Engineering Materials

This series provides topical information on innovative, structural and functional materials and composites with applications in optical, electrical, mechanical, civil, aeronautical, medical, bio- and nano-engineering. The individual volumes are complete, comprehensive monographs covering the structure, properties, manufacturing process and applications of these materials. This multidisciplinary series is devoted to professionals, students and all those interested in the latest developments in the Materials Science field, that look for a carefully selected collection of high quality review articles on their respective field of expertise.

More information about this series at http://www.springer.com/series/4288

Zhuohao Xiao · Shuangchen Ruan ·
Ling Bing Kong · Wenxiu Que ·
Kun Zhou · Yin Liu · Tianshu Zhang

MXenes and MXenes-based Composites

Processing and Applications

Springer

Zhuohao Xiao
School of Materials Science
and Engineering
Jingdezhen Ceramic Institute
Jingdezhen, Jiangxi, China

Ling Bing Kong
College of New Materials
and New Energies
Shenzhen Technology University
Shenzhen, Guangdong, China

Kun Zhou
School of Aerospace and Mechanical
Engineering
Nanyang Technological University
Singapore, Singapore

Tianshu Zhang
Anhui Target Advanced Ceramics
Technology Co. Ltd
Hefei, Anhui, China

Shuangchen Ruan
College of New Materials and New Energies
Shenzhen Technology University
Shenzhen, Guangdong, China

Wenxiu Que
Electronic Materials Research Laboratory,
School of Electronic and Information
Engineering
Xi'an Jiaotong University
Xi'an, Shaanxi, China

Yin Liu
School of Materials Science and Engineering
Anhui University of Science
and Technology
Huainan, Anhui, China

ISSN 1612-1317 ISSN 1868-1212 (electronic)
Engineering Materials
ISBN 978-3-030-59372-8 ISBN 978-3-030-59373-5 (eBook)
https://doi.org/10.1007/978-3-030-59373-5

This Springer imprint is published by the registered company Springer Nature Switzerland AG
The registered company address is: Gewerbestrasse 11, 6330 Cham, Switzerland

Preface

Renewable energies, environment protection and health engineering are the three hottest research topics all around the world. To solve these problems, the development of new materials is among the key solutions. For example, renewable energies, such as solar energies, wind energies and tide energies, require powerful storage technologies. Batteries and supercapacitors are electrochemical storage devices, in which electrodes play a significant role in determining their energy storage performances. Similar issues are also encountered in developing supercapacitors. High-performance materials are necessary in the synthesis and preparation of nanomaterials for applications in biosensing, nanomedicine and environmental remediation and so on.

MXenes have been drawn huge attentions in recent years, due to their specific 2D architectures, superb electrical conductivity and high mechanical strengths, which could find important applications in the abovementioned fields. Besides their pristine forms, MXenes have also been incorporated with other components to obtain hybrids and nanocomposites with enhanced or extra functionalities. Although reviews of articles have been available in various academic journals, they more often focused or emphasized one or limited number of aspects of the materials. This book is aimed to offer a relatively systematic overview on the progresses in synthesis, characterization and applications of MXenes.

After a brief introduction in Chap. 1, the progress in the preparation and properties of typical MXenes will be summarized in Chap. 2. In Chap. 3, the fabrication and characterization of MXenes-based hybrids and composites will be described and interpreted. The applications of MXenes in energy storages and conversions will be covered in Chap. 4, such as anodes of lithium-ion batteries, electrode materials of supercapacitors and storages of hydrogen. Other applications, such as biosensors, environmental remediating materials, piezoelectric devices and electromagnetic interference (EMI), shielding and absorption, will be presented in Chap. 5.

It is expected that this book can be used as a reference or handbook for senior undergraduates, postgraduates, researchers, R&D engineers, laboratory technicians, professors, application engineers, salespersons, etc., from universities, institute,

R&D departments or centers of companies, equipment/facility suppliers and so on, in the fields of materials science and engineering, chemical engineering, applied physics, nanomedicine, environmental engineering, health engineering, cosmetic engineering, to name a few.

Jingdezhen, China — Zhuohao Xiao
Shenzhen, China — Shuangchen Ruan
Shenzhen, China — Ling Bing Kong
Xi'an, China — Wenxiu Que
Singapore — Kun Zhou
Huainan, China — Yin Liu
Hefei, China — Tianshu Zhang

Contents

1 Introduction 1
References 2

2 Synthesis and Properties of MXenes 5
2.1 Introduction 5
2.2 Synthetic Strategies 6
2.2.1 HF Etching 7
2.2.2 Etching with Fluoride Salts 7
2.2.3 Other Methods 8
2.2.4 Large-Scale Synthesis 12
2.3 Materials Format 12
2.3.1 MXenes with Multilayers 12
2.3.2 MXenes with Single- or Few-Layers 51
2.3.3 Liquid Crystal MXenes 68
2.4 Key Properties 69
2.4.1 Electrical Properties 69
2.4.2 Chemical Properties 73
2.4.3 Thermal Properties 74
2.4.4 Mechanical Properties 76
2.4.5 Optical Properties 77
2.4.6 Magnetic Properties 81
2.5 Concluding Remarks and Perspectives 85
References 86

3 MXenes Based Composites and Hybrids 95
3.1 Introduction 95
3.2 MXene-Based Composites 96
3.3 MXenes-Based Hybrids 107
3.3.1 MXenes-Nanocarbons 107
3.3.2 MXenes-Metals 130

3.3.3 MXenes-Oxides 137
3.3.4 MXenes-Dichalcogenides 188
3.4 Conclusions and Outlook 197
References 198

4 Energy Related Applications 207
4.1 Introduction 207
4.2 Energy Storage and Conversion 207
4.2.1 Ion Batteries 208
4.2.2 Supercapacitors 258
4.2.3 Water Splitting 285
4.2.4 Hydrogen Storage 291
4.3 Conclusions and Outlook 294
References 295

5 Other Applications 303
5.1 Introduction 303
5.2 Environmental Remediating Agents 303
5.2.1 Adsorption of Heavy Metallic Ions 304
5.2.2 Removal of Organic Dyes 324
5.2.3 Elimination of Radionuclide Pollutants 335
5.2.4 Elimination of Gaseous Contaminants 337
5.2.5 Removal of Other Pollutants 338
5.2.6 Removal of Urea 339
5.3 Chemical/Biological Sensors 340
5.3.1 Electrochemical Sensors 340
5.3.2 Gas Sensors 348
5.3.3 Piezoelectric Sensors 362
5.3.4 Photoluminescence (PL) Sensors 365
5.3.5 Miscellaneous 368
5.4 Electrocatalytic Applications 369
5.5 EMI Shielding or Microwave Absorption 372
5.6 Transparent Conductive Films 386
5.7 Conclusions and Outlook 394
References 395

Chapter 1
Introduction

Two-dimensional (2D) materials have formed a new family of low-dimensional materials, which have drawn numerous attentions of the research community. Although various 2D materials have been proposed and synthesized, it is necessary to mention the titanium carbide, Ti_3C_2, which was first obtained by removing the Al atoms from the hexagonal ternary carbide, Ti_3AlC_2, through selective etching with aqueous hydrofluoric acid (HF) solution [1]. There are nearly one hundred similar ternary carbides and nitrides which are similar to Ti_3AlC_2, which have a general chemical formular of $M_{n+1}AX_n$, with M, A and X to stand for early transition metals, elements from the groups of IIIA or IVA and carbon/nitrogen, respectively, while n could be integer of 1–3 [2]. Moreover, the $M_{n+1}AX_n$ phases can be present as solid solutions, with different combinations of elements at the sites of M, A and X. As a result, the number of $M_{n+1}X_n$ should unlimited [3–16].

There are two formula units in each unit cell of the layer-structured hexagonal phases of $M_{n+1}AX_n$, with the M layers to be strongly adhered by the X atoms that are filled in between the octahedral sites, while the $M_{n+1}X_n$ layers are sandwiched by the A atom layers [17, 18]. As a consequence, the structures are of laminar architecture, thus having anisotropic characteristics. The M-X bond is a mixture of ionic, covalent and metallic behaviors, whereas the M-A is a pure metallic bond. The $M_{n+1}AX_n$ nanolayers are strongly bonded due to the bonding characteristics. In comparison, relatively weak van der Waals force is usually dominant the layer-structured materials, like graphite and transition metal dichalcogenide compounds (TMDs) [19]. As a result, they can be readily exfoliated through mechanical action to form 2D materials.

Owing to the difference in bonding properties, the strengths of the M-X and M-A interactions are different, so that the A layers can be taken away, thus forming $M_{n+1}X_nT_x$ layers, where T_x stands for surface functional groups, including =O, –OH and –F, which are linked to the M atoms on the surfaces generated during the etching reaction process. The layer thickness of the $M_{n+1}X_nT_x$ items is determined by the value of n, i.e., the number of the building blocks. They are single, two and three building blocks for n = 1, 2 and 3, respectively. This newly emerged group

Z. Xiao et al., *MXenes and MXenes-based Composites*, Engineering Materials,
https://doi.org/10.1007/978-3-030-59373-5_1

of materials are named as MXenes, in order to demonstrate the elimination of the component A from the initial compounds of $M_{n+1}AX_n$ and the 2D characteristic structure of graphene.

Besides the extensive studies on properties and applications of MXenes, the 2D materials have also been employed to form hybrids or composites, for a wide range of potential applications [20–22]. In this book, the advancement of MXenes and their nanohybrids and nanocomposites, in terms of synthesis, characterization and utilization. The synthesis and processing of representative MXenes will be covered in Chap. 2. In Chap. 3, the fabrication and characterization of MXenes-based hybrids and composites will be presented and discussed. The applications of MXenes in energy storage and conversion will be described in Chap. 4, such as anode materials of batteries, electrodes of supercapacitors, storage of hydrogen and so on. Other applications, including biosensing, environmental remediation, piezoelectric effects and electromagnetic interference (EMI) and shielding/absorption, etc., will be summarized in the last chapter.

Acknowledgements This work was supported by the National Natural Science Foundation of China (51762023 and 51962013), the Natural Science Foundation of Jiangxi, China (20192ACB20018), and Key R&D Program of Jiangxi Province (20171BBE50006, 20192ACB80007, and 20192ACB80004). Ling Bing Kong would like acknowledge Shenzhen Technology University (SZTU) for financial support through the Start-up Grant (2018) and grant from the Natural Science Foundation of Top Talent of SZTU (grant no. 2019010801002).

References

1. Naguib, M., Kurtoglu, M., Presser, V., Lu, J., Niu, J.J., Heon, M., et al.: Two-dimensional nanocrystals produced by exfoliation of Ti_3AlC_2. Adv. Mater. **23**, 4248–4253 (2011)
2. Barsoum, M.W.: MAX Phases: Properties of Machinable Ternary Carbides and Nitrides. Wiley (2013)
3. Anasori, B., Halim, J., Lu, J., Voigt, C.A., Hultman, L., Barsoum, M.W.: Mo_2TiAlC_2: A new ordered layered ternary carbide. Scripta Mater. **101**, 5–7 (2015)
4. Zhang, H.B., Zhou, Y.C., Bao, Y.W., Li, M.S., Wang, J.Y.: Intermediate phases in synthesis of Ti_3SiC_2 and $Ti_3Si(Al)C_2$ solid solutions from elemental powders. J. Eur. Ceram. Soc. **26**, 2373–2380 (2006)
5. Barsoum, M.W., El-Raghy, T., Ali, M.: Processing and characterization of Ti_2AlC, Ti_2AlN, and $Ti_2AlC_{0.5}N_{0.5}$. Metall. Mater. Trans. A.**31**, 1857–1865 (2000)
6. Handoko, A.D., Steinmann, S.N., Seh, Z.W.: Theory-guided materials design: two-dimensional MXenes in electro-and photocatalysis. Nanoscale Horizons. **4**, 809–827 (2019)
7. Jun, B.M., Kim, S., Heo, J., Park, C.M., Her, N., Jang, M., et al.: Review of MXenes as new nanomaterials for energy storage/delivery and selected environmental applications. Nano Research. **12**, 471–487 (2019)
8. Khazaei, M., Mishra, A., Venkataramanan, N.S., Singh, A.K., Yunoki, S.: Recent advances in MXenes: from fundamentals to applications. Curr. Opin. Solid State Mater. Sci. **23**, 164–178 (2019)
9. Khazaei, M., Ranjbar, A., Arai, M., Sasaki, T., Yunoki, S.: Electronic properties and applications of MXenes: a theoretical review. J. Mater. Chem. C. **5**, 2488–2503 (2017)

10. Lei, J.C., Zhang, X., Zhou, Z.: Recent advances in MXene: Preparation, properties, and applications. Front. Phys. **10**, 276–286 (2015)
11. Nan, J.X., Guo, X., Xiao, J., Li, X., Chen, W.H., Wu, W.J., et al.: Nanoengineering of 2D MXene-based materials for energy storage applications. Small 1902085 (2018)
12. Ronchi, R.M., Arantes, J.T., Santos, S.F.: Synthesis, structure, properties and applications of MXenes: current status and perspectives. Ceram. Int. **45**, 18167–18188 (2019)
13. Sun, Y.L., Meng, X., Dall'Agnese, Y., Dall'Agnese, C., Duan, S.N., Gao, Y., et al.: 2D MXenes as Co-catalysts in photocatalysis: synthetic methods. Nano-Micro Lett. **11**, 79 (2019)
14. Tang, H., Hu, Q., Zheng, M.B., Chi, Y., Qin, X.Y., Pang, H., et al.: MXene-2D layered electrode materials for energy storage. Progress Natural Sci. Mater. Inter. **28**, 133–147 (2018)
15. Verger, L., Xu, C., Natu, V., Cheng, H.M., Ren, W.C., Barsoum, M.W.: Overview of the synthesis of MXenes and other ultrathin 2D transition metal carbides and nitrides. Curr. Opin. Solid State Mater. Sci. **23**, 149–163 (2019)
16. Xiao, Z.B., Li, Z.L., Meng, X.P., Wang, R.H.: MXene-engineered lithium-sulfur batteries. J. Mater. Chem. A. **7**, 22730–22743 (2019)
17. Barsoum, M.W.: The $MN_{+1}A_XN$ phases: a new class of solids. Prog. Solid State Chem. **28**, 201–281 (2000)
18. Bai, Y.L., Srikanth, N., Chua, C.K., Zhou, K.: Density functional theory study of M(n+1)AX(n) phases: a review. Crit. Rev. Solid State Mater. Sci. **44**, 56–107 (2019)
19. Sun, Z.M., Music, D., Ahuja, R., Li, S., Schneider, J.M.: Bonding and classification of nanolayered ternary carbides. Phys. Rev. B **70**, 092102 (2004)
20. Naguib, M., Mochalin, V.N., Barsoum, M.W., Gogotsi, Y.: 25th anniversary article: MXenes: a new family of two-dimensional materials. Adv. Mater. **26**, 992–1005 (2014)
21. Xiao, Y., Hwang, J.Y., Sun, Y.K.: Transition metal carbide-based materials: synthesis and applications in electrochemical energy storage. J. Mater. Chem. A **4**, 10379–10393 (2016)
22. Kumar, P., Abuhimd, H., Wahyudi, W., Li, M.L., Ming, J., Li, L.J.: Review—Two-dimensional layered materials for energy storage applications. ECS J. Solid State Sci. Technol. **5**, Q3021–Q3025 (2016)

Chapter 2
Synthesis and Properties of MXenes

2.1 Introduction

MXenes, as a new group of two-dimensional (2D) structured materials, have attracted extensive attentions from all around the world. The 2D structures are obtained, as the element A is removed from the parent phase of $M_{n+1}AX_n$. In the unit cell of the layered hexagonal phases of, $M_{n+1}AX_n$, there are two formula units, in which the M layers are constructed through the close packing of the X atoms. The spaces in between the octahedral sites are filled with the X atoms, whereas the layers of a atoms serve to interleave the $M_{n+1}X_n$ layers [1]. The laminated structures exhibit strong anisotropic characteristics, because the M-X bond could have combined ionic, covalent and metallic natures, while the M-A bond is of just metallic behavior. The adjacent layers in the structures are strongly bonded, so that the exfoliation of $M_{n+1}AX_n$ is different from those of graphite and transition metal dichalcogenides (TMDs), in which the bonding is through the weak van der Waals interactions [2].

The strengths of the M-X and M-A are different, whereas the A layers have higher reactivity. Therefore, the A layers can be removed relatively easily, if suitable etching reagents are used. As a consequence, layer structured MXene, or $M_{n+1}X_nT_x$, is thus obtained. In addition, MXene usually contains surface-terminating functional groups, denoted as T_x, which include oxygen (=O), hydroxyl (–OH) and fluorine (–F). These functional groups are evidently formed during the etching process and linked to the M atoms in the MXenes. The thickness of $M_{n+1}X_nT_x$ is directly dependent on the value of n. For instance, for the values of $n = 1$, 2 and 3, they are single, double and triple octahedral blocks, respectively. More recently, MXene with $n = 4$ has also been synthesized, which has a formula of Mo_4VC_4, containing five atomic layers of transition metals [3].

The element A in the parent phases of $M_{n+1}AX_n$, which is selectively removed, is most likely Al in the materials reported in the open literature. The M site usually hosts transition metals, such as Ti, V, Cr, Nb, Ta, Zr, Mo, etc. The element A can also be taken away by using thermal etching at high temperatures, but the layered structures could be negatively affected or even totally destroyed [4]. In addition, both

Z. Xiao et al., *MXenes and MXenes-based Composites*, Engineering Materials,
https://doi.org/10.1007/978-3-030-59373-5_2

elements M and A can be etched through high temperature chlorination reaction, which results in porous products, known as carbide derived carbons (CDC), instead of layered structures [5]. In this regard, chemical etching is much preferred to develop 2D structured MXenes.

As the first 2D MXene, Ti_3C_2, was obtained by using selective etching of Al from the layered hexagonal ternary carbide, Ti_3AlC_2, with hydrofluoric acid (HF) solution to be the etching agent at room temperature (RT) [6]. In fact, the Ti_3AlC_2 belongs to the family of ternary carbides and nitrides, with a general chemical formula of $M_{n+1}AX_n$, in which M is an early transition metal, A is primarily an element from the groups of 13 or 14, X is C and/or N and $n = 1, 2$ or 3 [7]. Additionally, $M_{n+1}AX_n$ can also be solid solutions, with multiple components at the atomic sites of M, A and X, such as $(Mo_2Ti)AlC_2$, $Ti_3(Al_{0.5}Si_{0.5})C_2$, $Ti_2Al(C_{0.5}N_{0.5})$, from which $M_{n+1}X_n$ layered structures are formed, as A layer is selectively etched [8–10]. In addition, it has been theoretically predicted that ordered double-transition metal MXenes and ordered divacancy MXenes could also been present as stable phases [11, 12]. Moreover, they might be more eco-friendly, cost-effective, stable and much more active for the applications as hydrogen evolution reaction (HER) catalysts. In this respect, the 2D family is unlimited to allow for materials with various different and new functions. This chapter is aimed to elaborate the latest progress in synthesis, materials and properties of MXenes [13–18].

2.2 Synthetic Strategies

With the success in developing 2D layered $Ti_3C_2T_x$ with HF etching, various other methods were proposed and implemented. At the same time, various MXenes with new compositions have also been reported. In all cases, fluoride-containing agents formed the main stream in the exfoliation of transition metal carbides. Meanwhile, different surface properties and thus different functionalities are expected, if different strategies are employed. Therefore, it is important to classify and compare these synthetic methods.

Basically, there two strategies to synthesize 2D MXenes, i.e., (i) top-down method and (ii) bottom-up method [19, 20]. The top-down method is actually the exfoliation of bulk materials into few-layer or even single-layer sheets as mentioned above, whereas the bottom-up method is to grow or deposit such materials starting from atoms or molecules [21, 22]. For example, Xu et al. reported a chemical vapor deposition (CVD) method to obtain ultrathin 2D α-Mo_2C nanostructures with lateral sizes of up to 100 μm, which showed superconducting properties [23]. Comparatively, bottom-up methods are able to produce MXenes to be less defective and have large lateral sizes, due to the absence of defects that are usually observed in the top-down methods using strongly corrosive chemicals [6].

For the top-down method to prepare MXenes, $Ti_3C_2T_x$, the precursor Ti_3AlC_2 was synthesized by using the conventional ceramic processing, starting from Ti_2AlC and TiC. The two powders are mixed and then calcined at temperatures of >1200 °C in an inert environment in order to avoid oxidation of the components. Morphologies and dimensional properties of the precursors are highly dependent on the synthetic conditions, including properties of the raw materials, chemical compositions, calcination temperature, reaction time and atmosphere [24]. Significant progress has been made in developing approaches to prepare $Ti_3C_2T_x$, while various etching agents have been reported in the open literature, such as HCl-LiF, NH_4HF_2 and NH_4F [24–27].

2.2.1 *HF Etching*

The HF etching process to dissociate the M-A bond in $M_{n+1}AX_n$ phases with Al can be described with the following reactions:

$$M_{n+1}AlX_n + 3HF \rightarrow M_{n+1}X_n + AlF_3 + 1.5H_2, \quad (2.1)$$

$$M_{n+1}X_n + 2H_2O \rightarrow M_{n+1}X_n(OH)_2 + H_2, \quad (2.2)$$

$$M_{n+1}X_n + 2HF \rightarrow M_{n+1}X_nF_2 + H_2. \quad (2.3)$$

According to Eqs. (2.2) and (2.3), the surface terminations are formed due to the combination of –OH and –F during the etching of Al, resulting in the functionalization of the M layers [6].

The production yields, type of defects, and particle dimensions of the $Ti_3C_2T_x$ species, along with characteristics of the surface termination groups are strongly dependent on the etching process, such as concentration of HF, reaction temperature and reaction time, owing to the kinetically controlling behavior of the etching process [28–30]. The selective etching route has been applied to the development of various multi-layered MXenes, including Ti_2CT_x, $Ta_4C_3T_x$, $(V_{0.5}Cr_{0.5})_3C_2T_x$, $(Ti_{0.5}Nb_{0.5})_2CT_x$, $Ti_3CN_xT_x$, Nb_2CT_x, V_2CT_x, $Nb_4C_3T_x$, $Mo_2TiC_2T_x$, $Mo_2Ti_2C_3T_x$, $Cr_2TiC_2T_x$ and $(Nb_{0.8}Ti_{0.2})_4C_3T_x$ [11, 29, 31–33], to name a few.

2.2.2 *Etching with Fluoride Salts*

Despite the success in producing $Ti_3C_2T_x$ with HF etching, it is widely acknowledged that the special requirements to deal with HF could be problematic for large scale applications [34–36]. As a result, alternative milder and safer etching agents have been explored. For instance, Ghidiu et al. proposed a simple way to etch Ti_3AlC_2 through the reaction between HCl and fluoride salts (e.g., LiF) [26]. Normally, the HF-derived $Ti_3C_2T_x$ has a lattice parameter of $c = 2$ nm. In comparison, the LiF/HCl-produced $Ti_3C_2T_x$ exhibits larger interlayer distance of about 4 nm. This observation has been attributed to the possible intercalation of water molecules and/or cations into the $Ti_3C_2T_x$ layers which are hydrophilic in nature and negatively charged. Not only LiF-HCl is milder, but also the $Ti_3C_2T_x$ flakes produced in this way have much larger lateral sizes and less defects. Even more, single and/or double layers could be readily obtained. In addition, the product yield could approach 100%, which is much higher than that of the HF-etching route [6, 29, 37].

After that, different fluoride salts, such as NaF, KF and NH_4F, as well as acids, have been combined for such a purpose [30, 38–40]. During these processes, HF is formed due to the strong reaction between the fluoride salts and acids. Meanwhile, a wide range of cations, including Li^+, Na^+, K^+, Ca^{2+}, Al^{3+} and NH_4^+, have been employed to intercalate into the layers of Ti_3C_2. Even H_2O molecules could be simultaneously intercalated into the MXene layers, in the aqueous solutions of the fluoride salts. Therefore, materials with multiple new functions and characteristics, such as much larger interlayer distance and stronger surface adsorption capability, could be developed [26, 27, 41].

Besides powdery samples, film-like $Ti_3C_2T_x$ can also be obtained through the etching process. For example, Halim et al. epitaxially deposited Ti_3AlC_2 films first and then etched out Al to obtain $Ti_3C_2T_x$ thin films [22]. Such thin films could be used to study various interesting physical properties, such as optical modulation, electrical conductivity, magnetoresistance, superconductivity, etc. Microstructural analysis results indicated that the $Ti_3C_2T_x$ layers demonstrated highly ordered atomic structure, which was formed owing to the aid of a pre-coated TiC buffer layer. During the etching process, the etching agent was NH_4HF_2. At the same time, functional groups of $-NH_3$ and $-NH_4^+$ were intercalated into the interlayers of $Ti_3C_2T_x$, so that the interlayer spaces of the final products were effectively enlarged.

Additionally, Ti_3C_2 MXene has also been syntheized by selectively etching of silicon (Si) from the parent compound of titanium silicon carbide [42]. The Ti_3SiC_2 derived MXene could be produced at a relatively large scale with concentrations of 0.5–1.3 mg/mL, with which highly conductive flexible films were obtained. The MXene made in this way exhibited stronger resistance to oxidation as compared with those etched from Ti_3AlC_2. This finding opened a new door to develop MXenes.

2.2.3 *Other Methods*

As mentioned before, Xu et al. used a CVD method to deposit large area high quality 2D ultrathin α-Mo_2C crystal nanostructures [23]. The crystal layers had a thickness at nanometer scale, with a lateral size of up to 100 μm and pretty high stability under ambient conditions. The 2D nanostructures exhibited superconducting transitions and strong anisotropy at magnetic fields. It was found that the superconductivity had a strong dependence on thickness of the crystal nanostructures. This CVD process could be extended to other 2D TMC crystal materials.

A Cu foil with a dimension of 5×5 mm^2 was placed on top of a Mo foil with the same size, which was used as the substrates to deposit α-Mo_2C crystal nanostructures. The Cu/Mo bilayer substrates were heated to temperatures of >1085 °C in H_2 at a flow rate of 200 sccm in a tube furnace. Meanwhile, CH_4 was introduced into the reactor at flow rate of 0.35 sccm. The growth of 2D α-Mo_2C crystals was conducted at ambient pressure, for time durations of 2–50 min. If the Mo foil was replaced with W and Ta foils, WC and TaC nanocrystals would be obtained in a similar way. Lateral size of the α-Mo_2C crystal nanostructures could be readily tailored through controlling over the growth conditions.

An optical photograph of the α-Mo_2C nanosheet crystals that were grown at 1090 °C for 5 min is depicted in Fig. 2.1a. Interestingly, all the α-Mo_2C nanosheets adopted a regular shape, such as triangle, rectangle, hexagon, octagon, nonagon and dodecagon, implying that they were well-crystallized with high crystallinity. The average lateral dimension of the nanosheet crystals was 10 μm, while their thicknesses were in the range of 3–20 nm. After growth for 50 min, the lateral size of the α-Mo_2C nanosheet crystals could be up to 100 μm, while their thickness was less than 10 nm, as seen in Fig. 2.1b. Optical and atomic force microscope (AFM) images, of three representative hexagonal α-Mo_2C nanosheet crystals, with thicknesses of 6–11 nm, are shown in Fig. 2.1c–h, respectively. In addition, the nanosheet crystals all had a uniform thickness and a smooth surface. They had different colors on SiO_2/Si substrates under an optical microscope due to their different thicknesses. More importantly, all the α-Mo_2C nanosheet crystals were outstandingly stable in a wide range of solvents at room temperature. They experienced almost no change in structure when they were thermally treated in air at 200 °C.

A low-magnification high-angle annular dark-field (HAADF) STEM image of an individual hexagonal 2D Mo_2C nanosheet crystal is shown in Fig. 2.2a. EDS results revealed that the nanosheet crystal contained only two elements, i.e., Mo and C, with an atomic ratio to be close to 2:1, so the 2D phase had a near stoichiometric composition. Furthermore, the two elements of Mo and C were homogeneously distributed over the nanosheet, as demonstrated in Fig. 2.2b–d, confirming that the chemical composition was uniform. Zone axis selective area ED (SAED) pattern and convergent beam ED (CBED) pattern are shown in Fig. 2.2e, f, respectively, where the SAED pattern in Fig. 2.2e was along the [100] zone axis. Corresponding, the crystals had lattice parameters of $a = 0.477$ nm, $b = 0.598$ nm and $c = 0.521$ nm, which were all in agreement with the literature data.

Representative atomic-level HAADF STEM and bright-field (BF) STEM images of the α-Mo_2C nanosheet crystals are illustrated in Fig. 2.2g, h, respectively. The Mo atoms were closely packed to form a hexagonal structure. As revealed in Fig. 2.2h, the C atoms were located at the centre, while each C atom was surrounded by six Mo atoms in the view along the [100] direction. Furthermore, these ultrathin α-Mo_2C nanosheet crystals possessed very high quality, with almost no defects, disordering or impurities. In addition, all the crystals were grown along the [100] direction, i.e., in the perpendicular direction to their basal planes.

In summary, HF and the instantaneously formed HF through the reaction of NH_4HF_2 and LiF with strong acids exhibited slightly different effects the properties of the final Ti_3AlC_2, such as morphology, dimension, surface characteristic and defects. While HF etching led to small lateral sizes and more defects, the alternative etching agents resulted in relatively large sizes and less defects, but are much

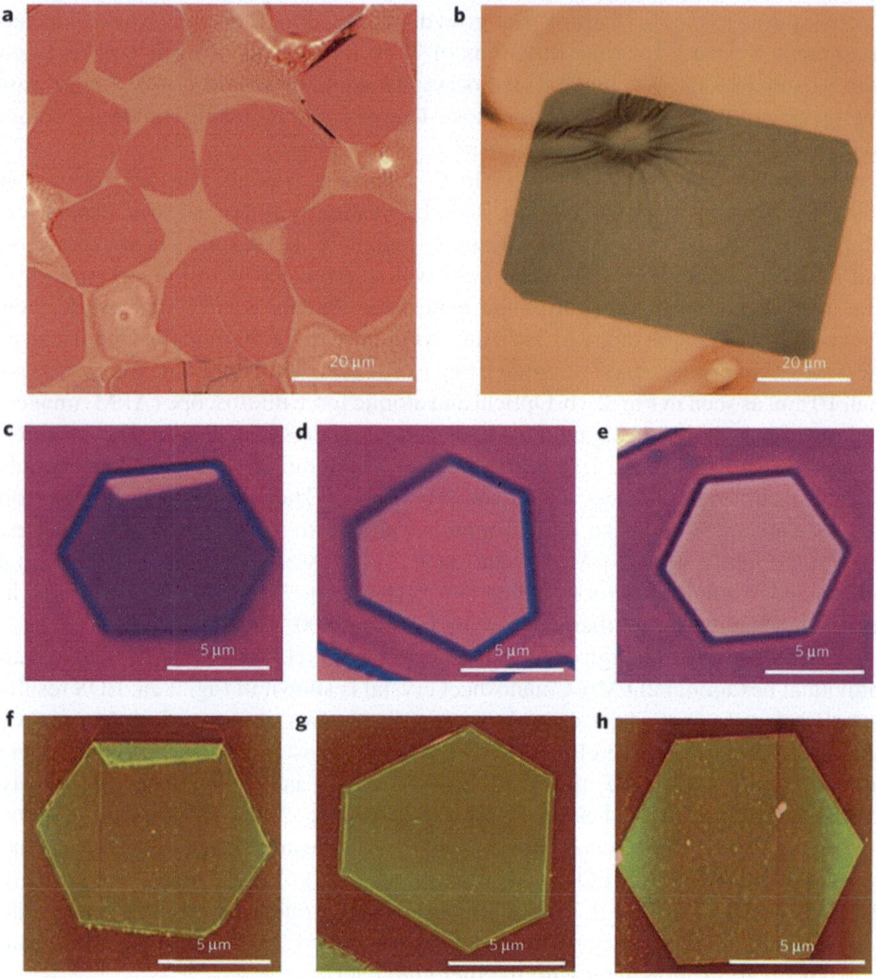

Fig. 2.1 **a** Representative optical image of the ultrathin α-Mo_2C nanosheet crystals on a Cu/Mo substrate, demonstrating the presence of different regular shapes. **b** Optical image of an individual octagonal α-Mo_2C nanosheet crystal with the longer side to be about 100 µm. **c–e** Optical images of three hexagonal α-Mo_2C nanosheet crystals with thicknesses of 6.7 nm (**c**), 8.2 nm (**d**) and 11.2 nm (**e**), which had been transferred onto a SiO_2/Si substrate, showing different colors for different thicknesses. **f–h** AFM images of three hexagonal α-Mo_2C nanosheet crystals of (**c–e**). Reproduced with permission from [23]. Copyright © 2015, Springer Nature Limited

time consuming. Moreover, dealing with HF needs special carefulness. It is reasonably expected more effective etching agents will be developed to prepare $Ti_3C_2T_x$ and other MXenes. However, bottom-up method is still in its infant stage, which deserves to be explored in the near future.

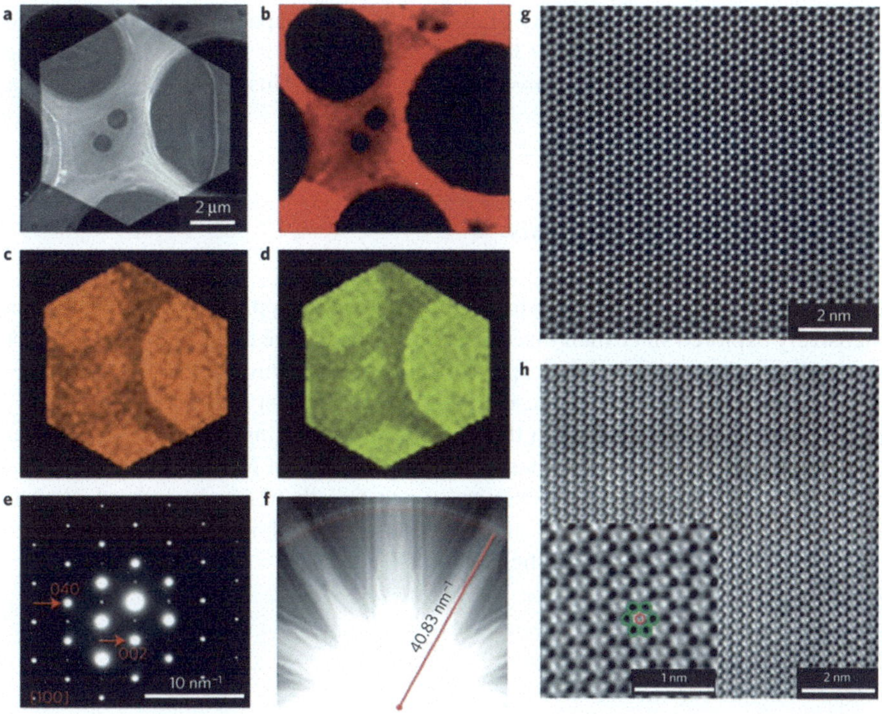

Fig. 2.2 **a** Low-magnification HAADF STEM image of an individual hexagonal α-MoC 2D crystal. **b–d** EDS elemental mappings of C–K (**b**), Mo–K (**c**) and Mo–L (**d**) lines across the sheet of (**a**), showing the homogeneous distribution of C and Mo. **e** SAED pattern along the [100] zone axis. **f** CBED pattern along the [100] zone axis, with the radius of the first-order Laue ring to 40.83 nm^{-1}. **g**, **h** Atomic-scale HAADF STEM (**g**) and BF-STEM (**h**) images of the α-Mo_2C nanosheet in (**a**). Reproduced with permission from [23]. Copyright © 2015, Springer Nature Limited

More recently, fabrication of MXenes by using water-free method [43] and solid–gas reaction strategy [44] has been reported. Due to their wide range potential applications, MXenes could be possibly used in fields where the absence of water is an essential requirement. To this end, Natu et al. reported an water-free to synthesize MXenes, by using organic polar solvents to dissolve ammonium dihydrogen fluoride [43]. The $Ti_3C_2T_z$ flakes obtained in this way were rich in fluorine terminational function groups. The $Ti_3C_2T_z$ MXenes etched in propylene carbonate exhibited promising electrochemical performances as anodes of Na-ion batteries, with capacity to be increased by 100% as compared with that of the one etched in aqueous solution. This achievement makes it possible to utilize MXenes for waster-sensitive applications.

Interestingly, Mei et al. developed a unique thermal reduction process to synthesize MXenes from sulfur-containing MAX compounds [44]. Because of their weak bonding in the compounds, S atoms tended to react with H_2 to produce H_2S, resulting in Ti_2C nanoflakes. The MXenes had outstanding performances as anode materials for LIBs, with a discharge capacity of 200 mAh/g. The capacity value was retained to be 70 mAh/g after cycling for 130 times at a high rate of 2.0 A/g, demonstrating a pretty high rate capability and strong cycling stability. This strategy can be readily extended to materials with other compositions.

2.2.4 Large-Scale Synthesis

Although MXenes have found a wide range of potential applications and have been extensively explored since their discoveries, it is a challenge to develop processes for large-scale production, which is the pre-requirement for industrialization of the materials. The latest development indicated that the preparation of 2D MXene $Ti_3C_2T_x$ can be easily scaled up basing on the currently used etching processes [45]. It has been experimentally confirmed that 1 and 50 g batches of MXene samples had no difference in morphology and properties, suggesting that large-scale production is not a critical issue for MXenes, simply because the etching reactions take place throughout the whole volume of the materials.

2.3 Materials Format

2.3.1 MXenes with Multilayers

Naguib et al. reported the first example of multi-layered $Ti_3C_2T_x$ from Ti_3AlC_2 powders through HF etching [6]. To prepare the starting powder of Ti_3AlC_2, Ti_2AlC and TiC commercial powders were mixed with a molar ratio of 1:1 through the conventional ball milling. The mixture was then calcined at 1350 °C for 2 h in the protection of Ar atmosphere. The calcined powder was soaked in an HF aqueous solution with a concentration of 50%, which was constantly stirred for 2 h at room temperature. Subsequently, the residue was removed from the supernatant by using filtration. Finally, Ti_3C_2 powder with layered nanostructures was obtained after the residue was thoroughly washed with DI water. Figure 2.3 shows a schematic diagram describing the etching steps to form the final Ti_3C_2 layered structures.

Figure 2.4a shows XRD patterns of the samples before after the etching reaction, together with the simulated one [6]. Obviously, Ti_3AlC_2 was formed due to the reaction of Ti_2AlC and TiC at the calcination temperature of 1350 °C. When the Ti_3AlC_2 powder was put into the HF solution, etching reaction started, which produced H_2 in the form of bubbles. The product was dispersed in methanol for 6 min, after which the diffraction peaks were weakened and broadened significantly. At the same time, a very broad peak was observed, indicating the occurance of amorphization in the sample. The exfoliation of the sample resulted in the disappearance of the diffraction

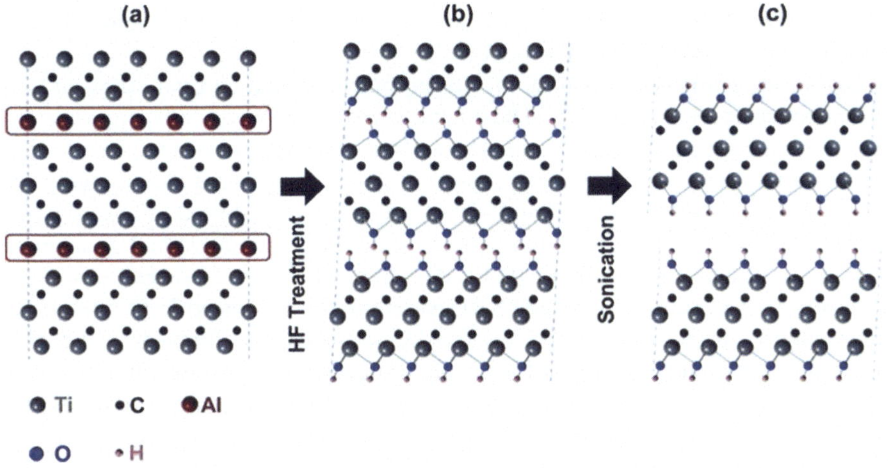

Fig. 2.3 Schematic diagram showing the exfoliation process of Ti_3AlC_2: **a** structure of Ti_3AlC_2, **b** Al atoms replaced by OH after the reaction with HF and **c** breaking of the hydrogen bonds and separation of the nanosheets after sonication in methanol. Reproduced with permission from [6]. Copyright © 2011, John Wiley & Sons

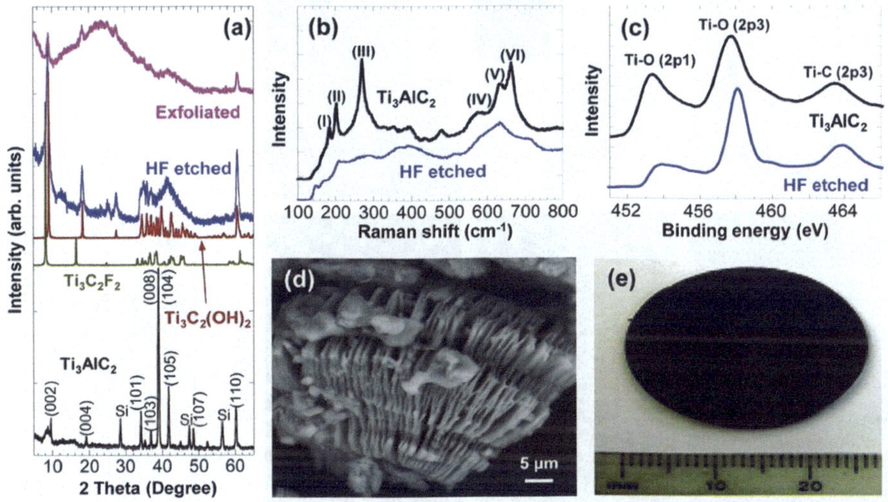

Fig. 2.4 Characterization results of the Ti_3AlC_2 before and after HF etching reaction. **a** XRD patterns of the Ti_3AlC_2 powders before the HF etching reaction, along with the simulated XRD patterns of $Ti_3C_2F_2$ and $Ti_3C_2(OH)_2$, measured XRD patterns of Ti_3AlC_2 after HF treatment and exfoliated nanosheets obtained after the sonication. **b** Raman spectra of Ti_3AlC_2 before and after the HF etching. **c** XPS spectra of Ti_3AlC_2 before and after the HF etching. **d** Representative SEM image of the sample after the HF etching. **e** Photograph of the disk with a diameter of 25 mm made of the etched and exfoliated powder with cold-pressing. Reproduced with permission from [6]. Copyright © 2011, John Wiley & Sons

peaks in the out-of-plane direction. Meanwhile, the formation of the nanostructure led to the broadening of the XRD peaks.

Due to the laminated structure, the etched powder could be easily pressed into disk or pellet, as seen in Fig. 2.4e. The thickness could be down to 300 μm, while the diameter could be as large as 25 mm. The middle XRD pattern in Fig. 2.4a revealed that almost all the non-basal plane peaks of Ti_3AlC_2 disappeared. Additionally, all the (00 *l*) peaks were broadened and weakened, while they shifted to lower angles at the same time. According to the Scherrer equation, the sample had an average crystal size of about 11 nm in the [000l] direction, which corresponded to roughly ten layers of $Ti_3C_2(OH)_2$. Simulated XRD patterns of hydroxylated $Ti_3C_2(OH)_2$ (red color pattern) and fluorinated $Ti_3C_2F_2$ (gold color pattern) well matched with that of the pressed sample (purple color pattern), suggesting that resultant Ti_3C_2 was functionalized to a certain degree. The pattern of powder was much similar to that of $Ti_3C_2(OH)_2$, implying its functionalization with OH groups.

It was estimated that hydroxylated (Fig. 2.5f) and fluorinated Ti_3AlC_2 structures would have expansions by 5% and 16%, respectively. Raman spectra of the Ti_3AlC_2 powders before and after the etching reaction are shown in Fig. 2.5b, also confirming

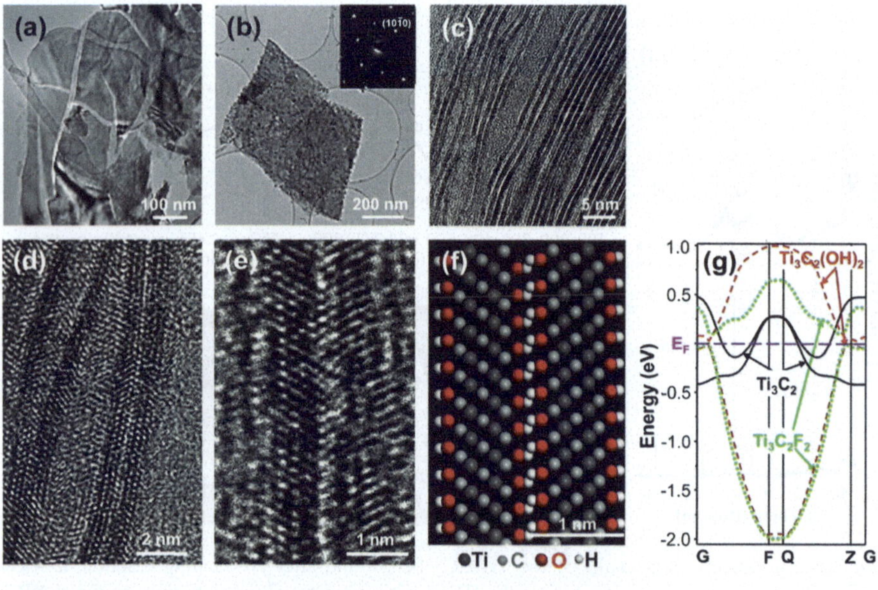

Fig. 2.5 Characterization results of the exfoliated MXene nanosheets: **a** TEM image of exfoliated 2D nanosheets with multiple layers, **b** TEM image of a single layer with the inset showing the SAED pattern that confirmed the hexagonal symmetry of the planes, **c** cross-sectional TEM image of single- and double-layer MXene nanosheets, **d** HRTEM image showing the separation of the individual nanosheets after the sonication, **e** HRTEM image of a bilayer $Ti_3C_2(OH)_xF_y$, **f** atomistic model of the layer structure shown in (**e**) and **g** calculated band structure of a single-layer MXene with –OH and –F surface terminations and without termination (Ti_3C_2), showing a transition from metal to semiconductor as a result of variation in the surface chemistry. Reproduced with permission from [6]. Copyright © 2011, John Wiley & Sons

the result of the XRD analysis. Figure 2.5c shows XPS spectra of the samples before and after the HF etching reaction. Both curves indicated the presence of the Ti–C and Ti–O bonds, confirming the generation of the etching product $Ti_3C_2(OH)_2$. The laminated structure of the product is clearly demonstrated by the SEM image depicted in Fig. 2.5d.

Representative TEM images of the exfoliated nanostructures are illustrated in Fig. 2.5a, b, indicating that the items were sufficiently thin so as to be transparent to electrons. The corresponding selected area electron diffraction (SAED), i.e., the inset in Fig. 2.5b, revealed that the planes were of hexagonal symmetry. The presence of Ti, C, O, and F was confirmed by the EDAX results. Cross-sectional TEM images are shown in Fig. 2.5c, d. A high-resolution TEM image and a simulated structure of two adjacent OH-terminated Ti_3C_2 sheets, are plotted in Fig. 2.5e, f, respectively. In addition, the nanosheets were flexible under the irradiation of electrons, when the samples were observed with TEM. DFT calculations also predict that the Electronic properties of the exfoliated Ti_3C_2 layers with the surface termination were calculated with DFT method, with result to be demonstrated in Fig. 2.5g.

The same authors used the HF etching route to exfoliate various other MAX derivatives, such as Ti_2AlC, Ta_4AlC_3, $(Ti_{0.5}Nb_{0.5})_2AlC$, $(V_{0.5}Cr_{0.5})_3AlC_2$ and Ti_3AlCN, evaluating the effects of concentration of the HF solution and time duration of the etching reaction [29]. The MXenes obtained in this study exhibited sheet resistances that were very close to those of multilayered graphene nanosheets, while the MXenes powders could be compacted into free-standing papers through a simple cold pressing without using any binders. Such MXenes free-standing sheets possessed hydrophilic surface characteristics.

The precursor Ti_2AlC was commercial product, while others were synthesized by using the solid-state reaction method. For example, Ta_4AlC_3 powder was prepared with the powders of Ta, Al and graphite (C). The mixture was thoroughly ball milled, followed by compaction and calcination at 1500 °C for 1 h in Ar. The calcined samples were used to obtain precursor powders. Similarly, TiNbAlC powder was from Ti, Nb, Al and C, whereas $(V_{0.5}Cr_{0.5})_3AlC_2$ powder was made of V, Cr, Al and C. The etching out of Al as confirmed by XRD results.

The Ti_2AlC powder was etched in10% HF solution for 10 h. Representative SEM images of the Ti_2AlC samples before and after the HF etching reaction are shown in Fig. 2.6a–c, confirming the nearly complete exfoliation of the individual particles [29]. In the exfoliated samples, the thin layers were distinctively separated from one another. After the aid of ultrasonication, 2D nanosheets were obtained, with an atomic ratio of Ti:C:F:O to be 39:19:20:22, as estimated by using energy-dispersive X-ray (EDX) micromapping. The absence of Al suggested the etching reaction as discussed above.

Ta_4AlC_3 could be exfoliated in 50% HF solution for 72 h. The effective and efficient exfoliation was confirmed by the XRD and SEM characterization results (Fig. 2.6d) [29]. It was observed that the XRD peak of (0002) of the sample after the etching reaction had a shoulder. Peak deconvolution results revealed two sub-peaks could be identified. The first one corresponded to a lattice parameter of $c \approx 3.034$ nm, with crystal size of about 38 nm, suggesting that there were about 12 unit-cells in

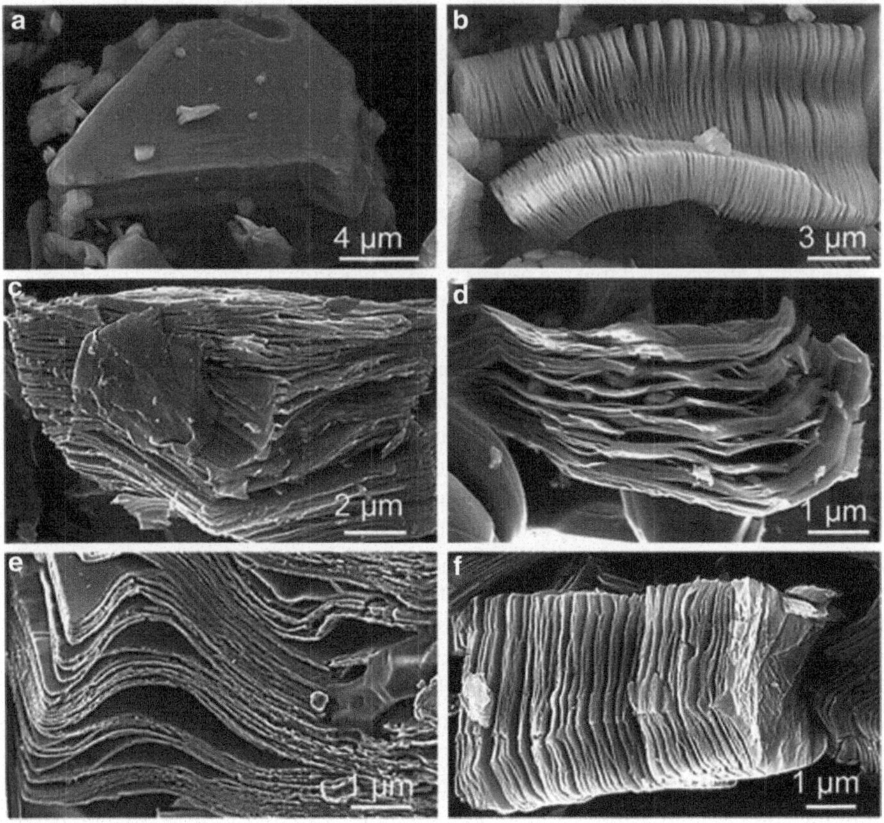

Fig. 2.6 SEM images of the layered structures: **a** Ti_3AlC_2 particle before etching, **b** Ti_3AlC_2 after HF etching, **c** Ti_2AlC after HF etching, **d** Ta_4AlC_3 after HF etching, **e** TiNbAlC after HF etching and **f** Ti_3AlCN after HF etching. Reproduced with permission from [29]. Copyright © 2012, American Chemical Society

the [0001] direction. The second peak implied a lattice parameter of $c = 2.843$ nm, corresponding to a crystal size of about 18 nm, meaning that there were about 6 unit-cells in the [0001] direction. The presence of the shoulder was possibly ascribed to the fact that the precursor powder Ta_4AlC_3 had two polymorphs, which however has not been confirmed yet.

Figure 2.7d shows a representative TEM micrograph of the Ta_4AlC_3 sample after HF etching reaction [29]. Figure 2.8a plots a HRTEM image of a Ta_4C_3 thin layer that had numerous nanometer-sized holes. The presence of C, Ta and O, together with a trace of F, was confirmed by the electron energy loss spectra (EELS), as seen Fig. 2.8b, while Al was absent. HRTEM and SAED are depicted in Fig. 2.8c, d, indicating that the basal planes of the MAX phases were still in crystalline state. The hexagonal structure of the Ta_4C_3 layers could be clearly identified. The values of d spacing were estimated to 0.269 nm and 0.155 nm for the lattice planes of $(01\bar{1}0)$ and

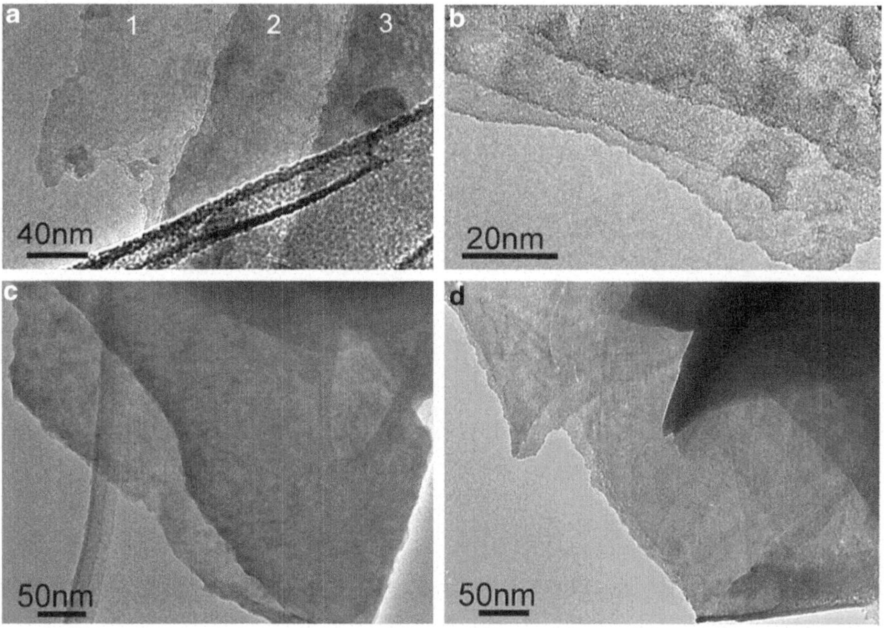

Fig. 2.7 TEM images of the layered structures: **a** Ti_3C_2 layers formed after HF etching of Ti_3AlC_2 at room temperature for 22 h, **b** Ti_3CN_x layers from Ti_3AlCN through HF etching, **c** layers from TiNbAlC after HF etching and **d** layers from Ta_4AlC_3 after HF etching. The numbers of the layers are indicated. Reproduced with permission from [29]. Copyright © 2012, American Chemical Society

($\bar{2}110$), respectively. Accordingly, the value of *a* lattice parameter was calculated to 0.31 nm, which was almost the same as the *a* lattice parameter of the Ta_4AlC_3 phase before exfoliation. In other words, the crystal structure of basal planes of MAX was retained in the MXenes.

Figure 2.6e shows SEM image of the TiNbAlC sample after etching reaction, clearly demonstrating the complete exfoliation of the precursor powder. A representative TEM image of the sample after ultrasonication is demonstrated in Fig. 2.7c. The thin sheets contained Ti, Nb, C, O and F, with an atomic ratio of 14:16:23:34:13, which was estimated from the EDX element mapping. A HRTEM image of an individual TiNbC layer is illustrated in Fig. 2.9a, along with the corresponding SAED as the inset, which clearly indicated the presence of hexagonal symmetry. The lattice parameter of ($10\bar{1}0$) planes was about 0.2606 nm, corresponding to an *a*-axis lattice constant of 0.301 nm, very close to the value of TiNbAlC (0.308 nm) reported in the literature [46]. The EELS results shown in Fig. 2.9b of the TiNbAlC sample after HF etching reaction confirmed its composition containing Ti, Nb, C, F and O, without the presence of Al.

After the concentration of the HF solution and etching reaction time duration were optimized, complete exfoliation of $(V_{0.5}Cr_{0.5})_3C_2$ was achieved, so that 2D

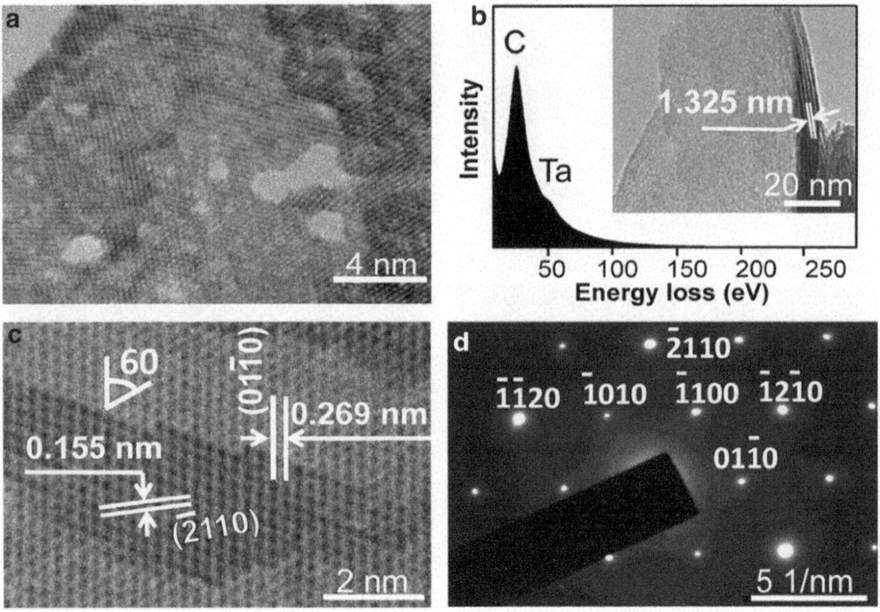

Fig. 2.8 HRTEM images and analytic results of the Ta_4AlC_3 after HF etching reaction: **a** TEM image, **b** EELS with a low-magnification TEM image (inset), **c** HRTEM of multilayer MXene and **d** SAED pattern. Reproduced with permission from [29]. Copyright © 2012, American Chemical Society

layered nanostructures were obtained. Figure 2.9c shows a HRTEM image of the exfoliated layer of $(V_{0.5},Cr_{0.5})_3C_2$ after HF etching reaction, whereas the hexagonal crystal structure was evidenced by the corresponding SAED shown as the inset. The a lattice parameter estimated according to the $(01\bar{1}0)$ reflections ($d \approx 0.286$ nm) was nearly the same as the value reported for $(V_{0.5}Cr_{0.5})_3AlC_2$ (i.e., $d \approx 0.289$ nm) [47]. The $(V_{0.5}Cr_{0.5})_3AlC_2$ after the HF etching reaction contained V, Cr, C, F and O, without the presence of Al, as confirmed by the EELS results, as observed in Fig. 2.9d.

Ti_3AlCN was fully exfoliated in 30% HF solution for 18 h. The 2D Ti_3CN_x layer nanostructures were composed of Ti, C, O and F, suggesting the surface termination of OH and/or F groups, which was evidenced by the EDX measurement results. At the same time, well exfoliation of the Ti_3CN_x layers were confirmed by SEM and TEM images, as noticed in Figs. 2.6f and 2.7b, respectively. All the exfoliated powders could be made into free-standing papers through could pressing without the requirement of glues or binders, with densities to be from 2.91 g/cm^3 of Ti_2C to 6.82 g/cm^3 of Ta_4C_3.

Zhao et al. reported a 2D Mxene, $Nb_4C_3T_x$, from Nb_4AlC_3 by using the HF etching at room temperature, for lithium ion battery applications [48]. High purity powders of Nb, Al and carbon black were used to synthesize Nb_4AlC_3, with at molar ratio of Nb:Al:C to 4:1.5:2.7. The powders were mixed, shaped and calcined at 1650 °**C** for

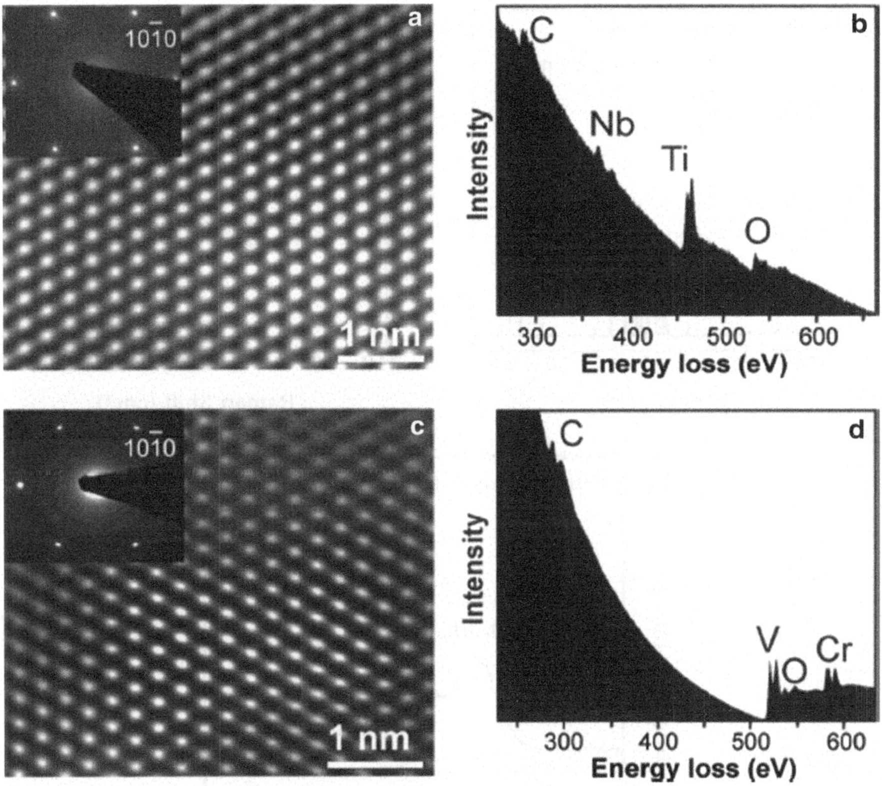

Fig. 2.9 **a** HRTEM of the TiNbAlC sample after HF etching reaction with the inset (top left) showing the SAED pattern. **b** EELS spectrum of the HF etched TiNbAlC. **c** HRTEM of the $(V_{0.5}Cr_{0.5})_3AlC_2$ sample after HF etching reaction with the inset (top left) showing the SAED pattern. **d** EELS spectrum the HF etched $(V_{0.5},Cr_{0.5})_3AlC_2$. Reproduced with permission from [29]. Copyright © 2012, American Chemical Society

4 h in Ar, at a heating rate of 5 °**C**/min. The calcined Nb_4AlC_3 powder was soaked in 49% HF for 140 h at room temperature with the aid of stirring. After reaction, the powder was washed thoroughly with DI water to a neutral pH level to form 2D $Nb_4C_3T_x$ powder.

Figure 2.10a shows XRD patterns of the Nb_4AlC_3 samples before and after the HF etching reaction. Before etching, the Nb_4AlC_3 precursor powder was of a single phase of the compound. After the etching reaction was finished, some of the diffraction peaks of the MAX phases were entirely absent, while others were significantly weakened. Meanwhile, a broadened peak of (002) at about $2\theta = 6°$ was present, which reflected the complete removal of Al. Raman spectroscopy of the 2D $Nb_4C_3T_x$ sample is illustrated in Fig. 2.10b The two peaks at about 222 cm^{-1} and 675 cm^{-1} confirmed the presence of the Nb–O and Nb-C vibrational bands, respectively [49]. The formation of the Nb–O band indicated surface functionalization of the 2D nanostructures

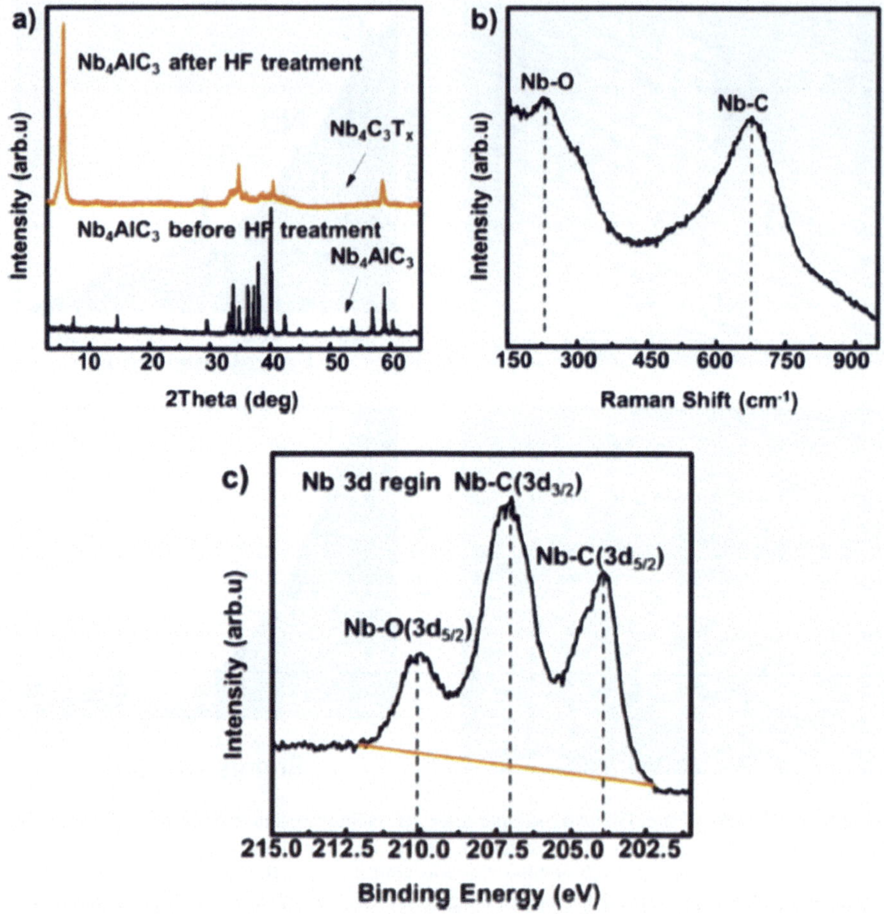

Fig. 2.10 **a** XRD patterns of the Nb_4AlC_3 samples before and after the HF etching reaction. **b** Raman spectrum of the 2D $Nb_4C_3T_x$ nanostructures. **c** XPS spectrum of the $Nb_4C_3T_x$ powder in the Nb 3d region. Reproduced with permission from [48]. Copyright © 2017, Elsevier

as Al component was etched out from the parent MAX. Typical XPS spectrum is demonstrated in Fig. 2.10c, further evidencing the presence of the Nb–O bands.

SEM images of the Nb_4AlC_3 samples before and after the HF etching reaction are plotted in Fig. 2.11a, b, respectively. The precursor Nb_4AlC_3 powder exhibited a clear layered structure. After it was completely etched with HF solution, typical MXene morphology was observed, which is very similar to that of thermally expanded graphite. The $Nb_4C_3T_x$ multilayer sheets had an accordion-like morphology. Figure 2.11c shows a TEM image of the 2D $Nb_4C_3T_x$ layered nanostructure, showing a typical lamellar morphology of MXene. At the same time, the hexagonal crystal structure stemmed from the parent MAX was supported by the SAED result, as seen in the inset in Fig. 2.11c. EDS elemental mapping results

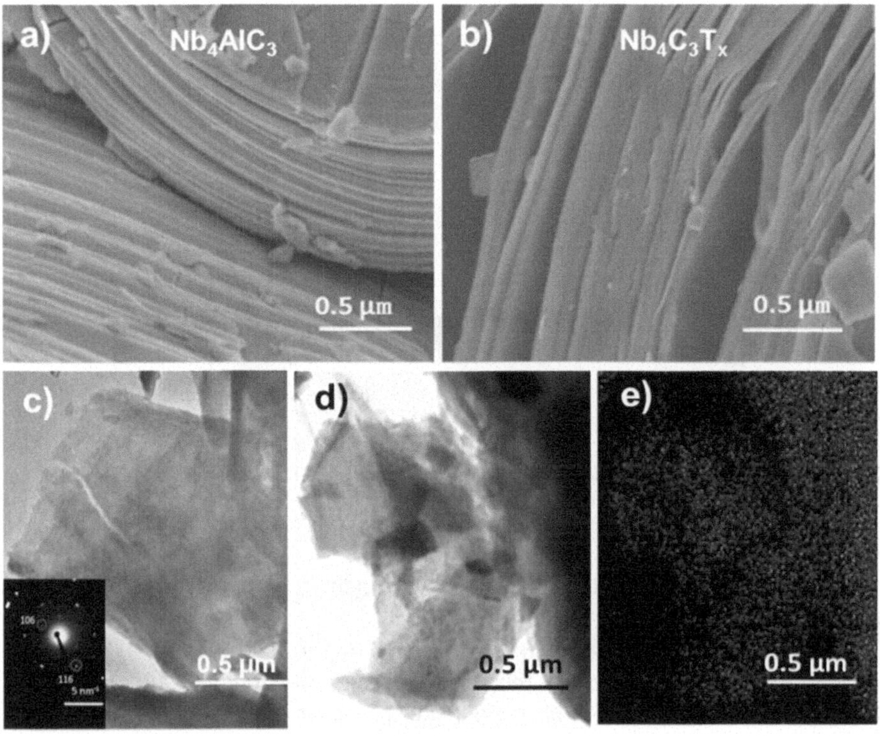

Fig. 2.11 **a** SEM image of the Nb_4AlC_3 with laminated structure. **b** SEM image of the multilayered $Nb_4C_3T_x$ nanoflakes. **c** TEM image of an individual $Nb_4C_3T_x$ nanoflake, with the inset showing its corresponding SAED pattern. **d** TEM image and **e** the corresponding elemental mapping of Nb in the area shown in (**d**). Reproduced with permission from [48]. Copyright © 2017, Elsevier

suggested that the distribution of the element Nb in the $Nb_4C_3T_x$ particles was very homogeneous. The 2D $Nb_4C_3T_x$ powder obtained in this way displayed promising electrochemical performances, when it was employed as the anode of lithium ion batteries.

The etching effect of HF has been applied to other elements different from Al. For example, Meshkian et al. obtained Mo_2CT_x layers by etching Ga from Mo_2Ga_2C with HF solution [50]. The precursor Mo_2Ga_2C thin films were prepared by using a DC magnetron sputtering method, with three elemental targets of Mo, Ga and C. Considering the relatively low melting point of Ga (~30 °C), a special strategy utilized to carry out the deposition, where the Ga target was in liquid state [51]. The deposition was conducted at 560 °C, with a base pressure of the chamber to be 1.5×10^{-7} Torr and 4.1×10^{-3} Torr after Ar gas was introduced. The Mo_2Ga_2C thin films were deposited on MgO(111) single crystal substrates that were pre-cleaned with acetone, ethanol and isopropanol with the aid of ultrasonication. The composition of the optimized sample was Mo:Ga:C of 39:43:18 in atomic ratio. The thin films were

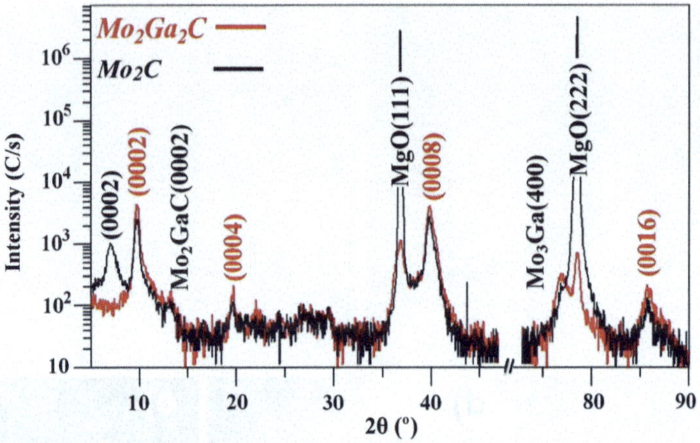

Fig. 2.12 XRD patterns of the Mo_2Ga_2C thin films before and after etching in 50% HF for 3 h at 50 °C. Reproduced with permission from [50]. Copyright © 2015, Elsevier

etched in 50% HF solution for 3 h at 50 °C. The etched samples thoroughly washed with DI water and ethanol.

Figure 2.12 shows XRD patterns of samples before and after the HF etching reaction. All the peaks ascribed to the Mo_2Ga_2C phase were retained after the sample was etched. In other words, the basal plane peaks of the Mo_2Ga_2C phase were maintained after the etching reaction. At the same time, an additional peak at ~7.1° in the etched sample suggested the formation of inter-layer spacing, corresponding to a lattice parameter of about 2.47 nm, i.e., Ga was etched out from the Mo_2Ga_2C phase, so that 2D Mo_2C MXene was developed. Moreover, the formation of the 2D layered nanostructures was also evidenced by the reduction in intensity of the (0002) plane. Additionally, a secondary phase Mo_3Ga was present due to its (400) diffraction peak in the XRD pattern. However, this phase was dissolved during etching reaction, such having no effect on the formation of the 2D Mo_2CT_x.

Figure 2.13a shows a typical TEM image of the etched sample, demonstrating the epitaxial growth of the layers on the MgO(111) single crystal substrate and the development of tilted grains in the topmost part of the film [50]. The tilted grains were well exfoliated, thus resulting in the 2D Mo_2C layered nanostructures. As seen in Fig. 2.13b, the effectiveness of the HF etching reaction was more distinguishable, in which obvious delamination of the exfoliated tilted grains was evidenced by the higher magnification image, confirming the complete removal of Ga. The titled grains were much easily etched, because they were readily attached by the etchant molecules. As the reactant entered the epitaxially grown thin films, the etching rate was decreased, because the Ga layer was protected by the MX layers.

Nevertheless, 2D Mo_2CT_x was finally developed through the HF etching reaction, as confirmed by the TEM images shown in Fig. 2.13a, as well as the XRD pattern discussed before. Figure 2.13c shows a higher magnification image from part of Fig. 2.13b, where the exfoliation of the layers was more obvious. However, there

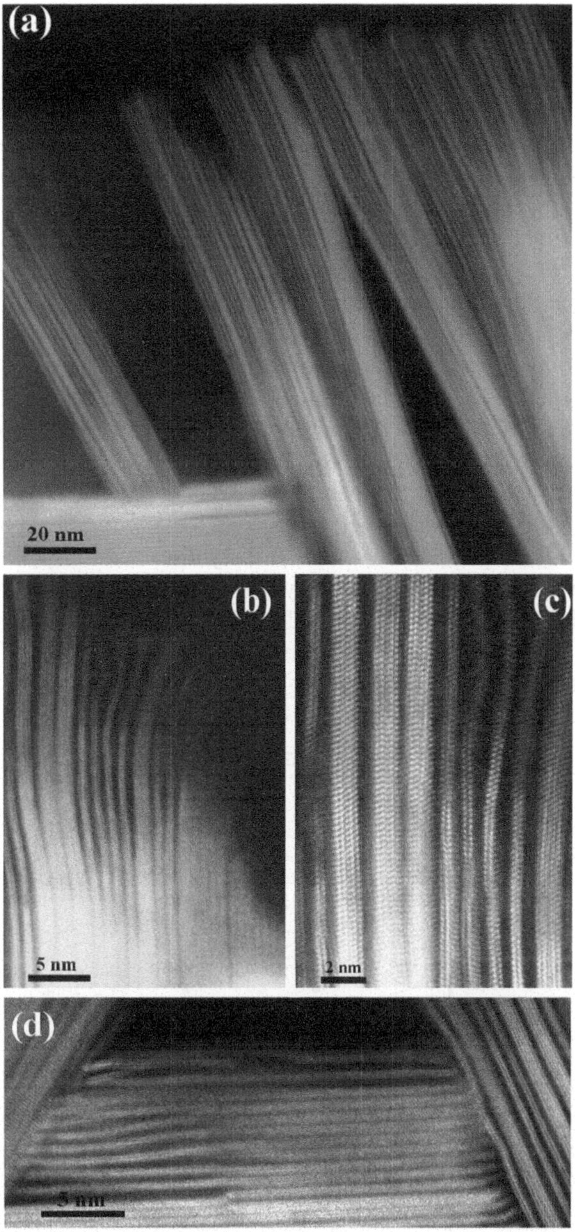

Fig. 2.13 TEM images of the Mo_2Ga_2C thin film after etching in 50% HF for 3 h at 50 °C: **a** overview TEM image revealing the formation of MXene 2D layered Mo_2C in the tilted grains, **b** high magnification image of selected tilted grains showing exfoliation of the layers, **c** high resolution image from the top part of (**b**) confirming the formation of the Mo_2C 2D layered nanostructures, with the left side of the image also revealing the difficulty of etching the Ga layers from the Mo_2GaC phase that was accompanied with the Mo_2Ga_2C phase, and **d** high magnification image showing exfoliation of the epitaxially grown thin films in the etched sample. Reproduced with permission from [50]. Copyright © 2015, Elsevier

were still Ga layers in the image (the left-hand side), which implied that removing Ga from Mo_2GaC MAX phase was more difficult than from Mo_2Ga_2C.

Zhou et al. developed the first Zr-containing layered $Zr_3C_2T_x$ by using the HF-etching protocol from $Zr_3Al_3C_5$, which was an alternative layered ternary and quaternary transition metal carbide beyond the $M_{n+1}AX_n$ phases [52]. $Zr_3Al_3C_5$ precursor powder was synthesized by using pulsed electric current sintering (PECS) technique, with desired Zr-Al-graphite powder mixture. Commercial element powders were mixed with a molar ratio Zr:Al:C of 3:3.2:4.8. The mixture was ball milled at a rate of 200 rpm. The milled powder was then dried at 50 °C for 24 h. The mixture was filled into a graphite die with a diameter of 40 and cold pressed at a pressure of 5 MPa. The green body, together with the die, was heated in a PECS facility to 1800 °C and kept for 20 min at a holding pressure of 30 MPa in Ar protection.

Ti_3AlC_2 powder was prepared similarly, but with a relatively low reaction temperature of 1280 °C. The $Zr_3Al_3C_5$ powder was etched in 50% HF solution at room temperature for 72 h, the Ti_3AlC_2 powder was reacted in the same HF solution for 40 h at room temperature. The as-obtained 2D $Ti_3C_2T_x$ and $Zr_3C_2T_x$ nanosheet powders were annealed at 1000 °C at a vacuum level of 2×10^{-2} Pa for 2 h.

Figure 2.14 shows that the as-synthesized $Zr_3Al_3C_5$ was nearly phase pure, with only a trace of ZrC impurity. As the $Zr_3Al_3C_5$ powder was put in 50% HF solution, the etching reaction proceeded, with bubbles of H_2 and CH_4 to be produced, forming Ti_3C_2 MXenes. As observed in Fig. 2.14, the diffraction peaks corresponding to the original $Zr_3Al_3C_5$ were tremendously reduced after the HF etching reaction. At

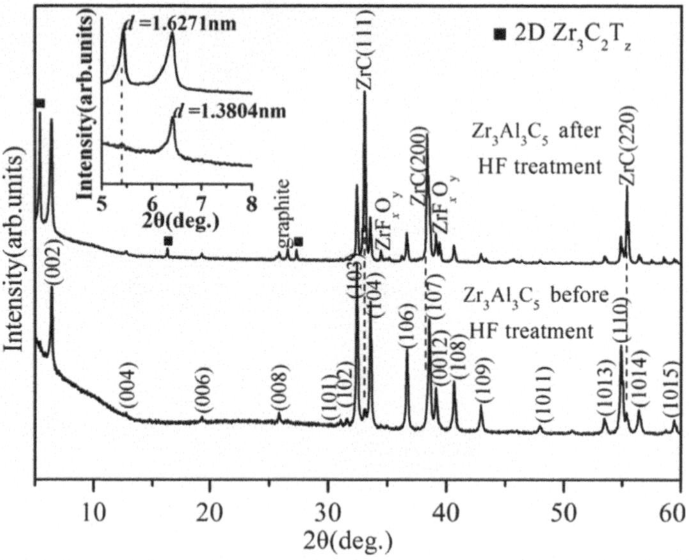

Fig. 2.14 XRD patterns of the $Zr_3Al_3C_5$ samples before and after the HF etching reaction. Reproduced with permission from [52]. Copyright © 2016, WILEY-VCH Verlag GmbH & Co. KGaA, Weinheim

the same time, the (0002) peak shifted to a lower angle of 5.43°, corresponding to a *c* lattice parameter of 3.253 nm, which was larger than the value of $Zr_3Al_3C_5$ (2.773 nm). Two additional (000l) peaks were present at 2θ of 16.32° and 27.24°. Instead the presence of broad peaks as discussed above for other HF etched MXenes, the shift in the low angle peak in this case suggested that spontaneous intercalation might have taken place, during the etching process. In addition, the ZrC impurity was inert to HF, so that its diffraction peaks were slightly increased after the HF etching reaction, because the peaks of $Zr_3Al_3C_5$ were reduced and ordered phase was formed.

SEM images of the sample after the HF etching are shown in Fig. 2.15a, b, demonstrating the exfoliation of the original particles along the basal planes. The accordion-like morphology of the exfoliated particles was similar to that observed for Ti_3AlC_2, as mentioned before. The presence of Ti, C, O and F was confirmed by EDS spectrum. Meanwhile, the content of Al was substantially diminished, confirming the removal of Al-C layer from the parent crystal structure.

In addition, the Zr/Al molar ratio of the etched sample was about 3.0:0.9, suggesting that the original $Zr_3Al_3C_5$ had been exfoliated by about 70%. Representative TEM images of the etched sample after sonication are shown in Fig. 2.15c–e,

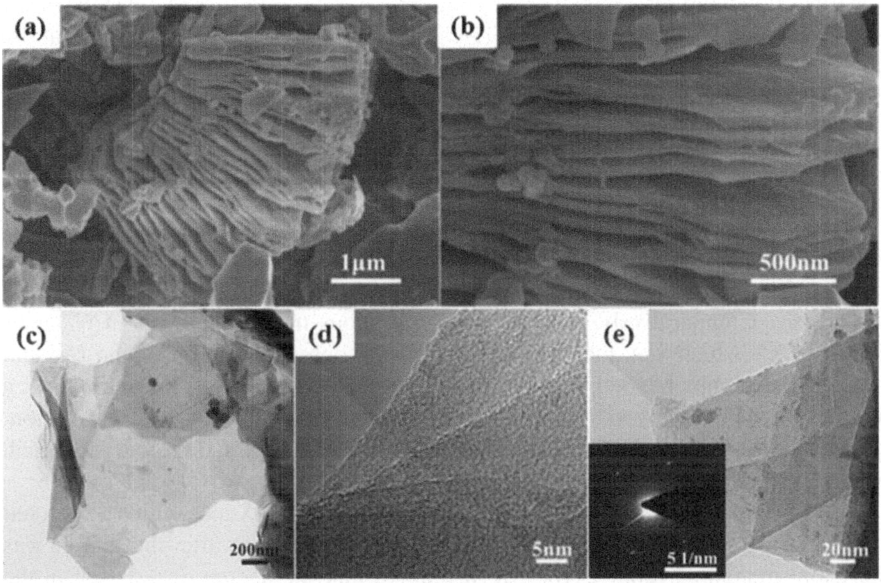

Fig. 2.15 **a, b** SEM images of the powders after HF etching reaction, showing the exfoliated grain to have an accordion-like structure. **c** Representative bright-field TEM images of the 2D $Zr_3C_2T_z$ after ultrasonication. **d** TEM image of the few-layered $Zr_3C_2T_z$. **e** TEM image of an individual scrolled $Zr_3C_2T_z$ nanosheet, with the inset to be a SAED pattern showing the reserved hexagonal symmetry of the basal plane. Reproduced with permission from [52]. Copyright © 2016, WILEY-VCH Verlag GmbH & Co. KGaA, Weinheim

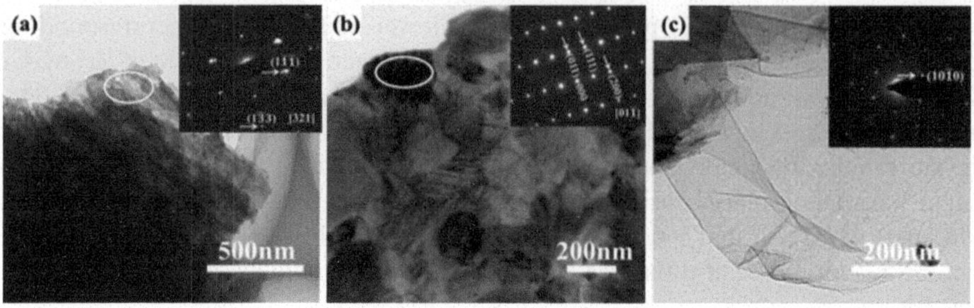

Fig. 2.16 **a**, **b** Representative bright-field TEM images of the vacuum annealed $Ti_3C_2T_z$ after ultrasonication, with insets to the SAED patterns corresponding to the circled areas. **c** Representative bright-field TEM image of the vacuum annealed $Zr_3C_2T_z$ after ultrasonication with the inset to be the SAED pattern of the observed flake. Reproduced with permission from [52]. Copyright © 2016, WILEY-VCH Verlag GmbH & Co. KGaA, Weinheim

revealing the thin morphology and layered structures stemmed from the exfoliation. EDS confirmed that the etched sample contained Zr, C, and O, without Al. The MXene nanosheets were highly flexible and foldable. The SAED (inset in Fig. 2.15e) indicated the hexagonal symmetry of the derived 2D MXene nanosheets.

Structural stabilities of 2D $Ti_3C_2T_z$ and $Zr_3C_2T_z$ MXenes, up to 1200 °C in Ar, were compared by the authors [52]. $Ti_3C_2T_z$ experienced a mass loss of 7% at the first stage over 50–200 °C, while an endothermic peak at 133 °C was observed in the DSC curve. In comparison, such weight loss for $Zr_3C_2T_z$ was only 4%, but up to 500 °C. Meanwhile, it had an endothermic peak at 71 °C in the DSC curve. In both cases, the first stage weight loss was caused by the removal of physically absorbed and intercalated water molecules.

At high temperatures, $Zr_3C_2T_z$ exhibited a broad exothermic peak at about 700 °C in the DSC curve, with almost no mass loss in the TG curve over 500–1200 °C. This exothermic thermal effect was probably associated with the crystal growth in the sample. Comparatively, $Ti_3C_2T_z$ had an obvious mass loss of about 6% over 200–800 °C, which was mainly attributed to the dehydroxylation of the residual isolated OH groups bonded with the metal-terminated surface. At the same time, a relatively large mass loss of about 14% was observed over 800–1000 °C. In addition, a relatively sharp exothermic peak was present at about 984 °C in DSC curves, which was ascribed to phase transition of the materials.

XRD results indicated that the diffraction peaks Ti_3C_2 MXenes with a hexagonal symmetry were absent, whereas only cubic TiC was observed, after the sample was thermally treated at 1000 °C for 2 h. It implied that the degree of ordering in the Ti_3C_2 MXenes in the *c* axis direction was decreased, while structural transformation might also take place in that case. The peaks of $Zr_3C_2T_z$ were still visible, although their intensity was reduced, reflecting the removal of the intercalated species, surface groups and decrease in ordering along *c* axis direction.

Figure 2.16a shows TEM image of the $Ti_3C_2T_z$ nanosheets, revealing the partial damage of their 2D nature and structural integrity. At the same time, structural transformation was observed, because cubic TiC_x was derived from the SAED pattern (inset in Fig. 2.16a). As seen in Fig. 2.16b, nanocrystals were observed on the $Ti_3C_2T_z$ nanoflakes, SAED pattern (inset in Fig. 2.16b) indicated that they were rutile TiO_2, implying that the nanosheets could be oxidized by the surface oxygen groups or residual O_2 in the vacuum. However, the nanosheets of $Zr_3C_2T_z$ were basically retained, after the high temperature annealing. As confirmed by the SAED pattern (inset in Fig. 2.16c), the original hexagonal symmetry was nearly unchanged. Therefore, it is safe to conclude that $Zr_3C_2T_z$ was thermally more stable than $Ti_3C_2T_z$.

The etching conditions, concentration of HF solution, time duration of reaction and etching temperature (if not room temperature), which have been reported in the open literature, are over wide ranges [31, 37, 53]. For example, $Ti_3C_2T_x$ was derived from Ti_3AlC_2 in 50% HF solution for 2 h at room temperature due to the complete exfoliation, while less stable Ti_2CT_x, with smaller values of n in $M_{n+1}AlX_n$, was formed from Ti_2AlC in 10% HF solution at the same temperature for the same time duration. Nb_2AlC has an M-A bond energy of about 1.21 eV, which is higher than that of Ti_2AlC (about 0.98 eV). As a result, a longer etching time of about 90 h and higher HF concentration of 50% are required to fully exfoliated Al from Nb_2AlC [29, 31, 54]. In some cases, excessive thermal treatment is necessary to promote the crystallization of the $M_{n+1}X_nT_x$ layers [29, 55]. Also, it is expected that the smaller the particle sizes, the fast the etching reaction will be. Therefore, the original $M_{n+1}AX_n$ powders are generally refined to increase the effectiveness and efficiency of the etching reactions.

As mentioned earlier, various less corrosive etching agents have been developed to replace HF. For example, Halim et al. exfoliated Ti_3AlC_2 epitaxial films with 1 M ammonium bifluoride (NH_4HF_2) solution at room temperature, which was less hazardous and milder etchant, thus requiring longer etching time duration [22]. At the same time, intercalation of ammonium items, such as $-NH_3$ and $-NH_4^+$, was observed during the etching of the Al layers.

Ti_3AlC_2 thin films were coated on sapphire substrates, with three elemental targets of Ti, Al and C, by using DC magnetron sputtering method [56]. The sputtering process was conducted in high purity Ar (99.9999%), at a constant pressure of 4.8 mbar. The sapphire substrates had *c*-axis orientation, i.e., Al_2O_3 (0001), with a dimension of 10×10 mm^2 and thickness of 0.5 mm. The substrates were preheated inside the chamber at the deposition temperature 780 °C for 1 h. During the deposition, after the Ti and C targets were sputtered for 5 s, the Al target was sputtered, thus forming a buffer layer of TiC (111) with thicknesses of 5–10 nm. After that, Ti_3AlC_2 layers were epitaxially grown.

The Ti_3AlC_2 films were first etched in 50% HF solution, where the samples with thicknesses of 15, 28, 43 and 60 nm were etched for times of 10, 15, 60 and 160 min, respectively. Secondly, they were etched in 1 M NH_4HF_2 for time durations of 150, 160, 420, and 660 min, respectively. Both etching processes were conducted at room temperature. After etching, the samples were thoroughly washed with DI water and ethanol. The exfoliation of the Ti_3AlC_2 films was confirmed by the XRD results. The *c* lattice parameter was increased from 1.86 nm for Ti_3AlC_2 to 1.98 nm for $Ti_3C_2T_x$, while it was 2.47 nm for the one etched with NH_4HF_2 ($Ti_3C_2T_x$-IC).

Figure 2.17 shows cross-sectional scanning TEM images of the as-deposited Ti_3AlC_2 films (a, d) and the ones etched with HF (b, e) and NH_4HF_2 (c, f). The presence of the TiC buffer layer and microstructural development of the etched samples were clearly demonstrated in the TEM images. The SAED patterns The out-of-plane epitaxial configuration was Ti_3AlC_2(0001)//TiC(111)//Al_2O_3(0001), as evidenced by the SAED pattern. The value of *c* lattice parameter for the Ti_3AlC_2, estimated from the SAED pattern and TEM images, was the same as that derived from the XRD pattern. This value for $Ti_3C_2T_x$ was also in a good agreement with the XRD one. In comparison, the *c* value of $Ti_3C_2T_x$-IC obtained from the SAED pattern was 2.1 ± 0.05 nm, which was much smaller than that from XRD (~2.5 nm). The authors attributed this observation to the deintercalation of the ammonium items caused by the TEM sample preparation and experiment.

The exfoliation caused an increase in *c* lattice and the separation of the MXene layers. As a result, the thicknesses of the films were slightly increased after the etching reaction. For example, the thickness of the as-deposited film was about 60 nm

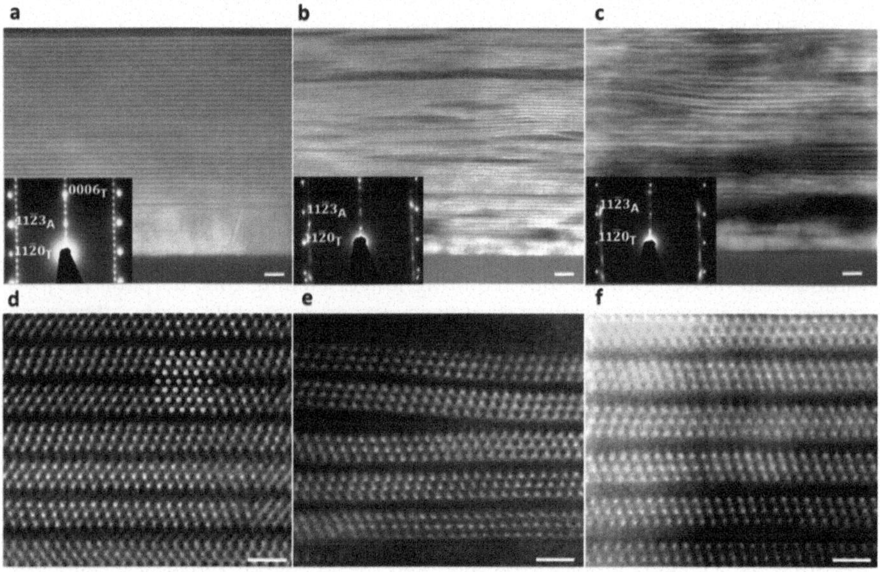

Fig. 2.17 Cross-sectional STEM images of the samples with a thickness of 60 nm, which was grown on sapphire substrate with a buffer layer of TiC: **a** Ti_3AlC_2, **b** $Ti_3C_2T_x$ and **c** $Ti_3C_2T_x$-IC films. The insets show SAED patterns of the film and the substrate, with the subscripts A and T to represent Al_2O_3 and Ti_3AlC_2, respectively. High-resolution STEM images of the films along the [$11\bar{2}0$] zone axis: **d** Ti_3AlC_2, **e** $Ti_3C_2T_x$ and **f** $Ti_3C_2T_x$-IC. The inset in panel (**d**) shows Ti, Al and C atoms in colors of yellow, gray and black, respectively. Scale bars for low resolution (**a**, **b** and **c**) and high-resolution (**d**, **e** and **f**) images are 5 and 1 nm, respectively. Reproduced with permission from [22]. Copyright © 2014, American Chemical Society

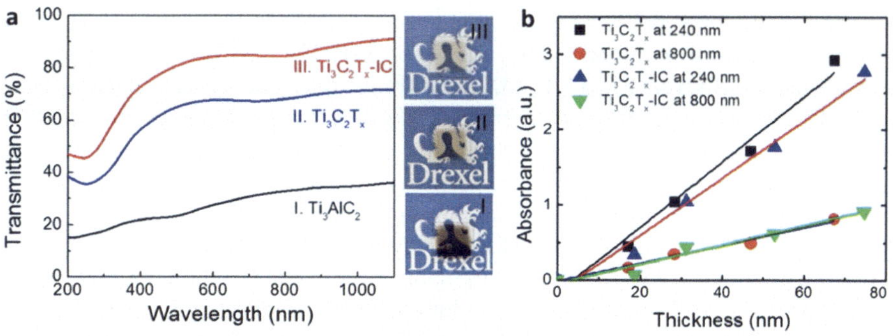

Fig. 2.18 **a** Transmittance spectra and photographs (on right) of the thin film samples with a thickness of 15 nm: (I) Ti_3AlC_2, (II) $Ti_3C_2T_x$ and (III) $Ti_3C_2T_x$-IC. **b** Light absorbance at wavelengths of 240 and 800 nm as a function of thickness for the $Ti_3C_2T_x$ and $Ti_3C_2T_x$-IC films. Reproduced with permission from [22]. Copyright © 2014, American Chemical Society

(Fig. 2.17a), while it became to be about 67.5 nm after exfoliation. The atomic layers in the $Ti_3C_2T_x$-IC sample (Fig. 2.17c, f) had spacings that were more uniform than those in the $Ti_3C_2T_x$ one (Fig. 2.17b, e). This phenomenon was ascribed to the milder characteristics of NH_4HF_2, as compared with stronger HF. Also, the intercalation of ammonia items could partly contribute to this effect.

Optical transparency of the films was increased after exfoliation. Both the $Ti_3C_2T_x$ and $Ti_3C_2T_x$-IC films exhibited higher transmittance than the original Ti_3AlC_2 one, as demonstrated in Fig. 2.18a and the insets. The transmittance values of the Ti_3AlC_2, $Ti_3C_2T_x$ and $Ti_3C_2T_x$-IC films were 30%, 70% and 90%, respectively. It was expected that the optical transmittance could be even higher if the TiC buffer layer was not used. Figure 2.18b shows absorbances of the films as a function of thickness. Almost a linear relationship was observed for all the films. The similarities in the transmittance curves and the linear relationship between absorbance and film thickness further proved that $Ti_3C_2T_x$ and $Ti_3C_2T_x$-IC should have similar microstructures.

Besides promising optical transparency, the etched films still exhibited metallic-like conductivity. The original Ti_3AlC_2 films were of metallic resistivity, with values in the range of 0.37–0.45 × 10^{-6} Ωm. The resistivity values of the $Ti_3C_2T_x$-IC films were higher than those of the ones etched with HF. For example, resistivities of the 28 nm thick $Ti_3C_2T_x$ and $Ti_3C_2T_x$-IC films were 2.3 and 5.0 μΩm, respectively. The conductivities of the films were closely related to the intercalated items, which implied that the MXenes could utilized for sensor applications.

Feng et al. studied the effect of temperature and etching time duration, when synthesizing $Ti_3C_2T_x$ in various bifluoride solutions with a concentration of 1 M, such as $NaHF_2$, KHF_2 and NH_4HF_2. The minimum etching time duration for the Ti_3AlC_2 to start to be exfoliated in the 1 M of bifluoride solutions at 60 °C was about 8 h. Based on the existence of corresponding by-products ($(NH_4)_3AlF_6$, NH_4AlF_4 and $AlF_3.3H_2O$ in a ratio of 10:1:1 for reaction with NH_4HF_2), probable reaction mechanism between Ti_3AlC_2 and bifluoride etchant solution has been proposed [57].

Wang et al. developed a simple method to prepare multi-layered $Ti_3C_2T_x$ from Ti_3AlC_2 together with either $NaHF_2$, KHF_2 or NH_4HF_2 [25]. The mechanisms to produce the multi-layered $Ti_3C_2T_x$ MXenes through the etching reaction were elucidated. The precursor Ti_3AlC_2 powder was synthesized by solid state reaction method in Ar. Etching solutions with a concentration of 1 M were prepared by dissolving the bifluorides DI water. For each experiment, 1 g Ti_3AlC_2 powder was dispersed in 100 ml etching solutions, which was constantly stirred at 60 °C for different time durations. The etched samples were thoroughly washed and dried.

Figure 2.19 shows XRD patterns of the samples derived from the Ti_3AlC_2 powder etched with different bifluoride etching solutions for different time durations. As seen in Fig. 2.19a, after etching with $NaHF_2$ at 60 °C for 2 h and 4 h, the main phase was still Ti_3AlC_2, whereas only very weak peaks of Ti_3C_2 were visible. After etching for 8 h, the intensity of the characteristic peak at 39.2° of Ti_3AlC_2 was largely reduced. As the etching time was continuously increased, the content of Ti_3C_2 was gradually increased. Therefore, MXene Ti_3C_2 was obtained after etching for 8 h with $NaHF_2$. Similarly, the etching time durations were 24 h and 12 h for etching agents of KHF_2 and NH_4HF_2, respectively.

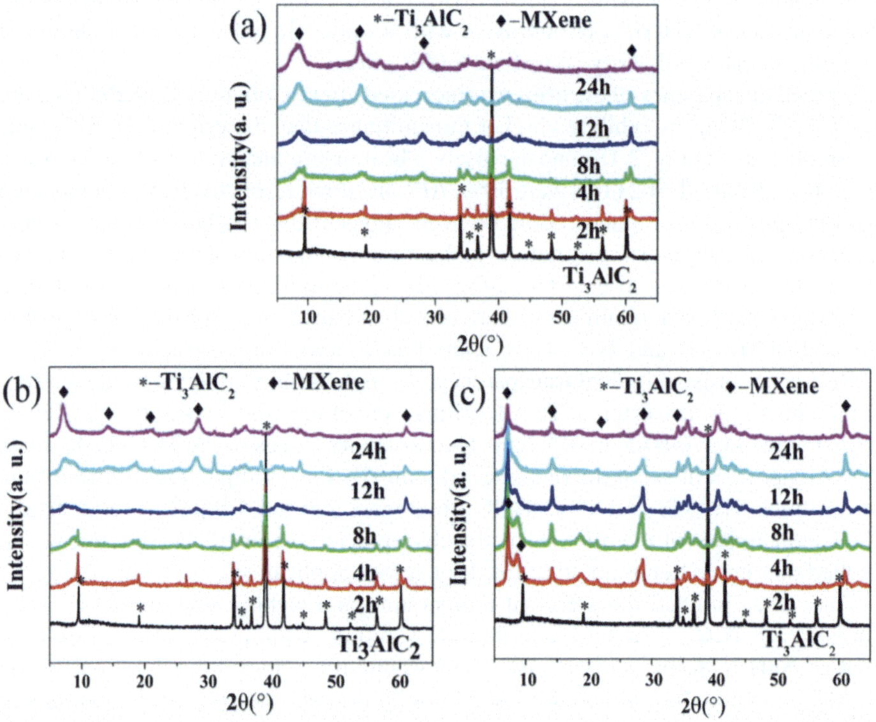

Fig. 2.19 XRD patterns of the original Ti_3AlC_2 and the etched products after etching in different solutions: **a** $NaHF_2$, **b** KHF_2 and **c** NH_4HF_2. Reproduced with permission from [25]. Copyright © 2017, Elsevier

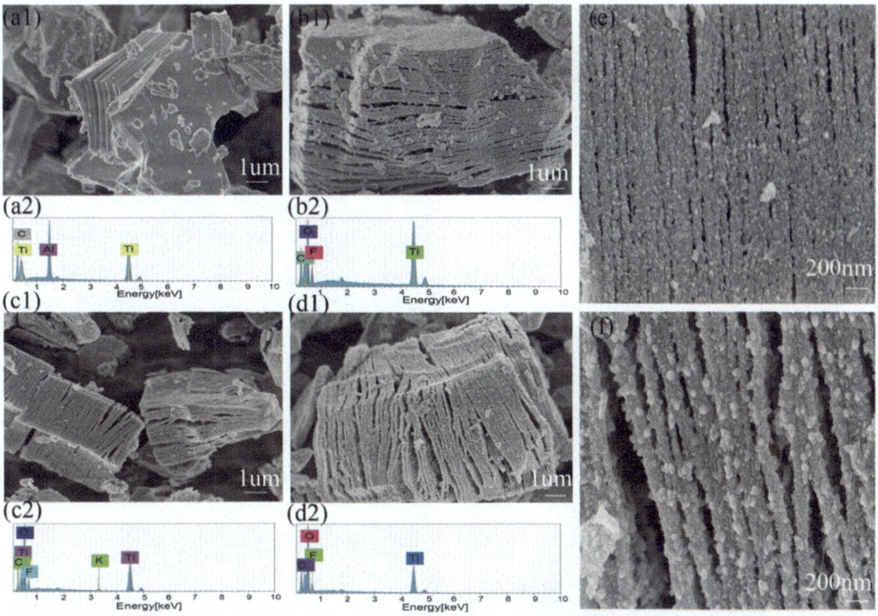

Fig. 2.20 SEM images and EDS spectra of the Ti_3AlC_2 and Ti_3C_2 samples: **a** Ti_3AlC_2, **b** samples exfoliated with $NaHF_2$, **c** samples exfoliated with KHF_2, **d** samples exfoliated with NH_4HF_2, **e** high magnification SEM image of Ti_3C_2 with KHF_2 and **f** high magnification SEM image of Ti_3C_2 with NH_4HF_2. Reproduced with permission from [25]. Copyright © 2017, Elsevier

Representative SEM and TEM images of original Ti_3AlC_2 and the Ti_3C_2 products are shown in Figs. 2.20 and 2.21. As observed in Figs. 2.20a1 and 2.21a, the precursor Ti_3AlC_2 had a typical layered structure, but with the layers to be strongly adhered together. The content of Ti, C and Al in the Ti_3AlC_2 was confirmed by the EDS spectra, as illustrated in Fig. 2.20a2. Crystal structure of the precursor was confirmed by the SAED pattern, which was plotted as the inset in Fig. 2.21a.

The SEM and TEM images of the Ti_3C_2 samples after exfoliation with $NaHF_2$ at 60 °C for 24 h are depicted in Figs. 2.20b1 and 2.21b. Obviously, lamellar structured Ti_3C_2 was generated, implying the successful exfoliation. The presence of Ti, C, O and F was confirmed by the EDS spectra, as seen in Fig. 2.20b2, while Al entirely absent. Meanwhile, the diffraction spots of the SAED pattern in the inset of Fig. 3.21b were not distinct, indicating the great reduction in crystallinity of etched product. Similar results were observed in the samples etched with KHF_2 and NH_4HF_2, which were demonstrated in Figs. 2.20c1 and 2.21c for KHF_2 and Figs. 2.20d1 and 2.21d for NH_4HF_2, respectively. For the sample made with KHF_2, a trace of K was observed in the EDS spectra, implying the formation of K^+ termination in the 2D Ti_3C_2. In addition, The Ti_3C_2 MXenes made with KHF_2 and NH_4HF_2 possessed relatively higher crystallinity.

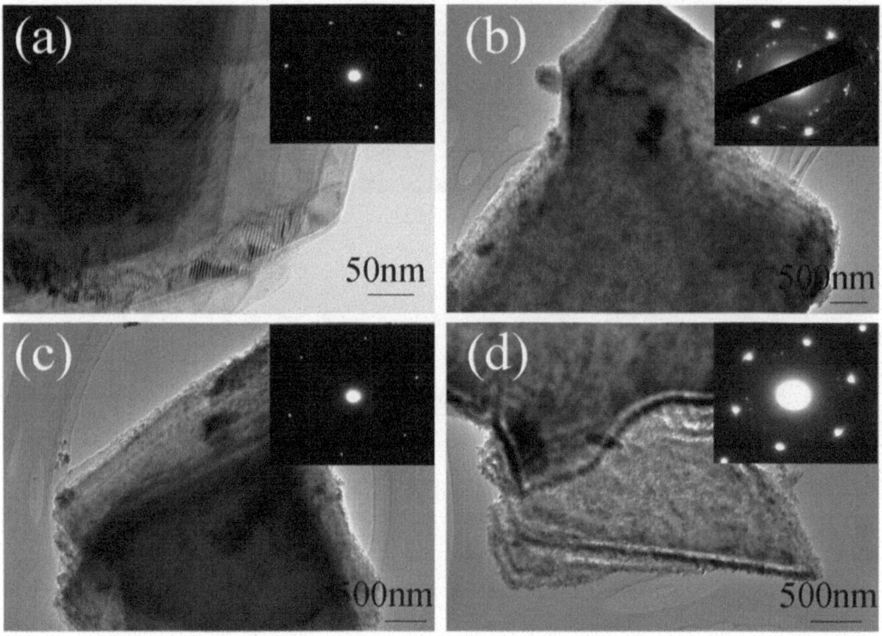

Fig. 2.21 TEM images of Ti_3AlC_2 and Ti_3C_2 samples exfoliated with different bifluorides, with the insets to be the corresponding SAED patterns: **a** TEM images of Ti_3AlC_2, **b** TEM images of Ti_3C_2 with $NaHF_2$, **c** TEM images of Ti_3C_2 with KHF_2 and **d** TEM images of Ti_3C_2 with NH_4HF_2. Reproduced with permission from [25]. Copyright © 2017, Elsevier

High magnification SEM images of the Ti_3C_2 made with KHF_2 and NH_4HF_2 are shown in Fig. 2.20e, f, respectively. The Ti_3C_2 nanostructures had a rough surface, while the edge of the layers was attached with small particles with a diameter of about 50 nm, which could be Ti_3C_2 that was formed during the sample preparation process. During the etching process, Ti_3AlC_2 and bifluoride reacted each other, removing Al to form layered structures. Because the fresh surface was not stable due to the high surface energy, small Ti_3C_2 particles would be formed to cover the large flakes.

To clarify the formation mechanism of the Ti_3C_2, the mixtures after reaction were directly dried without washing. According to XRD results, three byproducts were observed, which were Na_3AlF_6, $NaAlF_4$ and $AlF_3 \cdot 3H_2O$. The ratio was estimated from the main peak intensity of XRD patterns to 10:1:1. For $NaHF_2$, the main byproduct was Na_3AlF_6. Similarly, K_3AlF_6 and $(NH_4)_3AlF_6$ were for KHF_2 and NH_4HF_2, respectively. Also, during the direct drying process, $AlF_3 \cdot 3H_2O$ was formed from AlF_3 and H_2O to in.

As stated earlier, when Al was etched out from the precursor Ti_3AlC_2, surface groups, such as –F, –OH, =O, would terminate the broken bonds, thus leading further opening of the interplanar spacing of the Ti_3C_2 nanosheet structures. Therefore, the resultant Ti_3C_2 possessed a negatively charged surface. In this case, hydrated cations in the solution would be adsorbed onto the Ti_3C_2 layers through electrostatic interactions, so that the interplanar spacing of Ti_3C_2 was additionally increased. This assumption has been proved by various experimental results.

Ghidiu et al. reported a high-yield etching route to prepare $Ti_3C_2T_x$ from Ti_3AlC_2 by using 6 M hydrochloric acid (HCl), in which a small quantity of lithium fluoride (LiF) was dissolved [26]. After the as-etched $Ti_3C_2T_x$ was washed with DI water, a clay-like paste was obtained, due the intercalated water molecules and ions, of which could be conveniently shaped to desired dimensions and structures, so as to prepare free-standing films with flexibilities. Furthermore, the synthetic approach also was much fast and effectively avoided the use of hazardous concentrated hydrofluoric acid.

Ti_3AlC_2 powder (3 g) was dispersed in HCl solution with a concentration of 6 M solution which contained 1.98 g LiF in a volume of 30 ml. The reaction was carried out at 40 °C for 45 h, followed by filtration and thorough washing. The product together with a small amount of water was filtered through a cellulose nitrate membrane with a pore size of 0.22 μm. The resultant clay-like paste could be rolled in wet state as seen in Fig. 2.22a, leading to flexible free-standing films, as observed in Fig. 2.22c. The method could be easily scaled up without the limitation of dimension, given the flexibility of the paste. Also, the wet 'clay' paste could be moulded into different shapes as shown in Fig. 2.22d. It could be made into inks to for deposition or printing on various substrates. Furthermore, the product could experience rehydration, swelling when absorbing water and shrinking when dried, as illustrated in Fig. 2.22b.

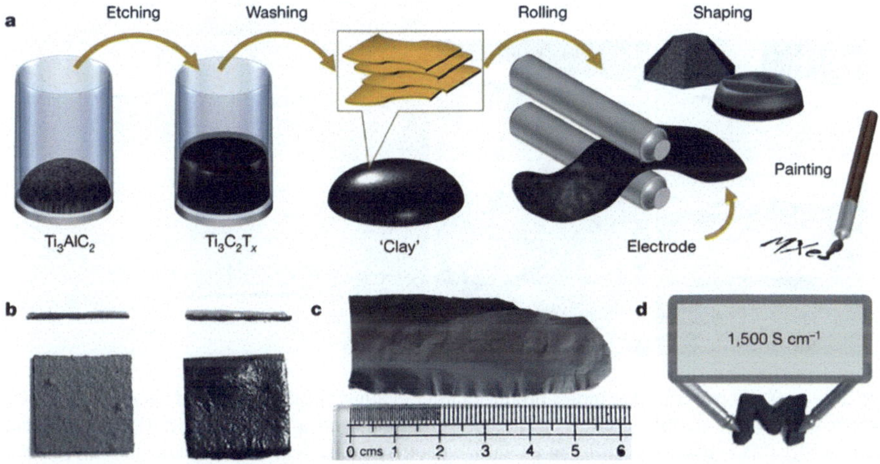

Fig. 2.22 **a** MAX phase was etched in the HCl/LiF solution (step 1), followed by thorough washing with water to remove reaction products and neutralize the pH level (step 2). The resultant sediment was similar to clay in nature, which could be rolled to produce flexible freestanding films (step 3), moulded and dried to yield conducting products with any desired shapes (step 4), or diluted and painted onto any substrate to form conductive coatings (step 5). **b** When the dried samples (left: cross-sectional and top views) were hydrated (right) they swelled, which then shrunk upon being dried. **c** Photograph of a rolled film. **d** Letter M (~1 cm) from the 'clay', which was conductive after dried. Reproduced with permission from [26]. Copyright © 2014, Springer Nature Limited

Figure 2.23a shows XRD patterns of the etched product, showing peaks with high intensity and strong sharpness, which was much different from those observed in HF etched MXene, as discussed previously. Moreover, the *c* lattice parameter was 2.7–2.8 nm, much larger than that of the HF derived $Ti_3C_2T_x$ ($c \approx 2.0$ nm). Even when the 'clay' was rolled into free-standing films, almost no change was encountered in the XRD patterns, indicating the high degree of ordering in the *c* direction (blue pattern in Fig. 2.23a).

Film thickness could be down to submicron and up to 100 μm. The particle shearing characteristic was demonstrated by the increase in intensity of the (110) peak at about 61°, as well as the SEM image shown in Fig. 2.23b. The multi-layered structure of the films was well revealed by the cross-sectional SEM images, as seen in Fig. 2.23e, f, while their flexibility was evidenced by the inset in Fig. 2.23e. The

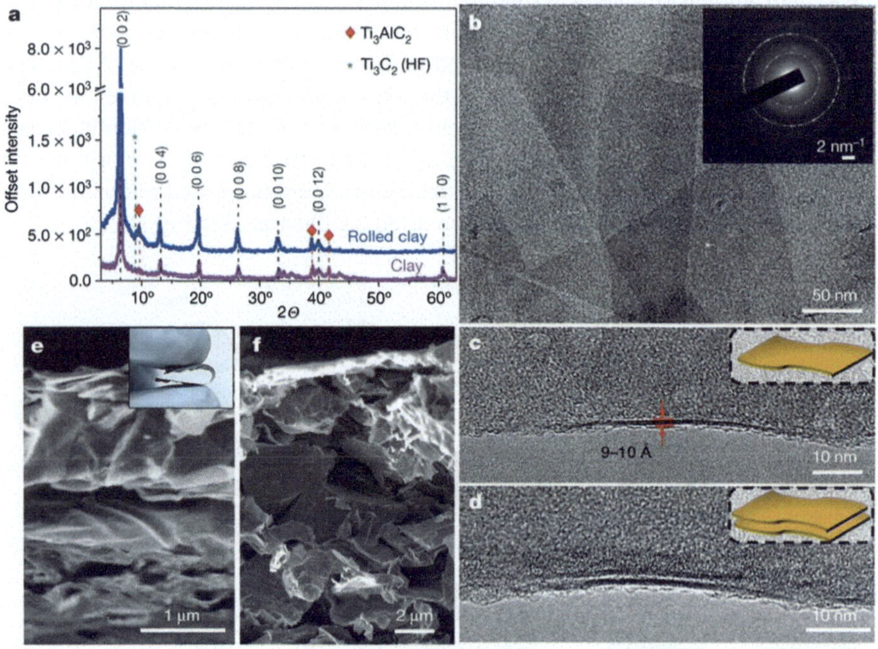

Fig. 2.23 **a** XRD patterns of samples produced after etching in LiF/HCl mixed solution. The pink pattern was for the multilayered $Ti_3C_2T_x$ nanostructure, with a sharp intense (0002) peak and higher-order (000l) peaks, resulting in a *c* lattice parameter of 2.8 nm and high degree of ordering in the *c* direction. The blue pattern is for the sample after being rolled into film with a thickness of 40 μm, demonstrating that the *c*-direction peaks were retained and the prominent (110) peak was absent, due to decrease in the degree of ordering in non-basal directions. **b** TEM image of a collect of flakes, with largest lateral size to be hundred nanometers, with the inset showing the overall SAED pattern. **c, d** TEM images of single- and double-layered nanoflakes, with the insets showing their sketches. **e** Fractured surface SEM image of a thick film of 4 μm made through rolling, showing shearing of layers, with the inset to illustrate flexibility of the film. **f** Fractured surface SEM image of a thicker rolled film (~30 μm), showing the alignment loss of the flakes in the film. Reproduced with permission from [26]. Copyright © 2014, Springer Nature Limited

contact angle of water on the rolled MXene film was as low as 21.5°, demonstrating its hydrophilic behavior.

The enlarge in *c* lattice parameter was also attributed to the fact that the interactions between the MXene layers was weakened, which was supported by the observation that multilayered particles could be easily delaminated by using sonication. As compared with the long-time delamination of f 4 h, sonication times of 0.5–1.0 h were sufficient to delaminate the product in this case. More importantly, the production yield of the flakes could be up to 45 wt%. Because the LiF/HCl mixed etching agent was much milder than HF, thus leading to nanoflakes that had much larger lateral sizes and less defects. MXenes papers obtained in this way exhibited promising electrochemical performances as electrode of supercapacitors.

Similarly, various other fluoride salts, including sodium fluoride (NaF), potassium fluoride (KF), caesium fluoride (CsF), tetrabutylammonium fluoride, calcium fluoride (CaF_2), have been combined with HCl or sulphuric acid (H_2SO_4), as the etching solutions to develop MXenes. It is expected that a wide range of MXenes would be available, with more interesting and useful properties, because of diverse modified surface chemistries and pre-intercalated ions. Successful examples include Nb_2CT_x, Ti_2CT_x, $Cr_2TiC_2T_x$, $Mo_2TiC_2T_x$, $Mo_2Ti_2C_3T_x$, Mo_2CT_x, $(Nb_{0.8}Zr_{0.2})_4C_3T_x$, etc. [11, 33, 58].

Yang et al. reported two groups of 2D Nb_4C_3-based solid solution MXenes, $(Nb_{0.8},Ti_{0.2})_4C_3T_x$ and $(Nb_{0.8}, Zr_{0.2})_4C_3T_x$, with *T* to be the surface termination [33]. The precursor MAX phases, $(Nb_{0.8}, Ti_{0.2})_4AlC_3$ and $(Nb_{0.8}, Zr_{0.2})_4AlC_3$, were synthesized through the conventional solid state reaction method. Intercalation of Li ions into the two materials made them suitable for potential applications in energy storage. Precursor MAX phases, including Nb_4AlC_3, $(Nb_{0.8}, Ti_{0.2})_4AlC_3$ and $(Nb_{0.8}, Zr_{0.2})_4AlC_3$, were prepared from commercial powders of Nb (99.5%, ≤44 μm), Al, Ti, and Zr (all 99.9%, ≤74 μm) and carbon (99%, ≤94 μm). The etching agents included HF (48–51 wt%), hydrochloric acid (HCl, 38 wt%) and lithium fluoride (LiF, 99%). The powders were mixed and hot pressed into dense and single-phase disks at 1700 °C for 1 h at a pressure of 30 MPa in the flow of Ar.

The $Nb_4C_3T_x$ was etched in 50% HF solution for 4 days at room temperature. The $(Nb_{0.8}, Ti_{0.2})_4C_3T_x$ was obtained by etching in 50% HF solution at 50 °C. To get $(Nb_{0.8}, Zr_{0.2})_4C_3T_x$, LiF/HCl mixed solution was used. The mixed solution was made by dissolving LiF in 12 M HCl solution. The $(Nb_{0.8},Zr_{0.2})_4AlC_3$ powder was then etched in the mixed solution for 168 h at 50 °C with the aid of constant stirring. 10 mL HCl solution contained 0.6 g of LiF.

XRD patterns of the Nb_4AlC_3, $(Nb_{0.8}, Ti_{0.2})_4AlC_3$ and $(Nb_{0.8}, Zr_{0.2})_4AlC_3$ powders, before and after the etching reactions, are shown in Figs. 2.24. As observed in Fig. 2.24a, the Nb_4AlC_3 was of single-phase characteristics. According to the positions of the (002) peaks for the three MAX phases, i.e., the inset I in Fig. 2.24a, the $(Nb_{0.8}, Ti_{0.2})_4AlC_3$ and Nb_4AlC_3 phases had a very close *c*-lattice parameter, whereas the value of the $(Nb_{0.8}, Zr_{0.2})_4AlC_3$ one was much larger relatively. As demonstrated by the inset II in Fig. 2.24a, the (106) peaks of the Ti- and Zr-containing MAXs were respectively at the right and left sides of that of Nb_4AlC_3. This was simply because

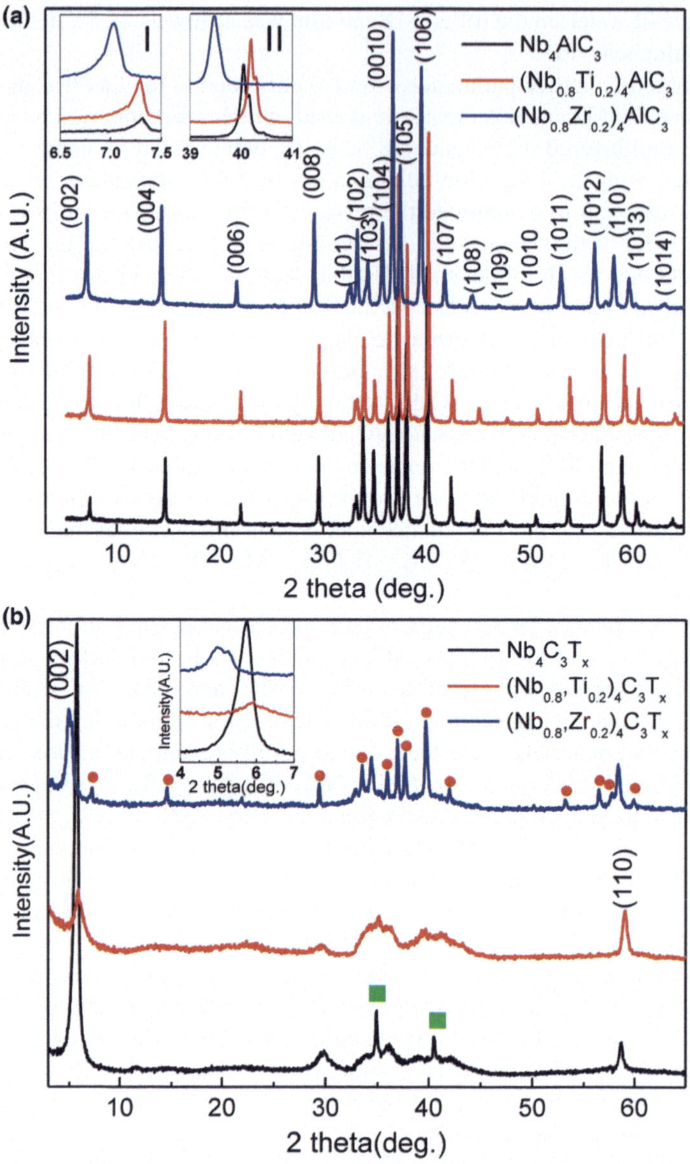

Fig. 2.24 XRD patterns of the three MAX samples: **a** before and **b** after the etching reaction, with the insets in (**a**) and (**b**) showing the partially enlarged XRD patterns. The peaks marked by green squares represent NbC as an impurity phase, while those marked by red circles represent the unreacted MAX phase in the MXene samples. Reproduced with permission from [33]. Copyright © 2016, John Wiley & Sons, Inc.

the atomic radius of Nb (145 pm) is just between those of the Ti (140 pm) and Zr (155 pm).

The Ti:Nb and Zr:Nb atomic ratios in $(Nb_{0.8},Ti_{0.2})_4AlC_3$ and $(Nb_{0.8},Zr_{0.2})_4AlC_3$ were 0.89:4.00, close to designed compositions. After the etching reactions, most of the diffraction peaks of the precursor compounds were absent or significantly broadened. Specifically, the (002) peak was largely weakened and shifted to lower angles, as demonstrated by the inset in Fig. 2.24b. This shift corresponded to an increase in the c-lattice parameters as discussed before. The content of Al was sharply reduced after the etching reactions. For instance, the Al:Nb ratio in the Nb_4AlC_3 was decreased from 0.64:4.00 before etching to 0.12:4.00 after etching. In addition, both the Ti:Nb and Zr:Nb ratios their respect etched products were also reduced, due mainly to then their dissolution during the etching reactions. The formation of the layered structures of MXenes are confirmed by the SEM and TEM images, as seen in Fig. 2.25. The variation in the content of different elements before and after the etching reactions was also supported by the XPS analysis results.

Halim et al. introduced a large-scale method to synthesize 2D Mo_2CT_x the ternary transition metal carbide Mo_2Ga_2C [58]. The effect of synthesis and delamination methods on nanoflake morphologies of the final products was systematically studied. The free-standing Mo_2CT_x films exhibited a semiconductor-like characteristic, with resistivity to increase by nearly one order of magnitude, as the temperature was

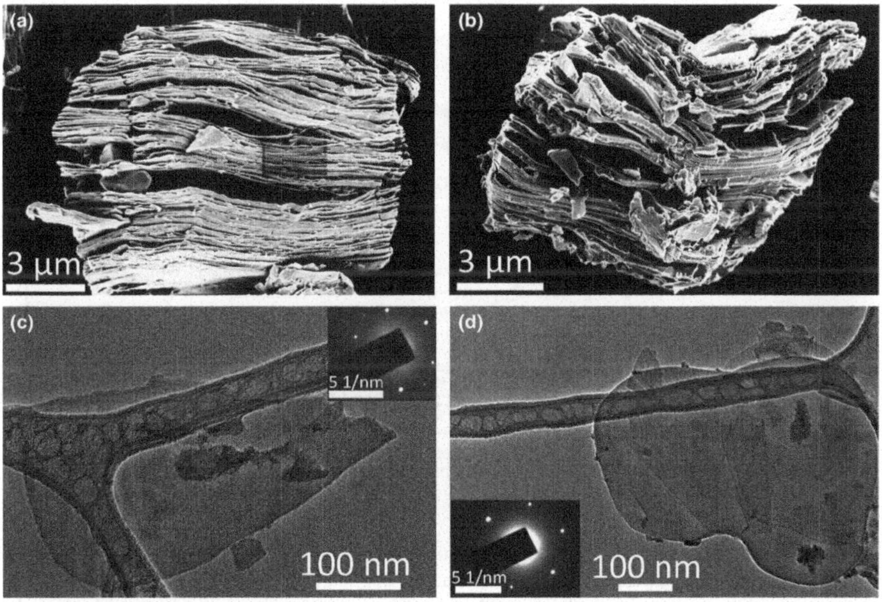

Fig. 2.25 Representative SEM images of the MXene powders: **a** $(Nb_{0.8}, Ti_{0.2})_4C_3T_x$ and **b** $(Nb_{0.8}, Zr_{0.2})_4C_3T_x$. TEM images of the **c** $(Nb_{0.8}, Ti_{0.2})_4C_3T_x$ and **d** $(Nb_{0.8}, Zr_{0.2})_4C_3T_x$ nanosheets, with the insets in (**c**) and (**d**) showing the corresponding SAED of the MXenes flakes. Reproduced with permission from [33]. Copyright © 2016, John Wiley & Sons, Inc.

decreased from 300 to 10 K. In contrast, to $Ti_3C_2T_x$ was metallic as stated earlier. The thick films made of the Mo_2CT_x MXene had promising electrochemical properties as electrodes of supercapacitors and lithium ion batteries.

Mo_2Ga_2C powder was prepared by using the solid liquid reaction of Mo_2C and Ga [59]. As seen earlier, Meshkian et al. used the same method to obtain Mo_2Ga_2C target that was sputtered onto substrates in the form of thin films, which were then etched to form MXene thin films [50]. Mo_2C powder and Ga were mixed at a molar ratio of 1:8. The mixture was sealed and heated at 850 °C for 48 h. The calcination was conducted for one more time at the same temperature for 16 h.

1 g Mo_2Ga_2C power was dispersed in 20 mL 12 m HCl solution for 2 d at room temperature in order to remove the unreacted Ga. Mixed etching solution was made by mixing 3 M LiF and 12 M HCl. 1 g Mo_2Ga_2C powder was soaked in 20 mL mixed etching solution at 35 °C for various time durations. The samples were then thoroughly washed with 1 M HCl solution and 1 M LiCl solution for three times each, following by further washing with DI water until the pH approached 6. The Mo_2Ga_2C power was etched with HF solution as a comparison, with the sample denoted as Mo_2CT_x. The delamination process to synthesize the MXenes is schematically illustrated in Fig. 2.26a [58].

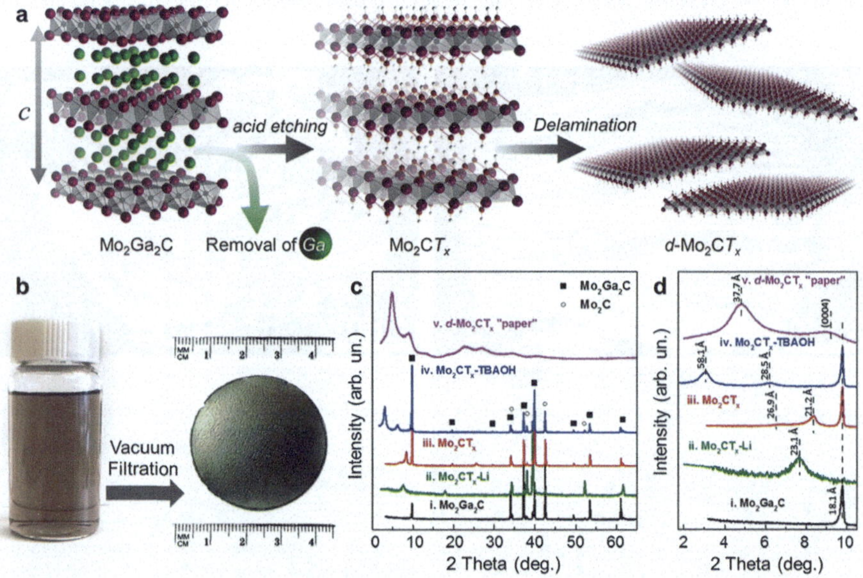

Fig. 2.26 **a** Schematic diagram of the delamination process to obtain Mo_2CT_x nanostructures. **b** Photographs of the diluted delaminated solution and the free-standing paper produced by filtering the delaminated solution through membranes. **c** XRD patterns: (i) Mo_2Ga_2C (black), (ii) Mo_2CT_x–Li (green), (iii) Mo_2CT_x (red), (iv) Mo_2CT_x (blue) intercalated with TBAOH and (v) (purple) free-standing paper after TBAOH intercalation. **d** Enlarged XRD patterns of those in (**c**). Reproduced with permission from [58]. Copyright © 2016, John Wiley & Sons, Inc.

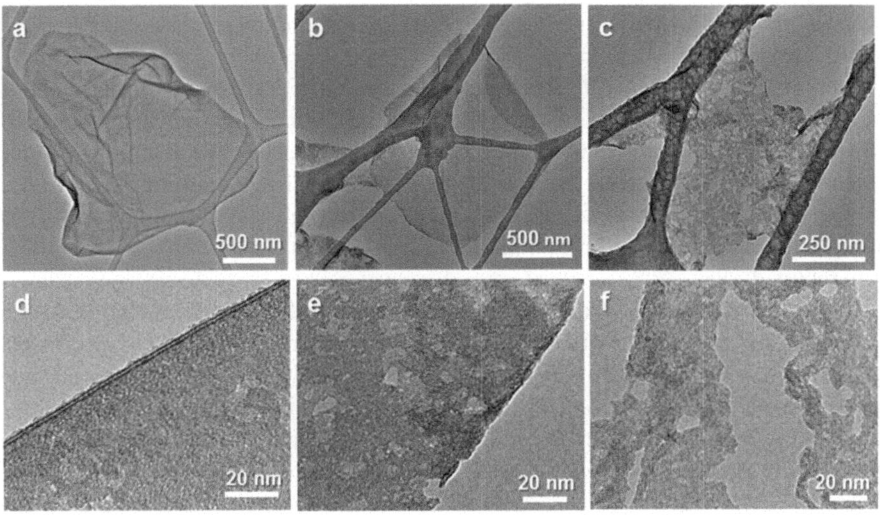

Fig. 2.27 Low magnification TEM images of individual flakes: **a** Mo_2CT_x–Li powders etched for 6 days and delaminated by manual shaking in water for 5 min, **b** Mo_2CT_x after intercalation with TBAOH and hand shaking for 5 min and **c** same as (**b**) but with sonication for 1 h in water. **d–f** High magnification TEM images of (**a–c**). Reproduced with permission from [58]. Copyright © 2016, John Wiley & Sons, Inc.

The sample etched with the LiF/HCl mixed etching solution and inserted with Li^+ ions was denoted as Mo_2CT_x–Li. XRD patterns of the Mo_2Ga_2C (i: black pattern in Fig. 2.26c, d) and Mo_2CT_x–Li (ii: green pattern in Fig. 2.26c, d). All the peaks of Mo_2Ga_2C were replaced by (000l) peaks of Mo_2CT_x–Li. The *c* lattice parameter was increased from the initial 1.81 nm to 2.31 nm. The Mo_2CT_x–Li nanoflakes through the etching for 16 days had pin-holes and other defects, while the one etched for 6 days possessed higher quality, which was almost defect-free, as seen in Fig. 2.27a. Although short etching time resulted in high quality nanoflakes, the production yield was also relatively low.

The Mo_2CT_x was further intercalated with tetrabutylammonium hydroxide (TBAOH) in water. The intercalated Mo_2CT_x powder was then delaminated in DI water either by using hand shaking or sonication, leading suspensions at a concentration of about 4 mg/mL, denoted as *d*-Mo_2CT_x. Free-standing paper sheets could be obtained by filtrating the suspensions through membranes, as illustrated in Fig. 2.27b.

XRD patterns of Mo_2Ga_2C powders after the etching reaction with HF (iii: red patterns in Fig. 2.26c, d) had two (0002) MXene peaks, along with a peak of the unreacted MAX. The two MXene peaks corresponded to *c* lattice constants of 2.12 nm and 2.69 nm, respectively. The observation was attributed to the intercalation of a single layer of water molecules into some of the Mo_2CT_x multilayer interspaces. At the same time, there were also some interlayer spaces were intercalated with two layers of water molecules. Similarly, the intercalation of TBAOH resulted in two

(0002) peaks (iv: blue patterns in Fig. 2.26c, d), corresponding to *c* lattice parameters of 2.85 nm and 5.81 nm, respectively.

In comparison, the Mo_2CT_x free-standing paper had a *c* lattice parameter of about 3.37 nm 37.7 Å (v: purple patterns in Fig. 2.26c, d), which was much smaller than 5.81 nm for the TBA intercalated sample. This corresponded to Δd of about 1.02 nm, which was comparable with the size of an individual TBA cation. In other words, the water molecules that were trapped in between the multilayers could be removed, during the delamination, filtration and drying processes, so that only TBA cations were retained.

As the XRD patterns of the free-standing paper and the multilayered powders (v: purple and ii: green patterns in Fig. 2.26c, d) were compared, it was found that the former were much purer, since the (0002) peak of the Mo_2Ga_2C precursor powders at about 9.5° was completely absent. In addition, the Mo_2CT_x–Li multilayered nanoflakes (ii: green patterns in Fig. 2.26c, d) had a *c* lattice constant that was longer than that of their Mo_2CT_x counterparts (iii: red patterns in Fig. 2.26c, d) by about 10%, because of the intercalation of the Li^+ cations in between the MXene layers.

TEM images of the Mo_2CT_x–Li nanoflakes after etching for 6 days are shown in Fig. 2.27a, d. The individual nanoflake had a lateral size of about 2 μm, with several macroscopic defects and pores. In comparison, the nanoflakes made after etching for 16 days had more defects as mentioned before. TEM images of the *d*-Mo_2CT_x nanoflakes are depicted in Fig. 2.27b, e, which exhibited more defects, as compared with the *d*-Mo_2CT_x–Li one. Defects were also observed in the nanoflakes after intercalation, as seen in Fig. 2.27c, f.

In summary, the samples that were relatively defect-free were suitable for electronic and optical applications, because electronic and optical properties are more sensitive to defects. However, the defective and nanoporous samples could also find their special applications, such as electrodes of energy storage devices and catalysts, because the defective and porous sites could serve as reactive centers to facilitate chemical and electrochemical reactions. Therefore, for different applications, the synthetic parameters of MXenes should be optimized according to different requirements on the final materials.

Liu et al. prepared 2D V_2C MXene from V_2AlC through etching with NaF and HCl mixed solution at 90 °C for 72 h [60]. The as-obtained V_2C MXene had purity of >90 wt% with a trace of $Na_5Al_3F_{14}$ and V_2AlC as the impurities, which was higher than that of the ones synthesized through the etching with HF solution at room temperature. The V_2C MXene also exhibited outstanding electrochemical performances as anode of lithium ion batteries. The precursor V_2AlC powder was synthesized from commercial powders of V, Al and C powders with an atomic ratio of 2:1.1:1 through solid state reaction at 1500 °C for 5 h in the flow in Ar. The as-prepared V_2AlC powder was dispersed in NaF and HCl mixed solution to etching Al at 90 °C for times of up to 168 h. The mixed solution was obtained by mixing 40 ml HCl (36–38 wt%) and 2 g NaF.

XRD results indicated that, after etching for 48 h, the diffraction peaks corresponding to V_2C MXene started to appear. The peak of V_2C shifted to lower angles, along with gradual peak broadening, with increasing etching time, which implied that the increase of inter-layer spacing was increased and the stacking thickness of the V_2C layers was decreased. At the same time, a small amount of $Na_5Al_3F_{14}$ was detected by the XRD analysis. With the proceeding of the etching process, the newly formed V_2C was dissolved in the etching solution. For example, if 2 g V_2AlC was used in the beginning, the quantities of the products were 1.85 g, 1.55 g and 0.93 g, after etching reaction for time durations of 48 h, 72 h and 120 h, corresponding to weight losses of 7.3%, 22.5% and 53.3%, respectively.

SEM results indicated that the precursor V_2AlC was fully exfoliated after etching reaction for 72 h, thus leading to 2D MXene nanosheets. As the etching reaction time was increased to 120 h, small particles were present near the 2D V_2C nanosheets, which were possibly the impurity phase $Na_5Al_3F_{14}$ that could not be washed away. Over the time range of 72–120 h, the separation between the V_2C MXene stacks was gradually enlarged, as the etching reaction time was prolonged. Further increasing the reaction time to 168 h resulted in MXenes with the surface that encountered corrosion and having more defects and impurity particles.

Besides various carbide and carbonitride MXenes, nitride MXenes have also been reported in the open literature, although their popularity is still much lower than that of the formers. However, nitride MXenes are potentially advantageous over the carbide ones in several aspects [61–64]. Generally, the conductivities of transition metal nitrides are much higher than those of carbides, which are thus more promising as electrodes in energy storage devices, such supercapacitors. Specifically, TiN has been acknowledged to show strong plasmonic behavior, implying that nitride MXenes could potentially find applications in transformation optics and metamaterials. Theoretical studies indicated that Ti_4N_3 properly functionalized with O, F, or OH groups exhibited a higher density of states (DOS) at the Fermi level ($N(E_F)$) than its Ti_4C_3 and Ti_3C_2 counterparts. Moreover, the $N(E_F)$ in increased with increasing value of n in the $Ti_{n+1}X_n$.

Ye et al. etched $Ti_4N_3T_x$ with HF solution at room temperature to get Ti_4AlN_3 nanosheets, which could be applied to develop micro-patterns [65]. The precursor Ti_4AlN_3 powder was prepared from powders of Ti, Al, and TiN, with a molar ratio of 1:1.2:2.05, through microwave sintering at 1250 °C for 0.5 h in Ar. For etching reaction, 3 g Ti_4AlN_3 powder was dispersed in 50 mL 20 wt% HF solution, which was conducted at room temperature for 30 h. The obtained powder was collected through centrifugation, followed by thorough washing and drying. The dried powder mixed alcohol with the aid of ultrasonication. After standing still for 24 h, the dispersion was subject to high speed centrifugation to remove the large particles, resulting in stable ink.

Too dilute HF solution (10 wt%) had no effect of exfoliation, while too high concentration (40 wt%) caused serious corrosion of the product. Therefore, the optimal HF concentration in this case was 20 wt% After etching reaction for 30 h, accordion-like structure was observed, indicating the proper exfoliation of the precursor, as seen in Fig. 2.28a. The etched product, Ti_3SiC_2, could be stably suspended in in water or alcohol, after the product was subject to appropriate ultrasonication.

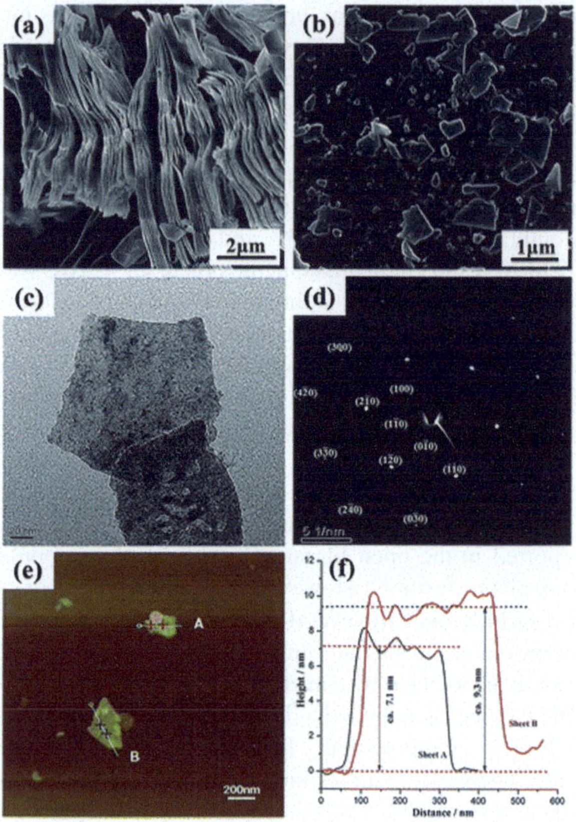

Fig. 2.28 **a** SEM image of the as-obtained Ti_4AlN_3 sample in 20 wt% HF solution. **b** Low magnification SEM image of the etched Ti_4AlN_3 powder after sonication for 6 h. **c** TEM image of newly formed MXene nanosheets and **d** the SAED pattern. **e** AFM image of the nanosheets and **f** the corresponding height measurement. Reproduced with permission from [65]. Copyright © 2015, The Royal Society of Chemistry

The nanosheets exhibited a wide size distribution, ranging from several nanometers to tens of micrometers, as demonstrated in Fig. 2.28b. Representative TEM and SAED pattern of the exfoliation product of Ti_3AlC_2 are shown in Fig. 2.28c, d. The nanosheets had a hexagonal crystalline structure which was inherited from the precursor Ti_4AlN_3. The contents of Ti and N elements were confirmed by the EDS results. The thickness of the exfoliated nanosheets was further examined by AFM, as illustrated in Fig. 2.28e, f. Experimentally, the as-exfoliated nanosheets possessed thickness of 7–9 nm, corresponding to 3–4 atomic layers of the Ti_4AlN_3 crystal unit in the *c* axis direction.

Figure 2.29 shows XRD patterns of the samples before and after the etching reactions. The single phase of Ti_4AlN_3 was confirmed by the XRD results. The relative intensity of the peaks corresponding to (00l) planes of Ti_4AlN_3 phase was increased, while the peaks of (104), (105), (106) planes were weaken. This was because the samples were lying on the silicon wafer with more basal plane of (00l) to be exposed to the X-ray, when compared with the samples in the form of powder. Although there was small amount of TiN presented in the as-exfoliated Ti_4AlN_3 nanosheets, they were most likely due to attachment on the surface of the Ti_4AlN_3

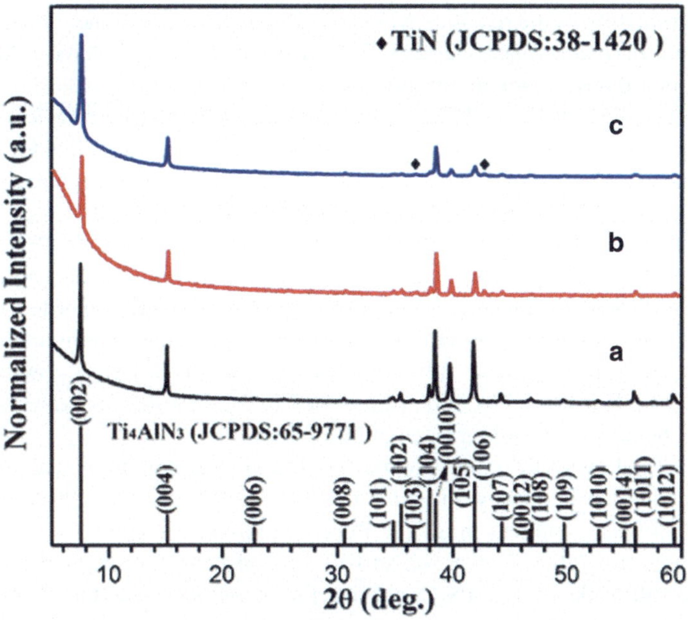

Fig. 2.29 XRD patterns of the Ti_4AlN_3 samples before (**a**) and after (**b**) etching reaction in 20 wt% HF solution, as well as the exfoliated sheet film (**c**). Reproduced with permission from [65]. Copyright © 2015, The Royal Society of Chemistry

nanosheets. Raman results also proved the presence of the Ti_4AlN_3 phase in the exfoliated samples.

The authors attributed this observation to the fact that a whole layer of Ti_4AlN_3 was etched in this case, rather than the removal of only Al during the etching reaction, which was different from those in other MXenes as discussed above. The result of the etching reaction is determined by relative reactivity of the elements. As the X atom was changed from C atom to N atom, the exfoliation reaction became different. In this case, the bond between Al atoms and the Ti–N sub-lattice was much stronger, so the inter-layer strength was too strong to prevent the formation of TiN_x phase in the HF etching solution. In contrast, the freedom of the Al atoms was relatively high in the Ti_3AlC_2 compound. In addition, Ti_2AlN possessed a similar periodic lattice structure to that of Ti_4AlN_3, except for the thinner Ti–N layer. The significance of this study offered a reference to develop other nitride nanosheets with different exfoliation behavior of the carbides.

Urbankowski et al. developed a method to obtain multilayered $Ti_4N_3T_x$ through heat treatment of the mixture with Ti_4AlN_3 powder and fluoride salt at 550 °C in Ar [66]. This route was different from all the previously reported ways to synthesize MXenes, e., selective etching in aqueous acidic solutions. In this case, molten fluoride salt was employed as the etching agent to remove Al from the precursor Ti_4AlN_3. Few-layered nanosheets and even monolayers of $Ti_4N_3T_x$ could be obtained through further delamination of the resultant MXenes. Density functional theory calculations indicated that bare and functionalized Ti_4N_3 were of metallic behavior. Moreover, bare Ti_4N_3 could have magnetic properties.

The precursor Ti_4AlN_3, or MAX, was synthesized by mixing powders of TiH_2, AlN and TiN, with a molar ratio of 2:1:2. These powders were ball milled for 14 h. The mixture was then compacted and calcined at 1275 °C for 24 h at a pressure of 70 MPa. The resultant Ti_4AlN_3 sample was crashed into powder with particles of <37 μm. Potassium fluoride (KF), lithium fluoride (LiF) and sodium fluoride (NaF), with a mass ratio of 0.59:0.29:0.12, were mixed to form fluoride salt. The composition was selected according to the eutectic point of ternary phase diagram, in order to achieve a sufficiently low melting temperature, at which the etching experiment could be conducted. The mixed fluoride salt was blended with the Ti_4AlN_3 powder with a mass ration of 1:1, followed by ball milling for 6 h. 2 g mixture was heated at 550 °C for 0.5 h, at heating and cooling rates of 10 °C/min.

Figure 2.30a shows XRD pattern of the Ti_4AlN_3 after the treatment with the molten salt that was not washed away, in which five fluoride phases were observed. It was found that Al was contained in all the fluoride phases, while no Ti was observed in them, suggesting that Al was selectively etched from the original compound. There were still peaks of Ti_4AlN_3, implying the incomplete reaction of the system. Meanwhile, two new peaks were present near the (002) peak of the precursor Ti_4AlN_3. The first one at about 6.3° was relatively broad, which was most likely to be (002) peak of a layered structure, corresponding to a *c* lattice parameter of 2.7 nm. In other words, $Ti_4N_3T_x$ could have been formed. The presence of another broad peak at 12.7°, which could correspond to (004) peak, further proved that the peak at 6.3° was from $Ti_4N_3T_x$.

The Al containing fluorides were dissolved with diluted sulfuric acid (H_2SO_4). After that, the sample was thorough washed with DI water, revealing the presence of $Ti_4N_3T_x$ and unreacted Ti_4AlN_3. The multilayered $Ti_4N_3T_x$ MXene was finally delaminated and the residual Ti_4AlN_3 was eliminated, by mixing the powder with tetrabutylammonium hydroxide (TBAOH), with the aid of strong ultrasonication and centrifugation. The delaminated $Ti_4N_3T_x$ flakes with small sizes were denoted as d-$Ti_4N_3T_x$.

After the conversion of MAX to MXene, the (001) peaks were broadened and shifted to lower angles, because the inter-layer spacing was enlarged, i.e., the *c* lattice constant was increased. The peak at 6.3° in the XRD pattern of Ti_4AlN_3 after the molten salt reaction was weakened and broadened. Similarly, the XRD pattern of the d-$Ti_4N_3T_x$ had such broadened peak. It was attributed to the shift of the (002) peak from 7.6° to 6.3°, corresponding to an increase in the *c* lattice parameter from 2.3 nm of Ti_4AlN_3 to 2.8 nm of $Ti_4N_3T_x$. Such an increase in *c* lattice parameter was mainly caused by two factors, (i) intercalation of ions during the molten salt reaction and (ii) water intercalation in between the layers during the delamination process.

Figure 2.30b shows SEM image of the sample after the molten salt reaction, confirming the transition of Ti_4AlN_3 to Ti_4N_3 multilayered structure. EDS spectra indicated that Al was still present, with a ratio of Ti:Al to be 4.0:0.9, which could be attributed to the Al containing fluorides. Although the delaminated $Ti_4N_3T_x$ particles

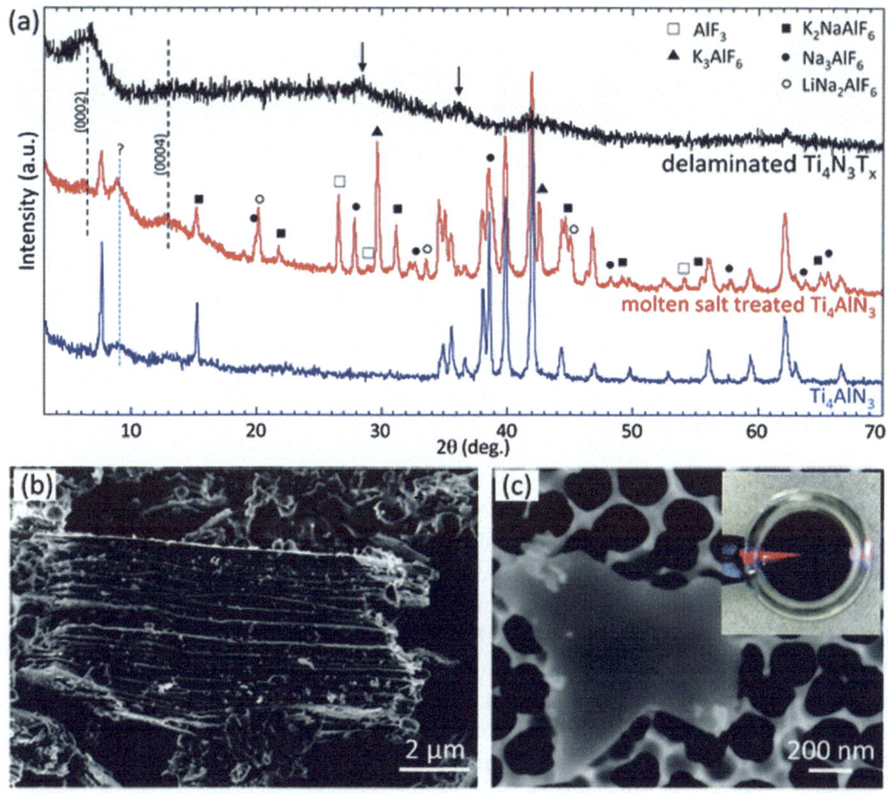

Fig. 2.30 **a** XRD patterns of the original Ti_4AlN_3, molten salt reacted Ti_4AlN_3 and delaminated $Ti_4N_3T_x$. **b** SEM image of the Ti_4AlN_3 after molten salt reaction at 550 °C for 0.5 h in the flow of Ar. **c** SEM image of few-layered $Ti_4N_3T_x$ particles on an alumina membrane, with the inset showing the Tyndall effect of the delaminated flakes dispersed in water. Reproduced with permission from [66]. Copyright © 2016, The Royal Society of Chemistry

were mostly at the scale of hundred nanometer, pretty large nanoflakes with a lateral size of close to 1 μm were occasionally observed, as illustrated in Fig. 2.30c. These large nanoflakes consisted of more layers. The colloidal nature of the suspension of the delaminated Ti_4N_3 nanoflakes in DI water was evidenced by the Tyndall effect, i.e., the inset in Fig. 2.30c.

Representative TEM images of the $Ti_4N_3T_x$ nanoflakes are shown in in Fig. 2.31. The absence of Al was confirmed by the EDX result, thus supporting the fact that the $Ti_4N_3T_x$ was formed through the molten salt etching reaction. SAED (top inset in Fig. 2.31b) of the $Ti_4N_3T_x$ nanoflakes, normal to the basal plane of (002) further provided that they were of single crystal characteristics, where the hexagonal symmetry of the original MAX phase was well maintained after the molten salt exfoliation process.

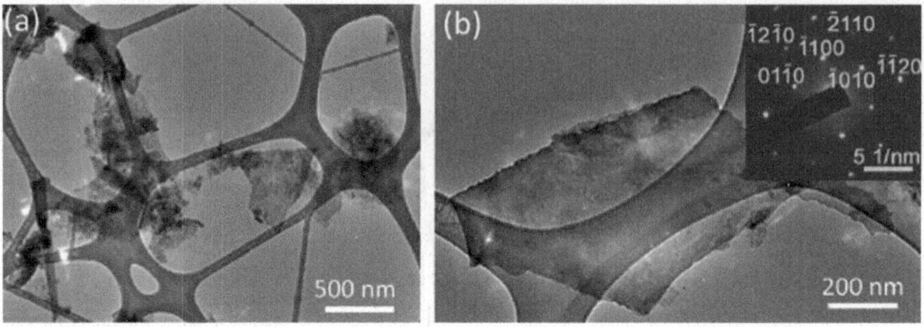

Fig. 2.31 TEM images of **a** several delaminated flakes and **b** an individual $Ti_4N_3T_x$ nanoflake. The inset in (**b**) was the SAED pattern of the $Ti_4N_3T_x$ nanoflake confirming the hexagonal symmetry of basal plane in the parent MAX phase. Reproduced with permission from [66]. Copyright © 2016, The Royal Society of Chemistry

Soundiraraju and George recently synthesized a 2D transition metal nitride of M_2X type MXene, Ti_2N [67]. Generally, it is more difficult to synthesize nitride-based MXenes ($M_{n+1}N_n$), because of the higher formation energy in $M_{n+1}AN_n$ and the lower stability of the $M_{n+1}N_n$ nanolayers in HF based etching solutions. In this study, the authors used a mixed solution of KF and HCl to achieve the selective etching of Al from Ti_2AlN (MAX). Then, multilayered Ti_2NT_x nanostructures were produced by sonicating the exfoliated samples in DMSO, combined with centrifugation, thus leading to few-layered Ti_2NT_x. The obtained Ti_2NT_x had surface-enhanced Raman scattering (SERS) effect, then it was made into paper, silicon and glass-based SERS substrates. A factor of 10^{12} was observed form Raman enhancement, as rhodamine 6G was used as the model compound with excitation at the wavelength of 532 nm. The simple Ti_2N (MXene) paper-based SERS substrate could be employed to detect trace level explosives with high sensitivity.

Ti_2AlN was prepared through high temperature solid state reaction method. It was etched with 2 g Ti_2AlN was put in 20 mL mixed solution of KF and HCl at room temperature for 3 h. The KF–HCl solution was made by dissolving 6 g KF in 100 mL 6 M HCl. After the etching reaction, the suspension was ultrasonicated at 40 °C for 1 h. Then, sample collection was carried out through centrifugation and washing. For delamination, 1 g powder was dispersed in 20 mL DMSO and sonicated for 1 h. Few-layered Ti_2N (FL-Ti_2NT_x) sample was collected accordingly. The FL-Ti_2NT_x product had a yield of 87%. HF (5 wt%), a HF–HCl mixture (5% HF + 100 mL 6 M HCl) and KF (6 wt%) were also tested for comparison.

The formation of Ti_2AlN phase was confirmed by Raman spectrum and XRD result, as illustrated in Fig. 2.32a, b, respectively. A trace of TiAl was detected in the XRD pattern as a second phase (Fig. 2.32b). The as-prepared Ti_2AlN had a *c* lattice constant of 1.359 nm. Figure 2.32c shows a representative SEM image of the layered structure of the MAX compound. The elemental composition, i.e., the presence of Ti, Al and N, was confirmed by the EDS spectrum.

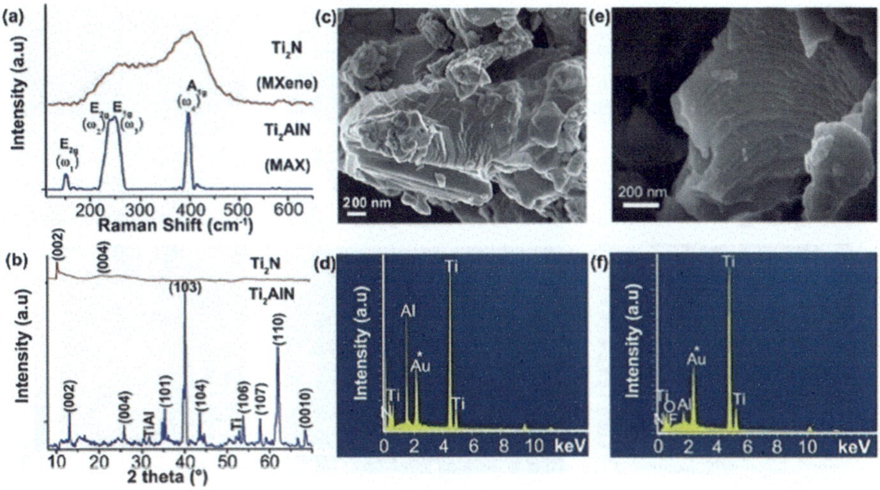

Fig. 2.32 **a** Raman spectrum of the Ti_2AlN (MAX) and Ti_2N (MXene) samples. **b** XRD pattern of the Ti_2AlN (MAX) and Ti_2N (MXene). **c**, **d** SEM image and EDS spectrum of Ti_2AlN (MAX). **e**, **f** SEM image and EDS spectrum of Ti_2N (MXene). Reproduced with permission from [67]. Copyright © 2017, American Chemical Society

When using 5% HF etching solution, oxides were formed on the surface together with the presence of fused morphology, after reaction for 1 day at room temperature. After etching reaction in HF–HCl mixture and KF solutions, the typical layered morphology of was damaged, along with formation of nanostructures on surface of materials. The white precipitate was titanium dioxide (anatase), as confirmed by Raman result. Therefore, the KF–HCl mixture solution was a promising etching agent to remove Al element.

After the removal of Al, the *c* lattice parameter was increased to 1.733 nm of Ti_2NT_x from the original 1.359 nm of Ti_2AlN. Obviously, the morphology of the Ti_2NT_x nanostructures prepared with the KF–HCl mixture etching agent was different from that of those MXenes exfoliated with HF solution, as observed in Fig. 2.32e. As discussed earlier, the delamination effect of the MXenes etched with LiF–HCl solution was not obvious, because of the intercalation of Li^+. Similarly, K^+ ions could also be intercalated into the layers of Ti_2NT_x. The elemental composition containing Ti and N, as well as O and F, was evidenced by the EDS spectrum, as illustrated in Fig. 2.32f.

Figure 2.33a shows a TEM image of the few-layered Ti_2NT_x, revealing transparent characteristics of the nanosheets, under the irradiation of electrons. Meanwhile, several wrinkled and curved parts were present at edges of the nanosheets in the TEM, FESEM, and AFM images, as demonstrated in Fig. 2.33a, c, f. SAED pattern is shown as the inset in Fig. 2.33a, confirming that the hexagonal symmetry of the original MAX nanoflakes was retained after the delamination process. Figure 2.33b shows HRTEM image of the sample, revealing profile of the cross section of stacked

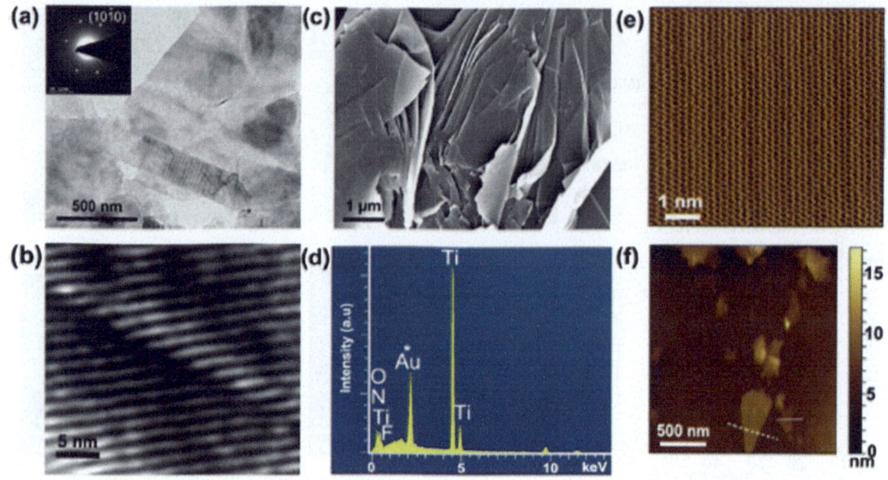

Fig. 2.33 Properties of the Ti_2NT_x (MXene) powder after delamination with DMSO: **a** low-magnification TEM image with the inset to be SAED pattern, **b** HRTEM image, **c** SEM image, **d** EDS spectrum of few-layered Ti_2NT_x, **e** STM image and **f** AFM topography image. Reproduced with permission from [67]. Copyright © 2017, American Chemical Society

layers in the few layered Ti_2NT_x and the spacing left in between the layers after the removal of Al. The delaminated Ti_2N exhibited stacked lamellar morphology, with lateral dimensions at micrometer scale, as seen in Fig. 2.33c.

Figure 2.33e shows representative scanning tunneling microscopic (STM) image of the sample, indicating the hexagonal close packing (top view) nature of titanium in the few-layered Ti_2NT_x. According to AFM topography image (Fig. 2.33f) and height profiles (Fig. 2.33g), the MXene sample had 1–3 layers. Specific surface areas were 1.03 m^2 g^{-1}, 12.6 m^2 g^{-1} and to 32.4 m^2 g^{-1}, for the original Ti_2AlN, exfoliated and intercalated multilayered Ti_2NT_x and delaminated few-layered Ti_2NT_x, respectively. With pellets of 13 mm in diameter and 2 mm in thickness, the Ti_2AlN and few-layered Ti_2NT_x had conductivities of 28,570 and 4950 S/cm, respectively, which further evidenced the removal of metallically bonded Al layers in the original Ti_2AlN compound.

Xie et al. reported a fluoride-free route to develop to etch Ti_3AlC_2 in 1 M sodium hydroxide (NaOH) solution at 80 °C for 100 h, combined with a subsequent hydrothermal treatment at 80 °C for 2 h in 1 M H_2SO_4 [68]. Although the etching reaction occurred at only partially on surface, this fluoride-free etching approach worth to be further explored. Formation process of Ti_3AlC_2 samples as supporting materials with the hydrothermal etching method is shown in Fig. 2.34. Firstly, Ti_3AlC_2 powder was treated in NaOH solution, which was then subject to hydrothermal treatment H_2SO_4. In this process, Al in the Ti_3AlC_2 surface layers was selectively etched, thus leading to surface exfoliated Ti_3C_2 layers terminated with OH groups, which was denoted as e-TAC. The e-TAC was loaded with Pt nanoparticles as a catalyst (Pt/e-TAC) for the oxygen reduction reaction (ORR).

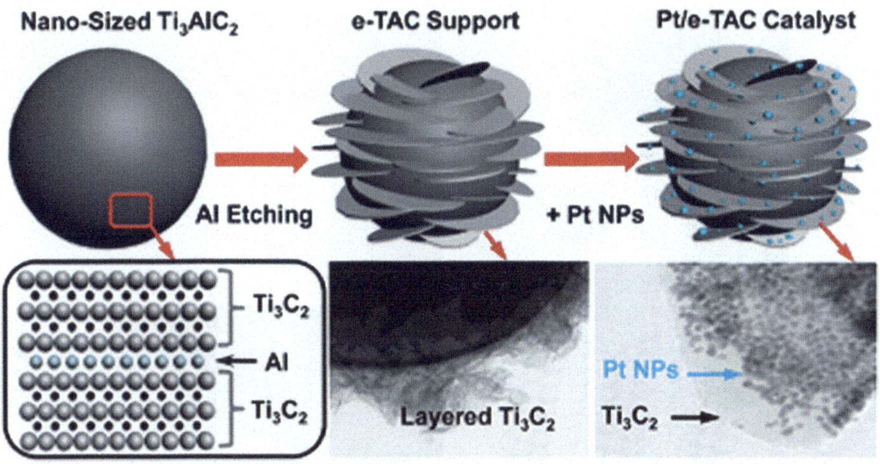

Fig. 2.34 Schematic diagram of the process to develop the Pt/e-TAC catalysts, together with representative TEM analysis results. Reproduced with permission from [68]. Copyright © 2014, The Royal Society of Chemistry

The e-TAC as a support to Pt nanoparticles had advantages over carbon and low conductive Ti-based composites, due to its high chemical stability and high conductivity. The high chemical stability could prevent the corrosion of the support materials, whereas the high conductivity ensured a smooth transport of electrons. Meanwhile, the interaction Pt nanoparticles into the e-TAC effectively increased the dispersing behavior of the well anchored Pt nanoparticles. Furthermore, the layer structure on surface of Ti_3C_2 served as a catalyst together with the Pt nanoparticles, through the electron transfer mechanisms from the e-TAC support to the Pt nanoparticles. In other words, the shift in the d-band structure of the Pt nanoparticles weakened the interactions between the Pt nanoparticles and the intermediate items, which are generally a second metal in the traditional Pt-M catalysts. As a result, the reduction of oxygen was largely enhanced, as compared with the conventional carbon-based support. In this regard, the hybrid Pt/e-TAC catalysts had potential to replace the conventional commercial Pt/C catalysts, in both reactivity and chemical stability.

XRD results indicated that the loading of Pt nanoparticles to Ti_3AlC_2 has not affect on its crystal structure, also confirming that the etching of Al only took place at the surface, due mainly to the outstanding chemical stability of Ti_3AlC_2 and the relatively weak etching capability of NaOH solution compared with HF solution. This observation was also confirmed by TEM analysis results, as shown in Fig. 2.35. The morphology of Ti_3AlC_2 is demonstrated in Fig. 2.35a. Figure 2.35b illustrates its laminated structure. A schematic diagram of the laminate structure is depicted in Fig. 2.35c. Figure 2.35d shows structure of the e-TAC sample with surface modification, where the thickness of the modification layer was about 0.1 μm. The Ti_3C_2 with a cicada wing-like morphology was attached onto the bulk Ti_3AlC_2 with Al layers that had been removed. Therefore, the Ti_3C_2 layer was more chemically stable than

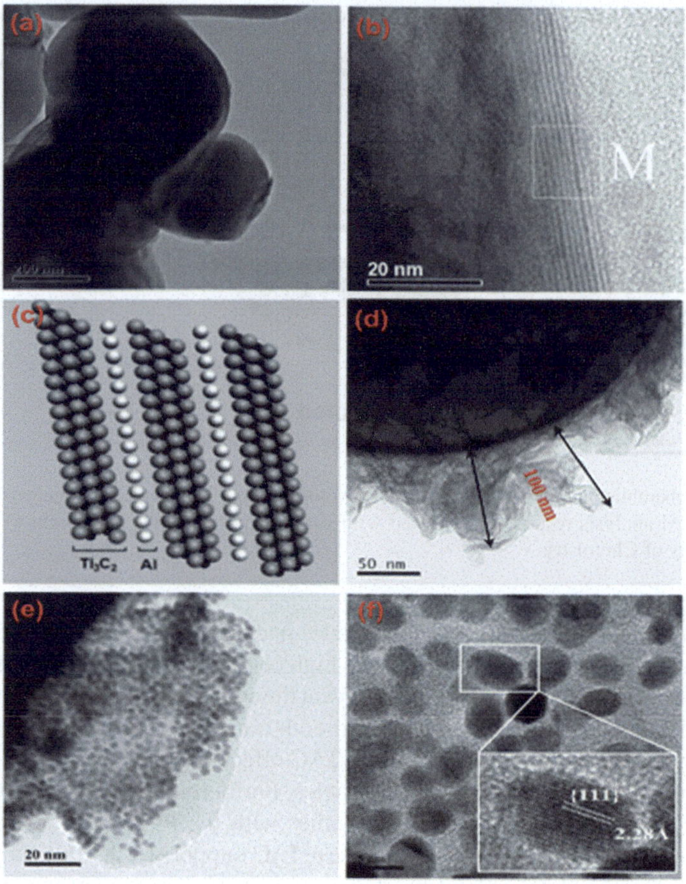

Fig. 2.35 **a** TEM image of the bulk Ti_3AlC_2. **b** HRTEM image of the Ti_3AlC_2 laminated structure. **c** Atomic configuration of square M in (**b**). **d** TEM image of the e-TAC sample. **e** TEM image of the Pt/e-TAC hybrid catalyst. **f** HRTEM image of the Pt/e-TAC. Reproduced with permission from [68]. Copyright © 2014, The Royal Society of Chemistry

the original Ti_3AlC_2 at both acidic and oxidizing conditions. The dominance of Ti and C on surface of e-TAC was also evidenced by the XPS results.

Nuclear magnetic resonance (NMR) spectroscopy, with high sensitivity to ^1H and ^{19}F nuclei, has been used to characterize the compositions of V_2CT_x and $Ti_3C_2T_x$ [69, 70]. For example, Harris et al. used ^1H NMR experiments to study HF-etched V_2CT_x [69]. They found that –OH groups was directly attached to V_2CT_x, with a layer of –H groups originated from H_2O to be bonded to the –OH surface. At the same time, ^{19}F NMR experiments suggested that the –F groups were also directly connected to the surfaces of V_2CT_x.

V_2AlC was ball milled before the etching reaction to form V_2CT_x, in order to reduce the particle sizes and thus enhance the penetration of the RF signals. Proton

NMR offers much strong signal intensity as compared with other NMR nuclides, due to its high natural abundances and the large magnetic moments of the ^{1}H nucleus. The ^{1}H NMR results of V_2CT_x could be well fitted with a summation of four sites. In most cases, the spectrum was composed of two sites at 6.5 and 85 ppm. The peak at 6.5 ppm was a typical chemical shift for water molecules inside the porous structures, implying that the materials adsorbed large amount of water. In contrast, the peak at 85 ppm was outside the range of the typical ^{1}H chemical shift, which was present as a very broad peak. These two characteristics of the 85 ppm peak could be observed, only when there were magnetic interactions with unpaired electrons. In particular, the large isotropic chemical shift was only visible, as the density of the unpaired electron was located at the ^{1}H nucleus. As a consequence, this kind of peak was attributed to the surface effect of MXene. Moreover, this effect was generally a Knight shift, because MXenes are usually of either metallic or low band gap semiconducting characteristics. Noting that the etching process is most likely conducted in aqueous solutions, the 85 ppm peak was safely ascribed the OH groups on the surface of MXenes.

The direct link of both the –OH and –F groups to the surfaces of $Ti_3C_2T_x$ was confirmed by Hope et al. [70]. Moreover, with quantitative NMR experiments demonstrated, the authors found that majority of the surface terminations were present as a mixture of –O and –F, in which the –OH was the minor one. In addition, the authors also observed that the proportion of the surface terminations was strongly dependent on the synthetic methods. For instance, HF exfoliated $Ti_3C_2T_x$ had more –F termination than the $Ti_3C_2T_x$ obtained through HCl/LiF etching process.

2.3.2 *MXenes with Single- or Few-Layers*

Once the strong bonds of M-A are replaced by relatively weak bonds, such as hydrogen bonds and van der Waals force, single- or few-layered MXene nanosheets could be obtained, by further intercalating and delaminating the multi-layered MXenes after exfoliation, as discussed above. For example, the as-obtained multi-layered $Ti_3C_2T_x$ and Mo_2CT_x through the etching reaction in HCl/LiF solution, could be completely and readily delaminated by additional ultrasonication in water [26, 58]. The effective delamination was ascribed to the intercalated water molecules and possibly cations in between the multi-layered MXenes nanostructures, which served as spacer and lubricant to allow the shearing movement of the layers.

Zhang et al. found that $Ti_3C_2T_x$ made with HCl-LiF etching solution had a larger *c* lattice parameter of about 3.1 nm. At the same time, the delamination ratio was enhanced from 6.5% to 29.2% after sonication for 60 min [71]. When using HCl-LiF to etch Ti_3AlC_2, there were two important parameters, i.e., (i) etching reaction temperature and (ii) washing solution. It was observed that an increase in etching temperature was positive to efficiency of the exfoliation step from Ti_3AlC_2 to multi-layered $Ti_3C_2T_x$ MXene. However, a more serious surface oxidation of was observed after the removal of Al. The optimal etching temperature was 35 °C, in terms of delamination ratio.

Furthermore, if ethanol was used to wash the exfoliated sample, instead of distilled water, the delamination ratio of the $Ti_3C_2T_x$ MXene could be largely enhanced, due to the co-intercalation behavior of ethanol molecules that are larger than water molecules. The $Ti_3C_2T_x$ nanoflakes could be well dispersed in water, with which flexible free-standing $Ti_3C_2T_x$ paper sheets could be obtained through filtration. The paper sheets had controllable thickness, hydrophilic surface and high conductivity. For instance, the sample with a thickness of 5 μm had a conductivity of 2×10^5 S/m. In addition, the delaminated $Ti_3C_2T_x$ MXene exhibited promising electrochemical performances as anode of lithium ion batteries, with reversible capacities of 226.3, 137.9, 102.0 and 47.9 mAh/g, at the current densities of 100, 300, 1000 and 3000 mA/g, respectively, which were much better than those of the Ti_3C_2 MXene obtained through the popular HF etching process.

6 M solution (10 ml). 0.666 g LiF dissolved in 10 ml 6 M HCl (diluted from concentration solution). 1 g Ti_3AlC_2 powder synthesized through high temperature solid state reaction method was immersed into to the etching solution. The reacted suspension was held at different temperatures for 1 day. Then, the mixtures were thoroughly washed with distilled water or ethanol, which were collected through centrifugation. 0.3 g exfoliated powder was dispersed in 150 ml deaerated water for delamination, which was triggered with strong sonication for 60 min in the flow of Ar, followed by centrifugation at 3500 rpm for another 60 min. As a result, stable MXene colloidal suspension of $Ti_3C_2T_x$ nanoflakes was obtained after vacuum filtration with polyethylene membranes.

According to XRD patterns, MXene (000l) peaks, such as (0002), were present for all samples, after etching in etching HCl-LiF solution for 1 day. Over the reaction temperature range of 30–55 °C, the intensity and sharpness of the (000l) peaks were gradually increased, while those of the peaks of Ti_3AlC_2 were decreased accordingly, suggesting that the contents of MXene and Ti_3AlC_2 were increased and decreased, respectively. XRD pattern of the sample etched at 55 °C had almost no peaks of Ti_3AlC_2, indicating that the original MAX had been completely exfoliated. The Al/Ti atomic ratio estimated from the EDS results was gradually decreased with increasing reaction temperature, which was in agreement with the XRD results. The 30 °C $Ti_3C_2T_x$ sample had a *c* lattice parameter of 2.575 nm and the values were 2.7–2.8 nm for the samples after etching reaction 35–55 °C, with the one at 35 °C having the largest *c* lattice constant of 2.777 nm. If the samples were washed with ethanol, the largest *c* constant was 3.099 nm for the 35 °C sample.

Figure 2.36 shows SEM images of the exfoliated items after etching reaction for different times, followed by washing with distilled water washed. Obviously, the degree of exfoliation was increased with increasing temperature of the etching reaction. All the samples exhibited typical accordion-like morphology, along with cross-sectional shear slips of the multi-layer MXenes nanostructures. This observation implied that the exfoliation of Ti_3AlC_2 through the etching with HCl-LiF was a temperature dependent kinetic process. As mentioned before, the MXenes etched

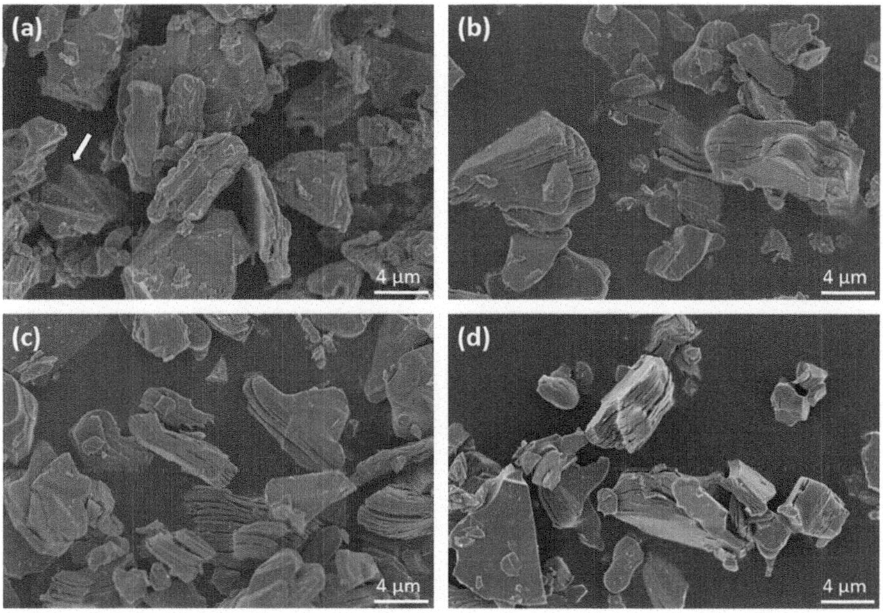

Fig. 2.36 SEM images of the exfoliated samples washed with distilled water after exfoliation at different temperatures: **a** 30 °C, **b** 40 °C, **c** 50 °C and **d** 55 °C. Reproduced with permission from [71]. Copyright © 2017, Elsevier

with HF had amorphous nature, with broadened (0001) peaks. In comparison, the HCl-LiF etching resulted in $Ti_3C_2T_x$ products with much better crystallization.

Figure 2.37 shows SEM images of the exfoliated samples after reaction at 35 °C, which were washed with distilled water and ethanol. The samples washed with distilled water seemed to be cleaner, whereas those washed with ethanol had impurity-like small particles with a regular columnar morphology. In addition, the latter exhibited well developed accordion-like structure and more distinctive exfoliation characteristics after washing with ethanol. This outcome was in good agreement with the XRD results as discussed above.

Figure 2.38 shows morphologies and microstructures of the $Ti_3C_2T_x$ free-standing papers. Thickness of the free-standing paper could be well controlled from several to tens of micrometers, by simply using the suspensions with different concentrations. Cross-sectional SEM images indicated that the papers had a well-aligned stacking structure, as seen in Fig. 2.38a. All the $Ti_3C_2T_x$ free-standing papers were highly flexible, which could be easily folded into different shapes without obvious damage, as demonstrated in Fig. 2.38d. The $Ti_3C_2T_x$ papers were highly conductive. Hydrophilic characteristics of the papers were confirmed by the small contact angle (25°) of the water droplet on surface of the $Ti_3C_2T_x$ paper.

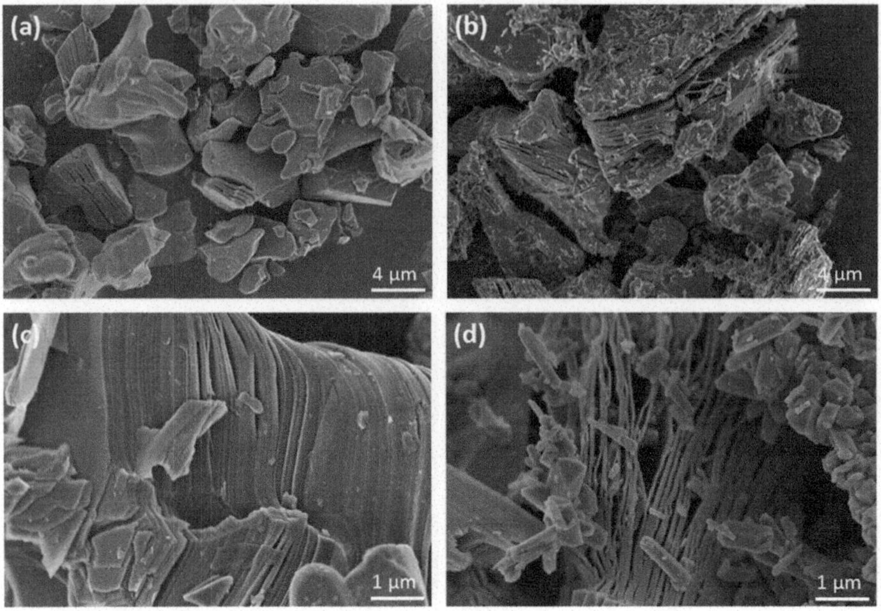

Fig. 2.37 SEM images of the exfoliated samples after etching reaction at 35 C for 24 h and washing with distilled water (**a**, **c**) and ethanol (**b**, **d**). Reproduced with permission from [71]. Copyright © 2017, Elsevier

Shahzad et al. developed a modified procedure with a high molar ratio of LiF:Ti_3AlC_2 to be 7.5:1 (instead of 5:1), so as to provide more Li^+ ions for intercalation [72]. Meanwhile, the authors used double molar ratio of HCl:LiF, in order to promote the etching behavior of aluminium. As a result, $Ti_3C_2T_x$ could be delaminated without the need of ultrasonication. A more concentrated HCl solution (9 M rather than 6 M) would have provided more protons (H^+) to be exchanged with the Li^+ ions.

Lipatov et al. compared qualities and properties of $Ti_3C_2T_x$ nanoflakes made with LiF-HCl mixed etching solution at 35 °C for 24 h [21]. Both the dimension and quality of the final $Ti_3C_2T_x$ nanoflakes were dependent on the molar ratios of LiF to Ti_3AlC_2 (5:1 and 7.5:1) in 6 M HCl solution. To be clearer, the two ways were denoted as Route 1 and Route 2. In Route 1, Ti_3AlC_2 powder was etched with LiF-HCl solution of 5:1, while Route 2 was the case of 7.5:1. In addition, the sample derived from Route 2 was not subject to sonication. Figure 2.39a shows a schematic diagram indicating the difference between two routes.

As seen in Fig. 2.39b, the nanoflakes made through Route 1 had lateral sizes of 200–500 nm. Meanwhile, there were incompletely exfoliated particles in this case. The Route 2 sample contained nanoflakes with much larger sizes, in the range of 4–15 μm, as demonstrated in Fig. 2.39d. Besides large sizes, they were also more uniform and had cleaner surfaces. At the same time, the nanoflakes had nearly same brightness in the SEM image, implying that their thicknesses were almost the same.

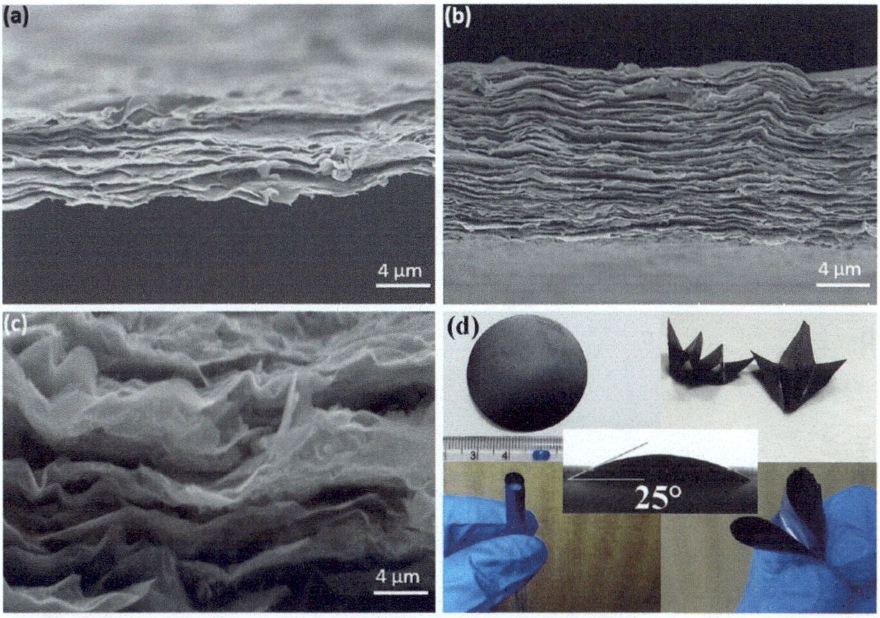

Fig. 2.38 **a**, **b**, **c** Cross-sectional SEM images of the MXene papers. **d** Photographs of the $Ti_3C_2T_x$ free-standing papers that could be folded into different shapes and wrapped onto a 4-mm-diameter glass rod without breakage, showing their outstanding mechanical flexibility, while the middle image shows the contact angle (25°) of a water droplet on surface of the paper, indicating its hydrophilic nature. Reproduced with permission from [71]. Copyright © 2017, Elsevier

TEM images of the $Ti_3C_2T_x$ nanoflakes made with Route 1 and Route 2 are shown in Fig. 2.39c, e respectively. According to the magnification images, it was clearly observed that shape, size and morphology of the $Ti_3C_2T_x$ nanoflakes were highly dependent on the synthetic parameters.

The $Ti_3C_2T_x$ nanoflakes through Route 1 were smaller in size and had uneven edges that were decorated with TiO_2 nanoparticles. High resolution TEM images revealed that the $Ti_3C_2T_x$ nanoflakes of Route 1 had pin holes. In comparison, the nanoflakes generated through Route 2 were larger in size, while they also exhibited well-defined and clean edges, without the presence of any visible pin holes. It was found that crystallographic shape of the MXene nanoflakes through Route 2 well followed that of the original MAX phase crystal, according to the TEM images. In other words, the laminar structure was not broken during the exfoliation process. HRTEM images indicated the hexagonal configuration of the atoms, demonstrating identical crystal structures of two nanoflakes. This observation was further evidenced by the SAED results. In summary, the $Ti_3C_2T_x$ nanoflakes through Route 2 had higher quality and larger size, as compared with those of Route 1.

Mechanical properties of the $Ti_3C_2T_x$ MXenes made with the two Routes were characterized by using MXene papers. The MXene papers were prepared by filtrating the suspensions through a PVDF membrane. Figure 2.40a shows photographs of the

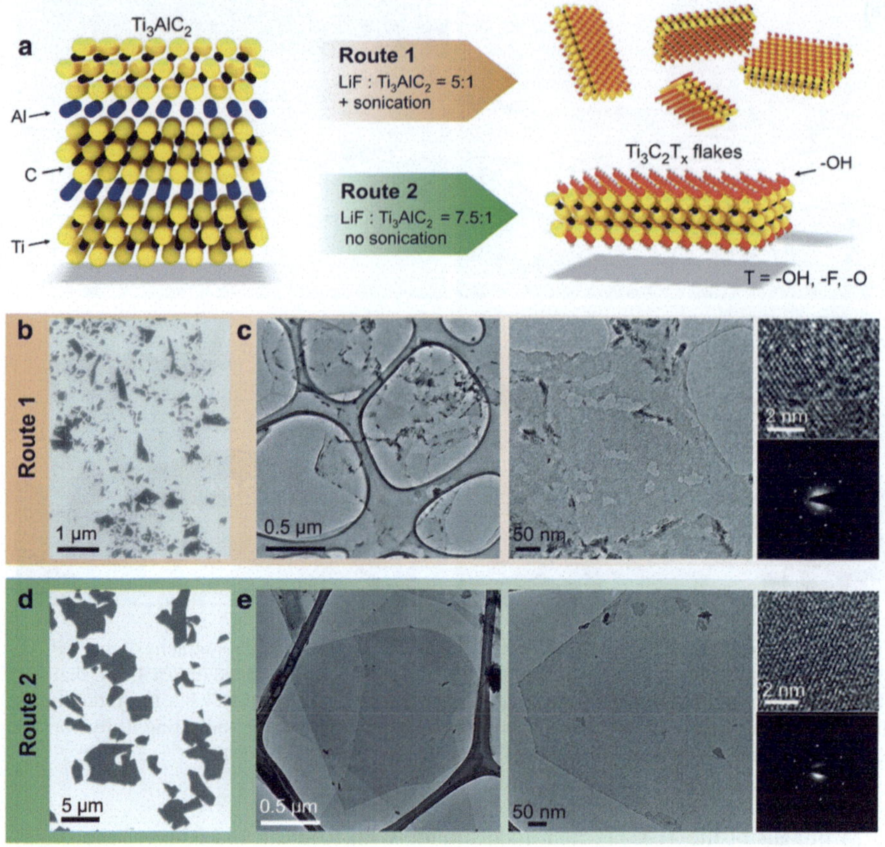

Fig. 2.39 Preparation and characterization of the $Ti_3C_2T_x$ flakes made through two different routes. **a** Schematic diagram showing the difference of Route 1 and Route 2. SEM (**b**) and TEM (**c**) images of the $Ti_3C_2T_x$ nanoflakes obtained through Route 1. SEM (**d**) and TEM (**e**) images of the $Ti_3C_2T_x$ flakes made with Route 2. The small panels in (**c**) and (**e**) were HRTEM images and SAED patterns of the monolayer $Ti_3C_2T_x$. Reproduced with permission from [21]. Copyright © 2016, John Wiley & Sons

two MXene papers. The MXene paper from the $Ti_3C_2T_x$ nanoflakes of Route 1 was relatively fragile, which could be easily crashed into powder, as seen in Fig. 2.40a. Comparatively, the $Ti_3C_2T_x$ paper of Route 2 was much stronger. Obviously, the higher mechanical strength of the Route 2 paper was attributed to the large sizes of the MXene nanoflakes.

Figure 2.40b shows XRD patterns of the two samples. The Route 2 $Ti_3C_2T_x$ sample had only (00*l*) peaks, suggesting that it had a layered structure and consisted of stacked nanoflakes, with an interplanar spacing of 1.242 nm. In comparison, XRD pattern of the Route 1 $Ti_3C_2T_x$ sample had only (001) peak, implying that the stacking of the MXene nanoflakes had a much lower degree of ordering. Meanwhile, peaks

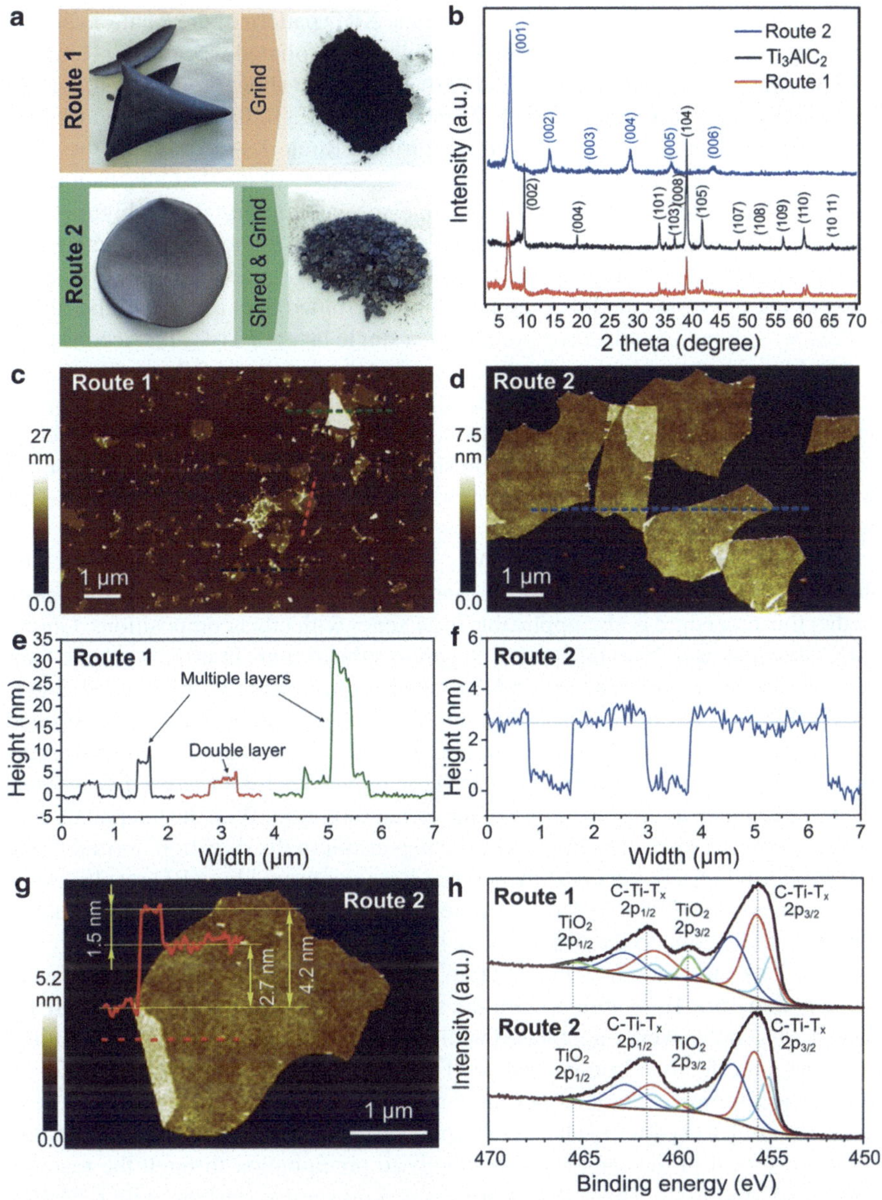

Fig. 2.40 **a** Photographs of the MXene papers from the powder synthesized by using two routes before and after grinding. **b** XRD patterns of the two $Ti_3C_2T_x$ samples shown in panel (**a**), together with that of the original Ti_3AlC_2. AFM images of the $Ti_3C_2T_x$ flakes deposited on Si/SiO_2: **c** Route 1 and **d** Route 2. **e** AFM height profiles measured along the dashed lines in (**c**). The colors of the height profiles in (**e**) correspond to the colors of the dashed lines in (**c**). **f** AFM height profile measured along the dashed line in (**d**). **g** AFM image of a folded $Ti_3C_2T_x$ nanoflake (Route 2) on Si/SiO_2 substrate, with the inset showing the height profile measured along the red dashed line. **h** Ti 2p XPS spectra of the two $Ti_3C_2T_x$ samples. Reproduced with permission from [21]. Copyright © 2016, John Wiley & Sons

of the original MAX phase were observed in the XRD pattern of the Route 1 sample, indicating that the MAX phase was not completely exfoliated when using Route 1.

AFM characterization results of the two samples are shown in Fig. 2.40c–g. As illustrated by the AFM images in Fig. 2.40c, d, the $Ti_3C_2T_x$ flakes obtained through Route 2 were much larger than those made through Route 1. According to the AFM height profile, along the blue dashed line in Fig. 2.40d, all the $Ti_3C_2T_x$ nanoflakes nearly had the same height of about 2.7 nm, as revealed in Fig. 2.40e. Figure 2.40f shows an AFM image of the Route 2 sample, revealing its monolayer characteristics. The $Ti_3C_2T_x$ nanoflake could be readily folded, as seen in Fig. 2.40g. The AFM height profile in the inset indicated that the height of the folded area relative to the rest of the flake was about 1.5 nm, corresponding to a single layer of $Ti_3C_2T_x$. According to the AFM height of the flake on the Si/SiO_2 substrate was about 2.7 nm (Fig. 2.40d, f). This increase in the height (thickness) of the nanoflakes was attributed to the adsorbed water molecules. In comparison, the nanoflakes of Route 1 had different thicknesses (Fig. 2.40c, e). Figure 2.40h shows XPS spectra in the Ti 2p region for the two MXenes, demonstrating their similarity in chemical composition, while the slight oxidation of the Route 1 sample was reflected at the same time.

Concurrent delamination of parent $M_{n+1}AX_n$ phases with HCl/LiF mixed etching solution was only observed for $Ti_3C_2T_x$ and Mo_2CT_x. It is thus necessary to explore whether this procedure is also applicable to MXenes with other compositions. Ghidiu et al. conducted experimental studies on cation substitutions in order to check their effect on the interlayer spacing of HF-etched multi-layered $Ti_3C_2T_x$ [73]. Interestingly, Li^+ ions together with a dynamic layer of H_2O molecules could resulted in HF-etched $Ti_3C_2T_x$ multi-layered structures with a high degree of ordering, thus demonstrating a clay-like (hydrated) behaviour, only when LiCl was used as an intercalant source that was present as part of the etchant solution. With the intercalation of H_2O molecules, a discontinuous structural expansion in the direction normal to the basal plane of $Ti_3C_2T_x$ occurred, which had been confirmed by XRD results. Also, different intercalated cations led to different structural variations upon exposure to relative humidity, due to the difference in hydration enthalpy.

Figure 2.41 shows a schematic diagram for the material process. Ti_3AlC_2 powder was etched in 10% HF solution, with the samples to be denoted as $Ti_3C_2T_x$ HF10. The etched samples were intercalated with Li by immersing it in 1 M LiCl solution, but it was not effective. In comparison, if LiCl was mixed in the etching solution, with a concentration of 5 molar per mole of Ti_3AlC_2, intense and sharp (002) peak was present in XRD pattern, which corresponded to a *c* lattice parameter of 2.45 nm. LiF was detected in the samples, due mainly to precipitation through the reaction between HF and LiCl. This LiF impurity was removed by washing with 6 M HCl solution. After that, the sample was suspended in 1 M LiCl for 24 h to retain the Li ions in the structure, so that the intercalation of H^+ could be effectively prevented. The sample obtain in this way was denoted as Li-$Ti_3C_2T_x$.

According to XRD patterns, the intercalation of Li^+ resulted in materials with higher degree of structural ordering. However, the Li^+ ions should be introduced during the etching reaction process, otherwise, no intercalation was achieved. For $Ti_3C_2T_x$ HF10, H_2O molecules were randomly distributed, together with varying

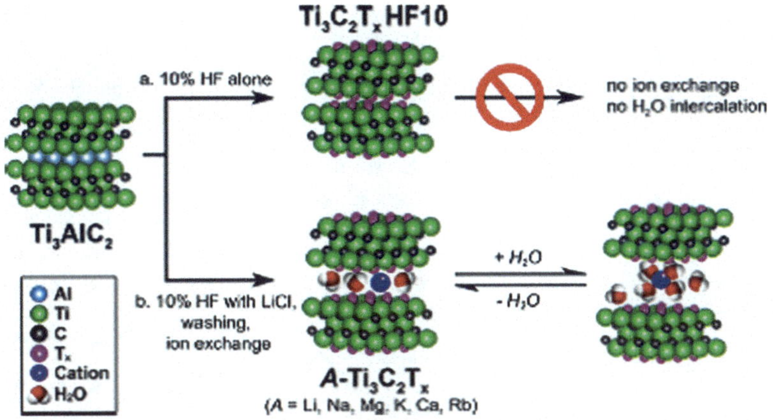

Fig. 2.41 Schematic diagram showing the material fabrication process. **a** Ti_3AlC_2 MAX phase etched with 10% HF solution to remove Al and yield MXene $Ti_3C_2T_x$ HF10 (with T standing for surface terminations, including O, OH, F, etc.). Intercalating cations or water molecules were observed. **b** Etching with 10% HF mixed with LiCl. After the excessive salts were washed out, a wide range of intercalated MXenes denoted as *A*-$Ti_3C_2T_x$ could be obtained, where *A* is an intercalated cation, through ion exchange. The materials were co-intercalated with H_2O reversibly, which caused variation in the basal spacing. Reproduced with permission from [21]. Copyright © 2016, American Chemical Society

surface group interactions, the basal spacing was varied significantly. However, in Li-$Ti_3C_2T_x$, the dynamic H_2O molecules and the Li^+ ions formed a stable layer to fix the interlayer spacings of the MXene nanosheets, thus having higher degree of crystallization. When the LiCl in the solution was replaced by other chlorides, such as 1 M NaCl, 1 M KCl, 1 M RbCl, 0.5 M $MgCl_2$ and 0.5 M $CaCl_2$, more intercalated samples, i.e., Na-$Ti_3C_2T_x$, K-$Ti_3C_2T_x$, Rb-$Ti_3C_2T_x$, Mg-$Ti_3C_2T_x$, and Ca-$Ti_3C_2T_x$ could be readily developed, opening a new way to obtain materials with new properties.

Mashtalir et al. reported a high yield method to delaminate accordion-like $Ti_3C_2T_x$ nanostructures into a mixture of single and few-layered $Ti_3C_2T_x$ nanosheets, with relatively narrow size distribution, through the intercalation with dimethyl sulphoxide (DMSO) at room temperature for 18 h, followed by sonication in DI water for 6 h [18]. Hydrazine monohydrate (HM, $N_2H_4{\cdot}H_2O$) dissolved in *N*,*N*-dimethylformamide (DMF) and urea could also be used for such purpose, which was evidenced by the shift in XRD peaks towards lower angles. Meanwhile, the authors found that solvents, such as thiophene, ethanol, acetone, tetrahydrofuran, formaldehyde, chloroform, toluene, hexane and DMF, could be intercalated into the $Ti_3C_2T_x$ nanosheets.

The intercalation process of MXene is shown schematically in Fig. 2.42a. According to XRD results, after intercalation with HM and/or DMF, the (0001) peaks were still there, while they were shifted to lower angles, as shown in Fig. 2.42b, c. The original compound had a *c* lattice parameter of 1.95 nm, which was nearly unchanged after various post-intercalation steps, such as drying. After intercalation

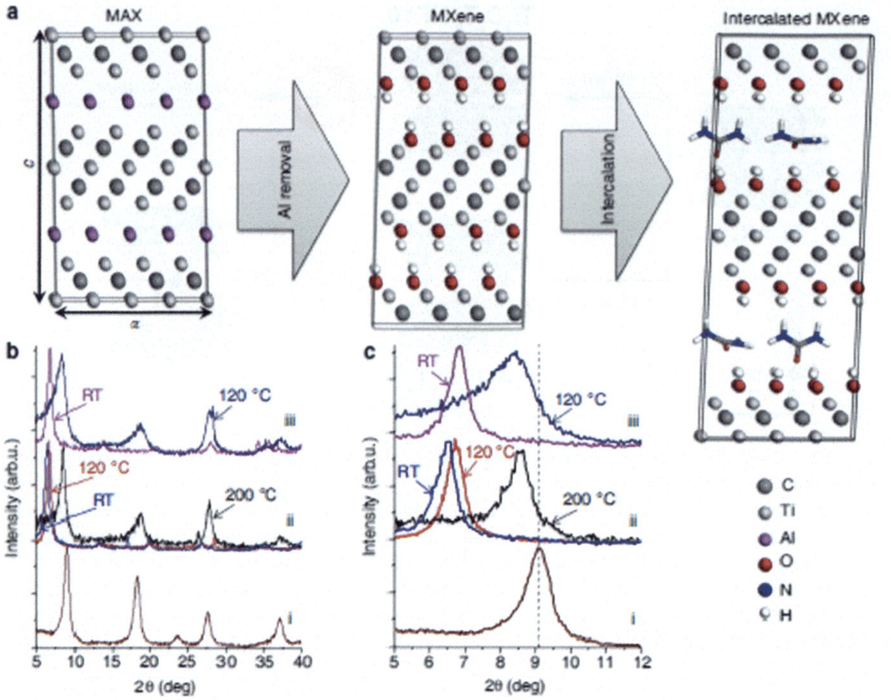

Fig. 2.42 Intercalation of MXenes. **a** Schematic diagram describing the synthesis and intercalation of f-Ti_3C_2: removal of Al layer and intercalation with urea. **b** XRD patterns of f-Ti_3C_2: (i) original compound, (ii) after HM in DMF treatment, washed with DMF, (iii) after HM treatment, washed with ethanol and dried at different conditions. **c** Zoom-ins of (b) in the range of 5–12° for the (0002) peaks. Reproduced with permission from [18]. Copyright © 2013, Macmillan Publishers Limited

with HM and HM in DMF at 80ºC for 1 day, the *c* lattice parameter was increased to 2.548 and 2.68 nm, respectively. The larger increase in *c* lattice parameter for the latter sample was attributed to a synergistic effect of HM and DMF.

If the HM intercalated sample was thermally treated at 120 °C in vacuum, the *c* lattice parameter was decreased to 2.06 nm, implying that the intercalation was a reversible process. In comparison, the sample intercalated with HM–DMF encountered a relatively small decrease, corresponding to smaller increase in the diffraction angles, as seen in Fig. 2.42b, ii). In other words, it was more difficult to de-intercalate sample after intercalation with the combination of HM–DMF than the one intercalated with only HM, which was most likely related to the higher boiling point of DMF (153 °C), as compared with that of HM (114 °C). After the f-Ti_3C_2 obtained through the intercalation with HM–DMF was heated at 200 °C in vacuum, the *c* lattice parameter was decreased to 2.01 nm. The effect of intercalation was also confirmed by XPS results.

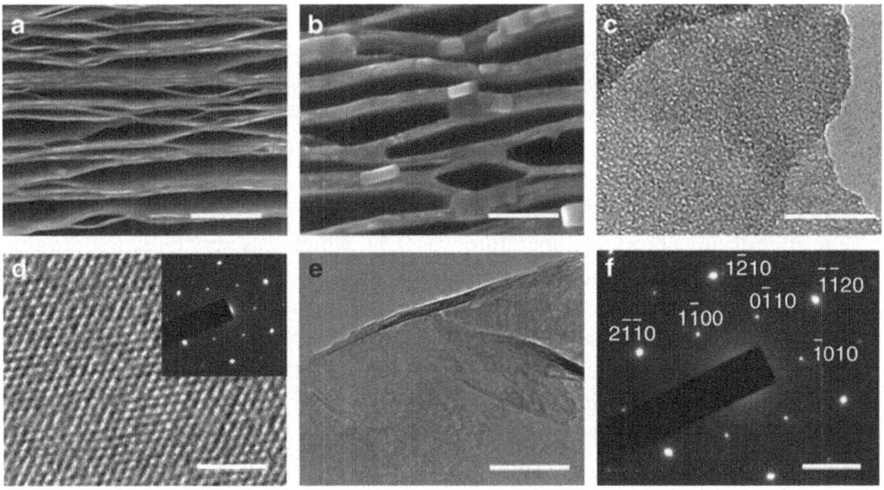

Fig. 2.43 SEM images before (**a**) and after (**b**) the intercalation of f-Ti_3C_2 with HM–DMF at 80 °C for 24 h. TEM (**c**) and HRTEM (**d**) images before intercalation with the inset showing the SAED pattern. TEM image (**e**) and SAED pattern (**f**) of intercalated f-Ti_3C_2. Scale bars: **a**, **b** 200 nm, **c** 10 nm, **d** 2 nm, **e** 40 nm and **f** 4 nm^{-1}. Reproduced with permission from [18]. Copyright © 2013, Macmillan Publishers Limited

SEM images of the f-Ti_3C_2 samples before and after the intercalation with HM–DMF at 80 °C for 1 day are demonstrated in Fig. 2.43a. It was found that the exfoliation state of MXene was retained after the intercalation, so that thickness of the multilayer structures was increased, as revealed in Fig. 2.43b. This was attributed to the gluing effect to combine the individual monolayers, leading to lamellas with thicknesses of 20–50 nm. The XRD peaks corresponding to *c* lattice parameter of 19.5 Å, as illustrated in Fig. 2.42b, i, disappeared, indicating that the completeness of the intercalation.

TEM images and the corresponding SAED patterns of the f-Ti_3C_2 sample that was intercalated with HM–DMF at 80 °C for 1 day are shown in Fig. 2.43e. At the same time, the hexagonal structure of the basal planes was almost unchanged after the intercalation process, as compares with the inset in Fig. 2.43d, f) The interplanar spacing values in the intercalated f-Ti_3C_2 were 2.648 Å for $(0\underline{1}10)$ and 1.540 Å for $(0\underline{2}10)$, thus leading to *a* lattice parameter to be 0.3057 nm, which in accordance with the *a* lattice parameters of f-Ti_3C_2 before intercalation and the original Ti_3AlC_2. This intercalation process was also applicable to other MXenes, such as Ti_3CN and TiNbC.

Electrical resistivities of the non-intercalated MXenes and MXenes intercalated with HM were examined. It was found that all the intercalated samples exhibited higher resistivity than those non-intercalated MXenes, which was primarily attributed to the increase in the *c* lattice parameter after they were intercalated. The magnitudes of increase in the resistivity were different for the MXenes with different degrees of intercalation, which was partially ascribed to the different numbers of the MXene atomic layers. For example, M_3X_2 with five atomic layers experienced an increase in the resistivity by about one order of magnitude. Comparatively, M_2C compounds with three atomic layers encountered a high increase by two orders of magnitude.

The experimental value of *c* lattice parameter for the f-MXene intercalated with N_2H_4 was 2.548 nm, corresponding to an MD-derived N/C ratio of about 0.39, which implied that an almost continuous monolayer of N_2H_4 molecules inside the interlayer space of the MXenes. According to XPS spectra, the N/C ratio in the HM-intercalated MXene after drying at room temperature was 0.37, which was very close to the value of about 0.4 predicted by the MD simulations, as a monolayer of N_2H_4 molecules was formed. In addition, the theoretical and experimental XRD patterns were also in agreement with one another.

The same authors used inelastic neutron scattering (INS) technique to characterize both as-exfoliated and HM intercalated multi-layered $Ti_3C_2T_x$ [74]. This technique is highly sensitive to ^1H atom. It was found that the intercalation of HM into $Ti_3C_2T_x$ resulted in a decrease in the contents of water and the –OH and –F terminations. In addition, the intercalation of DMSO was hindered by its efficiency when delaminating $Ti_3C_2T_x$. Also, complete removal of solvents with high boiling points was a challenge. Moreover, if the DMSO molecules were retained, the lamellar thickness would be increased, which pulled the delaminated nanosheets together, so as to increase the thickness. The underlying mechanism that DMSO only interacted with multi-layered $Ti_3C_2T_x$ has not been clarified.

Naguib et al. developed a simple method for large scale production through the delamination of HF-etched multi-layered Ti_3CNT_x, V_2CT_x and Nb_2CT_x [75]. Basic organic molecules with relatively large sizes, including tetrabutylammonium hydroxide (TBAOH), choline hydroxide and *n*-butylanime, were used for intercalation, which was carried out at room temperature, combined with mechanical agitation, either manual shaking or ultrasonication. In this case, the surface chemistry of the final MXenes was tremendously changed, because the –F terminations were taken away. The intercalation with TBAOH has also been applied to multi-layered $Ti_4N_3T_x$, which led to a mixture of single- and few-layered MXene nanosheets [66].

When the basic organic molecules reacted with the MXenes, they were intercalated with increased d-spacing. Stable colloidal suspensions were readily obtained after hand shaking the intercalated MXenes in water, due to the formation of delaminated nanoflakes. The method could be scaled up by simply increasing the volume of the reaction system. More importantly, this approach was applicable to MXenes with various compositions.

Figure 2.44a shows XRD patterns of Ti_3CNT_x samples before and after the intercalation reaction with TBAOH for different mixing time durations. After intercalation in the TBAOH solution for 2 h, the (0002) peak of Ti_3CNT_x was shifted from $2\theta = 8.26°$ to 4.57°, corresponding to an increment in the *c* lattice parameter from 2.14 nm to 3.86 nm. After intercalation for 4 h, the (0002) peak was further downshifted to $2\theta = 4.5°$, leading to *c* lattice parameter of 3.92 nm. Further increase in intercalation time to 21 h caused no more increase in the *c* lattice parameter. Therefore, 4 h intercalation was the optimal condition for Ti_3CNT_x.

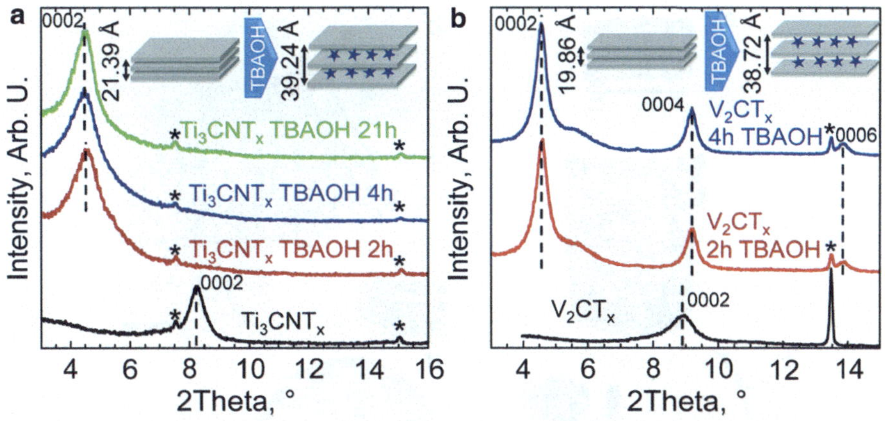

Fig. 2.44 **a** XRD patterns of the Ti_3CNT_x samples before (black pattern at bottom) and after the intercalation in TBAOH solution for 2, 4 and 21 h (red, blue and green, respectively). **b** XRD patterns of the V_2CT_x samples before (black pattern at bottom) and after the intercalation in TBAOH solution for 2 and 4 h (red and blue patterns, respectively). The peaks denoted by * were from the unreacted original MAX phase, with $Ti_4Al(C,N)_3$ in (**a**) and V_2AlC in (**b**). The insets are schematic diagrams of the MXene unit cells before and after the intercalation in TBAOH, with their corresponding change in the *c* lattice parameter. Reproduced with permission from [75]. Copyright © 2015, The Royal Society of Chemistry

XRD patterns of the V_2CT_x samples are shown in Fig. 2.44b. In this case, after reaction for 2 h in TBAOH solution, the (0002) peak shifted from $2\theta = 8.9°$ to 4.6°, which corresponded to *c* lattice parameter increased from 1.99 nm to 3.86 nm. Similarly, as the reaction time was increased to 4 h, the downshift of 2θ was as small as 0.04°, corresponding to a *c* lattice parameter of 3.872 nm. The total increase in *d*-spacing was about 0.94 nm. Meanwhile, (0004) and (0006) peaks were present after the intercalation in TBAOH solution.

As the TBAOH intercalated V_2CT_x powder was dispersed in water, a black colloidal suspension was obtained, after hand shaking. The suspension was still black even after it was centrifuged, as seen in Fig. 2.45a. Colloidal characteristics of the suspension were evidenced by the Tyndall effect, as illustrated by the inset in Fig. 2.45b. More importantly, the two samples intercalated in the solutions of choline hydroxide and n-butylamine were also black colloidal suspensions, which were similar to that of the TBAOH intercalated sample.

A representative TEM image of the delaminated V_2CT_x nanoflake after the intercalation with TBAOH is shown in Fig. 2.45b. According to electron energy loss spectroscopy (EELS) analysis results, the V_2CT_x nanoflake was composed of V, C and O. Most of the V_2CT_x nanoflakes had pin holes on them, which could be produced during the intercalation in the TBAOH solution or after the delamination process in water. As a matter of fact, the holes were mainly formed after the delamination process, because the XRD peaks of the TBAOH intercalated sample were very sharp and strong, confirming a less defective structure.

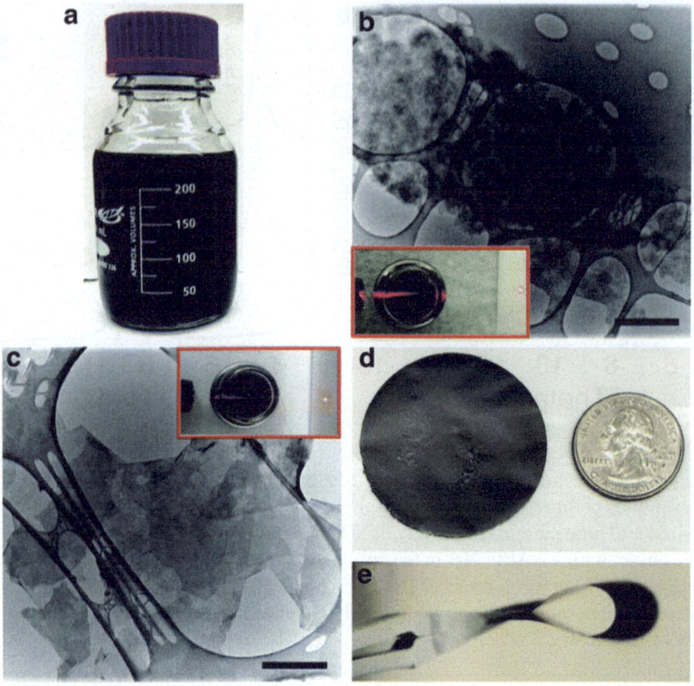

Fig. 2.45 **a** Photograph of a glass bottle containing the delaminated V_2CT_x in water after centrifuging. **b** TEM image of the delaminated V_2CT_x nanoflakes, with the inset showing the Tyndall effect of the suspension. **c** TEM image of the delaminated Ti_3CNT_x nanoflakes, with the inset showing Tyndall effect. Scale bars in (**b**) and **c** are 1 μm. **d** Photograph of the free-standing binder-free Ti_3CNT_x paper obtained through membrane filtration of the colloidal suspension of the delaminated Ti_3CNT_x. Photograph of the same paper shown in (**d**), demonstrating their flexibility. Reproduced with permission from [75]. Copyright © 2015, The Royal Society of Chemistry

The V_2CT_x colloidal suspension experienced a variation in color from the initial black to yellow in two days, which was attributed to the formation of pentavalent V in the forms of VO_2^+ or VO_3^-). Along with the color change, pH value of the suspension was reduced from about 7 to about 4. Moreover, the Tyndall effect was absent, which implied that the V_2CT_x nanoflakes were disassociated. In fact, the delaminated V_2CT_x nanoflakes were dissolved in water. Noting that vanadium carbide is stable, while vanadium oxide is not stable in water, the items experienced an oxidation process before dissolution, due to the presence of dissolved oxygen in water.

After pre-treatment in TBAOH for 4 h, the Ti_3CNT_x sample was sonicated in water for 20 min and then centrifuged, the delaminated Ti_3CNT_x in the colloidal suspension was separated from the non-delaminated settled particles, thus leading to a black colloidal suspension. Similarly, colloidal suspension could also been obtained from the sample intercalated with TBAOH for 21 h. As illustrated by the inset in Fig. 2.45c, Tyndall effect was observed in the diluted colloidal suspension. The formation of delaminated Ti_3CNT_x nanoflakes was confirmed by TEM observing results. As seen in Fig. 2.45c, there were numerous small nanoflakes laying on larger ones. EELS results indicated the presence of Ti, C, N and O.

Free-standing binder-free flexible Ti_3CNT_x papers could be derived from the colloidal suspension of the delaminated MXenes through the conventional filtration through membranes, as demonstrated in Fig. 2.45d, e. However, the papers made through membrane filtration of the delaminated V_2CT_x colloidal suspension were immediately disassociated, once they were dried in air, which was attributed to the instability of the vanadium oxide derived from the vanadium carbide.

The great decrease in the content of F and the increase in the level of O after the delamination process were confirmed by EDS results. For example, the as-synthesized Ti_3CNT_x powder had an atomic ratio of Ti:O:F to be 3.0:1.5:2.0, which became 3.0:3.6:0.8 after the delamination process. The as-obtained V_2CT_x powder possessed an atomic ratio V:O:F of 2.0:1.1:1.0. After the delamination treatment, the atomic ratio became to be 2.0:3.1:0.0. In other words, the F termination groups were nearly entirely removed. Theoretical prediction indicated that the stability of F termination was lower than that of O and OH [76]. Therefore, it is easily understood that the F termination was prone to be replaced by OH at basic conditions. The replacement of F by O/OH could be utilized to control the surface chemistry of MXenes, so as to manipulate their electrochemical or other performances for given applications.

Mashtalir et al. utilized delaminated isopropylanime (i-PrA) to intercalate HF-etched multi-layered Nb_2CT_x in DI water at room temperature for 18 h, followed by gentle sonication in deaerated DI water [77]. The surface of the HF-etched multi-layered Nb_2CT_x was negatively charged and acidic in nature, so that ammonium cations ($R\text{-}NH_3^+$) in i-PrA dissolved in water could be intercalated, through the electrostatic interaction [41, 77]. The i-PrA molecule has three alkyl groups, which was sufficiently small, so that the steric hindrance was readily overcome, thus leading the intercalation to form the Nb_2CT_x inter-layers. As a consequence, the Nb_2CT_x layers were pushed apart, because the interlayer interactions were effectively weakened. As this process was applied to Ti_3CT_x, multi-layered Ti_3CT_x could be partially delaminated after intercalation for 18 h without sonication. However, for $Nb_4C_3T_x$, much longer time of up to two weeks was necessary to achieve sufficient delamination [78].

Wang et al. attempted to intercalate Na^+ ions between HF-etched multi-layered $Ti_3C_2T_x$, by using diluted NaOH solution [79]. The experiment was carried out at room temperature for 2 h, which was followed by employing sulfanilic acid diazonium salts as surface modifier. The treatment was conducted in an ice bath for 4 h, followed by gentle sonication. The intercalation of Na^+ ions greatly weakened the interlayer cohesive strength, so that aryl diazonium salt molecules could be accommodated. Because diazonium salts are soluble in water, they can be easily cleaned if they are not reacted, when compared with their organic counterparts. The grafting of aryl groups onto $Ti_3C_2T_x$ made it very easy for dispersion. This effect could be used to further modify the surfaces, so that the number of the active sites on the surfaces could be increased, since *p*-phenyl-SO_3H was formed as reactive functional groups, from the aryl diazonium salts [79, 80].

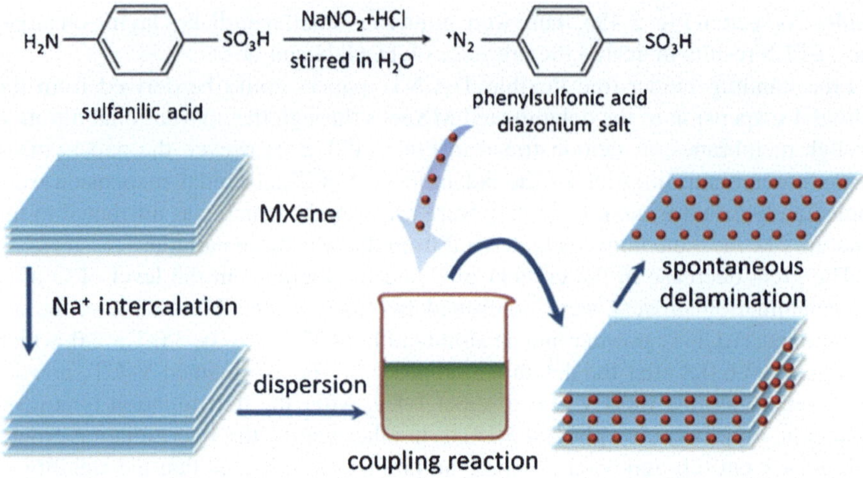

Fig. 2.46 Schematic diagram showing delamination process of the surface modified MXene multilayers with aryl diazonium salts. Reproduced with permission from [79]. Copyright © 2016, Elsevier

Figure 2.46 shows a schematic diagram that describes the delamination process. In this case, Na^+ was intercalated into the multilayered MXene Ti_3C_2, whose surface was further modified with sulfanilic acid diazonium salts. Due to the intercalation of Na^+ ions, the electrical conductivity was increased, thus leading to outstanding electrochemical performances of the materials. Also, the interlayer spacing was enlarged, so that the interlayer cohesive strength was weakened, where aryl diazonium salts could be accommodated in turn. As the surface of the MXenes was modified with diazonium salts, a small number of *p*-phenyl-SO_3H groups were attached onto the MXene surfaces, which expanded the interlayer spacing and weakened the bonds between the Ti_3C_2 layers. Therefore, the multilayered MXene Ti_3C_2 was delaminated.

As compared with other methods, this delamination approach had several advantages. Firstly, because diazonium salts are soluble in water, when they were used as the surface modifiers, it is easier to remove them as compared with other organic solvents or organic bases, after the modification process was finished. As a result, this the delamination procedure was much simpler and more efficient. Secondly, as the surfaces of the MXene nanoflakes were grafted with the aryl groups, their dispersibility in water was greatly enhanced, making them more useful in applications to fabricate MXene-based nanocomposites. Lastly, as a functional group of the aryl diazonium salts, *p*-phenyl-SO_3H had very high reactivity, which thus served as active sites on the surface of MXene, so that there could be various new surface chemical properties.

Figure 2.47a shows SEM image of the multilayered MXene Ti_3C_2 with a few stacked nanosheets, which had a size of several micrometers. After their surface was modified with the diazonium items, MXene-SO_3H nanoflakes were formed,

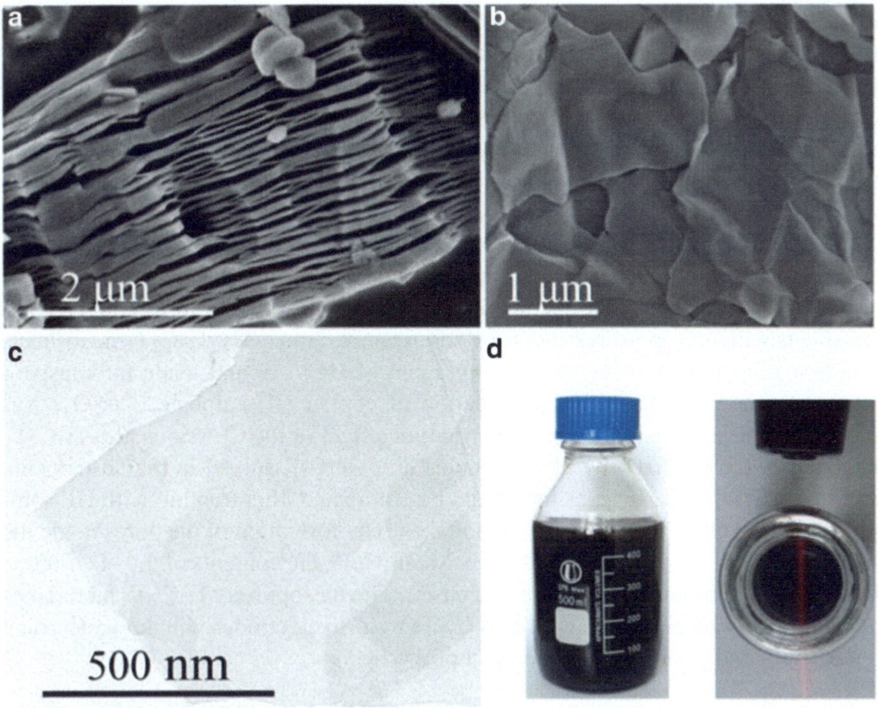

Fig. 2.47 **a** SEM image of the multilayered MXene Ti_3C_2. **b** SEM image and **c** TEM image of the surface-modified and delaminated Ti_3C_2 nanoflakes. **d** Photographs of a glass bottle and **e** Tyndall scattering effect of the surface-modified and delaminated Ti_3C_2 nanoflakes in water. Reproduced with permission from [79]. Copyright © 2016, Elsevier

which exhibited delamination profiles in the exfoliated MXene Ti_3C_2 domains, thus presenting a crumpled morphology, as demonstrated in Fig. 2.47b. Representative TEM image of the MXene-SO_3H nanosheets is shown in Fig. 2.47c. The delaminated nanoflakes were pretty transparent to electron beams, due to their relatively small thickness. Meanwhile, the sheet-like characteristic of the original MXene phase was kept roughly unchanged, even though their surfaces were modified with the aryl diazonium salts.

More importantly, the surface-modified MXene Ti_3C_2 could be re-dispersed in water, after it was washed with water and filtrated with membrane, thus forming a black stable colloidal suspension, as illustrated in Fig. 2.47d. The process could be easily scaled up, making it suitable for large scale industrial applications. Figure 2.47e shows a photograph of the MXene-SO_3H suspension, which had typical Tyndall effect, with a light beam from a laser pointer going through it. This confirmed its colloidal characteristics.

Zhang et al. first developed a method to exfoliate solid solution based $M_{n+1}AX_n$ nanosheets [81]. The resulting products had an average thickness of about 4 nm and lateral sizes in the range of 100–200 nm, after the solution was ultrasonicated in different organic solvents. It was found that only when the A layer in the $M_{n+1}AX_n$ phases contained different atoms that should follow the Hume-Rothery rules, the breaking of bonds between A and the MX layers was sufficiently efficient to achieve the exfoliation. However, this method had a relatively low yield production and produced only doped $M_{n+1}AX_n$ nanosheets, not MXenes. Therefore, it was difficult to completely convert the $M_{n+1}AX_n$ nanosheets to MXene and/or composite nanosheets. At the same time, it was only applicable to $Ti_3Si_{0.75}Al_{0.25}C_2$ and $Ti_3Al_{0.9}Si_{0.1}C_2$. The feasibility of this approach to other materials should be further explored and verified.

Ren et al. reported a wet-chemical etching reaction process to obtain $Ti_3C_2T_x$ nanosheets with mesoporous structure at room temperature [82]. $Ti_3C_2T_x$ nanosheets were first prepared by using the etching agent of HCl/LiF and made into suspension in water. After that, aqueous solution of copper (II) sulphate ($CuSO_4$) was mixed with the suspension. Due to the catalytic effect of the Cu^{2+} ions, the $Ti_3C_2T_x$ nanoflakes were oxidized by O_2 molecules that were dissolved in the suspension, resulting in TiO_2 nanoparticles that could be eliminated after treating with HF solution, so that highly porous $Ti_3C_2T_x$ nanoflakes. The formation of the pores made the $Ti_3C_2T_x$ nanoflakes to be more defective. Meanwhile, the contents of the –F, more = O and –OH functional groups were decreased. The mesoporous $Ti_3C_2T_x$ nanoflakes could be readily incorporated with MWCNTs to form electrodes with high electrical conductivity and strong electrochemical properties.

2.3.3 *Liquid Crystal MXenes*

Liquid crystals (LCs) are known as mesophases with high thermodynamic stability, which display both fluidity like liquids and structural ordering like crystals. The structures of LC phases exhibit very special ordering behaviors, while the ordering structures can be tailored by controlling concentration and temperature, applying shear forces and electric and magnetic fields, thus enabling a wide range of applications [83–85]. For example, Xia et al. realized vertical alignment of discotic lamellar liquid–crystal phase of $Ti_3C_2T_x$ by applying mechanical shearing action [86]. The films of vertically aligned $Ti_3C_2T_x$ possessed electrochemical energy storage effect that was not dependent on the film thickness up to 200 μm. This strategy tackled the problem encountered when using the traditional methods to fabricate electrodes based on 2D materials, where the electrochemical performances are limited if the films have too large thickness.

More recently, the authors experimentally observed for the first time self-assembled LC phases in aqueous $Ti_3C_2T_x$ MXene inks, without the involving LC additives, binders and stabilizers [87]. It was found that the transition concentration from isotropic phase to nematic one was dependent on the aspect ratio of the MXene nanoflakes. The nematic LC phases could be used to obtain fibers by using the conventional wet-spinning processes. The morphologies of the MXene fibers could be well controlled by simply adjusting the lateral size of the $Ti_3C_2T_x$ flakes in the inks, coagulation bath process and spinning parameters. The $Ti_3C_2T_x$ fibers exhibited a high electrical conductivity of 7750 S/cm, which is a record among the fibers based the nanomaterials reported in the open literature. Fiber supercapacitors assembled with the $Ti_3C_2T_x$ fibers possessed a volumetric capacitance of 1265 F/cm. In addition, the $Ti_3C_2T_x$ fibers could be utilized as heating elements. Moreover, similar nematic LC phases could be formed with other MXenes, such as $Mo_2Ti_2C_3T_x$ and Ti_2CT_x, in different organic solvents.

2.4 Key Properties

Because of their special structural feature and tailorable chemical compositions as discussed before, MXenes have been shown to exhibit rich physical, chemical, mechanical, electrical, electrochemical, catalytic and optical properties [32, 88, 89]. For instance, due to the metallic nature with three transition metal (Ti) atomic layers alternately in between two carbonic atomic (C) layers, 2D $Ti_3C_2T_x$ possessed very high electrical conductivity [31, 32, 90]. Furthermore, all the properties could be modified or even manipulated through controlling the surface terminations that are created during the exfoliation and intercalation processes. For example, it is possible to control over hydrophilicity and electronegativity of 2D $Ti_3C_2T_x$, so that it can be well dispersed in desired polar solvents [91, 92]. Meanwhile, synthetic conditions would play a significant role in determining the properties and performances of the final products [93]. Although all the properties have been mentioned directly or indirectly in the previous sections, it is necessary to put them together in order to maintain an integrity and have a comparison.

2.4.1 Electrical Properties

It has been theoretically predicted that MXenes with a monolayer structure have electrical conductivity like metals, with electron densities that are very close to the Fermi level, whereas the electronic properties of $Ti_3C_2T_x$ can be readily tailored through the surface functionalization [31, 58, 94]. In comparison, the electrical conductivity of $Ti_3C_2T_x$ is much higher metallic than that of most of the MXenes reported in the open literature. More importantly, it is dependent on the morphological profiles and surface properties, i.e., the quality of contact in between nanoflakes and the lateral size of the individual nanosheets [66, 95]. It has been reported that the $Ti_3C_2T_x$ nanoflakes with limited number of defects had an electrical conductivity of 6.5×10^{-3} S/cm, whereas the value could be less than 1000 S/cm if the $Ti_3C_2T_x$ sample contained a high concentration of defects, which is mostly observed when HF was used as the etching agent [32, 96].

Anasori et al. found that if Ti was replaced with Mo in the outer transition metal (M) layers of M_3C_2 and M_4C_3 MXenes, the 2D transition metal carbides would experience a transition from metallic behaviour to semiconducting-like one [94]. In the MXene structure, there were $n + 1$ layers of near-close packed M layers, with C or N to take the octahedral site in between them, thus exhibiting an arrangement of $[MX]_nM$. In the two newly synthesized ordered 2D double transition metal carbides MXenes, i.e., $M'_2 M'' C_2$ and $M'_2 M''_2 C_3$, where M′ and M″ represent two different early transition metals, which could be Mo, Cr, Ta, Nb, V and Ti. In this case, the M′ atoms only took the outer layers, while the M″ atoms were found in the middle layers. Therefore, the middle M″–C layers were sandwiched by the M′ layers.

According to X-ray atomic pair distribution function (PDF) analysis, it was evidenced that Mo atoms were in the outer layers of the $[MC]_nM$ structures for the MXenes of Mo_2TiC_2 and $Mo_2Ti_2C_3$. As compared with Ti_3C_2, the Mo-containing MXenes were not of metallic-like characteristics, because their resistance was increased as the temperature was decreased. Density functional theory (DFT) analysis results predicted that the Mo–Ti MXenes were semiconductors with narrow band gaps, when their surfaces were terminated with OH group. The theoretical predictions were confirmed by the experimental results, i.e., temperature dependencies of conductivities and magnetoresistances. In other words, both the electronic and optical properties of MXenes could be controlled by composition manipulation, thus opening a new door to explore these special materials for new applications.

The effect of Mo content on electronic properties of the MXenes was evaluated through by using the DC transport results. Figure 2.48a shows resistivity versus temperature curves of $Ti_3C_2T_x$ and the Mo-substituted MXenes. The levels of resistivity of the d-$Mo_2TiC_2T_x$ and d-$Mo_2Ti_2C_3T_x$ MXenes were lower than those of their multilayered counterparts. At the same time, the delaminated Mo-substituted MXenes had a resistivity that was higher than that of the multilayer $Ti_3C_2T_x$ by nearly one order of magnitude.

Figure 2.48b shows resistivity values of the multilayered $Mo_2TiC_2T_x$ (red, left-hand y-axis) and multilayered $Ti_3C_2T_x$ (black, right-side y-axis), in order for comparison. The difference in resistivity between the two samples was quite larger, which was about one order of magnitude, while the two samples both had a maxima at about 250 K. Such an observation was possibly attributed to the presence of water molecules and some ions inside the spaces of the MXene layers. In addition, the $Ti_3C_2T_x$ sample exhibited a metallic behaviour in the temperature range of 130–250 K, i.e., $d\rho/dT > 0$. However, for $Mo_2TiC_2T_x$, the resistivity was increased with decreasing temperature in the range of 10–250 K, i.e., $d\rho/dT < 0$. The resistivity of Ti_3C_2 was increased with decreasing temperature below about 130 K.

It was found that the transport behaviour of $Mo_2TiC_2T_x$ plotted in Fig. 2.48b could not be fitted by using the Arrhenius model, which is usually used to describe simple semiconductor behaviour. Moreover, no any hopping model could be applicable to the case. This suggested that interflake hopping processes might be responsible for the modest increases in resistivity as the temperature was decreased, as demonstrated in Fig. 2.48. Obviously, as the Mo layers sandwiched the Ti layers, i.e., in the Mo-substituted MXenes, the resistance was increased monotonically with decreasing temperature, while the resistance was decreased with decreasing temperature, before the substitution of Mo in the sample, as revealed in Fig. 2.48b.

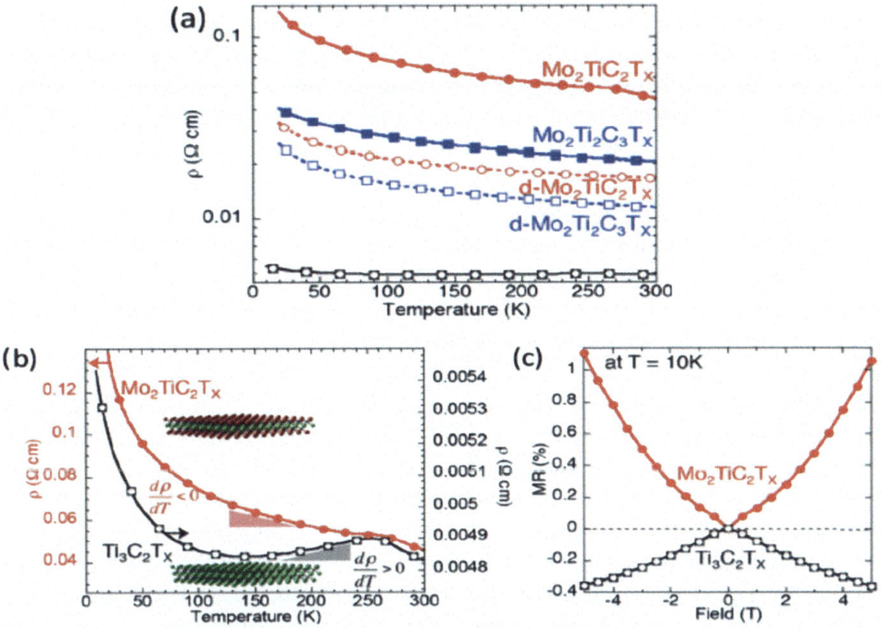

Fig. 2.48 **a** Temperature dependence of resistivity for $Mo_2TiC_2T_x$ (red), $Mo_2Ti_2C_3T_x$ (blue) and $Ti_3C_2T_x$ (black), with the dashed and solid lines to represent the delaminated films and the pressed multi-layered powders. **b** Resistivity curves of $Mo_2TiC_2T_x$ (red, left side *y*-axis) and $Ti_3C_2T_x$ (black, right-side *y*-axis), with the colored triangles (red and black) below the resistivity curves to demonstrate the differences in $d\rho/dT$. **c** Magnetic field dependent magnetoresistance measured at 10 K. Reproduced with permission from [94]. Copyright © 2016, Royal Society of Chemistry

Because the resistivity of two MXenes was both had increased at the temperatures of below 130 K, their magneto-resistance (MR) behaviours at 10 K were measured, in order to identify whether the mechanisms of observation was same or not. As seen in Fig. 2.48c, the two samples demonstrated an opposite behaviour in MR response in sign, implying that the transport mechanisms of $Ti_3C_2T_x$ and $Mo_2TiC_2T_x$ are distinctively different. The MR curves were measured at 10 K, which was within the range for the observation of $d\rho/dT < 0$. The negative MR of $Ti_3C_2T_x$ within the temperature range of $d\rho/dT < 0$ was readily ascribed to the weak localization. In comparison, the positive MR response and negative $d\rho/dT$ of $Mo_2TiC_2T_x$ over the whole temperature range confirmed that a different transport mechanism was present to control the conduction behaviour of the Mo-substituted MXene, as compared with $Ti_3C_2T_x$. The interesting observation was further elucidated by using DFT analysis in a detailed way [94].

Je et al. found that both the magnetic and electronic properties of Cr_2C with the functional groups of F and OH, i.e., $Cr_2CF(OH)$, which had a functional group at each side, could be theoretically predicted by using the properties of Cr_2C with only one type of functional group, which as either F or OH at the two sides [96]. Different from the conventional functionalized 2D materials which contained different functional groups, the functional group of $Cr_2CF(OH)$ at one sided possessed almost no influence on that at other side, in terms of the magnetic and electronic properties. This observation was attributed to the fact the two functional groups were geometrically similar, including the coordination number, length of bond and layer thickness. As a result, the properties of Cr_2CF_2 and $Cr_2C(OH)_2$ were nearly the same. Since the magnetic moments of the Cr atoms were bounded with the two functional groups, $Cr_2CF(OH)$ was of ferromagnetic characteristics. Also, $Cr_2CF(OH)$ had semiconducting behaviour, due to the atomic contributions of the energy levels of Cr_2CF_2 and $Cr_2C(OH)_2$ are. Therefore, the Cr_2C-based MXenes could be designed to have desired properties for spintronic and electronic applications.

Enyashin and Evanovskii studied the properties of MXenes, Ti_2C and Ti_3C_2, which could be functionalized with methoxy groups ($-OCH_3$), by using the density functional theory tight-binding method [97]. The calculation results revealed that the 2D-like geometry was well retained in the relaxed $Ti_2C(OCH_3)_2$ and $Ti_3C_2(OCH_3)_2$ structures, demonstrating their chemical stability, when they are modified, although their lattice parameters were related to the way of OCH_3 attachment.

According to the results of the DFT analysis, the stability of the possible configurations for $Ti_2C(OCH_3)_2$ and $Ti_3C_2(OCH_3)_2$ was elucidated. In both cases, if the methoxy groups were sitting above the hollow sites between the three neighboring carbon atoms and the tetrahedrally coordinated O atoms, the configuration was the most stable. However, if methoxy groups were directly connected to the Ti atoms and the double coordinated O atoms, the configuration was not stable.

Also, it was found that the formation energy of MXenes with both $-OCH_3$ and $-OH$ terminations was negative, suggesting that the MXenes could have non-uniform functionalization. Furthermore, the authors claimed that the most stable MXene derivatives were those that exhibited metallic-like conducting behaviors. In this case, the near-Fermi electronic bands were most likely due to the states of Ti 3d. According to the cross points of valence band and conduction band at the Fermi level, it was observed that some of the MXenes with metastable configurations possessed semimetal characteristics.

The authors also used MD simulations to predict the formation of new derivatives of MXenes, by considering the interface between the hydroxylated MXene layer of $Ti_2C(OH)_2$ and methanol molecules. It was highly possible that the CH_3OH molecules could be bonded with the Ti atoms at the surface of MXene through the OH groups. In addition, the hydroxylated MXenes displayed high protonating activity, which was confirmed by the simulation results. In this regard, these new materials could be used as promising catalysts. Moreover, the alkoxylated MXenes could be potential electrode materials for Li-ion batteries, because of the high Li-ion mobility and sufficiently high affinity to organic electrolytes, owing to the organic–inorganic hybrid nature.

2.4.2 Chemical Properties

Formation capability of highly stable colloidal suspensions is a critical requirement for various applications, especially in the forms of thin films [32, 98, 99]. In bulk phase of Ti_3AlC_2, the interlayer interactions are very strong, which are then largely weakened in $Ti_3C_2T_x$ after the exfoliation process. The weak interactions in between the $Ti_3C_2T_x$ nanosheets include van der Waals force related to O and/or F atoms and the hydrogen bonding force between the O/F atoms at one surface and the OH groups at the other surface, which promote the exfoliation of $Ti_3C_2T_x$ into few-layer or even single layer nanostructures [28].

In addition, the functional groups are usually randomly distributed on the surfaces of the nanosheets in the $Ti_3C_2T_x$ MXenes. Furthermore, the surface termination profiles are strongly dependent on the etching and delamination conditions. For example, if HF solutions with relatively low concentrations are used for exfoliation, the amount of –O termination will be much larger than that of –F group. Even more, $Ti_3C_2T_x$ exfoliated by using HCl-LiF etching agent contained water molecules bonded with –O/–OH terminations [96, 100]. Intercalation with cations, e.g., Li^+ ions, facilitated sliding of the $Ti_3C_2T_x$ nanosheets, which would further modify rheological properties of the suspensions [101–103]. Therefore, the intercalation of water molecules and cations could significantly enhance dispersion behavior of $Ti_3C_2T_x$ in various solvents.

As discussed before, $Ti_3C_2T_x$ MXenes delaminated through simple sonication in water usually have relatively low production yields. In comparison, multilayered $Ti_3C_2T_x$ generated with HCl-LiF etching agent could be easily delaminated through sonication, if intercalation items, such as water molecules and cations, are inserted in between the layers, which served as spacers [34, 104]. In this case, intercalation, with large organic molecules, including dimethyl sulfoxide (DMSO), urea, isopropyl amine, and so on, has been acknowledged to the most effective method to separate the strongly bonded layer structures of $Ti_3C_2T_x$ based MXenes [105].

Because of the formation of the surface terminations, the $Ti_3C_2T_x$ nanoflakes after delamination are of very strong hydrophilic characteristics [106, 107]. As a result, the $Ti_3C_2T_x$ MXenes processed in this way exhibited very high dispersion stability in aqueous suspensions, making it possible to develop $Ti_3C_2T_x$-based materials, such as thin films through deposition, free-standing papers through filtration and composite materials though blending with polymers [108, 109]. According to a systematic study on the dispersion behaviors of $Ti_3C_2T_x$ nanosheets in organic solvents, as presented earlier, it was found that the delaminated $Ti_3C_2T_x$ could be readily disperses in polar solvents, such as H_2O, ethanol, *N*-methy l-2-pyrrolidone, propylene carbonate, *N*,*N*-dimethylformamide (DMF) and dimethyl sulfoxide, simply owing to the compatible surface chemistry between $Ti_3C_2T_x$ and the organic solvents [91].

Expectedly, the solvents with weak polar properties, such as 1,2-dichlorobenzene, toluene and hexane, could not be used to for form stable suspensions. Besides intercalation with polar organic molecules [105, 110], other approaches, including mechanical vibration [111], and sonication [89], have been used to obtain delaminated

$Ti_3C_2T_x$ MXenes. The $Ti_3C_2T_x$ nanoflakes, with structures to be strongly dependent on the delamination processes, have thickness from one to several nanometers, with lateral sizes in the range from hundreds of nanometers to several micrometers [112].

2.4.3 *Thermal Properties*

Due to their chemical nature, MXenes like $Ti_3C_2T_x$ are prone to be oxidized, especially when H_2O and/or O_2 are encountered during the processing of the materials at high temperatures [113]. Therefore, thermal stability is an important parameter of $Ti_3C_2T_x$ for real applications. Generally, if $Ti_3C_2T_x$ is oxidized, the final product will be TiO_2 [114, 115]. For example, as flash oxidation of 2D $Ti_3C_2T_x$ occurred in air, a hybrid structure was formed, in which the thin sheets of the disordered graphitic carbon were decorated with TiO_2 nanoparticles. In a two-step oxidation process to synthesize $Ti_3C_2T_x$ MXene, the anomalous grain growth of TiO_2 could be well avoided, so that the original layered structure of MXene could be effectively retained. The consequence of oxidation is the loss of electrical conductivity and layered structure. Therefore, the oxidation of $Ti_3C_2T_x$ should be prevented as effectively as possible, even though certain unexpected structures and properties could be present as a result of the accident oxidation.

Various derivatives have been obtained from the $Ti_3C_2T_x$ precursor. For instance, Lorencova et al. synthesized graphene quantum dots (GQDs) with $Ti_3C_2T_x$ as the precursor by using a solvothermal process in DMF [116]. The authors found that the delaminated $Ti_3C_2T_x$ precursor was much more easily dispersed in DMF than the non-delaminated ones. Therefore, a higher product yield of GQDs was achieved by using the delaminated precursors. Moreover, Xu et al. obtained $Ti_3C_2T_x$-based QDs with different concentrations of titanium, by simply controlling the reaction temperature [117]. In a separate study, Rakhi et al. prepared sodium titanate (M-NTO, $NaTi_{1.5}O_{8.3}$) and potassium titanate (M-KTO, $K_2Ti_4O_9$) nanoribbons, through the simultaneous oxidation and alkalization of $Ti_3C_2T_x$, by using a hydrothermal reaction process [118]. The products could be promising candidates for biosensor applications.

More recently, Sarikurt et al. conducted theoretic study on Seebeck coefficients and lattice thermal conductivities of O terminated M_2CO_2, with M = Ti, Zr, Hf and Sc, by using density functional theory and Boltzmann transport theory [119]. The M_2CO_2 was monolayer MXene crystals, in two functionalization configurations, corresponding to model-II (MD-II) and model-III (MD-III). Thermoelectric figure-of-merit, zT, of the 2D materials was also evaluated by using two approaches. It was found that the structural model, i.e., adsorption site of O atoms on the surface of MXene, had a very strong effect on electronic and thermoelectric properties of the MXene crystals, implying that these materials could be exploited for thermoelectric applications.

Thermoelectric parameters, such as lattice thermal conductivity κ_l, Seebeck coefficient and zT values, were found to be variable by 40%, according to the structural models. For instance, the MD-III configuration had larger band gap, higher Seebeck coefficient and higher zT level and lower κ_l than the MD-II structure. This was simply because the MD-III structure possessed a highly flat valence band and reduced crystal symmetry, due to its larger band gap. The MD-III structure of Ti_2CO_2 and Zr_2CO_2 demonstrated the lowest thermal conductivity κ_l, as compared with Hf_2CO_2 and Sc_2CO_2 with same configuration. Among all the studied structures, Hf_2CO_2 with the MD-II model exhibited the highest κ_l, whereas Ti_2CO_2 and Zr_2CO_2 with MD-III configuration had the lowest κ_l. For example, the Ti_2CO_2 with the MD-II configuration had a band gap of 0.26 eV, but it would be 0.69 eV if MD-III structure was adopted. The optimized zT_{max} value of the MXenes could be as high as 1.1 through optimization of the structural model, which is also among the highest values of the conventional thermoelectric materials.

As mentioned earlier, the surface terminational groups allowed for tailoring the properties and performances of MXenes. This is especially important in the applications for heterogeneous catalysts, electrodes of electrochemical energy storage devices and the development of composite materials [120]. For instance, as the H_2O molecules and −OH groups were removed from MXenes, MXene–polymer composites could be readily fabricated, especially for the polymers that are hydrophobic. The presence of surface functional groups has a positive effect on the electrochemical performance of supercapacitors with aqueous electrolytes, whereas electrodes of batteries usually require nonterminated surface sites to avoid parasitic reactions [121]. In this regard, stabilities of the surface functional groups will play an important role in determining the properties and performances of the MXene materials.

Bulk transition-metal carbides and nitrides have relatively high thermal stabilities, with melting and decomposition temperatures to be in the range of 2000–3000 °C. To fabricate ceramics and metal matrix composites with MXenes, their thermal stability at high temperatures is a crucial factor. The thermal stability should be sufficiently high to prevent the decomposition of 2D MXenes, while the characteristics of the surface terminational groups are also very important [122]. However, these high temperature properties of MXenes still remained to be clarified.

Seredych et al. deeply studied thermal properties of three carbide MXenes ($Ti_3C_2T_x$, Mo_2CT_x, and Nb_2CT_x) with different surface terminational groups by using thermal gravimetric analysis and mass spectrometry technique (TA–MS) up to a high temperature of 1500 °C in He [123]. The surface terminational groups of hydroxyl (−OH), oxy (=O) and fluoride (−F), as well as the intercalated items, including salts and structural water molecules, could be detected and identified. Hydrophilic characteristics of the MXenes were dependent on the etching agents, e.g., concentration of hydrofluoric acid and composition of the acids. The post delamination parameters also had specific influence.

It was found that the content of structural water of the $Ti_3C_2T_x$ MXene was increased as the density of the O-containing surface terminational group. The thermal stability of $Ti_3C_2T_x$ was relatively higher, if the concentration of the HF solution was sufficiently low or mix etchant acids were used. Therefore, it was recommended to use H_2SO_4/HF or HCl/HF rather than HF, so that the number of defects could be minimized. As tetramethylammonium hydroxide (TMAOH) was employed to delaminated the MXenes, new items enriched with N were present on the surface of the MXene nanosheets. Furthermore, TMAOH could mitigate the difference of the $Ti_3C_2T_x$ samples obtained by using HF solutions with different concentrations, leading to free-standing films with nearly the same TA–MS results, suggesting that the surface chemistry of the $Ti_3C_2T_x$ could be modified by using the post-treatment. Therefore, thermal stability of MXenes is closely related to their chemical compositions and structures. In addition, the thermal stability of $Ti_3C_2T_x$ was higher than those of other MXenes.

2.4.4 Mechanical Properties

Kuturglu et al. were among the first to theoretically study the mechanical properties of 2D MXenes, including Ti_2C, Ti_3C_2, Ti_4C_3, V_2C, Cr_2C, Zr_2C, Hf_2C, Ta_2C, Ta_3C_2 and Ta_4C_3 [124]. The results revealed that the mechanical properties were much higher than those of multilayer graphenes when they had comparable thicknesses. For instance, MXenes exhibited higher elastic moduli when stretched along the basal planes, as compared with graphenes. More importantly, it was demonstrated that $T_3C_2T_x$ MXene, when blended with polyvinyl alcohol (PVA) to form $T_3C_2T_x$/polymer composites, had mechanical strength that was stronger as compared with its ow weight by fact of about 1.5×10^4 times [125].

Yorulmaz et al. used first principle calculation method to study the structures of Sc, Ti, Zr, Mo and Hf pristine MXenes with $n = 1$ and full surface termination of F and O [126]. The authors carefully evaluated the dynamical and mechanical stabilities of both pristine and fully terminated MXenes. In addition, they alto reported Raman and IR active mode frequencies of the materials, offering very useful reference for the experimental development of MXenes. It was found that all the carbides and nitrides of Sc, Ti, Zr, Mo and Hf pristine MXenes, except for Mo_2N, were highly stable and exhibited metallic conducting characteristics. Comparatively, the fully terminated carbide-based MXenes were dynamically and mechanically more stable than the nitride-based ones.

Meanwhile, structural, electronic, vibrational and mechanical properties of representative MXenes with and without full surface terminations were demonstrated. Electronic band opening in some of the MXenes was observed, which was dependent on the type of the materials (i.e., carbide or nitride) and the type of the functional groups. It was revealed that the MXene structures exhibited a large diversity in electronic and mechanical properties, were all the factors, including the type of the early transition metallic element, type of the X atom and type of the surface functional group, had an important role. The theoretical data could be used as a reference to select the composition of the MXenes with specific properties for desired applications.

2.4.5 Optical Properties

Magne et al. combined valence electron energy-loss (VEEL) spectroscopy in TEM and ab initio calculations to study the dielectric properties of multilayered (ML) 2D $Ti_3C_2T_2$ (T = OH or F) MXenes [127]. It was found that the ML − $Ti_3C_2T_2$ was similar to TiC, in terms of valence electron gas behaviors. At the same time, a clear interband transition was observed at the most stable site of the T functionalization groups, according to the VEEL spectral results. This characteristic was strongly dependent on the T-group localization at the surface of the MXene nanosheets. It had constantly an effect on optical properties of the ML nanostructures, resulting in variations in the optical conductivity in the middle of the visible spectrum by up to 40%. This strong dependence could be well used for thin films with optical transparent and sensing capability.

Electrically conducting thin films, based on 2D $Ti_3C_2T_x$ MXenes, have been prepared, by using a simple vacuum-free solution processible electro-hydrodynamic atomization method [128]. In order to deposit the $Ti_3C_2T_x$ thin films, delaminated $Ti_3C_2T_x$ powder was dispersed in ethanol to form suitable suspension for thin film deposition. Three groups of $Ti_3C_2T_x$ films, with thicknesses of 135 nm, 248 nm and 337 nm, were deposited on glass substrate, through controlling the deposition time, i.e., 20 min (S1), 40 min (S2) and 60 min (S3). After deposition, all the films were annealed at 400 °C for 2 h, in an inert environment. The samples S1, S2 and S3 had resistivities of 3.4×10^{-4} Ω·cm, 130×10^{-4} Ωcm and 210×10^{-4} Ωcm, respectively.

Meanwhile, the sample S1 exhibited an in-line optical transmittance of about 86.7% at the wavelength of 550 nm, as seen in Fig. 2.49a. Using the transmission spectra of Fig. 2.49a, both the absorption coefficient and optical gap (E_g) of the $Ti_3C_2T_x$ thin films could be calculated. Figure 2.49b shows the curves of $(\alpha h\upsilon)^2$ versus $h\upsilon$ of the three films, leading to their optical band gaps. Accordingly, the values of E_g were 4.05, 3.83 and 3.9 eV, for S1, S2 and S3, respectively.

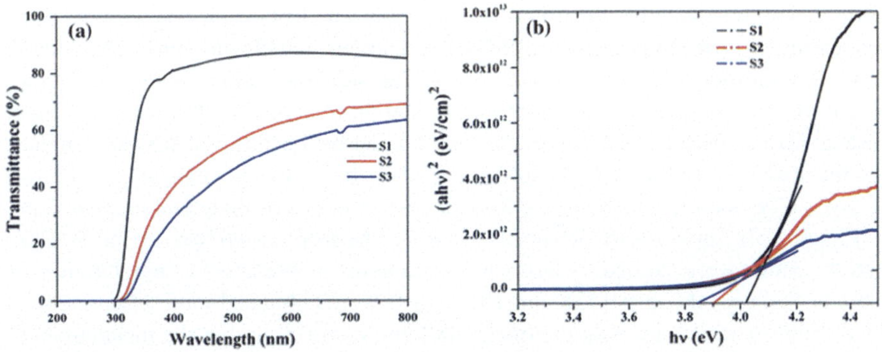

Fig. 2.49 **a** Transmittance spectra and **b** band gap determination curves of the S1, S2 and S3 $Ti_3C_2T_x$ films. Reproduced with permission from [94]. Copyright © 2016, Springer

Similarly, Marina et al. deposited $Ti_3C_2T_x$ delaminated monolayers that were used for the application of ultrathin transparent conductors [129]. The authors found that the electrical properties of the monolayer films were comparable with those of the films made of reduced graphene oxide. The $Ti_3C_2T_x$ films were deposited through a solution process. They displayed a very low sheet resistance of 437 Ω/□ and in-line transmittance of 77% at 550 nm. According to the field effect transistor measurement results, the films were confirmed to show metallic conduction characteristic, so that they could be used for applications as electrodes. Kelvin probe atomic force microscopy (PKAFM) examination indicated that the delaminated $Ti_3C_2T_x$ nanoflakes with –OH terminal groups had a work function of 5.28 eV. Therefore, it is concluded that these solution-processed $Ti_3C_2T_x$ thin films are promising candidates for next generation transparent conductive electrodes.

Sarycheva et al. found that $Ti_3C_2T_X$ could be employed to enhance Raman signal from organic dyes either on substrates or in solutions [130]. The MXene SERS substrates were prepared by using a spray coating method, which were then utilized to probe some commonly studied dyes. The enhancement factors were estimated to approach about 10^6. The SERS effect of MXene in aqueous solutions implied that it could be promising candidates for applications in biomedical or environmental areas, because the MXene could enhance the positively charged molecules with a high selectivity.

Figure 2.50a shows SERS enhancement (514 nm excitation) of Rhodamine 6G on the $Ti_3C_2T_X$ MXene substrates with different thicknesses that were controlled by adjusting the number of spray-coating passes. Very sharp R6G peaks could be observed for the samples made with 10^{-6} M solution drop-casted on the MXene substrate. Even at the dye solution concentration of 10^{-7} M, the vibrational peaks of R6G were still visible, especially for the main peaks, which implied that the $Ti_3C_2T_X$ substrates exhibited a high enhancement factor. It has been well known that the enhancement of Raman scattering of TiO_2 nanoparticles was attributed to the charge transfer absorption [131], which had a surface characteristic very similar to that of MXenes.

The dependence of the SERS behavior on the quality of the deposited $Ti_3C_2T_X$ was studied, with results to be illustrated in Fig. 2.50b. The R6G solutions with the same concentration were drop-casted on Si/SiO_2 substrates, which was spray-coated with $Ti_3C_2T_X$ nanosheets for 3, 10 and 20 times. The substrate spray-coated for 3 times demonstrated a higher enhancement factor than those coated with thicker and more homogeneous coatings. Basing on the relationship between the SERS performance and the properties of the $Ti_3C_2T_X$ substrates, it was found that the enhancement effect of $Ti_3C_2T_X$ was quite similar to the hot-spot effect in electromagnetic mechanism.

Figure 2.50c shows the SERS peaks of the 10^{-6} M drop-casted R6G on the $Ti_3C_2T_X$ islands, which were excited by lasers at wavelengths of 488 and 514 nm. Obviously, some peaks, e.g., those at 1528 and 1575 cm^{-1}, were enhanced more effectively by the excitation at 514 nm than the one at 488 nm, confirming that the mechanism of the enhancement was closely related to the interband transition or charge transfer. In addition, the R6G in the solution with a low concentration of 10^{-6} M could be detected by using the lasers at both 488 and 514 nm, even in the aqueous solution

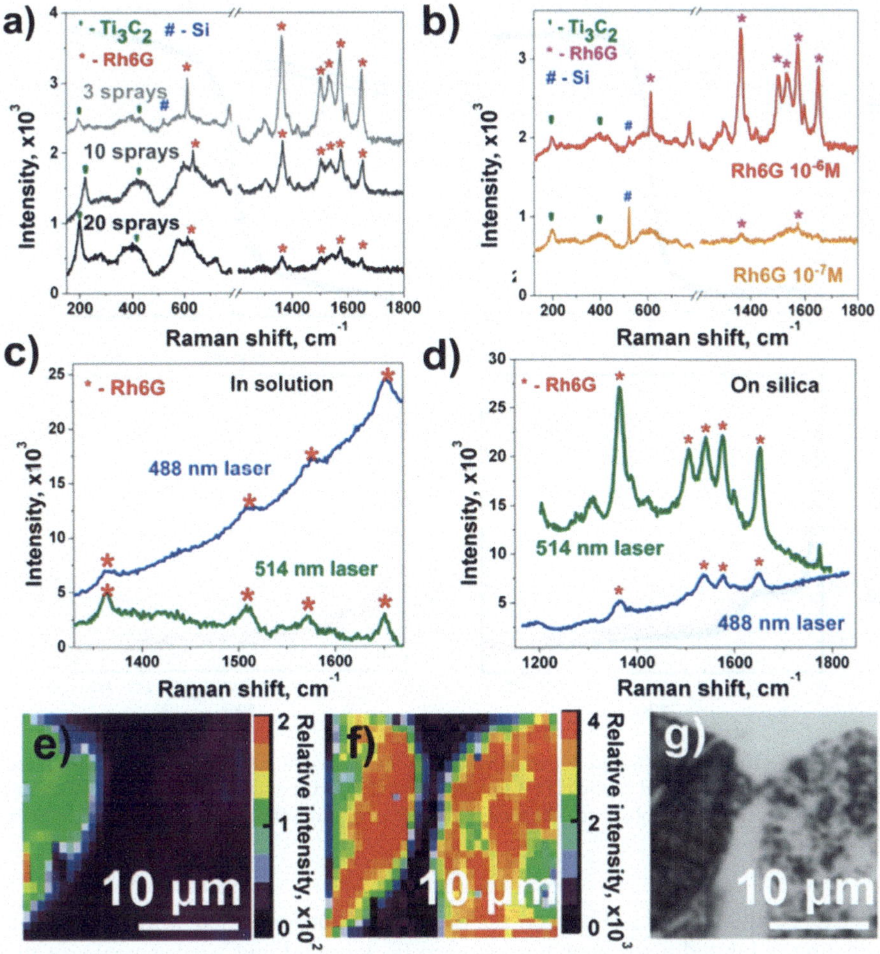

Fig. 2.50 **a** SERS enhancement (with 514 nm excitation) of Rhodamine 6G on $Ti_3C_2T_x$ MXene substrates with different thicknesses that were controlled by controlling the number of the spray-coating process. **b** SERS effect of R6G from the solutions with different concentrations after drop-casting on the MXene/SiO_2/Si substrate (three time of MXene spray-coating and 514 nm excitation). **c** Enhancement performances with the 488 and 514 nm lasers for (**c**) the aqueous colloidal solution and **d** that on MXene/SiO_2/Si substrate (3 times). Raman mapping profiles of the interstitial area of the same substrate, with color-coding of the intensity of the sum **e** the $Ti_3C_2T_x$ MXene peaks and **f** the R6G peaks (drop-casted on MXene from the solution with a concentration of 10^{-6} M). **g** Optical image of the Raman mapped area. Reproduced with permission from [130]. Copyright © 2017, American Chemical Society

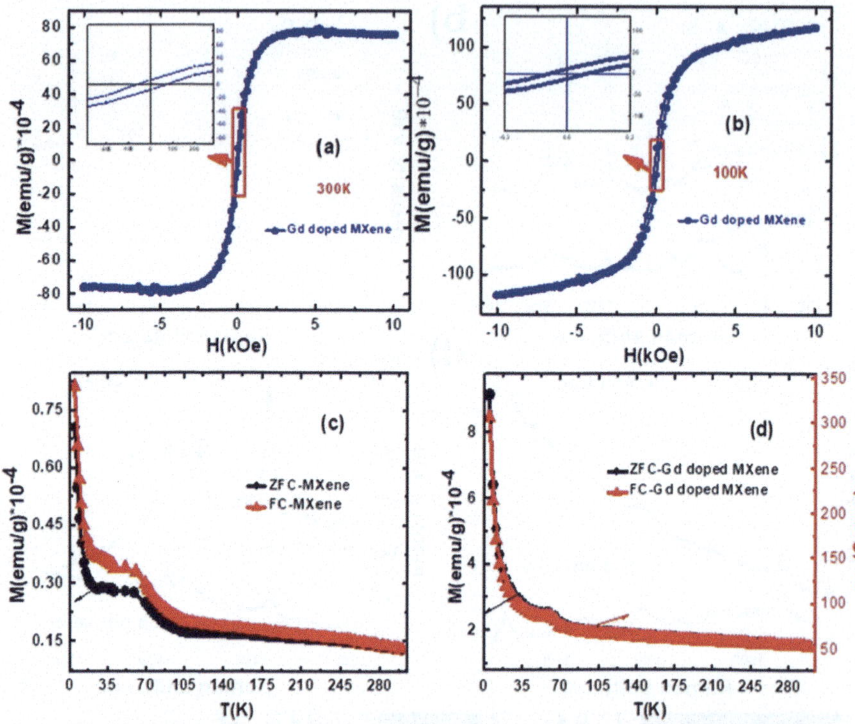

Fig. 2.51 M−H curves of the Gd-doped MXene phase measured at 300 K (**a**) and 100 K (**b**). Zero field cooling (ZFC) and field cooling (FC) curves recorded at 1 kOe of the pristine MXene phase (**c**) and the Gd-doped MXene phase (**d**). Reproduced with permission from [132]. Copyright © 2020, Elsevier

of $Ti_3C_2T_x$ at 1 mg/mL. However, the SERS spectrum obtained with the 488 nm excitation exhibited a stronger fluorescence background and weaker peaks of the dye, as compared with those with the excitation at 514 nm. This observation was probably ascribed to the shift of the excitation to higher energy level, as compared with the absorption peak, as shown in Fig. 2.50a. Figure 2.50d shows the results of the samples on the Si/SiO_2 substrates, from which the SERS enhancement of R6G could be estimated. The enhancement factors are 1.2×10^6 and 5.3×10^5 for the 488 nm and 514 nm excitations, respectively.

The distribution of the R6G molecules from the solution with a concentration of 10^{-6} M on the surface of the MXene islands obtained after spray-coating for 3 times on the Si/SiO_2 substrate was studied, by using the Raman intensity mapping with the excitation at 514 nm. Normalized false rainbow-color coded images are illustrated in Fig. 2.50e, f. The result of Fig. 2.50e suggested that the sample possessed a thicker coverage, i.e., there were more flakes and thus the surface was more homogeneous and flat, thus having fewer hot-spots. The intensity of the sum of the two peaks with excitation at 614 cm^{-1} and 776 cm^{-1} of the R6G from the SERS was higher at the edges of the MXene island, as demonstrated in Fig. 2.50f, evidencing the effect of hot-spots in the Raman signal enhancement phenomenon. The experimental results were in a good agreement with the optical image of the same area and the analysis outcome, as revealed in Fig. 2.50g, b respectively.

2.4.6 Magnetic Properties

Doping has been shown to be an effective way to develop MXenes with strong magnetic properties. For example, Rafiq et al. reported magnetic properties of MXene consisting of Ti_3C_2 and Gd^{+3} dopant, which was prepared by using a simple coprecipitation wet-chemical process [132]. The Gd-doped Ti_3C_2 had a bandgap of 1.93 eV, while the pristine MXene possessed a value of 2.06 eV. Both the pristine and Gd-doped MXene phases exhibited magnetic characteristics at room-temperature. Due to the presence of the spin-up and spin-down electrons of Ti atoms at both the top and bottom Ti layers, the pristine MXene demonstrated an asymmetric spatial spin-density of states (s-DOS). The spin-up state was strengthened by more than two time at room temperature after the MXene was doped with Gd^{+3}, simply because the more unpaired electrons were available near the Fermi level.

The Gd^{+3} doped MXene was expressed as $Ti_{3(1-x)}Gd_xC_2(OH)_2$ with a ratio of 1:1. To synthesize the Gd-doped MXene, Ti_3C_2 powder was dispersed in deionized water at a concentration of 0.4 mg/ml, with the aid of sonication at room temperature. $Gd(NO_3)_3{\cdot}6H_2O$ was dissolved in a mixed solvent of ethylene glycol and acetic acid was a ratio of 1:1, forming solution with a concentration of 0.01 M. The MXene suspension and the Gd-solution were mixed. After sonication for 0.5 h, the mixture was continuously heated for 3 h with stirring. The final product was thoroughly cleaned with deionized water followed by drying at 100 °C for one day.

Magnetic hysteresis loops of the Gd-doped MXene recorded at 300 and 100 K are illustrated in Fig. 2.51a, b. The pristine and the Gd-doped MXenes displayed saturation magnetization values of 22×10^{-4} emu/g and 80×10^{-4} emu/g, respectively, at the magnetic field of 10 kOe at room temperature. Meanwhile, the doped sample had a coercivity of 0.089 kOe and a remnant magnetization of 0.23×10^{-4} emu/g at room temperature. The undoped and doped MXene phases had saturation magnetization values of 3.53×10^{-4} emu/g and 120×10^{-4} emu/g, respectively. The slope of the curves at 100 K was increased with respect to that of ones recorded at 300 K. At the same time, the saturation magnetization of the pristine MXene at 100 K was reduced by about five times as compared with that at 300 K.

However, the Gd-doped sample exhibited an increase in saturation magnetization by about more than forty times, as the temperature was reduced from 300 to 100 K. This is because the thermal fluctuation of the net spin moment was minimized and the overlapping in wave function of the unpaired electrons was enhanced, thus leading to an increase in saturation magnetization. Therefore, owing to the doping with Gd^{+3}, the MXene phase exhibited increased in both the saturation and remnant magnetizations, while the coercive field was decreased, since the Gd^{+3} cation is ferromagnetic in nature with more unpaired electrons.

Zero-field cooled (ZFC) and field cooled (FC) curves of the samples are depicted in Fig. 2.51c, d, which were measured at a magnetic field of 1 kOe. As the temperature was increased from 5 to 300 K, ferromagnetic to paramagnetic phase transition was observed. The values of both the FC and ZFC curves were always beyond zero, suggesting strong interaction among the unpaired electrons. As a resut, net magnetization was observed. In addition, for the pristine MXene phase, the FC and ZFC values were almost the same. However, the Gd-doped sample, the FC values were always much higher than ZFC ones. For instance, the maximum values were 340×10^{-4} emu/g and 8.2×10^{-4} emu/g in the cases of FC and NFC, respectively. Meanwhile, the magnetization-temperature behaviors of the two samples were slightly different. Also, magnetic moment tended to decline with increasing temperature in the FC and ZFC curves. At low temperatures, the FC and ZFC curves were differentiated, as revealed in Fig. 2.51c, d. This observation was attributed to the formation of secondary phases, such as TiC and TiO_2, as evidenced by XRD measurement results.

It was concluded that stronger magnetic fields were required to align the magnetic moment in the pristine MXene phase. In comparison, it was much easier for the magnetic moments to be aligned in the Gd-doped sample, which needed relatively lower magnetic fields. The FC/ZFC curves also revealed that both the magnitude and stability of the magnetic moments in the Gd-doped MXene were enhanced, as compared with those in the pristine MXene. Figure 2.52 shows a schematic diagram demonstrating the synthesis process and possible mechanism for the room temperature ferromagnetic behaviour of the MXene phase due to the doping with Ga^{3+}.

Si et al. employed a hybrid density functional theory to study potential magnetic properties of Cr_2C-based MXenes [133]. The ferromagnetic effect was originated from the itinerant *d* electrons of Cr, leading to Cr_2C MXenes to show possible intrinsic half-metallic behaviors, with a half-metallic gap to be up to 2.85 eV. It was also found that there was a ferromagnetic–antiferromagnetic transition, corresponding to a metal–insulator transition in the Cr_2C MXenes, closely related to the surface

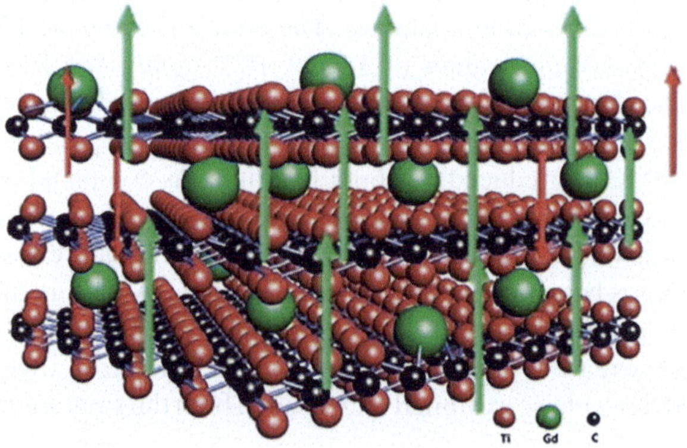

Fig. 2.52 Schematic diagram illustrating the mechanism of the enhanced magnetic behaviour in the Gd-doped MXene phase. Reproduced with permission from [132]. Copyright © 2020, Elsevier

functional groups of –F, –OH, –H or –Cl. In addition, the energy gaps of the antiferromagnetic insulating state were also dependent on what functional group was. The ferromagnetic–antiferromagnetic and metal–insulator transitions were attributed to the localization of the *d* electrons of Cr, which was induced by the surface functionalization effect. Therefore, the magnetic and electronic properties of the Cr_2C-based MXene could be tuned by controlling the surface functionalization, thus resulting in materials for applications in nanoscale spintronic and electronic devices.

Figure 2.53 shows energy gaps of four MXenes with different functional groups, which possessed similar lattice structures and antiferromagnetic ground states. They were all semiconductors, with energy gaps to be 1.43 eV, 1.76 eV, 2.56 eV and 3.49 eV, for $Cr_2C(OH)_2$, Cr_2CH_2, Cr_2CCl_2 and Cr_2CF_2, respectively. Obviously, the conduction of Cr_2C MXene could be altered from metallic to semiconducting and the bandgap could be predetermined by selecting certain functional groups. Most significantly, Cr_2CH_2 had a direct bandgap, whose conduction band minimum and valence band maximum were positioned at the Γ point, as given by the inset in Fig. 2.53. Because a direct gap corresponds to efficient light emission, the Cr_2CH_2 MXene could be a potential candidate for the applications as the active materials in optoelectronic devices. Finally, it is worth mentioning that the bandgap of the Cr_2C MXene could be continuously varied by using mixed functional groups.

Gao et al. Cr_2C-based MXene phase with magnetic properties and a high Néel temperature of 400 K [134]. Totally compensated antiferromagnetic order with zero magnetization and fully spin polarized semiconducting behavior were thought to be the magnetic properties. By using density functional theory calculations, it was predicted that, if the MXenes were asymmetrically functionalized, forming

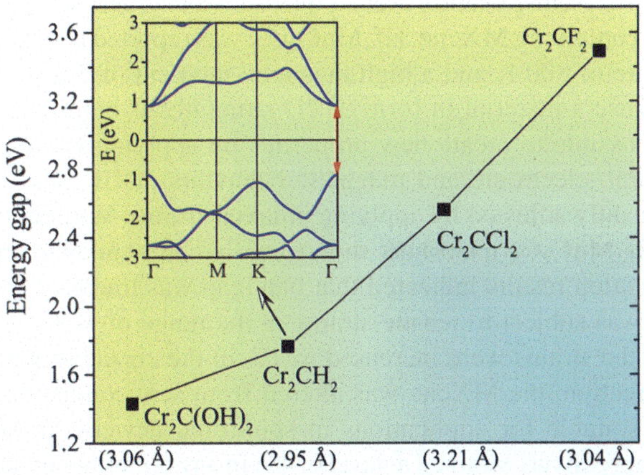

Fig. 2.53 Energy gaps of the Cr_2CT_2 MXenes terminated with different functional groups (T = F, OH, H or Cl). The numbers in the brackets in the *x*-axis are lattice constants of the corresponding Cr_2CT_2 MXene phases, with the inset showing the band structure of Cr_2CH_2 with the direct gap at the Γ point indicated by the red arrows in which the Fermi level was set to be zero. Reproduced with permission from [133]. Copyright © 2015, American Chemical Society

Janus Cr_2C–Cr_2CXX', with X/X′ = H, F, Cl, Br, OH, such magnetic characteristics would be present. The valence and conduction bands of the MXene phases were composed of opposite spin channels, thus leading to bipolar magnetic semiconduction behaviros with zero magnetization. The MXene phases, including Cr_2CFCl, Cr_2CClBr, Cr_2CHCl, Cr_2CHF and Cr_2CFOH, all had a high Néel temperature as mentioned above. More importantly, electron or hole doping strategies could be used to control the spin carrier orientation. Meanwhile, the transition from bipolar magnetic semiconducting state to half-metallic antiferromagnetic state could be well managed in a similar way. The bandgap of the Janus MXene phases could be tailored by selecting two groups of chemical elements or functional groups to decorate the two surfaces of the nanosheets. The spin-polarized semiconduction with zero magnetization was well retained as the MXene phases were transferred onto SiC (0001) substrates, making it possible to develop spintronic devices by incorporating with the silicon processing technology.

He et al. conducted theoretical studies on magnetic properties of Mn_2CT_2 (T = F, Cl, OH, O and H) MXenes, by using density functional theory calculations and Monte Carlo simulations [135]. It was revealed that the Mn_2CT_2 MXene with functional groups, which had a charge of -1, i.e., T = F, Cl and OH, displayed ferromagnetic ground state, whereas those with other functional groups were nonmagnetic in case that both surfaces of the nanosheets were symmetrically functionalized. Specifically, the results predicted that Mn_2CF_2 MXene was intrinsically of half-metallic characteristics, with a Curie temperature of as high as 520 K, a relatively wide half-metallic gap of 0.9 eV and a magnetic anisotropy of 24 μeV. Therefore, this Mn_2CF_2 MXene had potential applications to develope spintronic devices. In addition, functionalization with Cl and OH resulted in Mn_2CT_2 MXene with quantitative variation in magnetic behaviour, while O and H led to qualitative alternation.

Another Mn-containing MXene, $Hf_2MnC_2O_2$, was reported to exhibit a quite high Curie temperature of 800 K and a high magnetic moment of 3 μ_B in each formula unit [136]. Because a material in form of 2D structures is more elastic than in the bulk one, it is possible to create new properties by applying strains. Accordingly, both the structural, electronic and magnetic behaviors of $Hf_2MnC_2O_2$ monolayer could be significantly adjusted by applying uniaxial strains. Without the application of strains, the $Hf_2MnC_2O_2$ nanosheet showed an indirect bandgap semiconducting behavior. Calculation results indicated that bandgap was transited from indirect to direct one, if it was subject to tensile strains in the range of 1–3% in the armchair direction. Once the strains were increased to 7% in the zigzag direction and 9% in the armchair direction, the MXene was altered from semiconductor to half-metal, thus making it suitable for applications in spintronic devices. In addition, if the MXene was compressively strained at level of 4% in any direction, metallic behaviour was obtained. Furthermore, the application of tensile strains could raise the Curie temperature of the magnetic MXene. Specifically, the Curie temperature could reach 1200 K if an 8% strain was applied, which is of special interest when the materials are used in high temperature environment.

MXenes, $Sc_2C(OH)_2$ and Sc_2CO_2, were largely different in terms of structure and properties [137, 138]. Sc_2CO_2 displayed characteristics of indirect semiconductors.

The surface –O functional groups were present at the top sites of the bottom Sc atoms on one side, while the O atoms occupied the top sites of the central C atoms on the other side. In comparison, $Sc_2C(OH)_2$ was a direct semiconductor, with the –OH functional groups to be at the top sites of the bottom Sc atoms equally on the two sides of the MXene nanosheets. In addition, experimental results had been shown that the contents of –OH and –O functional groups could be readily tailored [76].

By using first-principle density functional theory (DFT), Zha et al. mechanistically revealed the process of dehydrogenation-hydrogenation was actually the removing-adding process of the surface H atoms between $Sc_2C(OH)_2$ and Sc_2CO_2. It was attempted to clarify the structural and compositional evolutions between the two MXene phases. With three supercells, i.e., $2 \times 2 \times 1$, $3 \times 3 \times 1$ and $4 \times 4 \times 1$, intermediate phases, $Sc_2C(OH)_xO_{2-x}$, are achieved, with the content of H (x) to be variable in the range of 0.0625–1.94. It was observed that discrepancy in both the number of the H atoms and the spacing in between the two on the two sides of the nanosheets tended to be minimized. The occupation sites of all the O, C and Sc atoms were kept unchanged as $Sc_2C(OH)_2$ was entirely converted to Sc_2CO_2. However, if the value of x was insufficiently low, the atoms in the carbon layer would be rearranged.

The electronic properties were dependent on the rearranged configurations. As the H atoms were randomly distributed at the two sides of the nanosheets and the values of x are in the range of 0.188–0.812, bipolar magnetic semiconductors (BMSs) were present. It was predicted that the $Sc_2C(OH)_xO_{2-x}$ MXene phases could have intrinsic spin-polarized semiconducting behaviors, which could be realized for real applications to fabricate devices, as long as of the materials are prepared in the form of nanoflakes. The authors also believed that similar effect could be applied to more MXenes and different functional groups.

2.5 Concluding Remarks and Perspectives

The newly emerged 2D nanostructures contained early transition metal carbides, carbonitrides and nitrides, which are produced by using various fluorides as etching agents, at different processing conditions. The most effective and efficient processing paths to generate layered structures of carbides and carbonitrides include the etching reactions with HF and HCl/LiF agents, with which multi-layered MXene nanosheets could be produced, with variable morphologies and surface chemical properties. Even though the bonding force is relatively weak and the HF etched multilayered MXenes are slightly difficult to separate into independent nanolayers, it is quite hard to obtain single- or few-layered MXenes, if the systems are just treated with ultrasonication, while the intercalating agents are not used.

In addition, if the composition of the HCl/LiF etching agent is altered, by increasing the molar ratio of LiF to Ti_3AlC_2 from 7.5:1 to 5:1, with other processing parameters to be fixed, delaminated $Ti_3C_2T_x$ nanoflakes could be achieved, without the requirement of ultrasonication. Furthermore, the productivity can be increased by simply raising the concentration of the HCl etching solutions. The delaminated $Ti_3C_2T_x$ nanoflakes obtained in this way have lateral sizes in the range of 4–15 μm, with well-developed clear edges and without the presence of obvious pin-holes. Also, the MXenes could be intercalated with various cations, it is possible to further use them to fabricate composite materials with various new functionalities and hence different new applications.

Due to their special 2D nanostructures, the MXenes have various exciting mechanical, physical and chemical properties, as predicted by theoretical studies and evidenced by experimental results. However, the theoretical investigations and experimental practices are usually conducted by different research groups, which have no close interactions. As a result, the reported data have not been compared and mutually verified. Therefore, it is suggested to the theorists and experimentalists to closely collaborate in future studies of this new family of 2D nanomaterials. Additionally, the reports on magnetic properties of MXenes are mainly based on theoretical predictions, while experimental studies need further explorations.

Acknowledgements This work was supported by the National Natural Science Foundation of China (51762023 and 51962013), the Natural Science Foundation of Jiangxi, China (20192ACB20018), and Key R&D Program of Jiangxi Province (20171BBE50006, 20192ACB80007, and 20192ACB80004). Ling Bing Kong would like acknowledge Shenzhen Technology University (SZTU) for financial support through the Start-up Grant (2018) and grant from the Natural Science Foundation of Top Talent of SZTU (grant no. 2019010801002).

References

1. Barsoum, M.W.: The $MN_{+1}AX_N$ phases: a new class of solids. Prog. Solid State Chem. **28**, 201–281 (2000)
2. Sun, Z.M., Music, D., Ahuja, R., Li, S., Schneider, J.M.: Bonding and classification of nanolayered ternary carbides. Phy. Rev. B. **70**, 092102 (2004)
3. Deysher, G., Shuck, C.E., Hantanasirisakul, K., Frey, N.C., Foucher, A.C., Maleski, K., et al.: Synthesis of Mo_4VAlC_4 MAX phase and two-dimensional Mo_4VC_4 MXene with five atomic layers of transition metals. ACS Nano **14**, 204–217 (2020)
4. Emmerlich, J., Music, D., Eklund, P., Wilhelmsson, O., Jansson, U., Schneider, J.M., et al.: Thermal stability of Ti_3SiC_2 thin films. Acta Mater. **55**, 1479–1488 (2007)
5. Sun, Z.M.: Progress in research and development on MAX phases: a family of layered ternary compounds. Int. Mater. Rev. **56**, 143–166 (2011)
6. Naguib, M., Kurtoglu, M., Presser, V., Lu, J., Niu, J.J., Heon, M., et al.: Two-dimensional nanocrystals produced by exfoliation of Ti_3AlC_2. Adv. Mater. **23**, 4248–4253 (2011)
7. Barsoum, M.W.: MAX phases: properties of machinable ternary carbides and nitrides. Wiley (2013)
8. Anasori, B., Halim, J., Lu, J., Voigt, C.A., Hultman, L., Barsoum, M.W.: Mo_2TiAlC_2: a new ordered layered ternary carbide. Scripta Mater. **101**, 5–7 (2015)
9. Zhang, H.B., Zhou, Y.C., Bao, Y.W., Li, M.S., Wang, J.Y.: Intermediate phases in synthesis of Ti_3SiC_2 and $Ti_3Si(Al)C_2$ solid solutions from elemental powders. J. Eur. Ceram. Soc. **26**, 2373–2380 (2006)
10. Barsoum, M.W., El-Raghy, T., Ali, M.: Processing and characterization of Ti_2AlC, Ti_2AlN, and $Ti_2AlC_{0.5}N_{0.5}$. Metall. Mater. Trans. A **31**, 1857–1865 (2000)

11. Anasori, B., Xie, Y., Beidaghi, M., Lu, J., Hosler, B.C., Hultman, L., et al.: Two-dimensional, ordered, double transition metals carbides (MXenes). ACS Nano **9**, 9507–9516 (2015)
12. Cheng, Y.W., Dai, J.H., Zhang, Y.M., Song, Y.: Two-dimensional, ordered, double transition metal carbides (MXenes): a new family of promising catalysts for the hydrogen evolution reaction. J. Phys. Chem. C **122**, 28113–28122 (2018)
13. Naguib, M., Mochalin, V.N., Barsoum, M.W., Gogotsi, Y.: 25th anniversary article: MXenes: a new family of two-dimensional materials. Adv. Mater. **26**, 992–1005 (2014)
14. Lei, J.C., Zhang, X., Zhou, Z.: Recent advances in MXene: preparation, properties, and applications. Front. Phy. **10**, 276–286 (2015)
15. Xiao, Y., Hwang, J.Y., Sun, Y.K.: Transition metal carbide-based materials: synthesis and applications in electrochemical energy storage. J. Mater. Chem. A **4**, 10379–10393 (2016)
16. Kumar, P., Abuhimd, H., Wahyudi, W., Li, M.L., Ming, J., Li, L.J.: Review—two-dimensional layered materials for energy storage applications. ECS J. Solid State Sci. Technol. **5**, Q3021–Q3025 (2016)
17. Ng, V.M.H., Huang, H., Zhou, K., Lee, P.S., Que, W.X., Xu, J.Z., et al.: Recent progress in layered transition metal carbides and/or nitrides (MXenes) and their composites: synthesis and applications. J. Mater. Chem. A **5**, 3039–3068 (2017)
18. Mashtalir, O., Naguib, M., Mochalin, V.N., Dall' Agnese, Y., Heon, M., Barsoum, M.W., et al.: Intercalation and delamination of layered carbides and carbonitrides. Nature Commun. **4**, 1716 (2013)
19. Mei, J., Liao, T., Kou, L.Z., Sun, Z.Q.: Two-dimensional metal oxide nanomaterials for next-generation rechargeable batteries. Adv. Mater. **29**, 1700176 (2017)
20. Pomerantseva, E., Gogotsi, Y.: Two-dimensional heterostructures for energy storage. Nature Energy **2**, 17089 (2017)
21. Lipatov, A., Alhabeb, M., Lukatskaya, M.R., Boson, A., Gogotsi, Y., Sinitskii, A.: Effect of synthesis on quality, electronic properties and environmental stability of individual monolayer Ti_3C_2 MXene flakes. Adv. Electr. Mater. **2**, 1600255 (2016)
22. Halim, J., Lukatskaya, M.R., Cook, K.M., Lu, J., Smith, C.R., Näslund, L.A., et al.: Transparent conductive two-dimensional titanium carbide epitaxial thin films. Chem. Mater. **26**, 2374–2381 (2014)
23. Xu, C., Wang, L.B., Liu, Z.B., Chen, L., Guo, J.K., Kang, N., et al.: Large-area high-quality 2D ultrathin Mo_2C superconducting crystals. Nat. Mater. **14**, 1135–1141 (2015)
24. Alhabeb, M., Maleski, K., Anasori, B., Lelyukh, P., Clark, L., Sin, S., et al.: Guidelines for synthesis and processing of two-dimensional titanium carbide ($Ti_3C_2T_X$ MXene). Chem. Mater. **29**, 7633–7644 (2017)
25. Wang, L.B., Zhang, H., Wang, B., Shen, C.J., Zhang, C.X., Hu, Q.K., et al.: Synthesis and electrochemical performance of $Ti_3C_2T_x$ with hydrothermal process. Electron. Mater. Lett. **12**, 702–710 (2016)
26. Ghidiu, M., Lukatskaya, M.R., Zhao, M.Q., Gogotsi, Y., Barsoum, M.W.: Conductive two-dimensional titanium carbide 'clay' with high volumetric capacitance. Nature **516**, 78–81 (2014)
27. Li, H.Y., Hou, Y., Wang, F.X., Lohe, M.R., Zhuang, X.D., Niu, L., et al.: Flexible all-solid-state supercapacitors with high volumetric capacitances poosted by solution processable MXene and electrochemically exfoliated graphene. Adv. Energy Mater. **7**, 1601847 (2016)
28. Wang, X.F., Shen, X., Gao, Y.R., Wang, Z.X., Yu, R.C., Chen, L.Q.: Atomic-scale recognition of surface structure and intercalation mechanism of Ti_3C_2X. J. Am. Chem. Soc. **137**, 2715–2721 (2015)
29. Naguib, M., Mashtalir, O., Carle, J., Presser, V., Lu, J., Hultman, L., et al.: Two-dimensional transition metal carbides. ACS Nano **6**, 1322–1331 (2012)
30. Sang, X.H., Xie, Y., Lin, M.W., Alhabeb, M., Van Aken, K.L., Gogotsi, Y., et al.: Atomic defects in monolayer titanium carbide ($Ti_3C_2T_x$) MXene. ACS Nano **10**, 9193–9200 (2016)
31. Naguib, M., Halim, J., Lu, J., Cook, K.M., Hultman, L., Gogotsi, Y., et al.: New two-dimensional niobium and vanadium carbides as promising materials for Li-ion batteries. J. Am. Chem. Soc. **135**, 15966–15969 (2013)

32. Ghidiu, M., Naguib, M., Shi, C., Mashtalir, O., Pan, L.M., Zhang, B., et al.: Synthesis and characterization of two-dimensional Nb4C_3 (MXene). Chem. Commun. **50**, 9517–9520 (2014)
33. Yang, J., Naguib, M., Ghidiu, M., Pan, L.M., Gu, J., Nanda, J., et al.: Two-dimensional Nb-based M_4C_3 solid solutions (MXenes). J. Am. Ceram. Soc. **99**, 660–666 (2016)
34. Ran, J.R., Gao, G.P., Li, F.T., Ma, T.Y., Du, A.J., Qiao, S.Z.: Ti_3C_2 MXene co-catalyst on metal sulfide photo-absorbers for enhanced visible-light photocatalytic hydrogen production. Nature Commun. **8**, 13907 (2017)
35. Wang, J., Tang, J., Ding, B., Malgras, V., Chang, Z., Hao, X.D., et al.: Hierarchical porous carbons with layer-by-layer motif architectures from confined soft-template self-assembly in layered materials. Nature Commun. **8**, 15717 (2017)
36. Xu, B.Z., Zhu, M.S., Zhang, W.C., Zhen, X., Pei, Z.X., Xue, Q., et al.: Ultrathin MXene-micropattern-based field-effect transistor for probing neural activity. Adv. Mater. **28**, 3333–3339 (2016)
37. Mashtalir, O., Naguib, M., Dyatkin, B., Gogotsi, Y., Barsoum, M.W.: Kinetics of aluminum extraction from Ti_3AlC_2 in hydrofluoric acid. Mater. Chem. Phys. **139**, 147–152 (2013)
38. Boota, M., Anasori, B., Voigt, C., Zhao, M.Q., Barsoum, M.W., Gogotsi, Y.: Pseudocapacitive electrodes produced by oxidant-free polymerization of pyrrole between the layers of 2D titanium carbide (MXene). Adv. Mater. **28**, 1517–1522 (2016)
39. Hantanasirisakul, K., Zhao, M.Q., Urbankowski, P., Halim, J., Anasori, B., Kota, S., et al.: Fabrication of $Ti_3C_2T_x$ MXene transparent thin films with tunable opoelectronic properties. Adv. Electr. Mater. **2**, 1600050 (2016a)
40. Liu, F.F., Zhou, A.G., Chen, J.F., Jin, J., Zhou, W.J., Wang, L.B., et al.: Preparation of Ti_3C_2 and Ti_2C MXenes by fluoride salts etching and methane adsorptive properties. Appl. Surf. Sci. **416**, 781–789 (2017)
41. Lukatskaya, M.R., Mashtalir, O., Ren, C.E., Dall' Agnese, Y., Rozier, P., Taberna, P.L., et al.: Cation intercalation and high volumetric capacitance of two-dimensional titanium carbide. Science **341,** 1502–1505 (2013)
42. Alhabeb, M., Maleski, K., Mathis, T.S., Sarycheva, A., Hatter, C.B., Uzun, S., et al.: Selective etching of silicon from Ti_3SiC_2 (MAX) to obtain 2D titanium carbide (MXene). Angew. Chem. Inter. Edn. **57**, 5444–5448 (2018)
43. Natu, V., Pai, R., Sokol, M., Carey, M., Kalra, V., Barsoum, M.W.: 2D $Ti_3C_2T_z$ MXene synthesized by water-free etching of Ti_3AlC_2 in polar organic solvents. Chem **6**, 616–630 (2020)
44. Mei, J., Ayoko, G.A., Hu, C.F., Sun, Z.Q.: Thermal reduction of sulfur-containing MAX phase for MXene production. Chem. Eng. J. **395**, 125111 (2020)
45. Shuck, C.E., Sarycheva, A., Anayee, M., Levitt, A., Zhu, Y.Z., Uzun, S., et al.: Scalable synthesis of $Ti_3C_2T_x$ mXene. Adv. Eng. Mater. **22**, 1901241 (2020)
46. Salama, I., El-Raghy, T., Barsoum, M.W.: Synthesis and mechanical properties of Nb_2AlC and (Ti, Nb)2AlC. J. Alloy. Compd. **347**, 271–278 (2002)
47. Zhou, Y.C., Meng, F.L., Zhang, J.: New MAX-Phase compounds in the V-Cr-Al-C system. J. Am. Ceram. Soc. **91**, 1357–1360 (2008)
48. Zhao, S.S., Meng, X., Zhu, K., Du, F., Chen, G., Wei, Y.J., et al.: Li-ion uptake and increase in interlayer spacing of Nb_4C_3 MXene. Energy Storage Mater. **8**, 42–48 (2017)
49. Dryza, V., Addicoat, M.A., Gascooke, J.R., Buntine, M.A., Metha, G.F.: Threshold photoionization and density functional theory studies of the niobium carbide clusters Nb_3C_n ($n = 1 - 4$) and Nb_4C_n ($n = 1 - 6$). J. Phys. Chem. A **112**, 5582–5592 (2008)
50. Meshkian, R., Näslund, L.A., Halim, J., Lu, J., Barsoum, M.W., Rosen, J.: Synthesis of two-dimensional molybdenum carbide, Mo_2C, from the gallium based atomic laminate Mo_2Ga_2C. Scripta Mater. **108**, 147–150 (2015)
51. Petruhins, A., Ingason, A.S., Dahlqvist, M., Mockute, A., Junaid, M., Birch, J., et al.: Phase stability of $Cr_{n+1}GaC_n$ MAX phases from first principles and Cr_2GaC thin-film synthesis using magnetron sputtering from elemental targets. Phy. Status Solidi-Rapid Res. Lett. **7**, 971–974 (2013)

52. Zhou, J., Zha, X.H., Chen, F.Y., Ye, Q., Eklund, P., Du, S.Y., et al.: A two-dimensional zirconium carbide by selective etching of Al_3C_3 from nanolaminated $Zr_3Al_3C_5$. Angew. Chem. **128**, 5092–5097 (2016)
53. Chang, F.Y., Li, C.S., Yang, J., Tang, H., Xue, M.Q.: Synthesis of a new graphene-like transition metal carbide by de-intercalating Ti_3AlC_2. Mater. Lett. **109**, 295–298 (2013)
54. Sun, Z., Li, S., Ahuja, R., Schneider, J.M.: Calculated elastic properties of M2AlC (M=Ti, V, Cr, Nb and Ta). Solid State Commun. **129**, 589–592 (2004)
55. Liao, L.Q., Jiang, S., Zhou, S.S., Wei, X.T., Duan, C.K., Yin, M., et al.: Enhanced upconverted emission in Yb3+/Er3+ co-doped GdPO4 transparent glass ceramics. Optics Commun. **326**, 130–133 (2014)
56. Frodelius, J., Eklund, P., Beckers, M., Persson, P.O.A., Hogberg, H., Hultman, L.: Sputter deposition from a Ti_2AlC target: Process characterization and conditions for growth of Ti_2AlC. Thin Solid Films **518**, 1621–1626 (2010)
57. Feng, A.H., Yu, Y., Wang, Y., Jiang, F., Yu, Y., Mi, L., et al.: Two-dimensional MXene Ti_3C_2 produced by exfoliation of Ti_3AlC_2. Mater. Des. **114**, 161–166 (2017)
58. Halim, J., Kota, S., Lukatskaya, M.R., Naguib, M., Zhao, M.Q., Moon, E.J., et al.: Synthesis and characterization of 2D molybdenum carbide (MXene). Adv. Func. Mater. **26**, 3118–3127 (2016)
59. Hu, C., Lai, C.C., Tao, Q., Lu, J., Halim, J., Sun, L., et al.: Mo_2Ga_2C: a new ternary nanolaminated carbide. Chem. Commun. **51**, 6560–6563 (2015)
60. Liu, F.F., Zhou, J., Wang, S.W., Wang, B.X., Shen, C., Wang, L.B., et al.: Preparation of high-purity V_2C MXene and electrochemical properties as Li-ion batteries. J. Electrochem. Soc. **164**, A709–A713 (2017)
61. Zhong, Y., Xia, X.H., Shi, F., Zhan, J.Y., Tu, J.P., Fan, H.J.: Transition metal carbides and nitrides in energy storage and conversion. Adv. Sci. **3**, 1500286 (2016)
62. Morel, A., Borjon-Piron, Y., Lucio Porto, R., Brousse, T., Belanger, D.: Suitable conditions for the use of vanadium nitride as an electrode for electrochemical capacitor. J. Electrochem. Soc. **163**, A1077–A1082 (2016)
63. Naik, G.V., Schroeder, J.L., Ni, X.J., Kildishev, A.V., Sands, T.D., Boltasseva, A.: Titanium nitride as a plasmonic material for visible and near-infrared wavelengths. Opt. Mater. Express **2**, 478–489 (2012)
64. Xie, Y., Kent, P.R.C.: Hybrid density functional study of structural and electronic properties of functionalized $Ti_{n+1}X_n$ ($X = C$, N) monolayers. Phy. Rev. B. **87**, 235441 (2013)
65. Ye, Q., Xiao, P., Liu, W.L., Chen, K., Chen, T., Xue, J.M., et al.: Exploring the potential of exfoliated ternary ultrathin Ti_4AlN_3 nanosheets for fabricating hybrid patterned polymer brushes. RSC Adv. **5**, 70339–70344 (2015)
66. Urbankowski, P., Anasori, B., Makaryan, T., Er, D.Q., Kota, S., Walsh, P.L., et al.: Synthesis of two-dimensional titanium nitride Ti_4N_3 (MXene). Nanoscale **8**, 11385–11391 (2016)
67. Soundiraraju, B., George, B.K.: Two-dimensional titanium nitride (Ti_2N) MXene: synthesis, characterization, and potential application as surface-enhanced raman scattering substrate. ACS Nano **11**, 8892–8900 (2017)
68. Xie, X.H., Xue, Y., Li, L.J., Chen, S.G., Nie, Y., Ding, W., et al.: Surface Al leached Ti_3AlC_2 as a substitute for carbon for use as a catalyst support in a harsh corrosive electrochemical system. Nanoscale **6**, 11035–11040 (2014)
69. Harris, K.J., Bugnet, M., Naguib, M., Barsoum, M.W., Goward, G.R.: Direct measurement of surface termination groups and their connectivity in the 2D MXene V_2CT_x using NMR spectroscopy. J. Phy. Chem. C **119**, 13713–13720 (2015)
70. Hope, M.A., Forse, A.C., Griffith, K.J., Lukatskaya, M.R., Ghidiu, M., Gogotsi, Y., et al.: NMR reveals the surface functionalisation of Ti_3C_2 MXene. Phy. Chem. **18**, 5099–5102 (2016)
71. Zhang, T., Pan, L.M., Tang, H., Du, F., Guo, Y.H., Qiu, T., et al.: Synthesis of two-dimensional $Ti_3C_2T_x$ MXene using HCl + LiF etchant: enhanced exfoliation and delamination. J. Alloy. Compd. **695**, 818–826 (2017)

72. Shahzad, F., Alhabeb, M., Hatter, C.B., Anasori, B., Soon, M.H., Koo, C.M., et al.: Electromagnetic interference shielding with 2D transition metal carbides (MXenes). Science **353**, 1137–1140 (2016)
73. Ghidiu, M., Halim, J., Kota, S., Bish, D., Gogotsi, Y., Barsoum, M.W.: Ion-exchange and cation solvation reactions in Ti_3C_2 MXene. Chem. Mater. **28**, 3507–3514 (2016)
74. Mashtalir, O., Lukatskaya, M.R., Kolesnikov, A.I., Raymundo-Pinero, E., Naguib, M., Barsoum, M.W., et al.: The effect of hydrazine intercalation on the structure and capacitance of 2D titanium carbide (MXene). Nanoscale **8**, 9128–9133 (2016)
75. Naguib, M., Unocic, R.R., Armstrong, B.L., Nanda, J.: Large-scale delamination of multilayers transition metal carbides and carbonitrides "MXenes" . Dalton Trans. **44**, 9353–9358 (2015)
76. Xie, Y., Naguib, M., Mochalin, V.N., Barsoum, M.W., Gogotsi, Y., Yu, X.Q., et al.: Role of surface structure on Li-ion energy storage capacity of two-dimensional transition-metal carbides. J. Am. Chem. Soc. **136**, 6385–6394 (2014)
77. Mashtalir, O., Cook, K.M., Mochalin, V.N., Crowe, M., Barsoum, M.W., Gogotsi, Y.: Dye adsorption and decomposition on two-dimensional titanium carbide in aqueous media. J. Mater. Chem. A **2**, 14334–14338 (2014)
78. Mashtalir, O., Lukatskaya, M.R., Zhao, M.Q., Barsoum, M.W., Gogotsi, Y.: Amine-assisted delamination of Nb_2C MXene for Li-ion energy storage devices. Adv. Mater. **27**, 3501–3506 (2015)
79. Wang, H.B., Zhang, J.F., Wu, Y.P., Huang, H.J., Li, G.Y., Zhang, X., et al.: Surface modified MXene Ti_3C_2 multilayers by aryl diazonium salts leading to large-scale delamination. Appl. Surf. Sci. **384**, 287–293 (2016)
80. Mahouche-Chergui, S., Gam-Derouich, S., Mangeney, C., Chehimi, M.M.: Aryl diazonium salts: a new class of coupling agents for bonding polymers, biomacromolecules and nanoparticles to surfaces. Chem. Soc. Rev. **40**, 4143–4166 (2011)
81. Zhang, X.D., Xu, J.G., Wang, H., Zhang, J.J., Yan, H.B., Pan, B.C., et al.: Ultrathin nanosheets of MAX phases with enhanced thermal and mechanical properties in polymeric compositions: $Ti_3Si_{0.75}Al_{0.25}C_2$. Angew. Chemie Inter. Edn. **52**, 4361–4365 (2013)
82. Ren, C.E., Zhao, M.Q., Makaryan, T., Halim, J., Boota, M., Kota, S., et al.: Porous two-dimensional transition metal carbide (MXene) flakes for high-performance Li-Ion storage. Chem. Electr. Chem. **3**, 689–693 (2016)
83. Ravnik, M., Alexander, G.P., Yeomans, J.M., Zumer, S.: Three-dimensional colloidal crystals in liquid crystalline blue phases. Proc. Natl. Acad. Sci. U.S.A. **108**, 5188–5192 (2011)
84. Yuan, Y., Martinez, A., Senyuk, B., Tasinkevych, M., Smalyukh, I.I.: Chiral liquid crystal colloids. Nat. Mater. **17**, 71–79 (2018)
85. Nakayama, M., Kajiyama, S., Kumamoto, A., Nishimura, T., Ikuhara, Y., Yamato, M., et al.: Stimuli-responsive hydroxyapatite liquid crystal with macroscopically controllable ordering and magneto-optical functions. Nature Commun. **9**, 568 (2018)
86. Xia, Y., Mathis, T.S., Zhao, M.Q., Anasori, B., Dang, A., Zhou, Z.H., et al.: Thickness—independent capacitance of vertically aligned liquid-crystalline MXenes. Nature **557**, 409–412 (2018)
87. Zhang, J.Z., Uzun, S., Seyedin, S., Lynch, P.A., Akuzum, B., Wang, Z.Y., et al.: Additive-free MXene liquid crystals and fibers. ACS Central Sci. **6**, 254–265 (2020)
88. Zhu, J., Ha, E.N., Zhao, G.L., Zhou, Y., Huang, D.S., Yue, G.Z., et al.: Recent advance in MXenes: a promising 2D material for catalysis, sensor and chemical adsorption. Coord. Chem. Rev. **352**, 306–327 (2017)
89. Li, R.Y., Zhang, L.B., Shi, L., Wang, P.: MXene Ti_3C_2: An effective 2D light-to-heat conversion material. ACS Nano **11**, 3752–3759 (2017)
90. Zhang, X., Lei, J.C., Wu, D.H., Zhao, X.D., Jing, Y., Zhou, Z.: A Ti-anchored Ti_2CO_2 monolayer (MXene) as a single-atom catalyst for CO oxidation. J. Mater. Chem. A **4**, 4871–4876 (2016)
91. Naguib, M., Come, J., Dyatkin, B., Presser, V., Taberna, P.L., Simon, P., et al.: MXene: a promising transition metal carbide anode for lithium-ion batteries. Electrochem. Commun. **16**, 61–64 (2012)

92. Gao, G.P., O'Mullane, A.P., Du, A.J.: 2D MXenes: a new family of promising catalysts for the hydrogen evolution reaction. Acs Catal. **7**, 494–500 (2017)
93. Li, Z.Y., Wang, L.B., Sun, D.D., Zhang, Y.D., Liu, B.Z., Hu, Q.K., et al.: Synthesis and thermal stability of two-dimensional carbide MXene Ti_3C_2. Mater. Sci. Eng. B **191**, 33–40 (2015)
94. Anasori, B., Shi, C.Y., Moon, E.J., Xie, Y., Voigt, C.A., Kent, P.R.C., et al.: Control of electronic properties of 2D carbides (MXenes) by manipulating their transition metal layers. Nanoscale Horizons **1**, 227–234 (2016)
95. Li, J.X., Du, Y.L., Huo, C.X., Wang, S., Cui, C.: Thermal stability of two-dimensional Ti_2C nanosheets. Ceram. Int. **41**, 2631–2635 (2015)
96. Je, M.Y., Lee, Y.B., Chung, Y.C.: Structural stability and electronic properties of multi-functionalized two-dimensional chromium carbides. Thin Solid Films **619**, 131–136 (2016)
97. Enyashin, A.N., Ivanovskii, A.L.: Structural and electronic properties and stability of MXenes Ti_2C and Ti_3C_2 functionalized by methoxy groups. J. Phy. Chem. C **117**, 13637–13643 (2013)
98. Ren, J.W., Antonietti, M., Fellinger, T.P.: Efficient water splitting using a simple Ni/N/C paper electrocatalyst. Adv. Energy Mater. **5**, 1401660 (2015)
99. Ma, T.Y., Cao, J.L., Jaroniec, M., Qiao, S.Z.: Interacting carbon nitride and titanium carbide nanosheets for high-performance oxygen evolution. Angew. Chem. Int. Ed. **55**, 1138–1142 (2016)
100. Seh, Z.W., Fredrickson, K.D., Anasori, B., Kibsgaard, J., Strickler, A.L., Lukatskaya, M.R., et al.: Two-dimensional molybdenum carbide (MXene) as an efficient electrocatalyst for hydrogen evolution. ACS Energy Lett. **1**, 589–594 (2016)
101. Shein, I.R., Ivanovskii, A.L.: Graphene-like titanium carbides and nitrides $Ti_{n+1}C_n$, $Ti_{n+1}N_n$ ($n = 1, 2$, and 3) from de-intercalated MAX phases: first-principles probing of their structural, electronic properties and relative stability. Comput. Mater. Sci. **65**, 104–114 (2012)
102. Zhang, X., Zhao, X.D., Wu, D.H., Jing, Y., Zhou, Z.: High and anisotropic carrier mobility in experimentally possible Ti_2CO_2 (MXene) monolayers and nanoribbons. Nanoscale **7**, 16020–16025 (2015)
103. Zou, G.D., Liu, B.Z., Guo, J.X., Zhang, Q.R., Fernandez, C., Peng, Q.M.: Synthesis of nanoflower-shaped MXene derivative with unexpected catalytic activity for dehydrogenation of sodium alanates. ACS Appl. Mater. Interfaces **9**, 7611–7618 (2017)
104. Li, X.J., Fan, G.Y., Zeng, C.M.: Synthesis of ruthenium nanoparticles deposited on graphene-like transition metal carbide as an effective catalyst for the hydrolysis of sodium borohydride. Int. J. Hydrogen Energy **39**, 14927–14934 (2014)
105. Kim, S.J., Naguib, M., Zhao, M.Q., Zhang, C.F., Jung, H.T., Barsoum, M.W., et al.: High mass loading, binder-free MXene anodes for high areal capacity Li-ion batteries. Electrochim. Acta **163**, 246–251 (2015)
106. Wu, R.Y., Du, H.F., Wang, Z.Y., Gao, M.X., Pan, H.G., Liu, Y.F.: Remarkably improved hydrogen storage properties of $NaAlH_4$ doped with 2D titanium carbide. J. Power Sources **327**, 519–525 (2016)
107. Xie, X.H., Chen, S.G., Ding, W., Nie, Y., Wei, Z.D.: An extraordinarily stable catalyst: Pt NPs supported on two-dimensional $Ti_3C_2X_2$ ($X = OH$, F) nanosheets for oxygen reduction reaction. Chem. Commun. **49**, 10112–10114 (2013)
108. Zhang, Z.W., Li, H.N., Zou, G.D., Fernandez, C., Liu, B.Z., Zhang, Q.R., et al.: Self-reduction synthesis of new MXene/Ag composites with unexpected electrocatalytic activity. ACS Sustain. Chem. Eng. **4**, 6763–6771 (2016)
109. Lin, H., Wang, X.G., Yu, L.D., Chen, Y., Shi, J.L.: Two-dimensional ultrathin MXene ceramic nanosheets for photothermal conversion. Nano Lett. **17**, 384–391 (2017)
110. Dall'Agnese, Y., Lukatskaya, M.R., Cook, K.M., Taberna, P.L., Gogotsi, Y., Simon, P.: High capacitance of surface-modified 2D titanium carbide in acidic electrolyte. Electrochem. Commun. **48**, 118–122 (2014)
111. Wang, L., Xie, X.J., Imazato, S., Weir, M.D., Reynolds, M.A., Xu, H.H.K.: A protein-repellent and antibacterial nanocomposite for Class-V restorations to inhibit periodontitis-related pathogens. Mater. Sci. Eng. C-Mater. Biol. Appl. **67**, 702–710 (2016)

112. Zhou, J., Zha, X.H., Zhou, X.B., Chen, F.Y., Gao, G.L., Wang, S.W., et al.: Synthesis and electrochemical properties of two-dimensional hafnium carbide. ACS Nano **11**, 3841–3850 (2017)
113. Rakhi, R.B., Ahmed, B., Anjum, D.H., Alshareef, H.N.: Direct chemical synthesis of MnO_2 nanowhiskers on transition-metal carbide surfaces for supercapacitor applications. ACS Appl. Mater. Interfaces **8**, 18806–18814 (2016)
114. Wang, F., Yang, C.H., Duan, C.Y., Xiao, D., Tang, Y., Zhu, J.F.: An organ-like titanium carbide material (MXene) with multilayer structure encapsulating hemoglobin for a mediator-free biosensor. J. Electrochem. Soc. **162**, B16–B21 (2015)
115. Wang, F., Yang, C.H., Duan, M., Tang, Y., Zhu, J.F.: TiO_2 nanoparticle modified organ-like Ti_3C_2 MXene nanocomposite encapsulating hemoglobin for a mediator-free biosensor with excellent performances. Biosens. Bioelectron. **74**, 1022–1028 (2015)
116. Lorencova, L., Bertok, T., Dosekova, E., Holazova, A., Paprckova, D., Vikartovska, A., et al.: Electrochemical performance of $Ti_3C_2T_x$ MXene in aqueous media: towards ultrasensitive H_2O_2 sensing. Electrochim. Acta **235**, 471–479 (2017)
117. Liu, H., Duan, C.Y., Yang, C.H., Shen, W.Q., Wang, F., Zhu, Z.F.: A novel nitrite biosensor based on the direct electrochemistry of hemoglobin immobilized on MXene-Ti_3C_2. Sensors Actuators B-Chem. **218**, 60–66 (2015)
118. Rakhi, R.B., Nayuk, P., Xia, C., Alshareef, H.N.: Novel amperometric glucose biosensor based on MXene nanocomposite. Sci. Reports. **6**, 36422 (2016)
119. Sarikurt, S., Cakir, D., Keceli, M., Sevik, C.: The influence of surface functionalization on thermal transport and thermoelectric properties of MXene monolayers. Nanoscale **10**, 8859–8868 (2018)
120. Handoko, A.D., Fredrickson, K.D., Anasori, B., Convey, K.W., Johnson, L.R., Gogotsi, Y., et al.: Tuning the basal plane functionalization of two-dimensional metal carbides (MXenes) to control hydrogen evolution activity. ACS Appl. Energy Mater. **1**, 173–180 (2018)
121. Hu, M.M., Hu, T., Li, Z.J., Yang, Y., Cheng, R.F., Yang, J.X., et al.: Surface functional groups and interlayer water determine the electrochemical capacitance of $Ti_3C_2T_x$ MXene. ACS Nano **12**, 3578–3586 (2018)
122. Zhang, C.F.J., Pinilla, S., McEyoy, N., Cullen, C.P., Anasori, B., Long, E., et al.: Oxidation stability of colloidal two-dimensional titanium carbides (MXenes). Chem. Mater. **29**, 4848–4856 (2017)
123. Seredych, M., Shuck, C.E., Pinto, D., Alhabeb, M., Precetti, E., Deysher, G., et al.: High-temperature behavior and surface chemistry of carbide MXenes studied by thermal analysis. Chem. Mater. **31**, 3324–3332 (2019)
124. Kurtoglu, M., Naguib, M., Gogotsi, Y., Barsoum, M.W.: First principles study of two-dimensional early transition metal carbides. MRS Commun. **2**, 133–137 (2012)
125. Ling, Z., Ren, C.E., Zhao, M.Q., Yang, J., Giammarco, J.M., Qiu, J.S., et al.: Flexible and conductive MXene films and nanocomposites with high capacitance. Proc. Natl. Acad. Sci. **111**, 16676–16681 (2014)
126. Yorulmaz, U., Ozden, A., Perkgoz, N.K., Ay, F., Sevik, C.: Vibrational and mechanical properties of single layer MXene structures: a first-principles investigation. Nanotechnology. **27**, 335702 (2016)
127. Magne, D., Mauchamp, V., Celerier, S., Chartier, P., Cabioc'h, T.: Spectroscopic evidence in the visible-ultraviolet energy range of surface functionalization sites in the multilayer Ti_3C_2 MXene. Phys. Rev. B **91**, 201409 (2015)
128. Hantanasirisakul, K., Zhao, M.Q., Urbankowski, P., Halim, J., Anasori, B., Kota, S., et al.: Fabrication of $Ti_3C_2T_x$ MXene transparent thin films with tunable optoelectronic properties. Adv. Electr. Mater. **2**, 1600050 (2016b)
129. Mariano, M., Mashtalir, O., Antonio, F.Q., Ryu, W.H., Deng, B.C., Xia, F.N., et al.: Solution-processed titanium carbide MXene films examined as highly transparent conductors. Nanoscale **8**, 16371–16378 (2016)
130. Sarycheva, A., Makaryan, T., Maleski, Satheeshkumar, E., Melikyan, A., Minassian, H., et al.: Two-dimensional titanium carbide (MXene) as surface-enhanced Raman scattering substrate. J. Phys. Chem. C **121**, 19983–19988 (2017)

131. Yang, L.B., Jiang, X., Ruan, W.D., Zhao, B., Xu, W.Q., Lombardi, J.R.: Adsorption study of 4-MBA on TiO_2 nanoparticles by surface-enhanced Raman spectroscopy. J. Raman Spectrosc. **40**, 2004–2008 (2009)
132. Rafiq, S., Awan, S., Zheng, R.K., Wen, Z.C., Rani, M., Akinwande, D., et al.: Novel room-temperature ferromagnetism in Gd-doped 2-dimensional $Ti_3C_2T_x$ MXene semiconductor for spintronics. J. Magn. Magn. Mater. **497**, 165954 (2020)
133. Si, C., Zhou, J., Sun, Z.M.: Half-metallic ferromagnetism and surface functionalization-induced metal-insulator transition in graphene-like two-dimensional Cr_2C crystals. ACS Appl. Mater. Interfaces **7**, 17510–17515 (2015)
134. He, J.J., Lyu, P.B., Sun, L.Z., Garcia, A.M., Nachtigall, P.: High temperature spin-polarized semiconductivity with zero magnetization in two-dimensional Janus MXenes. J. Mater. Chem. C **4**, 6500 (2016)
135. He, J., Lyu, P., Nachtigall, P.: New two-dimensional Mn-based MXenes with room-temperature ferromagnetism and half-metallicity. J. Mater. Chem. C **4**, 11143–11149 (2016)
136. Siriwardane, E.M.D., Karki, P., Loh, Y.L., Cakir, D.: Strain-spintronics: Modulating electronic and magnetic properties of $Hf_2MnC_2O_2$ MXene by uniaxial strain. J. Phys. Chem. C **123**, 12451–12459 (2019)
137. Khazaei, M., Arai, M., Sasaki, T., Chung, C.Y., Venkataramanan, N.S., Estili, M., et al.: Novel electronic and magnetic properties of two-dimensional transition metal carbides and nitrides. Adv. Func. Mater. **23**, 2185–2192 (2013)
138. Zha, X.H., Luo, K., Li, Q.W., Huang, Q., He, J., Wen, X.D., et al.: Role of the surface effect on the structural, electronic and mechanical properties of the carbide MXenes. EPL **111**, 26007 (2015)

Chapter 3
MXenes Based Composites and Hybrids

3.1 Introduction

Similar to other nanostructures, such as graphene and CNTs, MXenes have also been explored as components to form nanocomposites or nanohybrids, in order to obtain properties that are improved as compared with the individual components or have new properties that are not available in the respective components. Although there are differences between composites and hybrids, they will be differentiated in this book. As an example, when the multiple-layered MXene nanostructures were delaminated through intercalation process, paper-like structures with relatively high porosity were obtained. Such paper-like structures could be incorporated with CNTs or conducting polymers to produce nanocomposites [1–6]. The nanocomposites exhibited higher electrochemical performances, due to the enhanced electrical conductivities. The enhancement in electrical conductivity was attributed to the generation of extra electron conduction pathways. Meanwhile, the interlayer spacings of the MXenes were effectively enlarged, so that both the access and transportation of the electrolyte ions were boosted. Furthermore, because polymeric binders or conductive additives were not necessary, the problems of dead volumes and interfaces, which are harmful to the electrochemical performances, would be absent.

Another strategy to open up the interlayer spaces of MXenes is to attach cations onto the surfaces of the nanolayers. In order for this, the surface chemistries of the MXenes should be tailored [7–10]. As a result, the dominant functional groups on the surfaces would be altered, thus making the attachment of specific cations to be more effective and efficient. Moreover, once the interlayer spaces were enlarged, the next processing steps, like oxidations and self-assembling, would become more flexible. As a consequence, hybridized properties and hence enhanced performances could be achieved.

Attempts have also been made to functionalize the MXenes, so as to create new fillers/inclusions or matrices to allow the development of nanocomposites or nanohybrids with enhanced properties and performances [11–19]. For instance, the surface functional groups of MXenes can be removed using thermal annealing in

Z. Xiao et al., *MXenes and MXenes-based Composites*, Engineering Materials,
https://doi.org/10.1007/978-3-030-59373-5_3

Ar atmosphere, so that the interlayer surfaces are passivated, which would ensure more possibilities of modification. Similarly, hydrothermal oxidizing reaction has been employed to increase the degree of crystallization of MXenes, as well as their production capabilities. At the same time, morphologies and compositions can be adjusted to produce various MXenes/oxides nanohybrids.

In nanocomposites, the conductive MXenes matrix and the inclusions/fillers should be intimately contacted each other, so that the materials are sufficiently homogeneous to ensure the designed performances. In fact, achieving hierarchically multiple structures with intimate contacts in between the components and maintaining homogeneous distribution of the fillers/inclusions in the nanocomposites or nanohybrids should be balanced in practice. However, if the MXenes are utilized as both precursors of the cations and the supportive matrices, the contact issue could be readily solved, but number of cations available from the MXenes is still very limited.

3.2 MXene-Based Composites

Conductive flexible free-standing $Ti_3C_2T_x$/polyvinyl alcohol (PVA) composite were first reported by using vacuum filtration method, from colloidal suspensions of delaminated HF-etched multi-layered $Ti_3C_2T_x$ MXene with DMSO as the intercalant [20]. The PVA used in the study had a molecular weight of $M_r = 115{,}000$. In this type of $Ti_3C_2T_x$/PVA composites, the PVA matrix strongly interacted with the negatively charged $Ti_3C_2T_x$ nanosheet fillers through hydrogen bonding. Additionally, the $Ti_3C_2T_x$ nanolayers were highly aligned along the in-plane direction. As the content of PVA was increased from 10 to 60 wt%, tensile strength of the $Ti_3C_2T_x$/PVA composites, with reference to the pristine $Ti_3C_2T_x$ sample, was increase by about 34% to nearly 100%. Accordingly, electrical conductivity of the composites was sacrificed. Therefore, the mechanical and electrical properties should be balanced for different applications.

In a separate study, $Ti_3C_2T_x$/PVA composites were fabricated and evaluated in terms of thermal properties of MXene by using temperature-dependent Raman spectroscopy and polarized-laser power-dependent Raman spectroscopy [21]. It was found that the incorporation of PVA greatly improved the thermal stability of $Ti_3C_2T_x$, where the thermal coefficient of the $E_g{}^1$ mode was reduced from −0.06271 to −0.03357 cm^{-1}/K. This reduction was ascribed to the fact that strong Ti–O bonds were formed between the MXene and PVA polymer. At the same time, the thermal conductivities of both the pristine $Ti_3C_2T_x$ and the $Ti_3C_2T_x$/PVA composites exhibited pretty high values, which were 55.8 and 47.6 W/m K, respectively.

Microstructure of the pristine $Ti_3C_2T_x$ sample is illustrated in Fig. 3.1a, b, with typical layered structure of MXenes. With the presence of PVA, the MXene nanosheets were stuck, as shown in Fig. 3.1c, due to the bonding between the hydroxyl groups in PVA and the –OH functional surface of MXene. Also, Ti–O bonds were formed at the surface, according to the XPS characterization. As seen in Fig. 3.1d, the

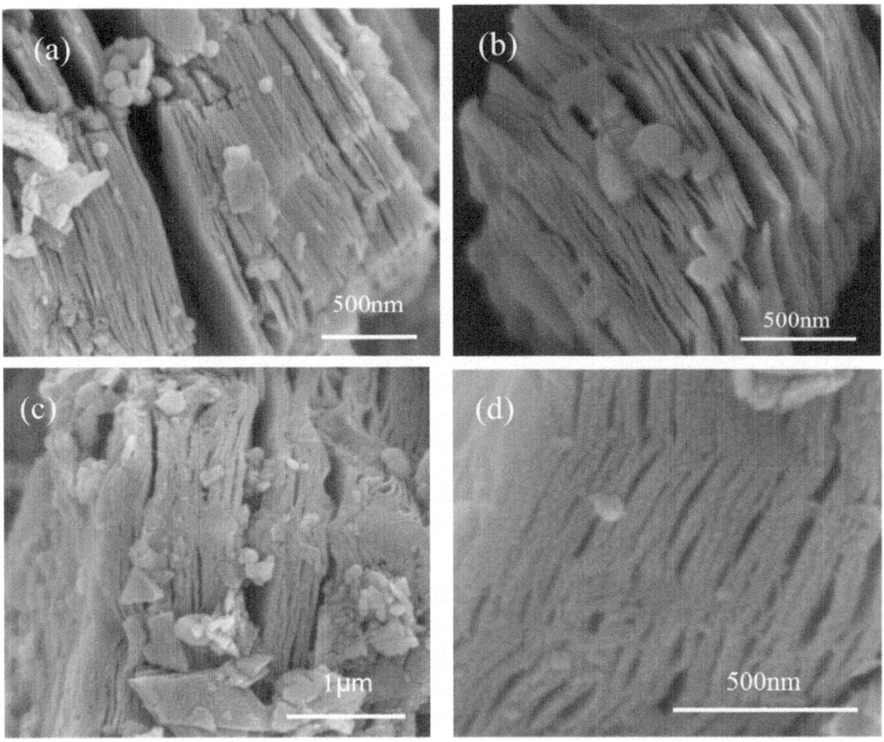

Fig. 3.1 **a** SEM image of $Ti_3C_2T_x$ MXene, **b** enlarged view of **a**, **c** SEM image of $Ti_3C_2T_x$/PVA, and **d** enlarged view of **c**. Reproduced with permission from [21]. Copyright © 2018, American Chemical Society

$Ti_3C_2T_x$/PVA composites displayed a densified layer structure. TEM images indicated that the interlayer spacing of the $Ti_3C_2T_x$ MXene was enlarged slightly after the incorporation of PVA. The presence of the polymer also reduced the uniformity of the MXene nanosheets in the composites.

TGA curves of the bare $Ti_3C_2T_x$ and $Ti_3C_2T_x$/PVA composite in nitrogen are shown in Fig. 3.2a, b, respectively. For pure $Ti_3C_2T_x$, three weight variation steps were observed, corresponding to the evaporation of water molecules adsorbed on the surface of $Ti_3C_2T_x$, release of bonding water and functional groups and the decomposition of $Ti_3C_2T_x$ MXene, respectively. In comparison, TGA curve of the $Ti_3C_2T_x$/PVA composite had one more weight change step, which over the temperature range of 220–480 °C, due to the decomposition of PVA. The composite sample contained about 12.7% PVA. However, the decomposition temperature of $Ti_3C_2T_x$ of in the composite sample was higher than that of pure $Ti_3C_2T_x$, suggesting that the thermal stability of $Ti_3C_2T_x$ was increased when it was incorporated with PVA to form composites.

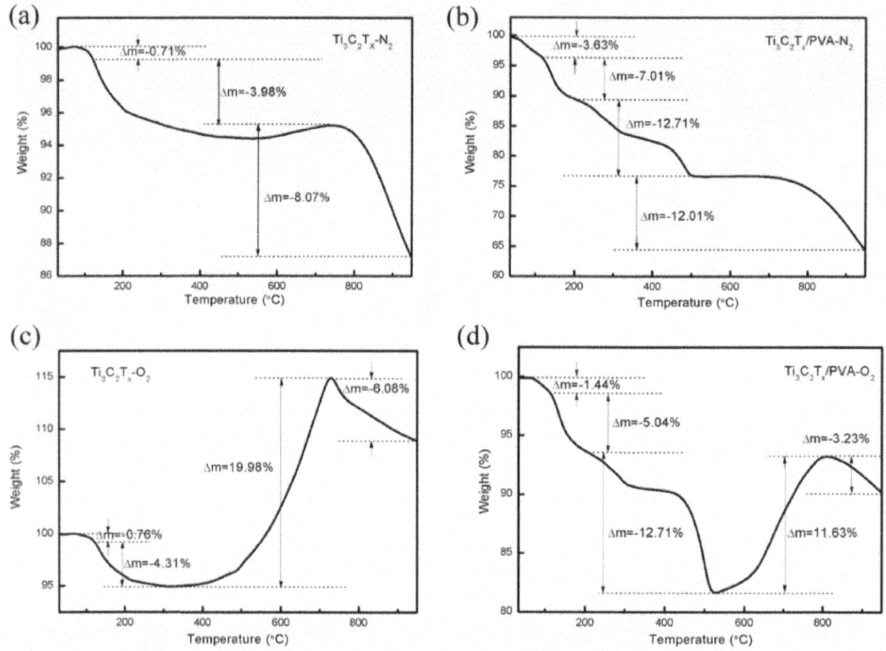

Fig. 3.2 TG curves of **a** $Ti_3C_2T_x$ and **b** $Ti_3C_2T_x$/PVA composites in nitrogen and **c** $Ti_3C_2T_x$ and **d** $Ti_3C_2T_x$/PVA composites in oxygen. Reproduced with permission from [21]. Copyright © 2018, American Chemical Society

TGA curves of the two samples in oxygen are plotted in Fig. 3.2c, d. The pure $Ti_3C_2T_x$ experienced four weight change steps. As the temperature was increased from 322 to 729 °C, the sample exhibited a weight increase by 20%, which was readily ascribed to the oxidation of $Ti_3C_2T_x$. Therefore, the oxidation temperature of $Ti_3C_2T_x$ was near 322 °C, while the decomposition temperature was about 729 °C. Similarly, the composite sample one more weight variation step, i.e., five weight change steps. However, the oxidation temperature was 530 °C, whereas the decomposition temperature was about 811 °C. Both the oxidation and the decomposition temperatures of the composite sample were increased, as compared with those of the pristine $Ti_3C_2T_x$, confirming the positive effect of PVA on thermal stability of $Ti_3C_2T_x$ MXene.

$Ti_3C_2T_x$ MXene-reinforced PVA nanofibers were prepared by using electrospinning method [22]. In these nanocomposites, the delaminated MXene nanosheets acted as fillers to obtain outstanding performances, due to the special characteristics of the 2D MXenes, including high aspect ratio, hydrophilic behavior and metallic conductivity. Meanwhile, cellulose nanocrystals (CNC) were also employed to enhance mechanical strength of the nanofibers as a comparison. The nanofibers had diameters in the range of 174–194 nm, which could be controlled by adjusting the ratios of PVA, CNC and MXene. It was observed that room temperature elastic modulus was increased from 392 MPa for bare PVA fibers to 855 MPa for those reinforced with CNC and MXene. More importantly, the PVA based composite nanofibers with 0.14 wt% $Ti_3C_2T_x$ displayed an electrical conductivity of 0.8 mS/cm, which as among the highest values of similar composites made with other methods. Due to their greatly enhanced mechanical and electrical performances, the $Ti_3C_2T_x$/CNC/PVA composite nanofibers could find applications in energy harvesting and storage.

Date palm leaves were used to prepare CNCs. Dried leaves were made into powder, which was dispersed in 1 M ammonium persulfate solution. The suspension was heated at 60 °C overnight with constant stirring, thus releasing CNCs. The CNCs were collected through repeated centrifugation/washing process, resulting in white powder. The sample was slowly titrated with 1 M NaOH till a pH value of about 7. ML-$Ti_3C_2T_x$ suspension was made with 70% ethanol/water. PVA solution with a concentration of 15 wt% was made by dissolving PVA powder in distilled water. Both the $Ti_3C_2T_x$ and CNCs suspensions were mixed with the PVA solution with the aid of ultrasonication in ice bath. The mixtures were used to obtain composite nanofibers by using electrospinning.

Highly porous MXene/PVA composite foams were made of few-layered Ti_2CT_x (f-Ti_2CT_x) nanosheets and PVA, by using a simple freeze-drying process [23]. The composite foams exhibited excellent EMI shielding effects, with theoretical specific shielding effectiveness levels of up to 5136 dB cm^2/g^{-1}, with the loading of the MXene inclusion to be as low as 0.15 vol%. Since the reflection effectiveness (SE_R) was <2 dB, the EMI shielding performance was dominated by the absorption effect of the materials. In this case, the high impedance matching, the multiple structures of the pores, the internal reflections and the multiple polarization effects were synergistically responsible for the superior absorption effect and hence the overall EMI shielding behavior.

$Ti_3C_2T_x$/polyacrylamide (PAM) nanocomposites were prepared from $Ti_3C_2T_x$/PAM mixed solution by using casting method on Teflon, which were then naturally dried at room temperature [24]. In these nanocomposites, the $Ti_3C_2T_x$ nanosheets were homogeneously distributed in the PAM matrix. The $Ti_3C_2T_x$ was intercalated with DMSO, so as to interact with the PAM molecules effectively to form solutions, with the few-layered $Ti_3C_2T_x$ nanosheets to be uniformly dispersed and randomly oriented. The as-prepared nanocomposites were highly flexible.

Specifically, the sample with 6 wt% (1.7 vol%). MXene was semiconducting, with an electrical conductivity of 0.033 S/m, which was largely higher than that of pure PAM. As the content of $Ti_3C_2T_x$ was increased to 31 wt%, the sample showed metal-like conduction characteristics, suggesting that it was over the percolation threshold. Electrical conduction properties of the nanocomposites are shown in Fig. 3.3. Such nanocomposites could be used as electromagnetic interference (EMI) shielding materials and electrodes of various electronic devices and electricity storage devices. Also, it was expected that the presence of $Ti_3C_2T_x$ could increase mechanical properties of PAM.

$Ti_3C_2T_x$, etched with HF and intercalated with DMSO, was used to form nanocomposites with hydrophilic polyethyleneimine (PEI) as the polymer matrix, by using solution filtration method [25]. $Ti_3C_2T_x$/PEI was mixed with trimesoyl chloride

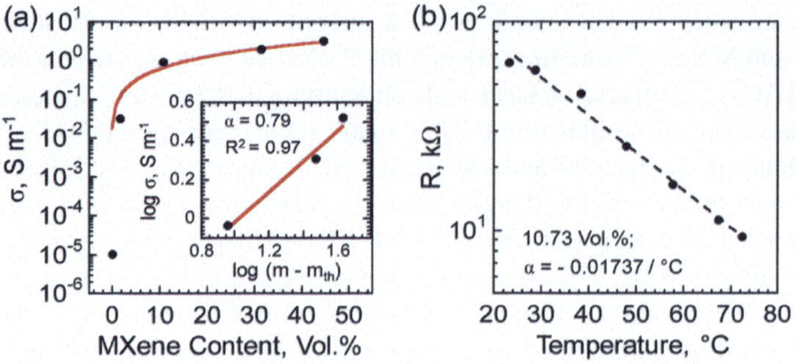

Fig. 3.3 **a** Conductivity of the MXene–PAM nanocomposite films as a function of MXene volumtric concentration. Solid line represents theoretical fit to the experimental data. Inset illustrates the power law dependence of conductivity above the percolation threshold, m_{th}. **b** Temperature dependence of resistance of the 10.7 vol% (31 wt%) MXene–PAM nanocomposite film. Reproduced with permission from [24]. Copyright © 2016, Royal Society of Chemistry

(TMC) and n-hexane solution to form suspensions, which were filtrated through polyacrylonitrile (PAN) ultrafiltration membrane to obtain membranes. After annealing at 60 °C for 2 h, the residual solvent was eliminated, while cross-linking polymerization took place. The $Ti_3C_2T_x$ nanosheets were homogeneously in the PEI matrix, due to the hydrogen bondings.

Poly(ethylene oxide)(PEO)/MXene nanocomposites were fabricated in order to modify the structural and crystallization properties [26]. MXene $Ti_3C_2T_x$ were incorporated as nanofillers to form polymer nanocomposites by using solution blending method. Thermal behaviors of the nanocomposites were studied in terms of both non-isothermal and isothermal crystallization characterizations. Figure 3.4 shows non-isothermal DSC scan curves of the nanocomposites during first cooling and second heating processes. It was found that the addition of $Ti_3C_2T_x$ promoted the crystallization of PEO initially at relatively low concentrations of MXene. However,

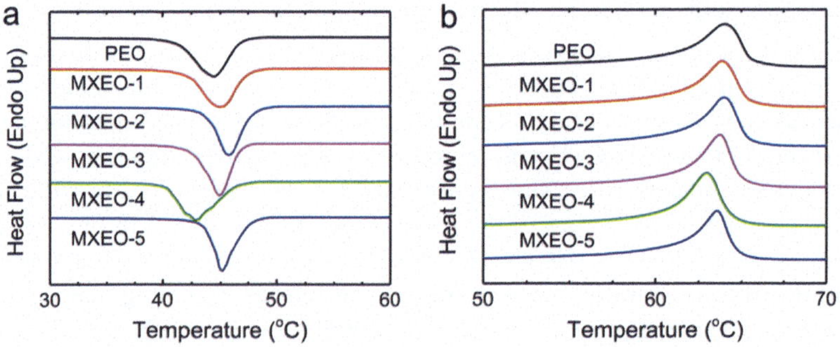

Fig. 3.4 Non-isothermal DSC scan curves of the nanocomposites during **a** first cooling and **b** second heating. Both scans were conducted at the rate of 10 °C/min. Reproduced with permission from [26]. Copyright © 2016, Elsevier

a high content of PEO could delayed the crystallization of the polymer. The rate of crystallization was maximized at 0.5 wt% MXene, which could be understood by considering the competition of nucleation and confinement effect of the 2D nanofillers.

Delaminated Ti_3C_2 MXene nanoflakes were incorporated with poly(acrylic acid) (PAA), poly(ethylene oxide) (PEO), poly(vinyl alcohol) (PVA), and alginate/PEO to develop nanofibers by using electrospinning technique [27]. It was observed that a small quantity of delaminated Ti_3C_2 (1% w/w) could have high effect on structure and performances of the nanofibers. For instance, Ti_3C_2 could influenced the properties of the polymer solutions, thus having an effect on average diameters of the nanofiber. Representative SEM images and fiber diameter distribution characteristics of the samples are demonstrated in Fig. 3.5. The overall effect was the most pronounced in the $Ti_3C_2T_x$/PEO solution, whose viscosity and conductivity were increased by 11% and 73.6%, respectively. Also, the crystallization of Ti_3C_2/PEO was enhanced, while that Ti_3C_2/PVA was slightly weakened.

PAA (9% w/v) was dissolved in DMF/water mixed solvent with a volume ratio of 3:7, PEO (5% w/v) in ethanol (EtOH)/water a volume ratio of 4:6 and PVA (9% w/w) in EtOH/water with a volume ratio of 1:9, which were all at 90 °C for 12 h. The polymer solutions were mixed with $Ti_3C_2T_x$ suspensions with a concentration of 1% w/w, aided with stirring for 12 h at room temperature. Then, $Ti_3C_2T_x$/alginate

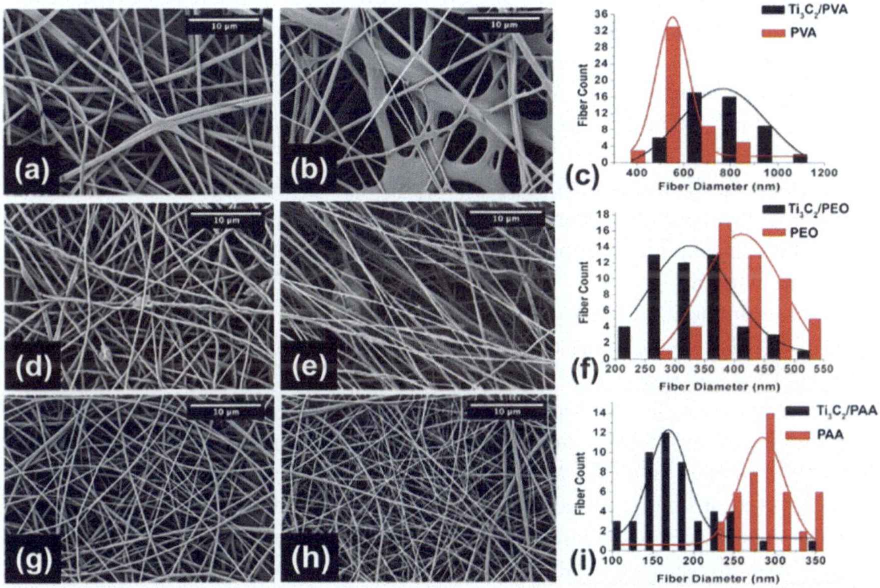

Fig. 3.5 SEM images of the electrospun nanofibers: **a** PVA, **b** $Ti_3C_2T_x$/PVA, **d** PEO, **e** $Ti_3C_2T_x$/PEO, **g** PAA and **h** $Ti_3C_2T_x$/PAA. Fitted fiber diameter distribution characteristics of **c** PVA, **f** PEO and **i** PAA and their respective nanocomposites. Reproduced with permission from [27]. Copyright © 2017, John Wiley & Sons

was dissolved in glycerol/water with a volume ratio of 2:1, thus leading to PEO concentration of 1% w/w. Meanwhile, PEO was dissolved in water resulting in a weight ratio of PEO/alginate to be 1:1. Finally, the mixture of $Ti_3C_2T_x$/PEO/alginate was stirred for one day. Polymer samples were prepared in a similar way for the purpose of comparison.

For electrospinning experiment, the $Ti_3C_2T_x$/PVA solution was pumped at a flow rate of 0.75 ml/h, while a voltage of 20 kV was applied, at a needle-to-collector distance of 10 cm. Electrospinning of the $Ti_3C_2T_x$/PEO solution was conducted at a flow rate of 0.5 ml/h, a voltage of 15 kV and a distance of 15 cm. For the $Ti_3C_2T_x$/PAA solution, the three parameters were 0.25 ml/h, 20 kV and 15 cm. Similarly, those for the $Ti_3C_2T_x$/alginate/PEO were 0.25 ml/h, 15 kV and a 25 cm. A 21 gauge needle was used for all electrospinning processes. All the experiments were carried out at room temperature with a relative humidity of <30%.

Poly(2-(dimethylamino)ethyl methacrylate) (PDMAEMA) was grafted onto multi-layered V_2CT_x nanosheets, in order to obtain V_2CT_x/PDMAEMA composites, through self-initiated photografting and photopolymerization (SIPGP) [28]. The composites were formed, due to the interaction between the positively charged PDMAEMA and the negatively charged V_2CT_x. Surface characteristics of the V_2CT_x/PDMAEMA composites were varied from hydrophilic to hydrophobic behaviour, as the samples were thermally stimulated. In addition, the conductivity could be altered by CO_2. Figure 3.6 shows schematic diagram of the fabrication process to develop the V_2CT_x/PDMAEMA composites. In this case, PDMAEMA was grafted on the photo-active group functionalized surface of V_2C through the step of SIPGP, leading to hybrids of MXene which were sensitive to the stimulation of CO_2 and temperature.

An in-situ method was reported to prepare mono- and co-doped MXene($Ti_3C_2T_x$)/poly(3,4-ethylenedioxythiophene) (PEDOT) composites (MPTs) for electrode applications [29]. The co-doped-MPTs had a lamellar microstructure, which promoted the transport of both electron and ions, thus leading to enhanced electrochemical properties. The doped MPTs based electrodes possessed a potential window of 1.8 V. It was found that the 1,5-naphathalene disulfonic acid/anthraquinone-2-sulfonate (AQS) co-doped MPTs exhibited a specific capacitance of 323 F/g, which was higher than that of the 1,5-naphathalene disulfonic acid/$H_3PMo_{12}O_{40}$ co-doped MPTs. Moreover, the former had an ideal capacitive characteristic, due to the co-doping effect and the Faradaic reaction of AQS.

Figure 3.7 shows SEM images of the pristine MXene, PEDOT, N-MPT, NP-MPT (NP0.4-MPT) and NA-MPT(NA0.6-MPT). The pristine MXene exhibited typical layered structure, as seen in Fig. 3.7a. Pure PEDOT had submicro-sized to micrometer-sized stacked lamellar structure, as observed in Fig. 3.7b. As revealed in Fig. 3.7c–e, the surface of MXene was coated with submicro-sized PEDOT with a plate-like morphology, after the reaction with EDOT monomer. By comparing Fig. 3.7c–e, it was found that the PEDOT plates had smaller sizes in the co-doped MPTs than in the 1,5-NDA doped MPTs, which was closely related to the addition of $H_3PMo_{12}O_{40}$ and AQS.

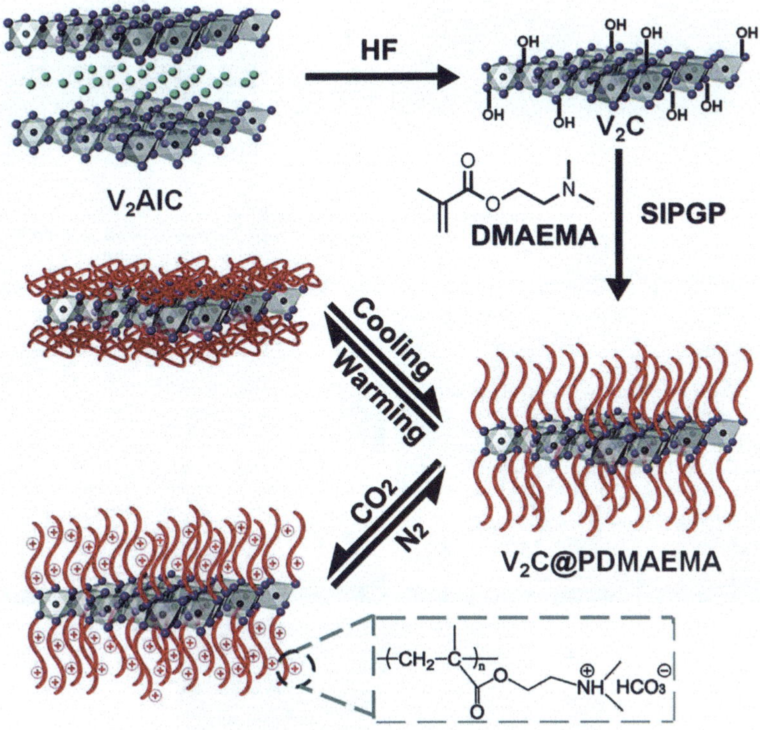

Fig. 3.6 Schematic diagram showing fabrication steps of the V_2C/PDMAEMA smart hybrids. C atoms are denoted as off-white color, V atoms are denoted as purple color which occupy every other elemental atomic plane and Al atoms are denoted as blue color which occupy every third plane. For simplicity, only two layers of V_2C are illustrated. Reproduced with permission from [28]. Copyright © 2015, Royal Society of Chemistry

A simple process was develop to obtain free-standing films with a self-assembled layered structure, in which aligned and doped polypyrrole (PPy) and C_4H_4NH were inserted into the spacings of single-layered $Ti_3C_2T_x$ MXene, as shown schematically in Fig. 3.8 [3]. To prepare the filtration solution, pyrrole (Py) was dispersed in the suspensions of HCl/LiF etched $Ti_3C_2T_x$. Due to the acidic nature of $Ti_3C_2T_x$, PPy molecules could be facilitated to form long chains.

According to SEM images, it was observed the binder-free PPy/$Ti_3C_2T_x$ (1:2) film had a thickness of 13 μm, with the entire film to be consisting of highly orientated MXene nanosheets. XRD patterns indicated that the pristine $Ti_3C_2T_x$ and the PPy/$Ti_3C_2T_x$ (1:2) films had (0002) peaks at 6.4° and 3.6°, respectively, suggesting that the interlayer spacing of $Ti_3C_2T_x$ was enlarged by about 1 nm. Because PPy is amorphous, its XRD pattern had a broad diffraction peak centered at about 25°, due to the scattering effect of the PPy molecular chains. Therefore, XRD pattern of the PPy/$Ti_3C_2T_x$ (1:2) sample in the range of 20°–35° was almost the same as that of pure $Ti_3C_2T_x$ film, without the presence of any extra diffraction peaks.

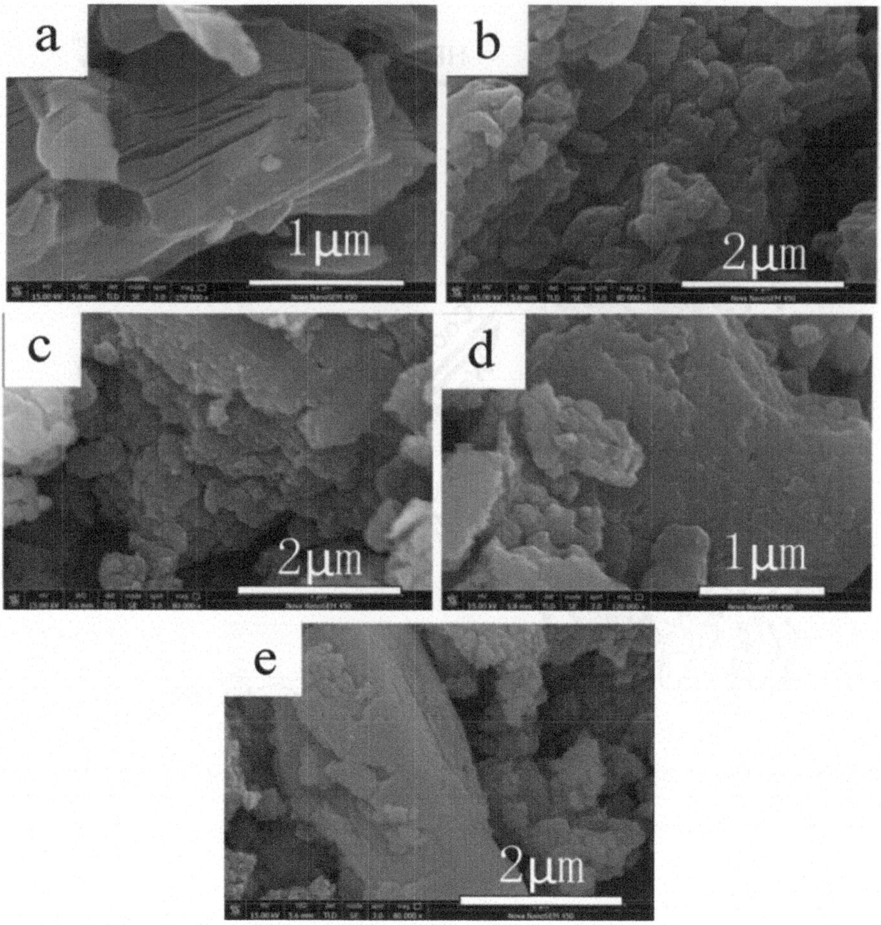

Fig. 3.7 SEM images of the samples: **a** MXene, **b** PEDOT, **c** N-MPT, **d** NP-MPT and **e** NA-MPT. Reproduced with permission from [29]. Copyright © 2018, Elsevier

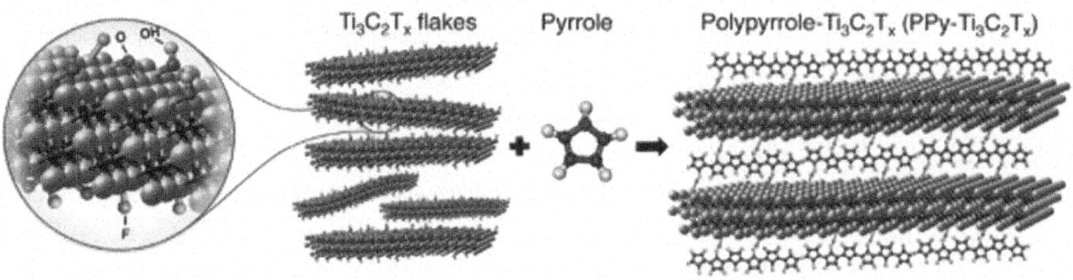

Fig. 3.8 Schematic diagram of pyrrole polymerization with MXene. The terminating groups on the latter facilitated the polymerization reaction. Reproduced with permission from [3]. Copyright © 2016, John Wiley & Sons

It was believed that the functional groups of –O, –OH and –F on the surface of $Ti_3C_2T_x$ nanosheets might have played a significant role during the formation of the hybrid between PPy and the MXene, because hydrogen bonds could be formed with the N–H groups of PPy molecules. As a result, the polymer chains were automatically aligned, while the random aggregation was effectively prevented. Meanwhile, the PPy molecules were doped by –F, thus leading to an increase in electrical conductivity. This study presented a simple approach that achieved intercalation, alignment and metal-free polymerization of Py with the presence of $Ti_3C_2T_x$. At optimized content of PPy, conductivity of the composites was maximized. The composite films had outstanding performances as the electrodes of supercapacitors.

An in-situ polymerization method was reported to prepare polyaniline (PANI) decorated Ti_3C_2-based composites, as electrodes of supercapacitors with enhanced electrochemical performances [30]. It was found that the decoration of PANI could increase the capacitive effect of Ti_3C_2 due to the dual effect of increased electrical conductivity and enhanced surface wetting behavior of the MXenes. The key function was related to the amino groups of PANI molecules. As a result, the ion transport was boosted, while the number of active sites for faradaic reaction was increased. The PANI-Ti_3C_2 can exhibited a maximum specific capacitance of 164 F/g at 2 mV/s, which was higher than that of the electrode based only on Ti_3C_2. Furthermore, the PANI-Ti_3C_2 composite electrode had a strong cycling stability, with 96% capacitance retention after cycling for 3000 times.

Similarly, an oxidant-free in-situ polymerization process was utilized to fabricate PANI-$Ti_3C_2T_x$ composites to serve as electrode of supercapacitors with large thicknesses [31]. The PANI molecules were attached on the surface of MXene nanosheets, thus leading to free-standing films with different thicknesses. Typical MXene/PANI based electrodes demonstrated gravimetric and volumetric capacitances of 503 F/g and 1682 F/cm^3, respectively. Because of the high accessibility, the electrochemical performances of the composite electrodes were limited by their thickness or mass loading. At a thickness of 90 μm and a mass loading of 23.82 mg/cm^2, the specific capacitance of the sample was 336 F/g, corresponding to a volumetric capacitance of 888 F/cm^3.

To coat PANI onto MXene nanosheets, 0.3 L aniline monomer (99+%) was dissolved in 20 mL HCl solution with a concentration of 1 M. The aniline/HCl solution mixed with $Ti_3C_2T_x$ water suspension, under constant stirring for 10 min. During the mixing process, PANI molecules were adsorbed onto the surface of the MXene nanosheets and thus precipitated. After that, the mixtures with different ratios of MXene to aniline were kept at 4 °C for 5 h, followed by centrifugation at 3500 rpm for 1 min. The collections were thoroughly washed with DI water. Finally, $Ti_3C_2T_x$/PANI composite samples were dispersed in 30 mL DI water, followed by filtration through Celgard® membranes, resulting in filter free-standing MXene/PANI composites sheets.

Figure 3.9 shows schematic diagrams illustrating fabrication routes of both the MXene and the composite samples. As the aniline solution was mixed with the MXene suspensions, the aniline monomor molecules were adsorbed onto surface of the MXene nanosheets, while the latter was immediately crumpled, because the low

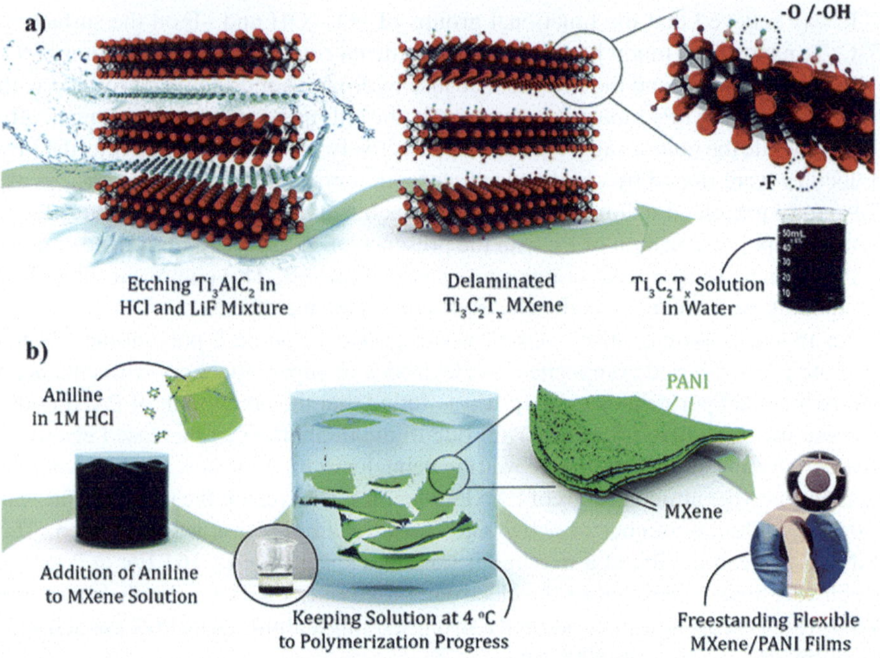

Fig. 3.9 **a** Schematic diagrams showing synthetic route and fabricate step of the samples: **a** $Ti_3C_2T_x$ MXene and **b** $Ti_3C_2T_x$/PANI composite electrode. Reproduced with permission from [31]. Copyright © 2018, Royal Society of Chemistry

pH value of the mixture caused flocculation of the composite material, as shown in Fig. 3.9a. The aniline HCl solution was slightly yellow in color, while supernatant of the MXene/aniline mixture was clear, as observed in Fig. 3.9b. As the $Ti_3C_2T_x$/aniline mixture was stored at 4 °C, polymerization of aniline monomors took place. It was found the optimal ratio of $Ti_3C_2T_x$ to aniline was 1:2, in terms of electrochemical performances.

In the $Ti_3C_2T_x$/PANI composites, the *c*-lattice parameter of the MXene was increased by about 0.15 nm, while the *d*-spacing was enlarged by about 0.075 nm, due to the insertion of the PANI molecules. The increased experimental value of the *d*-spacing was much larger than that theoretical calculation (0.018 nm). Also, the interlayer spacing was increased with increasing content of PANI. For instance, as the ratio of MXene to aniline was 1:500, the *d*-spacing could be 0.61 nm. Roughly, the presence of the PANI molecules led to an decrease in the content of –O groups in the MXene nanosheets.

$Ti_3C_2T_x$MXene/polyaniline (PANI) composites with different contents of PANI were prepared for microwave absorption applications [32]. The as-prepared MXene/PANI composites were further incorporated with paraffin matrix for the characterization of electromagnetic properties. At the optimal content of PANI, the paraffin based composites displayed a maximum reflection loss of −56.3 dB at the frequency of 13.8 GHz and the thickness of 1.8 mm. Additionally, the effective absorption bandwidth of −10 dB (>90%) was shifted from X-band (8–12.4 GHz) to Ku-band (12.4–18 GHz), as the tuneable thickness of the composite was increased from 1.5 to 2.6 mm. The outstanding microwave absorption behavior of the composites could be readily attributed to the multilayered structure of the MXene, special dielectric properties of $Ti_3C_2T_x$ MXene and PANI, as well as the synergistic effect of the two components.

PANI@TiO_2/$Ti_3C_2T_x$ composites were developed by using hydrothermal reaction process together with in-situ polymerization, as electrode of supercapacitors [33]. In this type of composites, the $Ti_3C_2T_x$ MXene nanosheets acted as active materials, because of their relatively large surface area. Meanwhile, they formed frameworks with more sites for insertion/extraction of the electrolyte ions. Both the PANI molecules and TiO_2 particles could increase electrochemical performances of the composite electrode, because they have typical pseudocapacitive effect. At the same time, surface area of the composites was effectively increased. The composite-based electrodes displayed a specific capacitance of 188.3 F/g at 10 mV/s and 435.4 F/cm^2 at 0.5 A/g in KOH solution, which were much higher than those of the TiO_2/$Ti_3C_2T_x$-based one.

3.3 MXenes-Based Hybrids

3.3.1 MXenes-Nanocarbons

In order to increase electrical conductivity of $Ti_3C_2T_x$, highly conductive carbon nanofibers (CNFs) were grown in between the MXene nanosheets as a linker by using a CVD method [5]. The CNFs bridged the $Ti_3C_2T_x$ nanosheets so as to create conductive pathways, which were perpendicular to the nanosheet planes, while not blocking the electrolyte ions. To synthesize the $Ti_3C_2T_x$-CNFs composites, cobalt (II) nitrate ($Co(NO_3)_2$ catalyst precursor and poly(vinyl pyrrolidone (PVP) dispersant were incorporated with $Ti_3C_2T_x$ first. After that, the samples were calcined at 600 °C for 30 min in Ar, during which acetylene was introduced as the carbon source to form CNFs. Finally, the Co catalyst was eliminated through chemical reaction. The synthetic process is shown schematically in Fig. 3.10. For different mass ratios of PVP to Co salt, the Ti_3C_2/CNF composite samples were denoted as Ti_3C_2-CNF-1–2, Ti_3C_2-CNF-1–5, Ti_3C_2-CNF-1–10, Ti_3C_2-CNF-1–15 and Ti_3C_2-CNF-1–20.

Figure 3.11 shows SEM and TEM images of the as-obtained $Ti_3C_2T_x$ and Ti_3C_2/CNF composite samples. As expected, the pristine $Ti_3C_2T_x$ had typical accordion-like morphology, as revealed in Fig. 3.11a. Because of the hydrophilic behavior of the $Ti_3C_2T_x$ nanosheets, the impregnated Co salt was uniformed coated. As a result, the catalytic Co nanoparticles were uniformly distributed on the surfaces of the $Ti_3C_2T_x$ nanosheets. After CVD reaction, the surface functional groups of –F and –OH were eliminated due to the high temperature treatment in Ar/H_2 atmosphere.

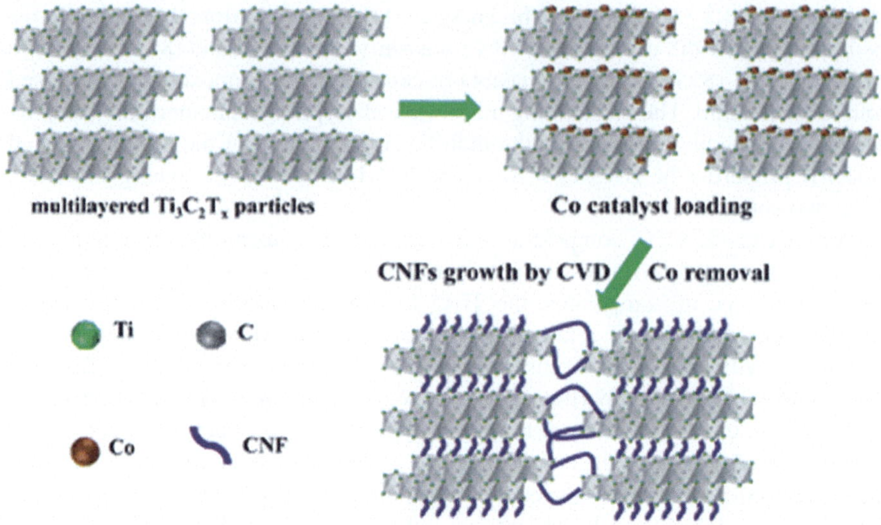

Fig. 3.10 Schematic diagram demonstrating synthetic process of the MXene-CNFs composites. Reproduced with permission from [5]. Copyright © 2015, Royal Society of Chemistry

Instead, CNFs were present inside the interlayer spacings of the MXene nanosheets, as illustrated in Fig. 3.11b. Significantly, the network structural characteristic of MXene was well retained.

As seen in Fig. 3.11c, TEM image of the Ti_3C_2-CNF-1–10 sample indicated that CNFs were visible in between and surrounding MXene particles. According to HRTEM image, the multilayered structure of the Ti_3C_2 was unaltered during the CVD reaction process, as demonstrated in Fig. 3.11d. Figure 3.11e shows HRTEM image of an individual CNF, which had a diameter of about 20 nm. A small hole was present at the end of the CNF, owing to the removal of the Co nanoparticle. Therefore, the CNFs were grown following the tip-growth mechanism. At the same time, TiO_2 nanoparticles with an anatase phase and an average size of 20 nm were formed, as depicted in Fig. 3.11f, indicating oxidation of the MXene phase during the CVD process.

In this case, the retention of the MXene characteristic network structure was critical to the electrochemical performances as the composites were used as the electrodes of LIBs. The partial oxidation of the Ti_3C_2 phase to anatase TiO_2 was ascribed to the presence of oxygen during the CVD reaction. The oxygen could be derived from the –O and –OH functional groups and that related to the Co salt. In addition, XRD results suggested that CNFs were in a amorphous or weak crystalline state.

Free-standing $Ti_3C_2T_x$–C nanofiber mats were prepared by using electrospinning technique, which were then used as electrodes of supercapacitors [34]. The suspensions for electrospinning were derived from $Ti_3C_2T_x$ MXene nanosheets that were dispersed in polyacrylonitrile (PAN), followed by carbonization. In these special

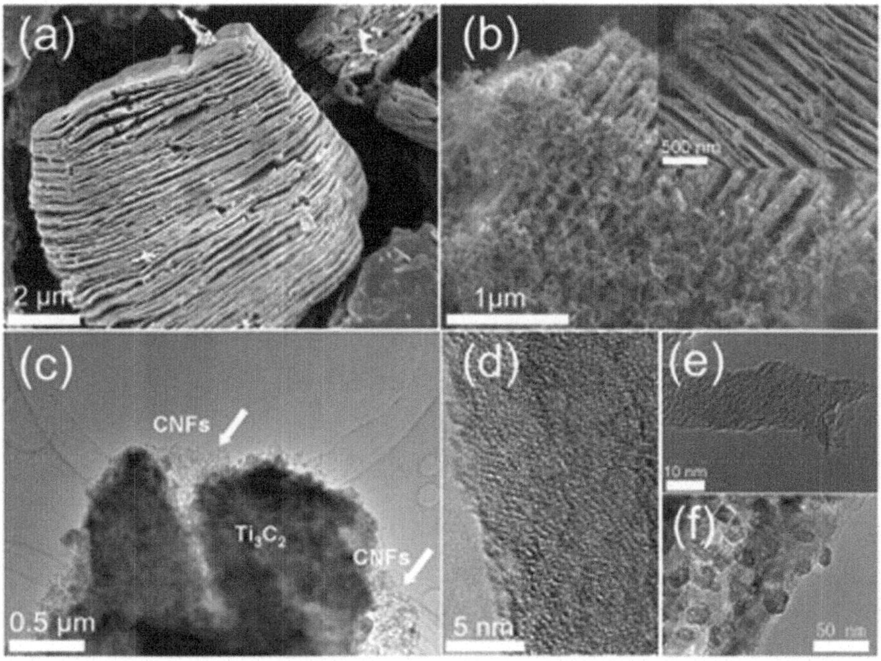

Fig. 3.11 Microstructures of the $Ti_3C_2T_x$ and Ti_3C_2/CNF composite samples: **a** SEM images of the pristine $Ti_3C_2T_x$ nanosheets, and **b** SEM images of the Ti_3C_2-CNF-1–10 sample, with the inset showing the CNFs grown in between the interlayer spaces of the Ti_3C_2 nanosheets, **c** TEM image of the Ti_3C_2-CNF-1–10 sample, **d** HRTEM images of a Ti_3C_2 flake, **e** TEM image of the CNF and **f** TEM image of the TiO_2 particles on the Ti_3C_2 nanosheets. Reproduced with permission from [5]. Copyright © 2015, Royal Society of Chemistry

composites, the multilayered $Ti_3C_2T_x$ nanosheets were homogeneously distributed inside the carbon nanofibers (CNFs). The nanofiber mats were obtained after spinning without the use of any binder or additive.

The highest weight ratio of MXene to PAN was 2:1, corresponding to 35 wt% $Ti_3C_2T_x$ in the final nanofiber mats. At optimal conditions, the $Ti_3C_2T_x$-CNFs based electrodes exhibited an areal capacitance of 205 mF/cm^2 at the scan rate of 50 mV/s, which was higher than that of pure carbonized PAN nanofiber mats by nearly three times. The composite based electrodes also demonstrated higher electrochemical performances than those based on electrospun nanofiber mats coated with $Ti_3C_2T_x$ by using spray-coating method.

Figure 3.12 shows schematic diagrams illustrating fabrication processes of the $Ti_3C_2T_x$ MXene nanosheets and the $Ti_3C_2T_x$-CNFs mats. The minimum concentration of PAN required to ensure uniform bead-free nanofibers was 8 wt% PAN. $Ti_3C_2T_x$ MXene/PAN suspensions used for electrospinning were made with DMF as the solvent, because DMF has a relatively high vapor pressure. However, the conductivity of $Ti_3C_2T_x$ MXene was heavily damaged when it was suspended in DMF, since

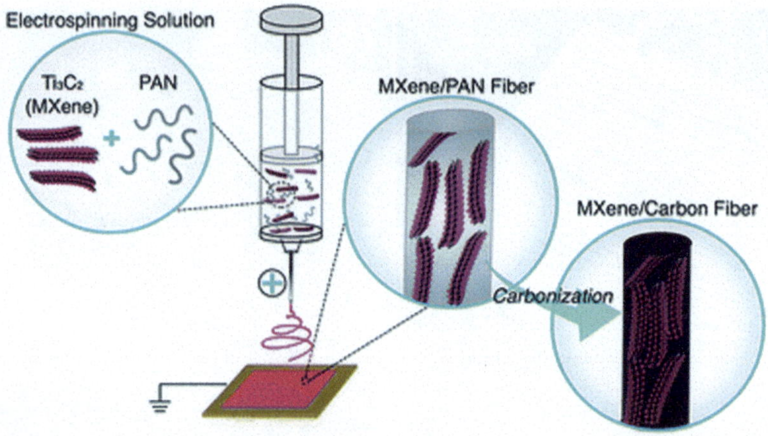

Fig. 3.12 Schematic of the production of $Ti_3C_2T_x$ MXene/carbon nanofibers via electrospinning. MXene/PAN solutions were prepared by dispersing delaminated MXene flakes in DMF and mixing with PAN. The solution was loaded into a syringe capped with a needle. A high voltage was applied to the needle tip and a nonwoven mat of MXene/PAN composite fibers was collected on a stationary copper plate. The fibers were stabilized and carbonized to form MXene/carbon nanofibers. Reproduced with permission from [34]. Copyright © 2019, Royal Society of Chemistry

the relative large DMF molecules could block the inter-nanosheet electron transport [2, 35]. In order to tackle this problem, the authors suggested to use organic solvents with smaller molecules to disperse the MXene nanosheets.

Figure 3.13 shows SEM images of the $Ti_3C_2T_x$ MXene composite nanofibers, with the content of MXene to be up to 16 wt% in the electrospinning solutions. The nanofibers were uniform in diameter, without the presence of beads. The nanofiber mats were free-standing with sufficiently high mechanical flexibility to withstand bending or twisting. As the content of MXene in the solutions was increased from 0 to 16 wt%, the resultant nanofiber mats experienced a continuous variation in color. For instance, the pure PAN fiber mat was white fiber, while those from the solutions with 10 and 16 wt% MXenes were black.

It was found that average diameter of the nanofibers was gradually increased with increasing content of $Ti_3C_2T_x$ MXene in the precursor solutions, as shown in Fig. 3.13f. The pure PAN nanofiber had an average diameter of 179 nm, whereas the 16 wt% $Ti_3C_2T_x$ fibers were much thicker, with an average diameter of 958 nm. Accordingly, the fibers derived from the solutions with relatively low content of MXene were less uniform in diameter, simply because of their small thickness. For example, MXene nanoflakes could be seen outside the fiber surface, as observed in Fig. 3.13b. In comparison, the 16 wt% sample contained MXene nanoflakes that were well embedded in the fibers and distribute in the direction of the fiber axis, with uniform diameters and smooth surfaces.

According to TEM characterization results, the $Ti_3C_2T_x$ MXene nanosheet were of single- to few-layered structures, which were well within the body of the fibers.

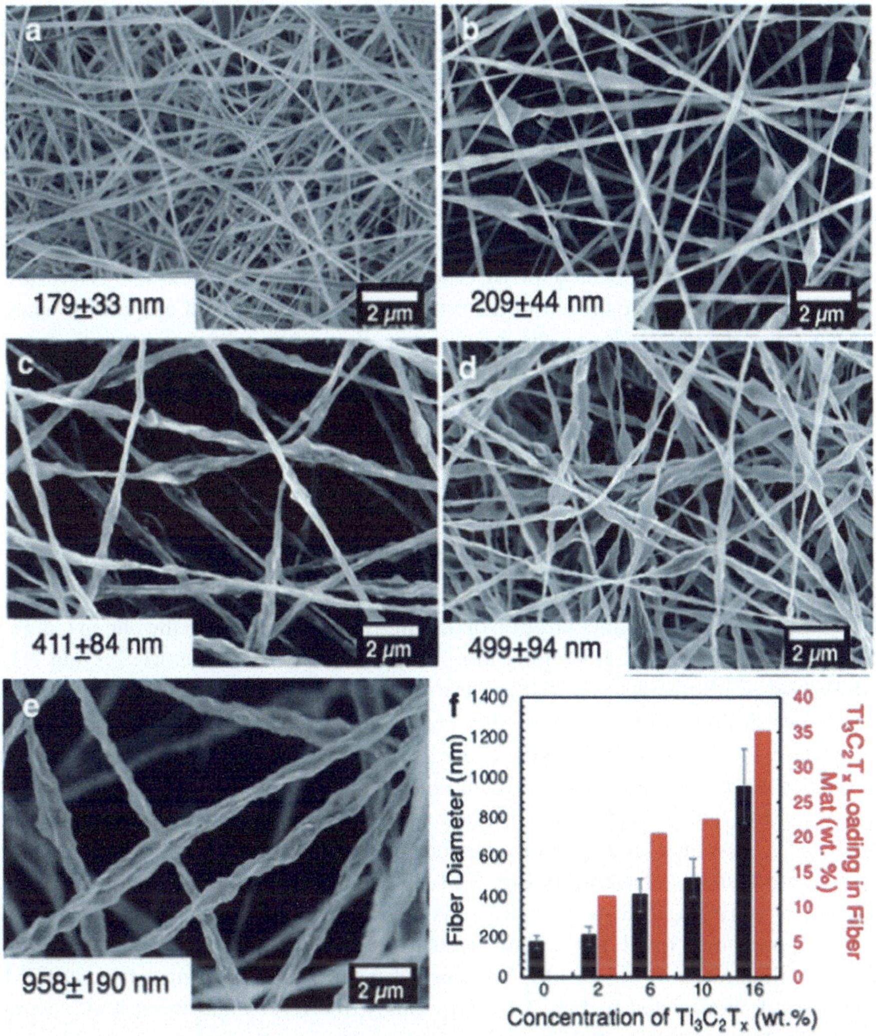

Fig. 3.13 Morphologies of the electrospun $Ti_3C_2T_x$ MXene/PAN fibers. SEM images of the electrospun PAN fibers with different contents of $Ti_3C_2T_x$ MXene in the electrospinning solutions: **a** 0 wt%, **b** 2 wt%, **c** 6 wt%, **d** 10 wt% and **e** 16 wt%. **f** Bar chart of average fiber diameters and the estimated loading of MXene in the final fiber mats. Reproduced with permission from [34]. Copyright © 2019, Royal Society of Chemistry

Meanwhile, the MXene nanoflakes were highly oriented along the axis of the fibers. XRD results indicated that the inter-layer spacing of the MXene was almost the same before and after the incorporation with PAN. In addition, the appearance and morphology of the fibers were not affected by the thermal carbonization, with the diameters to be slightly decreased.

A multi-step vacuum filtration process was used to fabricate $Ti_3C_2T_x$/CNTs composite films, which were flexible and free-standing with a sandwich-like structure [2]. The composite films could contain $Ti_3C_2T_x$ up to 95 wt%. In the sandwich-like $Ti_3C_2T_x$/CNT composite films, the $Ti_3C_2T_x$ nanosheets were highly oriented. The ordered structure led to high electrical conductivities of the films, with an optimal value of 385 S/cm, which was higher than that of the randomly mixed $Ti_3C_2T_x$/CNT samples. The increased electrical conductivity was attributed to the enhanced connectivity of the MXene nanosheets offered by the highly conductive CNTs. Also, the presence of the CNTs prevented the MXene nanosheets from restacking. As a result, the $Ti_3C_2T_x$/CNTs composite exhibited promising electrochemical properties, when it was used as the electrodes of supercapacitors.

A similar work was reported by Yan et al. to prepare CNT-Ti_3C_2 composites as electrodes of supercapacitors [36]. The composite based electrodes had a volumetric capacitance of 393 F/cm^3 at a scan rate of 5 mv/s in KOH electrolyte solution, together with high cycling stabilities. The MXene (Ti_3C_2) was treated with dimethyl sulfoxide (DMSO), followed by centrifugation. The Ti_3C_2 powder was dispersed in DI water, with ultrasonication for 2 h and 6 h, resulting in samples that were denoted as ex-Ti_3C_2 and d-Ti_3C_2, respectively. CNTs with diameters of 20–40 nm were dispersed to form suspensions with a concentration of 0.3 mg/ml. Then, the d-Ti_3C_2 suspension with a concentration of 0.3 mg/ml was mixed with the one of CNTs, leading to d-Ti_3C_2/CNTs, at weight ratios of d-Ti_3C_2 to CNTs to be 6:1, 2:1 and 1:1, with the aid of ultrasonication. After filtration with polytetraflouroethylene (PTFE) membranes, followed by drying at 70 °C, to obtain d-Ti_3C_2/CNTs composites.

Figure 3.14 shows SEM images of the Ti_3C_2 samples. The Ti_3C_2 had a typical 2D layered structure. After the treatment with strong ultrasonication, the number of the layers was reduced, while the surface roughness of the Ti_3C_2 nanosheets was increased, as revealed in Fig. 3.14b. Further ultrasonication led to breaking of the Ti_3C_2 nanosheets, as observed in Fig. 3.14c. It was found that the pulverized Ti_3C_2 nanosheets were prone to restack to form aggregation. However, the presence of the CNTs prevented the d-Ti_3C_2 nanosheets to agglomerate, leading to well isolated Ti_3C_2, as illustrated in Fig. 3.14d. Such effect was further confirmed by the TEM images, as demonstrated in Fig. 3.15. Obviously, the CNTs and the d-Ti_3C_2 were interweaved in the d-Ti_3C_2/CNTs composite.

An electrostatic self-assembly approach was reported to obtain flexible free-standing composite films of $Ti_3C_2T_x$/MWCNTs and Mo_2CT_x/MWCNTs, for the applications as electrodes of supercapacitors [4]. The composites films could be made because of the negatively charged $Ti_3C_2T_x$ and Mo_2CT_x nanosheets and the positively charged MWCNTs. To ensure the success of the self-assembly process, the MWCNTs should be grafted with cetyltrimethylammonium bromide (CTAB). Otherwise, the self-assembly was impossible, instead, the $Ti_3C_2T_x$ nanoflakes and the MWCNTs were coated on the filter papers in a disordered manner. The $Ti_3C_2T_x$/CNT paper-based electrodes displayed optimal volumetric capacities of 421 mA h/cm^3 and 89 mA h/cm^3 at the current densities of 20 mA/g and 5000 mA/g, respectively. When the composite was used to assemble full sodium-ion cells with $Na_{0.44}MnO_2$, a high volumetric discharge capacity of 286 mA h/cm^3 could be readily achieved.

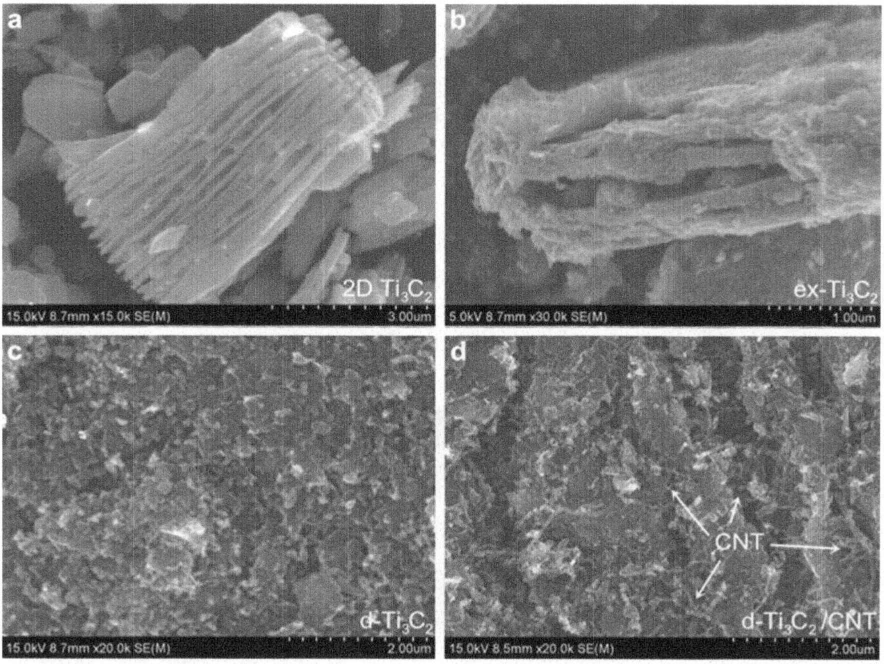

Fig. 3.14 SEM images of the samples: **a** 2D Ti_3C_2, **b** ex-Ti_3C_2, **c** d-Ti_3C_2 and **d** d-Ti_3C_2/CNTs. Reproduced with permission from [36]. Copyright © 2015, Elsevier

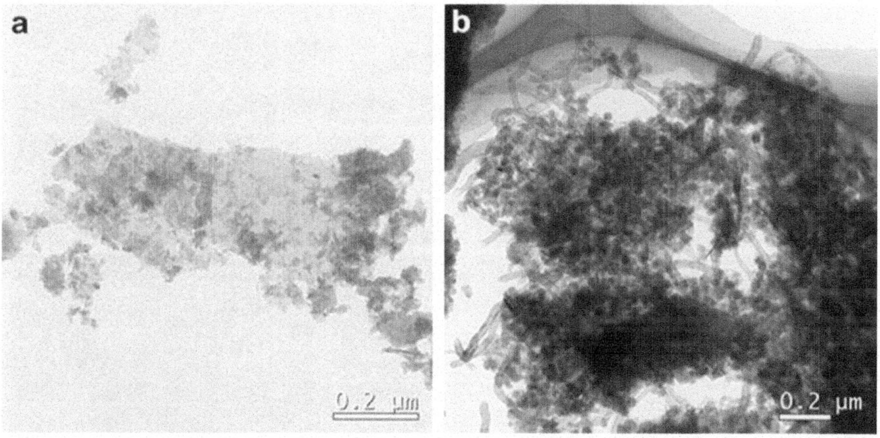

Fig. 3.15 TEM images of the samples: **a** d-Ti_3C_2 and **b** d-Ti_3C_2/CNTs. Reproduced with permission from [36]. Copyright © 2015, Elsevier

The filtration induced directional flow triggered the assembly of the $Ti_3C_2T_x$ nanosheets to have a layer-by-layer stacking. Top-view and cross-sectional SEM images of $Ti_3C_2T_x$ paper are shown in Fig. 3.16a, b. Without the incorporation of CNTs, the $Ti_3C_2T_x$ nanosheets were restacked closely to form dense films with relatively smooth surface profiles. The $Ti_3C_2T_x$/CNT-SDS composite papers possessed a similar surface morphology, as seen in Fig. 3.16c. In both cases, the $Ti_3C_2T_x$ nanosheets in the pure and the CNT-SDS samples were distributed in the suspensions in a random way, due to the absence of any interactions among the items. Therefore, the randomly distributed nanosheets would be settled down during the filtration process, thus leading to compact papers. For the CNT-SDS sample, the presence of the CNTs was unable to align the $Ti_3C_2T_x$ nanosheets.

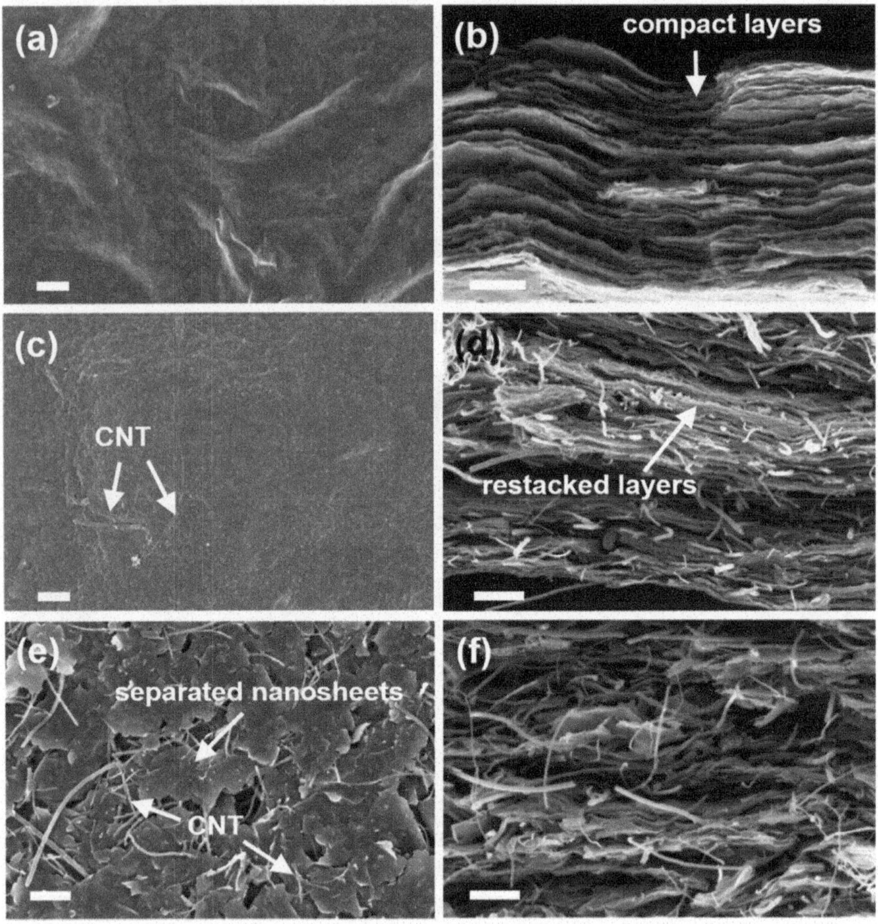

Fig. 3.16 Top view (*left panels*) and cross-sectional (*right panels*) SEM images of the samples: **a**, **b** pure $Ti_3C_2T_x$, **c**, **d** $Ti_3C_2T_x$/CNT-SDS and **e**, **f** porous $Ti_3C_2T_x$/CNT-SA. Scale bars: 1 μm. Reproduced with permission from [4]. Copyright © 2016, Elsevier

However, self-assembly was realized in the CNT-SA sample. As a result, CNTs, $Ti_3C_2T_x$ nanosheets were isolated by the CNTs. The inter-distribution of the $Ti_3C_2T_x$ nanosheets and the CNTs was homogeneous. In this case, the $Ti_3C_2T_x$ nanosheets were linked by the CNTs, resulting in a conductive network, as observed in Fig. 3.16e. In addition, a large number of pores were present in between the $Ti_3C_2T_x$ nanosheets, because of the spacing effect of the CNTs. The realization of such microstructure was attributed to the strong electrostatic interactions between the $Ti_3C_2T_x$ nanosheets and the CNTs, which had different charges. Representative cross-sectional SEM image of the $Ti_3C_2T_x$/CNT-SA film is shown in Fig. 3.16f. The CNTs were ordered sandwiched by the $Ti_3C_2T_x$ nanosheets.

$Ti_3C_2T_x$/CNTs flexible free-standing composite films were prepared by using vacuum filtration process as electrodes of electrochemical energy storage devices [37]. The $Ti_3C_2T_x$/CNT composite film-based electrode delivered a reversible capacity of 489 mA h/g at a current density of 50 mA/g, with a high cycling stability. A full-cell lithium-ion capacitor (LIC) was constructed, with the $Ti_3C_2T_x$/CNT film and activated carbon as the anode and cathode, respectively. An energy density of 67 Wh/kg and a capacity retention of 81.3% after 5000 cycles were achieved.

An in-situ CVD growth process was used to grow CNTs onto grown $Ti_3C_2T_x$ MXenes, forming $Ti_3C_2T_x$/CNTs composites for the applications in microwave absorption [38]. In the $Ti_3C_2T_x$/CNT composites, the 1D CNTs were homogeneously embedded inside the interlayer spaces of the 2D $Ti_3C_2T_x$ MXene nanoflakes. The composites demonstrated enhanced microwave absorption capability, as compared with the pristine MXene samples. Specifically, a minimum reflection coefficient of −52.9 dB, corresponding to 99.999% absorption, with a broad absorption bandwidth of 4.46 GHz, was achieved. The mass loading of MXene was as low as 35 wt%, while the thickness of the absorber was 1.55 mm. The authors attributed the outstanding microwave absorption effect to the presence of various absorption mechanisms, although further studies are required to support the explanations.

To facilitate the growth of CNTs, $Ti_3C_2T_x$ powder was dispersed in an aqueous solution containing 5 wt% nickel nitrate hexahydrate ($Ni(NO_3)_2 \cdot 6H_2O$) as the precursor of catalyst. After vacuum freeze-drying, the powders were subject to CVD reaction process at 800 °C in Ar/H_2, at flowing rates of 150 sccm and 150 sccm, respectively. As the temperature was heated to 800 °C, C_2H_4 was brought in the reactor as the source of carbon. The reaction was conducted for 30 min, at a flow rate of 40 sccm. The fabrication process is schematically illustrated in Fig. 3.17.

The pristine $Ti_3C_2T_x$ and the a-$Ti_3C_2T_x$ had similar multilayered structures, as observed in Fig. 3.18a, b. The CNTs were homogeneously embedded in between the $Ti_3C_2T_x$ nanoflakes. This was the reflection of the fact that the catalyst precursor was uniformly distributed in the samples. Ni^{2+} ions were interacted with the surface oxygen related functional groups of the MXene nanoflakes. Also, the $Ti_3C_2T_x$ particles were highly porous, with the pores to be produced during the freeze-drying process. As a result, the carbon source molecules could fill in to ensure the growth of the CNTs. The CNTs acted as a connector to link the $Ti_3C_2T_x$ nanosheets, while the multilayered structure was still retained, as demonstrated in Fig. 3.18d.

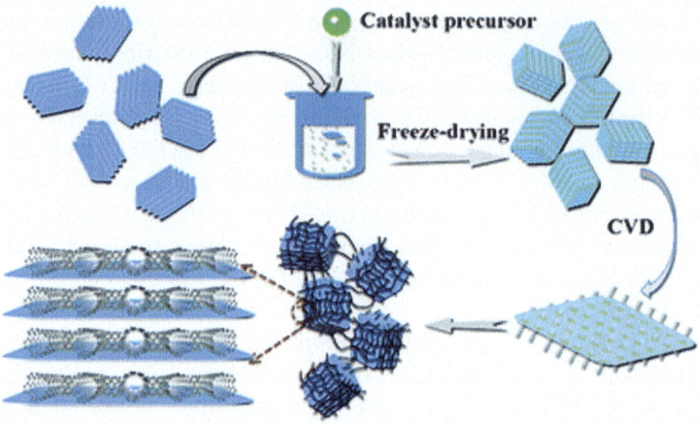

Fig. 3.17 Schematic diagram illustrating synthetic process of the $Ti_3C_2T_x$/CNTs composites with CVD method. Reproduced with permission from [38]. Copyright © 2017, Royal Society of Chemistry

The uniform distribution of the CNTs inside the MXene nanoflakes has also been confirmed by TEM results, as shown in Fig. 3.18e. The CNTs had diameters in the range of 9–45 nm. Ni catalyst particles were at the tips of the CNTs, implying that the growth of the CNTs followed the typical vapor–liquid–solid (VLS) mechanism. The reduction of Ni^{2+} ions resulted in Ni nanoparticles, onto which carbon atoms due to the pyrolysis of the carbon source were adsorbed. The continuous supply of the carbon source triggered the growth of the CNTs. According to the SAED pattern of the $Ti_3C_2T_x$/CNTs composite, the CNTs had a polycrystalline nature, but with a relatively low degree of crystallinity, which could be attributed to the presence of defects in the in-situ grown CNTs. Figure 3.18f shows a representative HRTEM image of the CNTs, indicating their multiwalled characteristics. In this case, the inner and outer diameters were 5 nm and 12 nm, respectively. XRD patterns indicated the formation of both anatase and rutile TiO_2, due to the partial oxidation of the MXene during the CNT growth.

A simple process was developed to fabricate $Ti_3C_2T_x$@CNTs 3D structured composites as electrodes of supercapacitor [39]. The $Ti_3C_2T_x$ nanosheets were covered with a thin layer of polydopamine (PDA), which acted as protection coating to ensure the mechanical strength and structural integrity of the MXene. Then, CNTs were t grown on the $Ti_3C_2T_x$ nanolayers by using an in-situ pyrolysis step with urea as the source of carbon. The Co nanoparticles from Co^{2+} salt were used as the catalyst for CNT growth. Figure 3.19 shows a schematic diagram of the fabrication procedure. Both the length and density of the CNTs could be varied by controlling the quantity of urea. The CNTs were vertically aligned, thus allowing the electrolyte ions to access and enlarging the interlayer space of the $Ti_3C_2T_x$ nanosheets. Electrodes based on the $Ti_3C_2T_x$@$CNTs_{6.0}$ composites exhibited an areal capacity of 924.9 mF/cm^2 at the scan rate of 2 mV/s, together with a strong cycling stability.

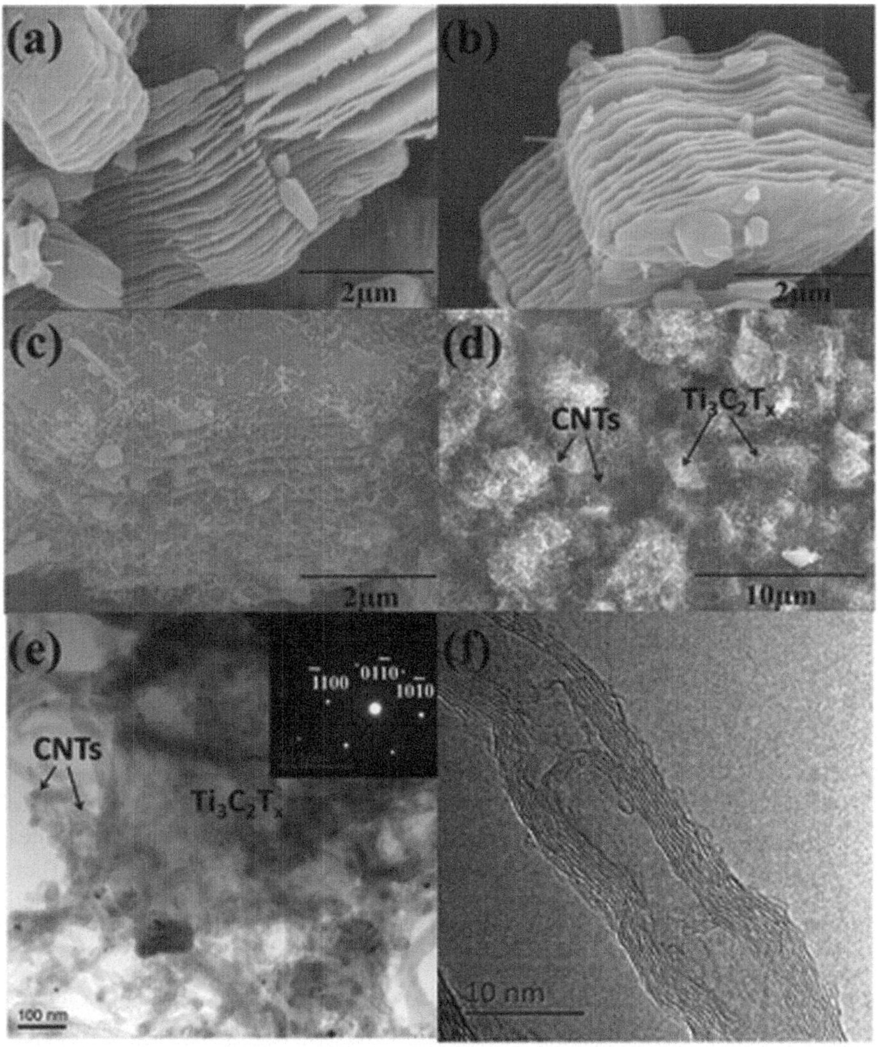

Fig. 3.18 SEM images of the samples: **a** $Ti_3C_2T_x$, **b** a-$Ti_3C_2T_x$ and **c**, **d** $Ti_3C_2T_x$/CNTs. **e** TEM images of the $Ti_3C_2T_x$/CNTs with the inset showing the corresponding SAED pattern. **f** HRTEM image of the CNTs. Reproduced with permission from [38]. Copyright © 2017, Royal Society of Chemistry

A microwave irradiation process was reported to grow CNTs to form CNTs@MXene composites, as electrodes of lithium-ion batteries [40]. MXenes that have been studied included Ti_3C_2, Ti_2C and V_2C. In the composites, multi-walled CNTs were attached on the MXene nanosheets to obtain porous network structures. The CNTs@Ti_3C_2 composites displayed the highest electrochemical performances,

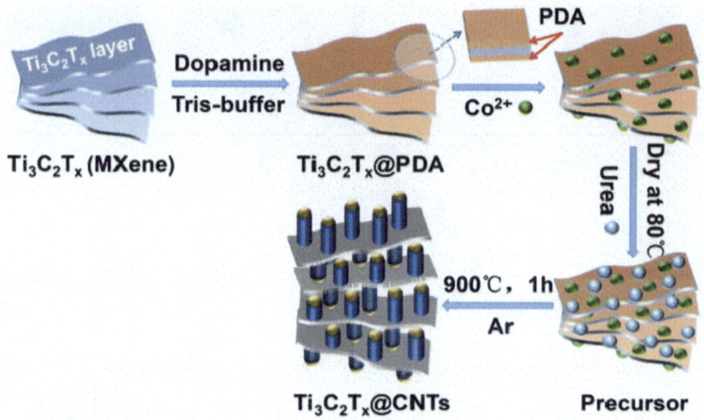

Fig. 3.19 Schematic diagram illustrating fabrication process of the multilayered and interconnected 3D $Ti_3C_2T_x$@CNTs composites. Reproduced with permission from [39]. Copyright © 2017, Elsevier

with specific capacities of 430 mA h/g and 175 mA h/g at 1 A/g and 10 A/g, respectively. MXene powders and ferrocene were mixed manually. The mixtures were then subject to microwave irradiation for designed time durations at 900 W. The reaction time was controlled to avoid extended oxidation of the MXenes and ensure the sufficient growth of the CNTs. The fabrication process of the CNTs@MXene composites is schematically illustrated in Fig. 3.20.

Another in-situ process was presented for the preparation of Co/N-CNTs@$Ti_3C_2T_x$ composites, as bifunctional electrocatalyst for oxygen reduction reaction (ORR) and oxygen evolution reaction (OER) [41]. In these composites, the CNTs were distributed on the surface of $Ti_3C_2T_x$ MXene nanosheets. There were strong interfacial coupling and electron transfer in the composites, which were responsible for the bifunctional electrocatalytic activities for ORR and OER in alkaline solutions. Figure 3.21 shows schematic diagram of the fabrication process. To synthesize the Co/N-CNTs@Ti_3C_2T composites, $Co(NO_3)_2 \cdot 6H_2O$, dicyandiamide (DCD) and $Ti_3C_2T_x$ powder were mixed with the aid of ethanol. After the ethanol was complete evaporated, the mixture was dried at 50 °C in vacuum overnight. Heat

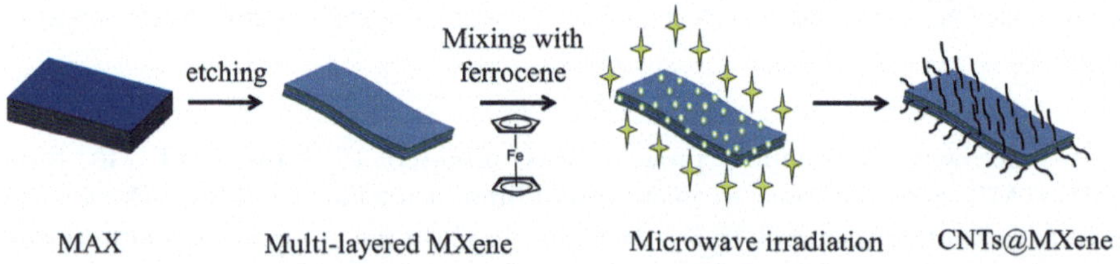

Fig. 3.20 Schematic diagram describing preparation steps of the CNTs@MXene composites. Reproduced with permission from [39]. Copyright © 2018, Royal Society of Chemistry

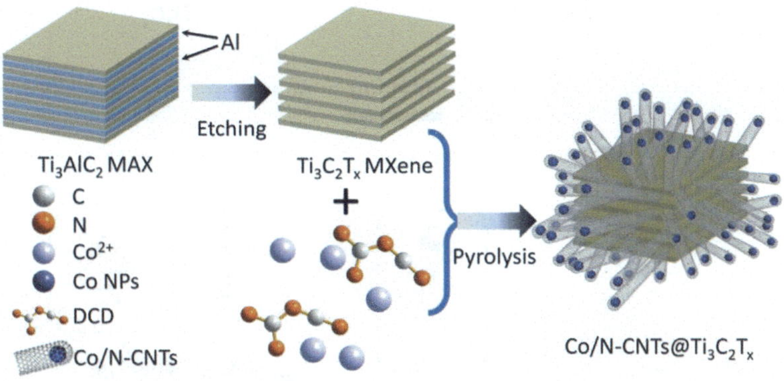

Fig. 3.21 Schematic diagram describing synthetic process of the Co/N-CNTs@$Ti_3C_2T_x$composites. Reproduced with permission from [41]. Copyright © 2018, Royal Society of Chemistry

treatment was conducted at 900 °C for 1 h in N_2. In this case, the DCD served as the source of both carbon and nitrogen, while Co^{2+} was reduced to Co nanoparticles as the catalysts for the growth of CNTs.

SEM images indicated that the $Ti_3C_2T_x$ nanosheets were entirely coated by the Co/N-CNTs, whereas the Co/N-CNTs were grown on the surface of the $Ti_3C_2T_x$ nanosheets. Therefore, the two components were firmly connected each other. According to TEM images, it was observed that the Co nanoparticles were encapsulated the N-CNTs, with an average diameter of 10 nm. Several reasons were responsible for the high electrocatalytic activities of the Co/N-CNTs@$Ti_3C_2T_x$ composites for ORR and OER, including (i) the encapsulation of the Co nanoparticles in CNTs leading to high electrical conductivity and high corrosion resistance, (ii) N doped CNTs offering more active sites for the adsorption of O_2 molecules, (iii) electron transport from Ti to C generating holes to increase the adsorption OH^- ions in alkaline solutions, (iv) N doped $Ti_3C_2T_x$ creating holes at the Ti sites to promote the adsorption of OH^- ions and (v) enhanced charge transportation due to the coupling of the CNTs and the $Ti_3C_2T_x$.

A relatively sophisticated process was employed to fabricate Co-CNTs/Ti_3C_2 composite for ORR applications [42]. In this case, a metal–organic-framework (MOF), ZIF-67, was grown on Ti_3C_2 nanolayers, which was then converted to Co nanoparticles. The Co-tipped CNTs/Ti_3C_2 composites had more Co–N/C active sites, with suitable carbon graphitization of carbon and proper surface area, thus leading to promising ORR activity. The catalytic efficiency of the Co-CNTs/Ti_3C_2 was close to that the commercial Pt/C catalyst, while the former displayed a much higher stability than the latter. Figure 3.22 shows a schematic diagram of the synthetic steps towards the Co-CNTs/Ti_3C_2 composite. Due to the functional groups of –O and –F, nucleation and growth of ZIF-67 took place effectively on the surface of the MXene nanosheets. The ZIF-67 polyhedra had sizes at the nanometer scale. Co-CNT/Ti_3C_2-60 composite was formed after the pyrolysis of ZIF-67/Ti_3C_2-60 in H_2/N_2.

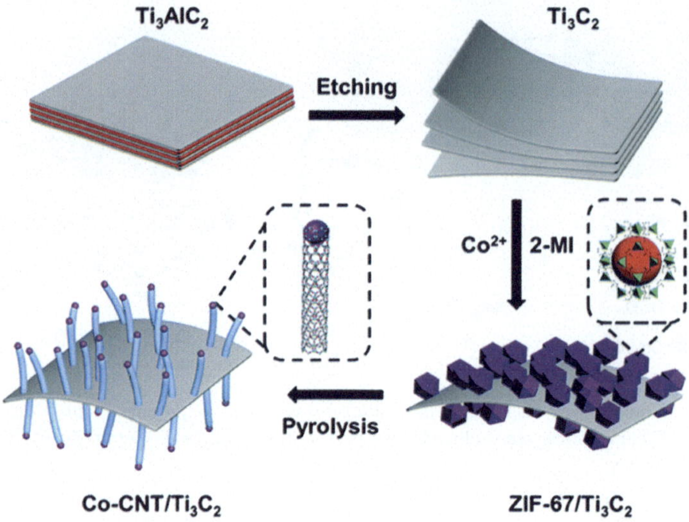

Fig. 3.22 Schematic diagram showing fabrication steps of the Co-CNT/Ti_3C_2 composites. Reproduced with permission from [42]. Copyright © 2019, Royal Society of Chemistry

After the Co-CNT/Ti_3C_2-60 and ZIF-67 were calcined at 800 °C, both metallic Co and graphitic carbon were present. According to the morphological characterization results, it was confirmed that the 2D structure was well retained, while CNTs were standing vertically on surface of the nanosheets. The Co nanoparticles sitting at the tips of the CNTs had an average size of 10 nm, while the CNTs had an average diameter of 15 nm. Both the C and N were homogeneously distributed inside the composite. The strong ORR activity of the Co-CNT/Ti_3C_2 was responsible by the increased active sites of Co–N/C at the tips of the CNTs, high degree of carbon graphitization and a sufficiently large surface area, so as to promote electron transfer and the diffusion of O_2 molecules.

An electrophoretic deposition (EPD) process was used to deposit Ti_3C_2/CNTs composite films on graphite papers, as electrodes of supercapacitors [43]. Specific capacitance of the electrodes based on the Ti_3C_2/CNTs composites was higher than those of pure Ti_3C_2 and CNTs by about 1.5 and 2.6 times, respectively. The Ti_3C_2/CNTs composite based electrodes also exhibited strong cycling stability. To prepare the EPD suspension, d-Ti_3C_2 powder and MWCNTs were dispersed in a mixture solvent of DI water and acetone with a volume ratio of 1:1. To conduct the EPD experiment, conductive graphite papers were used as the substrates of composite films and counter electrode. The film formation was lasted for 10 min, at a applied DC voltage of 8 V, with the two electrodes to be separated by 1 cm. At the optimal conditions, the Ti_3C_2/CNTs composite film had mass loadings of 3–5 mg.

Mo_2C MXene nanosheets with hydroxyl functional group prepared by by etching our the Sn layer of Mo_2SnC, which was further decorated with CNTs for the applications as electrodes of lithium-sulfur batteries [44]. Importantly, due to the presence of the hydroxyl functional group on the surface of the Mo_2C nanosheets, the shuttle effect of lithium polysulfides (LiPSs) was prevented, because of the strong interactions between the Mo atoms of the MXenes and the LiPSs. Furthermore, the incorporation of CNTs resulted in composite with increased specific surface area, enhanced electrical conductivity and buffered the volume variation during the discharging-charging process. In addition, sulfur atoms were strongly embedded inside the Mo_2C-CNTs porous network, thus ensuring high electrochemical behaviors when the composite was used as electrodes of lithium–sulfur batteries. A reversible capacity of 925 m Ah/g was achieved at 0.1 C.

Nb_2CT_x/MWCNTs composite papers were fabricated by using filtration method from suspensions of the component mixture [1]. In this case, multi-layered Nb_2CT_x was delaminated with i-PrA as the intercalant. MWCNTs dispersed in DI water with 0.03 mol/L sodium dodecylsulphate (SDS). Then, mono- or few-layered Nb_2CT_x nanosheets were added into the resultant suspensions, which were used to obtain Nb_2CT_x/MWCNTs composite papers through vacuum filtration. Flexible free-standing composites with the content of CNTs of up to 10 wt% could be made at room temperature. As anode of LIBs, the composite papers could deliver a specific capacity of 400 mAh/g at 0.5 C. In addition, a high volumetric capacitance of 325 F/cm^3 was achieved when examined in the configuration Li-ion capacitors.

$Ti_3C_2T_x$/graphene nanocomposites were fabricated by coating $Ti_3C_2T_x$ and graphene alternatively on glass substrates, in order to study their transport properties [45]. The $Ti_3C_2T_x$ and graphene layers were deposited by using an electrohydrodynamic atomization (EHDA) process. The content of graphene had an effect on structure and electrical conductivity of the composites. For example, the incorporation of graphene reduced surface roughness of the composite. Electrical conductivity and Hall carrier mobility were largely increased, with values of 9.5×10^4 S/cm and 54.58 cm^2/V s, respectively, as the content of graphene was as low as 2.5 wt% in the composite.

A solution route was used to prepare hybrid inks from $Ti_3C_2T_x$ MXene and graphene nanosheets, which were applied to fabricate flexible all-solid-state supercapacitors [46]. The graphene powder was obtained by using electrochemical exfoliation method. The concentrations of the MXene and graphene were 0.2 mg/mL and 0.15 mg/mL^1, respectively. Microsupercapacitor based on the composite exhibited an areal capacitance of 3.26 mF/cm^2 and a volumetric capacitance of 33 F/cm^3 at the scan rate of 2 mV/s. All-solid-state supercapacitor displayed a volumetric capacitance of 216 F/cm^3 at the current density of 0.1 A/cm^3. In this case, the $Ti_3C_2T_x$ nanosheets were uniformly distributed in between the graphene nanolayers, serving as active materials to boost the access of the electrolyte ions. Also, the MXene nanosheets with small sizes prevented the restacking of the graphene sheets.

Highly flexible conductive $Ti_3C_2T_x$/rGO composites were fabricated to be used as electrodes of supercapacitors [47]. The composites were obtained by using electrostatic self-assembly with positively charged rGO nanosheets due to the modification with poly(diallyldimethylammonium chloride) and negatively charged $Ti_3C_2T_x$ nanosheets. In the composites, the rGO nanosheets were sandwiched by the $Ti_3C_2T_x$

nanosheets. Therefore, both components were prevented from restacking. Meanwhile, interlayer spacing of the MXene was greatly enlarged, thus leading to high electrochemical performances. Specifically, electrodes based on $Ti_3C_2T_x$/rGO-5 wt% exhibited a volumetric capacitance of 1040 F/cm^3 at the scan rate of 2 mV/s, with a retention of 61% at 1 V/s after cycling for a long time. In addition, symmetric supercapacitor based on the composite displayed a volumetric energy density of 32.6 Wh/L.

Similarly, rGO/$Ti_3C_2T_x$ composites were fabricated by using vacuum filtration method, as electrodes of all-solid-state supercapacitors [48]. With the supercapacitors, energy conversion and storage devices (ECSDs) were fabricated through the integration with flexible thin-film solar cells by using the co-cathode method. The filtration suspensions were prepared with GO/$Ti_3C_2T_x$ dispersed in DI water. The resultant porous GO/$Ti_3C_2T_x$ films reduced at 300 °C in vacuum to form rGO/$Ti_3C_2T_x$ composites. Electrodes based on the rGO/$Ti_3C_2T_x$ composite film delivered volumetric and gravimetric capacitances of 370 F/cm^3 and 405 F/g, respectively, in 6 M KOH solution. With polyvinyl alcohol (PVA)/KOH gel electrolyte, all-solid-state supercapacitors derived from the composite possessed a energy density of 63 mW h/cm^3, corresponding to a power density of 0.06 W/cm^3.

Both the GO and $Ti_3C_2T_x$ suspensions had a concentration of 1 mg/mL, which were mixed according to the GO/$Ti_3C_2T_x$ mass ratios of 1:1, 2:1, 3:1 and 4:1. Filtration was conducted with cellulose membranes with pore of 0.22 μm, resulting in films denoted as GM-1, GM-2, GM-3 and GM-4, respectively. The samples were peeled off from the cellulose membranes and dried in air. After reduction at 300 °C for 5 h in vacuum, rGO/$Ti_3C_2T_x$ (RGM) composite films, denoted as RGM-1, RGM-2, RGM-3 and RGM-4, were obtained. Figure 3.23 shows a schematic diagram of the fabrication process of the composite films.

Because $Ti_3C_2T_x$ and GO nanosheets contained oxygen-related functional groups, making them hydrophilic in nature, so that they could be well dispersed in water, as demonstrated in Fig. 3.24a. The as-filtrated GO/$Ti_3C_2T_x$ composite films were black brown in color and mechanically flexible. After vacuum reduction, the rGO/$Ti_3C_2T_x$ composite films were black grey in color, while the flexibility was retained, as illustrated in Fig. 3.24b. Figure 3.24c shows a representative cross-sectional SEM image of the rGO/$Ti_3C_2T_x$ films, confirming their porous hierarchical microstructure. As the films were reduced in vacuum at 300 °C, the –O, –OH and other functional groups were eliminated, during which the graphene nanolayers were expanded, resulting in the porous structures. In addition, the incorporation of the $Ti_3C_2T_x$ nanosheets also blocked the restacking of the rGO nanosheets, as observed in Fig. 3.24d. Meanwhile, the rGO nanosheets linked the individual $Ti_3C_2T_x$ nanoparticles, as shown in Fig. 3.24e.

A wet-spinning method was developed to continuously fabricate MXene/graphene composite fibers, as flexible electrodes of supercapacitors [49]. Due to the synergistic effect of liquid crystal behavior of the graphene oxides and 2D structure of the MXene nanosheets, the wet-spinning process was realizable. In this case, the graphene oxides acted as liquid crystalline templates to guide the MXene nanosheets. The composite

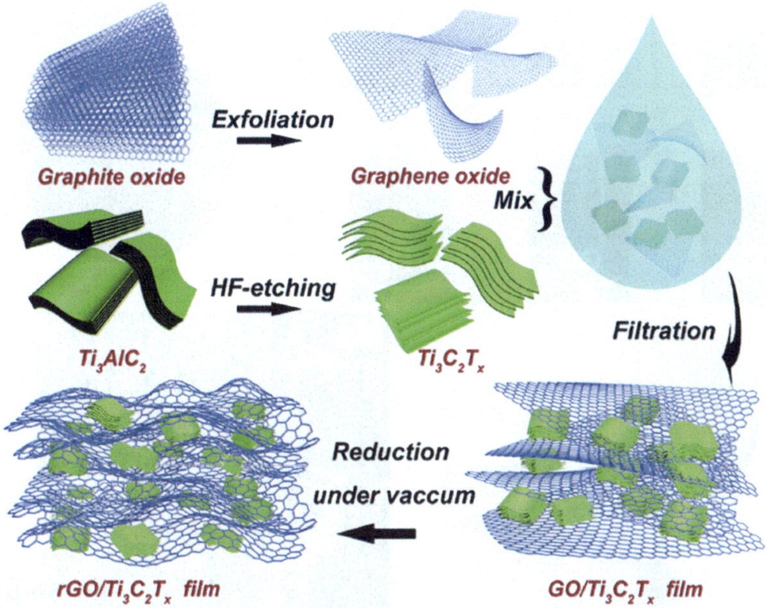

Fig. 3.23 Schematic diagram of the processing steps of the porous rGO/$Ti_3C_2T_x$ composite films. Reproduced with permission from [48]. Copyright © 2017, Royal Society of Chemistry

fibers could contain a high MXene level of 95 wt%. The highest electrical conductivity was 2.9×10^4 S/m, while a volumetric capacitance of 586.4 F/cm^3 could be achieved for the electrodes based on composite fibers.

GO nanosheets were dispersed into Ti_3C_2 solutions, followed by condensation to concentrations of 17–20 mg/mL through high speed centrifugation. The Ti_3C_2/GO mixtures were filled into a plastic syringe with a spinning nozzle having a diameter of 0.250 mm. Then, the mixtures were injected into a coagulation bath by using an injection pump at a rate of 50 μL/min. The coagulation solution was made of isopropanol and DI water, together with 5 wt% $CaCl_2$. The fibers were coagulated for 0.5 h, followed by washing and natural drying. The dried Ti_3C_2/GO composite fibers were reduced with a mixture of HI and acetic acid, with a ratio of HI to acetic acid to 2:3, at 90 °C for 12 h, thus leading to Ti_3C_2/rGO composite fibers. The fabrication process is shown schematically in Fig. 3.25a.

The as-obtained Ti_3C_2/GO fibers were dark black in color, due to the inclusion of the Ti_3C_2 nanosheets, as observed in Fig. 3.25b. In comparison, the GO fiber had a color of light brown, as seen in Fig. 3.25c. In the coagulation solution, the isopropanol served to solidify the Ti_3C_2 nanosheets, while water was to float the wet gel fibers due to its relatively high density. High content of IPA ensured a high rate of solvent exchange and rapid shrinkage of the composite fibers, as demonstrated in Fig. 3.25d, e. Comparatively, the MXene/GO-77 fiber with 77 wt% Ti_3C_2 exhibited a more crumpled surface and more dense core.

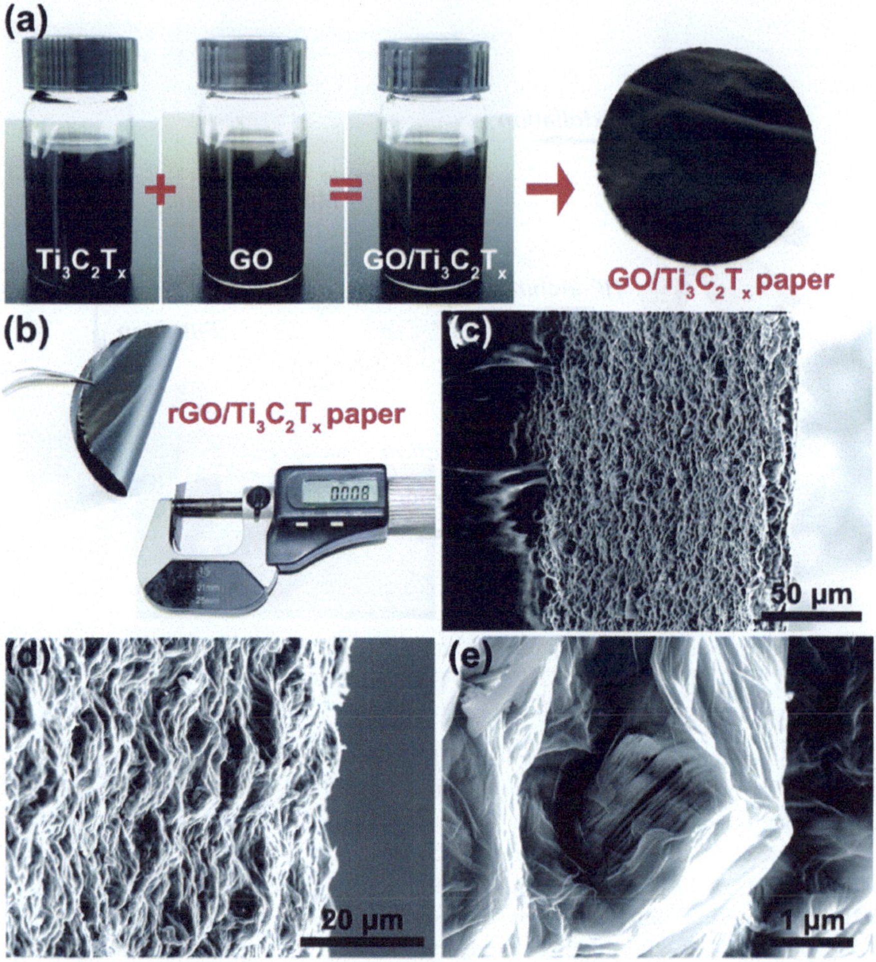

Fig. 3.24 **a** Photographs of the stable $Ti_3C_2T_x$–water, GO–water, and $GO/Ti_3C_2T_x$–water dispersions, together with a flexible free-standing $GO/Ti_3C_2T_x$ film (GM-3) with a diameter of 5 cm. **b** Photograph of a flexible $rGO/Ti_3C_2T_x$ composite film (RGM-3) with thickness measurement. **c**–**e** Cross-sectional SEM images of the flexible $rGO/Ti_3C_2T_x$ composite film at different magnifications. Reproduced with permission from [48]. Copyright © 2017, Royal Society of Chemistry

An electrolyte-induced self-assembly process was reported to fabricate 3D porous MXene/rGO composite films, as electrodes of LIBs [50]. The porous structure of the composite films ensured fast electrolyte ion transportation, while high electrical conductivity of the MXene was well retained. As a result, as they used as the electrodes LIBs, a high specific capacity of 335.5 mA h/g at 0.05 A/g, together with a high rate stability, was achieved. Also, the 3D porous MXene–rGO structure facilitated a fast reaction kinetics and thus high capacity. Figure 3.26 shows a schematic

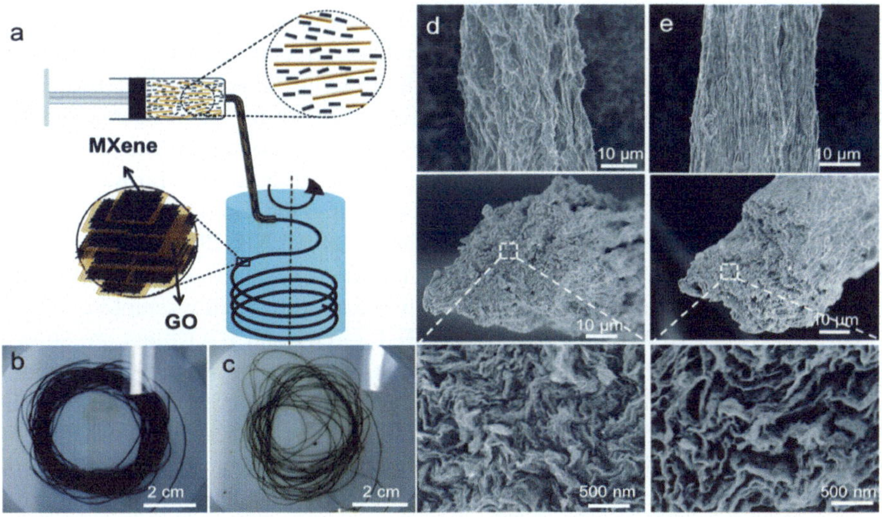

Fig. 3.25 **a** Schematic diagram of the wet-spinning process to fabricate the MXene/GO composite fibers. **b** Photographs of the MXene/GO-90 fiber and **c** pure GO fiber. SEM images of the MXene/GO-77 fiber coagulating from the precursor suspensions with different ratios of IPA:water: **d** 9:1 and **e** 1:3. Reproduced with permission from [49]. Copyright © 2017, Royal Society of Chemistry

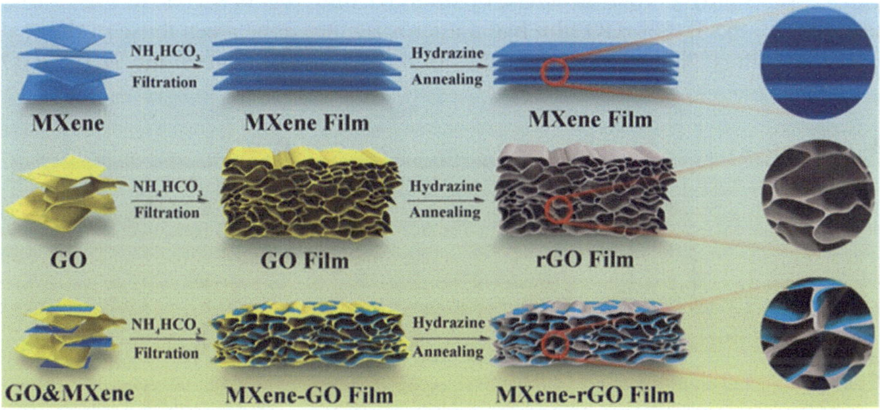

Fig. 3.26 Schematic illustration of the fabrication of the MXene, rGO, and MXene–rGO films. Reproduced with permission from [50]. Copyright © 2018, American Chemical Society

diagram of the fabrication processes for the two components and the MXene/rGO composite films.

To form the precursor solutions, GO and MXene dissolved in water and were mixed, with the aid of constant stirring for 1 h. Then, ammonium bicarbonate (NH_4HCO_3) was put into the mixture, so as to supply NH_4^+ ions to unbalance the electrostatic charges of the MXene and GO nanosheets. As a result, Ti_3C_2–GO composite clusters were formed, due to the linkage effect of the NH_4^+ ions to both the MXene and GO nanosheets. After vacuum filtration, 3D porous composite films were developed, in which the two components were strongly connected. The 3D porous structure was well retained because of the freeze-drying process. After that, Ti_3C_2–GO was reduced to Ti_3C_2–rGO with hydrazine vapor, so as to further increase electrical conductivity of the composite films. Finally, the samples were thermally treated at 200 °C for 2 h in Ar to eliminate the excessive NH_4HCO_3, H_2O, and the functional groups. The composite films were named as *x*–*y*-Ti_3C_2–rGO, with *x*–*y* to represent the weight ratio of $Ti_3C_2T_x$ to GO in the as-filtrated film samples.

Cross-sectional SEM images of Ti_3C_2, Ti_3C_2–rGO and rGO films are shown in Fig. 3.27a–c. The Ti_3C_2 film had a dense structure, with a thickness of about 2 μm, as seen in Fig. 3.27a. In contrast, the Ti_3C_2–rGO and rGO films were of 3D porous structure, with thicknesses of 30 and 250 μm, respectively. The difference in morphology between the two component films was attributed to their difference in state in their respective precursor solutions. The Ti_3C_2 nanoflakes were smaller in lateral size and more hydrophilic functional groups, thus leading to denser structure. In comparison, functional groups of GO nanosheets were most likely located at the edges, so that the efficiency of the NH_4^+ ions to link the GO nanosheets was relatively low. As a result, the GO film was highly porous.

High magnification SEM images of the three samples are shown in Fig. 3.27d–e. Multilayered structure of the Ti_3C_2 film was further confirmed, as illustrated in Fig. 3.27d. Highly porous structure of the rGO film was observed in Fig. 3.27f. As expected, the Ti_3C_2–rGO film had a microstructure in between those of the two component films. Nevertheless, the walls of the porous structure of the Ti_3C_2–rGO

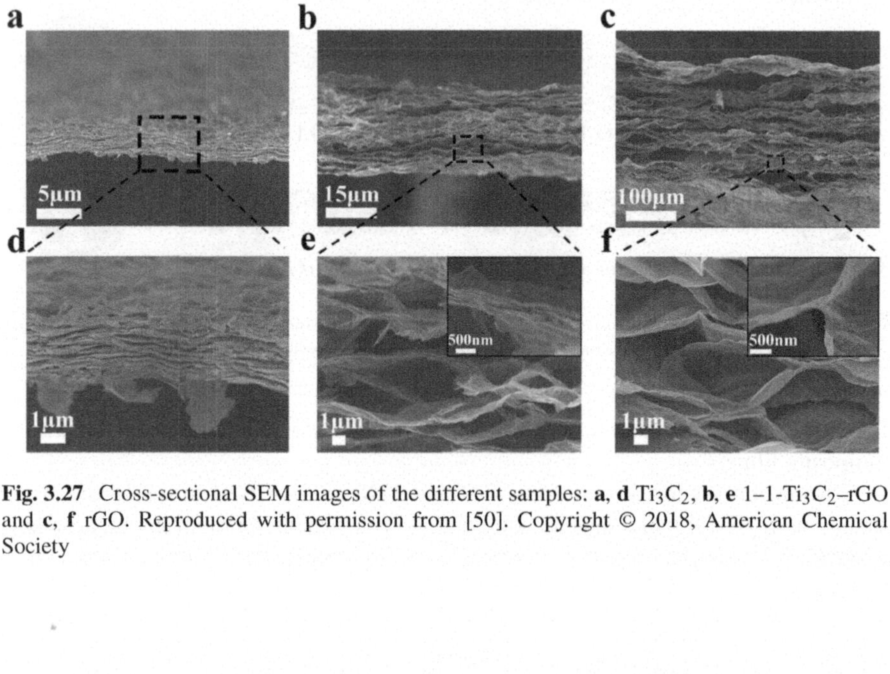

Fig. 3.27 Cross-sectional SEM images of the different samples: **a, d** Ti_3C_2, **b, e** 1–1-Ti_3C_2–rGO and **c, f** rGO. Reproduced with permission from [50]. Copyright © 2018, American Chemical Society

film was of multilayer characteristics, due to the presence of the MXene nanosheets. Therefore, the incorporation of the rGO nanosheets resulted in the formation of the porous structure, which was favourite the electrolyte ions to access the electrodes, while the electrical conductivity due to the MXene was not compromised, thus leading to promising electrochemical performances.

A similar method was reported to prepare flexible rGO/Ti_2CT_r composite films, as electrodes of LIBs [51]. In this case, the electrochemical activity of both the rGO and the Ti_2CT_r nanosheets was utilized. Moreover, the two components prevented their self-restacking. It was found that the electrochemical property of Ti_2CT_x was highly dependent on the conditions of thermal treatment. Electrodes based on the thermally treated rGO/Ti_2CT_r hybrid films had a specific capacity of 700 mAh/g at the current density of 0.1 A/g, with strong cycling rate stabilities.

A MXene/rGO hybrid aerogel was prepared for piezoresistive sensor applications [52]. The hybrid gel had an extremely low density and superstrong elastic properties with pressure-sensitive behviors. The MXene/rGO aerogel was highly porous due to the presence of the rGO nanosheets and highly conductive becuase of the $Ti_3C_2T_x$. Specifically, the MX/rGO aerogel based piezoresistive sensor exhibited a sensitivity of 22.56 per kPa, with short response times of <200 ms and high stability for >10,000 cycles. The same authors also used MXene/rGO hybrid aerogel as electrodes of microsupercapacitors [53]. The microsupercapacitors had an area specific capacitance of 34.6 mF/cm^2 at a scan rate of 1 mV/s^1, together with a high cycling stability cycling for up to 15,000 times. In addition, the micro-device had self-healing capability After five healing cycles, the specific capacitance was retained by 81.7%.

A combined process of hydrothermal assembly and freeze-drying was developed to produce highly conductive 3D $Ti_3C_2T_x$/graphene porous architectures, as electromagnetic interference (EMI) materials [54]. The hybrid aerogels displayed a cellular microstructure with strong alignment, with graphene nanosheets serving as the network and $Ti_3C_2T_x$ nanosheets forming the cell walls. An electrical conductivity of 1085 S/m was achieved at the optima conditions. Epoxy nanocomposites filled with the hybrid gels secured a conductivity of 695.9 S/m, which resulted in EMI shielding effectiveness of 50 dB at the X-band. Interestingly, the content of $Ti_3C_2T_x$ was only 0.74 vol%.

Fabrication process of the $Ti_3C_2T_x$/graphene porous composites is depicted schematically in Fig. 3.28. Due their strong gelating behavior, GO nanosheets were used to facilitate the gelation. $Ti_3C_2T_x$/GO mixed suspension was prepared for hydrothermal processing. To avoid the oxidization of $Ti_3C_2T_x$, the hydrothermal experiment was conducted at mild conditions. GO was reduced to rGO during the treatment. The interconnection of the rGO nanosheets resulted in the formation of 3D architectures. At the same time, the $Ti_3C_2T_x$ nanosheets were attached onto the wall surfaces of the rGO network, thus forming $Ti_3C_2T_x$/rGO hybrid hydrogels. If the content of $Ti_3C_2T_x$ was less than 80 wt%, all the nanosheets were engaged in the areogels. The $Ti_3C_2T_x$/rGO hybrid hydrogels were then subject to dialyzation in 10 vol% of ethanol aqueous solution, before they were freezed in liquid nitrogen. During the freezing process, the rGO and $Ti_3C_2T_x$ nanosheets were assembled guided

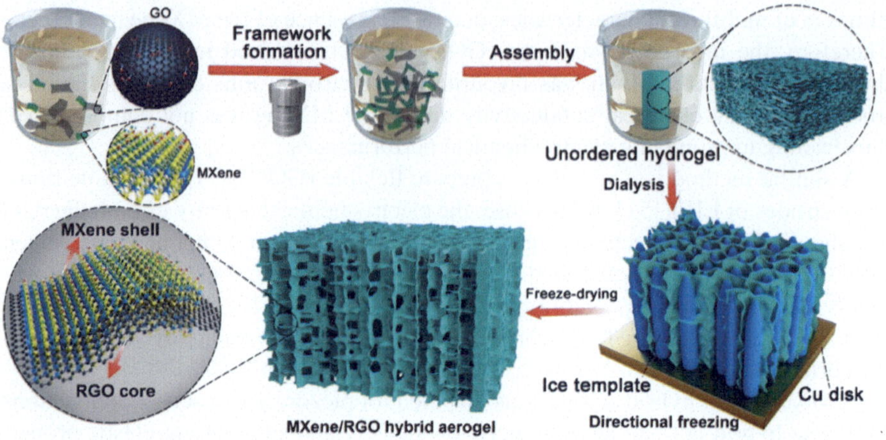

Fig. 3.28 Schematic diagram of the processing steps to fabricate the $Ti_3C_2T_x$/rGO gybrid aerogels through a combined process consisting of GO-assisted gydrothermal assembly, directional freezing and freeze-drying. Reproduced with permission from [54]. Copyright © 2018, American Chemical Society

by the ice cylinders. Eventually, cellular-like $Ti_3C_2T_x$/rGO hybrid hydrogels were obtained.

$Ti_3C_2T_x$/rGO hydrogel film was prepared for adsorbent applications [55]. The purpose was to extract 16 polycyclic aromatic hydrocarbons (PAHs) in river and tap water, before gas chromatography–tandem mass spectrometric analysis (GC–MS/MS) was applied. The $Ti_3C_2T_x$/rGO hydrogel was able to accumulate the PAHs in water, which then allowed for elution with *n*-hexane, separation with HP-5MS column and detection with mass spectrometer. The analytical result was linear in the range of 0.005–2.00 μg/L, with correlation coefficients of >0.995. The method had recoveries of 80.3–109.9%, with relative standard deviations (RSDs) to be <10.01%. The limit of detection was as low as 0.005 μg/L.

$Ti_3C_2T_x$ has also been incorporated with exfoliated graphitic carbon nitride (g-C_3N_4) nanosheets to form hybrids. For instance, a vacuum filtration method was used to obtain free-standing $Ti_3C_2T_x$/g-C_3N_4 hybrid films, as catalyst for oxygen-evolution reaction (OER) [56]. The outstanding catalytic effect was attributed to the Ti-N_x motifs, which served as electroactive sites, as well as the high surface area to offer high electrochemical double layer capacitances, due to the highly porous architecture. In addition, the hybrids had strong hydrophilic and porous characteristics, which ensured the surfaces to be highly wettable for the aqueous electrolytes, rapid transport of the reactants to the active sites and quick removal of the products.

Similarly, a Ti_2C/g-C_3N_4 hybrid was prepared for water splitting applications [57]. The enhanced water splitting effect was attributed to the strong interaction between the Ti_2C and g-C_3N_4. At the optimal content of 0.4 wt% Ti_2C, the hydrogen production rate was 47.5 μmol/h, which was increased by about 13 times as compared with that of g-C_3N_4. The hybrid samples were prepared by simply mixing melamine with Ti_2C in aqueous ethanol suspensions, followed by calcination. The mixtures were stirred at 85 °C to remove ethanol and then calcined at 550 °C for 4 h at a heating rate of 5 °C/min.

Ti_3C_2/g-C_3N_4 composites were obtained by using electrostatic self-assembly process, as photocatalysts to promote hydrogen evolution through visible light irradiation [58]. The incorporation of Ti_3C_2 significantly enhanced catalytic activity of g-C_3N_4 by more than 10 times. The great enhancement in photocatalytic effect of the Ti_3C_2/g-C_3N_4 composites was closely related to the 2D/2D heterostructures. Because the Ti_3C_2 and g-C_3N_4 nanosheets were closely contacted, the photogenerated charge carriers generated at the interfaces of Ti_3C_2/g-C_3N_4 could be effectively separated. Moreover, due to the high conductivity of Ti_3C_2 and the g-C_3N_4/Ti_3C_2 interface Schottky junction effect, the photoinduced electrons were transferring smoothly, while their recombination with the photogenerated holes was prevented.

To prepare the Ti_3C_2/g-C_3N_4 composites, g-C_3N_4 was dispersed in 0.5 M HCl aqueous solution protonation with the aid of ultrasonication. After thorough washing, the g-C_3N_4 suspension reached pH value of 4. S-Ti_3C_2 was dispersed in DI water to form Ti_3C_2 suspension. The two component suspensions were mixed according to the designed compositions. All the mixtures were centrifuged to precipitate Ti_3C_2/g-C_3N_4 composites, followed by drying at 80 °C for one day.

Ti_3C_2 nanoparticles, together with Pt nanoclusters, was used to increase the photoactivity of g-C_3N_4 for hydrogen production [59]. The Ti_3C_2 and Pt co-doped g-C_3N_4 displayed enhanced electrical conductivity and had more active sites, resulting in high photocatalytic activity for H_2 production. The production rate was 5.1 mmol/h g, which was higher than those of g-C_3N_4/Ti_3C_2 and g-C_3N_4/Pt by 3 and 5 times, respectively. The enhancement in the photocatalytic effect was attributed to the synergetic effect of the Ti_3C_2 and Pt nanoparticles, which led to inhibition charge carrier recombination and catalytic stability, due to the rapid electron transfer. Firstly, Ti_3C_2 and g-C_3N_4 were dispersed in DI water to form a mixed suspension, followed by overnight drying. Then, the g-C_3N_4/Ti_3C_2 mixtures were put into H_2PtCl_6 aqueous solution with 10 vol% methanol. After that, the suspensions were illuminated with a 300 W Xe lamp for 0.5 h, during which Pt^{4+} was reduced to Pt nanoclusters.

Figure 3.29 shows representative TEM images of the samples. Pure g-C_3N_4 had a lamellar structure, with folds and wrinkles at the edges, as observed in Fig. 3.29a. Average diameter of the Ti_3C_2 nanoparticles was 30 nm, as illustrated in Fig. 3.29b. The impregnation process had almost no effect on crystalline structures of both g-C_3N_4 and Ti_3C_2. The Ti_3C_2 nanoparticles were attached onto the g-C_3N_4 nanosheets, with a porous morphology, as demonstrated in Fig. 3.29c. HRTEM image revealed the lattice fringes with an inter-plane spacing of 0.35 nm, corresponding to the (006) plane of Ti_3C_2, as seen in Fig. 3.29d. Meanwhile, the Pt nanoclusters were uniformly distributed on g-C_3N_4, with an average diameter of 1.5 nm. The formation of the C_3N_4/Ti_3C_2/Pt ternary heterostructures was also clearly depicted by TEM images, as shown in Fig. 3.29e, f.

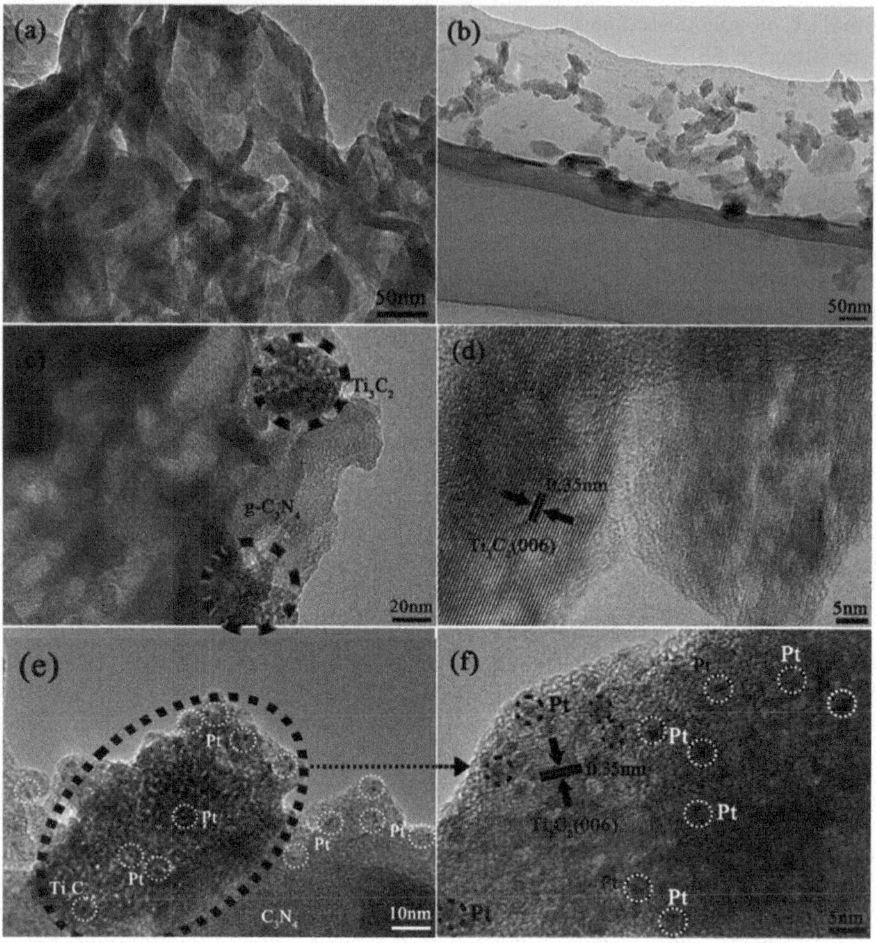

Fig. 3.29 TEM images: **a** g-C_3N_4, **b** Ti_3C_2 nanoparticles, **c** Ti_3C_2 loaded g-C_3N_4 and **e** g-$C_3N_4/Ti_3C_2/Pt$. HR-TEM images: **d** g-C_3N_4/Ti_3C_2 and **f** g-$C_3N_4/Ti_3C_2/Pt$. Reproduced with permission from [59]. Copyright © 2018, Royal Society of Chemistry

3.3.2 MXenes-Metals

Ti_3C_2/Ag nanocomposite has been prepared for LIBs applications [60]. The Ag nanoparticles were derived from $AgNO_3$ aqueous solution through direct reduction, in the presence of $Ti_3C_2(OH)_{0.8}F_{1.2}$. Electrodes based on the MXene/Ag nanocomposite had specific capacities of 310, 260 and 150 mAh/g, at 1 C, 10 C and 50 C, respectively. The outstanding electrochemical performances were attributed to the decreased interface resistance and the transition of Ti^{2+} to Ti^{3+} during the charge–discharge process.

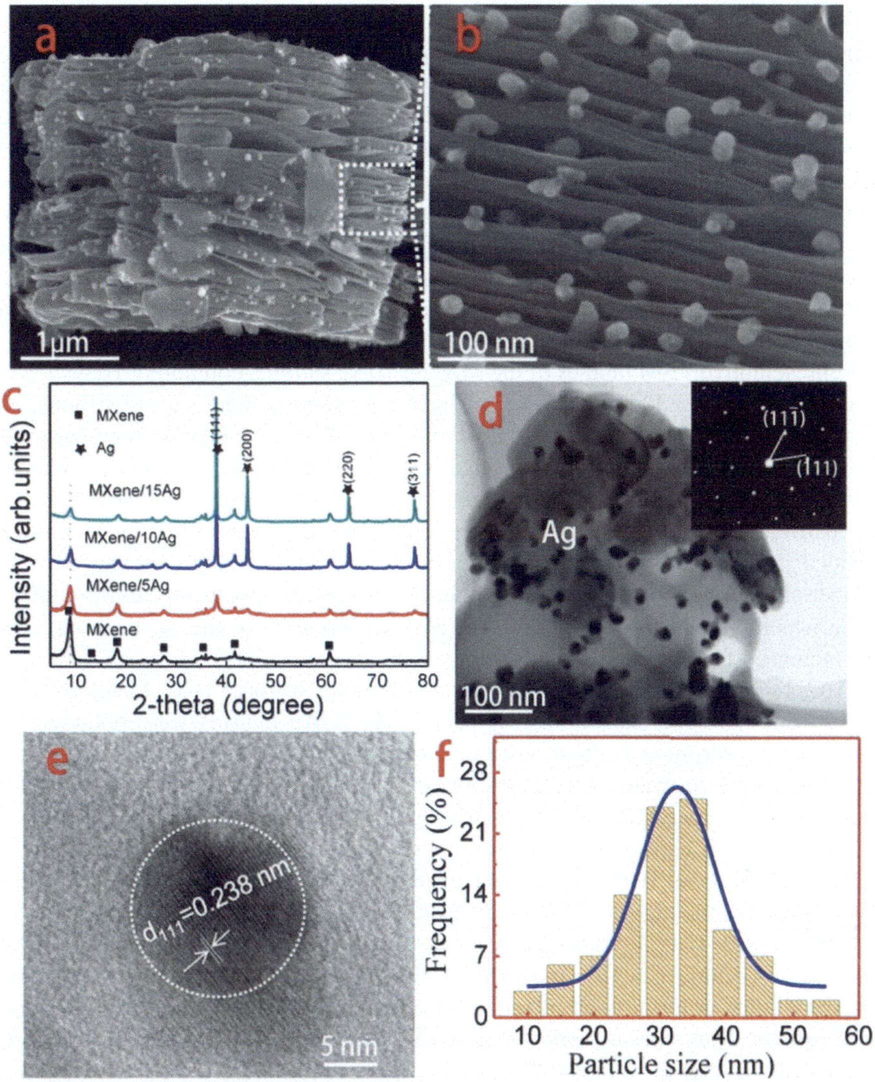

Fig. 3.30 **a** Representative SEM image of the layered MXene/10Ag sample. **b** High-magnification SEM image of a local area in panel (**a**). **c** XRD patterns of samples with different contents of Ag. **d** Representative TEM image of the MXene/10Ag sample, with the inset to be the SAED along the [1–10] direction. **e** HRTEM image of an isolated Ag particle. **f** Particle size distribution of the Ag nanoparticles. Reproduced with permission from [60]. Copyright © 2016, American Chemical Society

A representative SEM image of the nanocomposite is shown in Fig. 3.30a, indicating the retention of the layered structure of the MXene nanoparticles, which were decorated with numerous small Ag nanoparticles. There were also Ag particles to be present on the surfaces of the MXene, as demonstrated in Fig. 3.30b. Figure 3.30c shows XRD patterns of different samples. Obviously, the Ag nanoparticles possessed face-centered-cubic crystal structure, with lattice constant of 0.401 nm. As the content of Ag was increased, its diffraction peaks were gradually sharpened and increased, suggesting an increase in its crystallinity. Meanwhile, the (002) peak intensity of MXene was slightly weakened, implying a decrease in the degree of ordering in the *c*-axis direction, while the peak position was not shifted, suggesting no intercalation of Ag into the MXene nanolayers.

Representative TEM images of the MXene/10Ag sample, containing 10 wt% Ag, are depicted in Fig. 3.30d, e. It was found that the Ag nanoparticles were distributed on the surfaces of the MXene. The high degree of crystallization of the Ag nanoparticles was also confirmed by the clear selected area electron diffraction pattern (SAED) of the nanoparticle along the [1–10] zone axis. HRTEM image indicated that Ag nanoparticles were single crystals, with an average diameter of about 35 nm, as revealed in Fig. 3.30f. With increasing content of Ag, both the diameter and the number of the Ag nanoparticles were gradually increased. Specifically, as the content of Ag was increased from 5 to 15%, the average diameter of the Ag nanoparticles was increased from about 10 to >100 nm.

$Ti_3C_2T_x$ was also modified with Ag nanoparticles to form $Ti_3C_2T_x$/Ag hybrids as electrodes of supercapacitors [61]. The $Ti_3C_2T_x$/Ag hybrids were obtained by using vacuum filtration method, with mixed suspension of $Ti_3C_2T_x$ nanosheet and Ag nanoparticles in water. Electrodes based on the $Ti_3C_2T_x$/Ag hybrid films had a high specific surface area of 107 m^2/g, a high areal capacitance of 332.2 mF/cm at 2 mV/s, a high capacitance retention by 63.2% from 2 to 100 mV/s and a 87% cycling stability over 10,000 cycles. An even high areal capacitance of 1173 mF/cm was achieved at the highest mass loading of 15.0 mg/cm.

Delaminated $Ti_3C_2T_x$ suspensions were directly obtained after the etching of Ti_3AlC_2. To synthesize Ag nanoparticles, 1.0 mM $AgNO_3$ and 10.0 mM sodium dodecyl sulfate solution was mixed with 2.0 mM $NaBH_4$ solution at controlled temperatures. Then, the suspensions of $Ti_3C_2T_x$ and Ag nanoparticles, with concentrations of 1.2 mg/mL and 0.027 mg/mL, respectively, were mixed with the aid of ultrasonication, followed by vacuum filtration with a polytetrafluoroethylene separator membranes with an average pore diameter of 0.22 μm.

Urchin-shaped MXene-$Ag_{0.9}Ti_{0.1}$ bimetallic nanowire nanocomposites have been reported as photocatalysts for oxygen reduction reaction (ORR) [62]. The MXene-Ag nanocomposites were prepared by directly mixing $AgNO_3$ and alkalization-intercalated MXene (alk-MXene, $Ti_3C_2(OH/ONa)_2$) suspensions in the presence of polyvinylpyrrolidone (PVP). In this case, the use of PVP was to trigger the formation of fivefold nanotwined Ag seeds, thus leading to the development of Ag/Ti ($Ag_{0.9}Ti_{0.1}$) bimetallic nanowires. The bimetallic nanowires could facilitate a four-electron transfer process, with high current density and high stability. They also provided with more adsorption sites for oxygen molecules and shortened the diffusion path of the adsorbed oxygen molecules, thus offering high catalytic efficiency for ORR.

The MXene $Ti_3C_2(OH)_{0.8}F_{1.2}$ was of a multilayered structure, as seen in Fig. 3.31a. After alkalization-intercalation in 6 M NaOH for 2 h, layered alk-MXene $Ti_3C_2(OH/ONa)_2$ was obtained. The as-etched MXene had an XRD (0002) peak at about 9°, corresponding to *c* lattice of 0.975 nm, as shown in Fig. 3.31b. After alkalization-intercalation process, an additional peak at about 6.7° present, accompanied by widening in full width at half-maximum (FWHM), which implied that the *c* lattice was increased and the crystallinity was decreased [20]. During alkalization-intercalation process, –F functional groups were replaced by the –OH ones, while –ONa groups were formed due to ion exchange effect, which expanded the *c* lattice.

Figure 3.31c shows SEM image of the urchin-like structured nanocomposite. On surface of the alk-MXene, there were long wires (A) and dot-like particles (B). As observed in Fig. 3.31d, the XRD pattern of Ag had sharp diffraction peaks, indicating its high crystallinity nature. According to TEM images, it was found that the Ag nanowires possessed a relatively rough surface, as illustrated in Fig. 3.31d, e. The nanowires had a length of about 5 μm and a width of about 42 nm, with a chemical composition of Ag/Ti to be 9:1. The Ag nanoparticles had a fivefold twin morphology, with diameters of 20–50 nm, as revealed in Fig. 3.31f.

MXene/Ag hybrid has also been synthesized as membranes for ultrafast water purifications [63]. The MXene/Ag hybrid membranes contained Ag nanoparticles with concentrations of 0–35%. The Ag nanoparticles were made through self-reduction of $AgNO_3$ on the surface of MXene nanosheets. In this case, the 2D structure of the MXene ensured the formation of the hybrid membranes, while Ag^+ was reduced to Ag by MXene at the same time. The MXene/Ag membrane sample, with 21% Ag nanoparticles, thickness of 470 nm and average pore size of 2.1 nm, demonstrated the highest of 420 L/m^2 h bar. In comparison, the water flux achieved by using the pristine MXene membrane was 118 L/m^2 h bar. The optimal membrane also exhibited high rejection efficiencies and strong flux recovery capabilities for organic molecules. With the reference of hydrophilic polyvinylidene difluoride (PVDF) membrane, the 21% MXene/Ag hybrid membrane displayed a 99% *E. coli* growth inhibition, as compared with 60% for MXene membranes. In addition, the 21% MXene/Ag membrane showed much stronger rejection to organic foulants, such as bovine serum albumin (BSA) and methyl green (MG), as compared with the membranes reported in the open literature.

Ti_3C_2/Au composites were prepared to be combined with MOF-derived MnO_2/Mn_3O_4, in order to set up an electrochemical sensing platform to detect organophosphorus pesticides (OPs) [64]. When the 3D MnO_2/Mn_3O_4 hierarchical microcuboids with vertically aligned highly ordered nanosheets were incorporated with MXene/Au, a synergistic effect for signal amplification was observed, together with superb electrochemical performance, huge specific surface area and strong environmental biocompatibility. After optimization, the sensing platform, AChE-Chit/MXene/Au/MnO_2/Mn_3O_4/GCE was able to detect methamidophos over a wide range of concentrations of 10^{-12}–10^{-6} M, which was accompanied by high linearity.

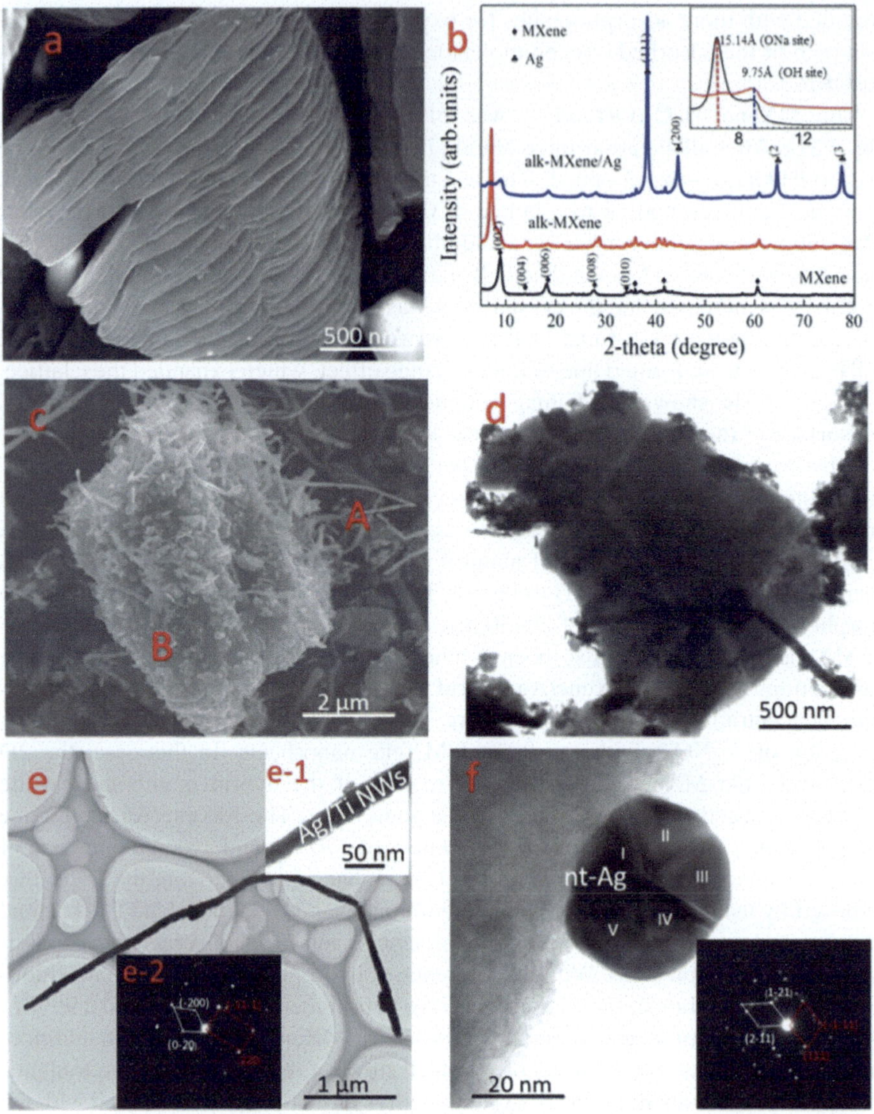

Fig. 3.31 **a** Representative SEM image of the layered alk-MXene. **b** XRD patterns of different samples, with the inset to be the enlarged low angle range. **c** SEM image of the MXene/NW-$Ag_{0.9}Ti_{0.1}$ nanocomposite, with A and B to represent a nanowire and nanosized particle, respectively. **d** TEM image of the MXene/NW-$Ag_{0.9}Ti_{0.1}$ nanocomposite. **e** High magnification TEM image of the nanowire A in panel (**c**), with a diameter of about 42 nm (**e-1**) and the corresponding SAED fringe (**e-2**) confirming the nanotwinned Ag structure **f** High-magnification TEM image of the nanoparticle B with a fivefold twin, with the inset corresponding to its SAED pattern. Reproduced with permission from [62]. Copyright © 2016, American Chemical Society

This method could be extended to other composites for different electrochemical sensor applications.

Ti_3C_2/Au nanohybrids were prepared for nanomedicine applications [65]. As the Au nanoparticles were deposited onto the surface of the Ti_3C_2 nanosheets, nanocomposites with enhanced stability and biocompatibility were realized, due to the thiol modification. Meanwhile, the near-infrared optical absorption was significantly increased. Because of the strong optical absorbance and X-ray attenuation, the Ti_3C_2/Au nanocomposites could be employed to build photoacoustic and computed tomography dual-modal imaging systems. In addition, the Ti_3C_2/Au nanocomposites exhibited a mild photothermal effect, which allowed for the improvement in tumor oxygenation, thus leading to highly efficient radiotherapy. More importantly, nanocomposites showed no long-term toxicity at the dose in that study.

Xie et al. [66] $Ti_3C_2T_x$ was combined with Au nanorods (AuNRs) to form MXene-AuNRs hybrids, with could be used as stable surface-enhanced Raman scattering (SERS) substrates with high sensitivity of organic pollutants. The AuNRs were evenly attached on the surface of the MXene nanolayers through electrostatic interactions, which offered numerous SERS hot-spots. The MXene-AuNR SERS substrates were highly sensitive and reproductive to various organic dyes, like rhodamine 6G, crystal violets and malachite greens. The limits of detection were 1×10^{-12} M, 1×10^{-12} M and 1×10^{-10} M for the three dyes correspondingly. The SERS substrates were also useful in the detection of more complicated pesticides and pollutants, with the limits of detection to be 1×10^{-10} M and 1×10^{-8} M, for thiram and diquat, respectively. Such MXene-AuNRs nanohybrids could be used for environmental protection and food safety examination.

$Ti_3C_2T_x$ was decorated with Pt nanoparticles as a platform for electrochemical sensing applications [67]. The GCE modified with the $Ti_3C_2T_x$/Pt nanocomposite had higher and more stable redox behavior than that deposited with pure $Ti_3C_2T_x$. Specifically, the H_2O_2 sensor based on $Ti_3C_2T_x$/Pt/GCE exhibited a limit of detection (LOD) of 448 nM. In addition, the $Ti_3C_2T_x$/Pt-based sensors could be used to detect various redox molecules, including ascorbic acid (AA), dopamine (DA), uric acid (UA) and acetaminophen (APAP), with high selectivities and LOD of nM level.

$Ti_3C_2T_x$/Pt nanocomposite has also been synthesized as potential catalysts for hydrogen evolution (HER) or oxygen reduction reaction (ORR) [68]. Factors including Pt precursor concentration, reduction time and additional reducing agents could influence the properties and performances of the nanocomposites. The sample with 14 wt% Pt possessed HER effect, with onset potential and current density of -75.9 mV and -10 mA/cm^2, respectively.

Similarly, with its stable multilayered structure and special physical properties, Ti_3C_2 was utilized as support of Pt nanoparticles [69]. The Pt/Ti_3C_2 nanocomposites exhibited electrocatalytic activities for methanol oxidation reaction (MOR) and hydrogen evolution reaction (HER). Specifically, the electrocatalytic activity for methanol oxidation, with a peak current density of 1.137 mA/cm^2, was much higher that of Pt/C. For HER in 0.5 M H_2SO_4 solution, a relatively low overpotential of 7 mV was observed at -10 mA/cm^2, much lower than that (17 mV) of Pt/C. The enhanced MOR and HER activities of the Pt/Ti_3C_2 nanocomposites were readily attributed to the stable layered structure and high electrical conductivity of the Ti_3C_2 MXene, along with the smooth electron transfer between the Pt nanoparticles and the MXene.

A facial self-reduction process was developed to prepare self-assembled MXene-Pd nanohybrids as high-efficient catalysts for nitro compounds and morin [70]. Pd nanoparticles (PdNPs) were directly anchored on the surface of the MXene nanolayers, with the compositions of the hybrids to be controlled by simply varying the reducing reaction time durations. The MXene-PdNPs20 sample demonstrated a high catalytic efficienty for 4-NP and 2-NA, with first-order rate constants to be 0.180/s and 0.089/s, respectively. The hybrid-based catalysts exhibited strong cycling stability, with high conversion efficiencies of 94% and 92% after cycling for 8 times for 4-NP and 2-NA, respectively. Such MXene-PdNPs nanohybrids could find applications in environmental management and health monitoring.

A one-step method was reported to hybridize Ag, Au and Pd nanoparticles with Ti_3C_2 nanosheets, as substrates for surface-enhanced Raman spectroscopy (SERS) [71]. For Ag and Au nanoparticles, theoretical SPR wavelengths were 440 nm and 558 nm, respectively, which were in a good agreement with the experimental results. For Pd nanoparticles, the predicted SPR wavelength was 160 nm for the planar geometry. In other words, a broad red-shifted SPR band peaked at 230 nm was observed, when the non-spherical Pd nanoparticles were coupled with MXene. The hybrid substrates showed pretty high SERS detection sensitivities of methylene blue (MB), with theoretical enhancement factors of up to 105.

To prepare hybrid suspensions, $Ti_3C_2T_x$ nanosheets were dispersed in DI water with the aid of strong ultrasonication. Then, aqueous solutions of $AgNO_3$, $HAuCl_4$ and $PdCl_2$ were mixed with the $Ti_3C_2T_x$ suspension, thus forming hybrids of Ag, Au, and Pd@$Ti_3C_2T_x$. The mixtures were centrifuged at 10,000 rpm for 5 min to obtain final hybrid samples. Figure 3.32 shows TEM images of the pristine $Ti_3C_2T_x$ and the hybrid samples. The MXene was composed of single- and few-layer nanoflakes, with lateral sizes of 2–3 μm. All the nanoparticles were directly decorated on the MXene nanosheets. The Ag and Au nanoparticles were of nearly spherical morphology, with sizes in the ranges of 10–70 nm and 40–50 nm, respectively. It was more difficult to control the size of the Ag nanoparticles, because of the high reactivity of the Ag^+ ion. In comparison, the Pd nanoparticles had a sheet-like planar morphology, probably because the reduction process of the Pd^{2+} ion was different from those of the Ag^+ and Au^{3+} ions.

MXene/DNA/Pd/Pt multiple component nanocomposites have been prepared as sensors to dopamine (DA) [72]. In this case, the MXene nanosheets were used as conductive matrix to host the Pd/Pt nanoparticles, which were gown through the aromatic hydrophobic bases of DNA on the surface of MXene. It was found that the DNA was crucial to the development of the nanocomposites, with which the Ti_3C_2 nanosheets were uniformly dispersed the growth of the metallic nanoparticles was ensured. The combination of the Pd/Pt nanoparticles and the Ti_3C_2 nanosheets resulted in an significant enhancement in the electrocatalytic activity of the nanocomposites to sense DA. Specifically, the nanocomposite based sensors displayed a linear

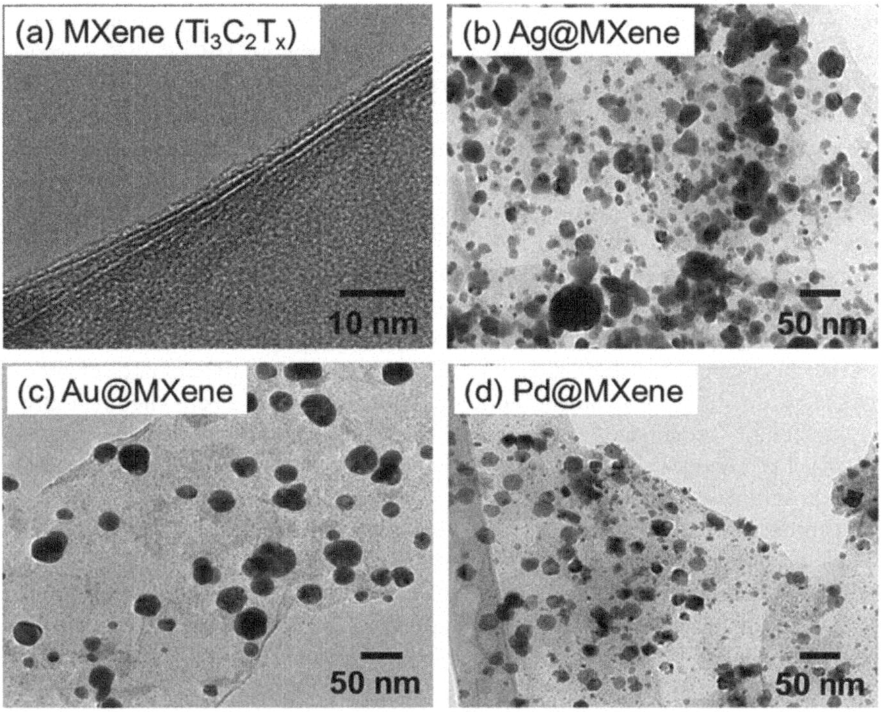

Fig. 3.32 HRTEM and TEM images of the samples: **a** High resolution TEM image of MXene nanosheets, **b** Ag@MXene, **c** Au@MXene and **d** Pd@MXene hybrid nanosheets. Reproduced with permission from [71]. Copyright © 2016, Springer Nature

amperometric sensitivity to DA in the concentration range of 0.2–1000 μM, with a relatively low limit of detection (LOD) of 30 nM and a high selectivity.

3.3.3 MXenes-Oxides

Due to the content of Ti, it is naturally expected that TiO_2 could be derived from Ti-based MXenes. The oxidation of $Ti_3C_2T_x$ in air, CO_2 and pressurized DI water to form TiO_2 has been confirmed [11]. After flash oxidization of $Ti_3C_2T_x$ in air at 1150 °C for 0.5 min, anatase TiO_2 nanocrystals were formed, with different morphologies and sizes, which were included in amorphous carbon nanosheet matrix, thus leading to TiO_2-C hybrid. If it was isothermally oxidized in CO_2 at 150–300 °C for 60 min or hydrothermally treated at temperatures of 150–250 °C, a small quantity of $Ti_3C_2T_x$ remained in TiO_2-C hybrid.

The oxidation mechanisms of $Ti_3C_2T_x$ have been studied by using in-situ TEM technique [73]. It was revealed that TiO_2 nanoparticles were present at the top and bottom surfaces of the Ti-layers. TiO_2 nanocrystals due to flash oxidation and slow heating were anatase and rutile, respectively. Similarly, as $Ti_3C_2T_x$ was partially oxidized with O_2 at a relatively low temperature of 200 °C for 40 min, $Ti_3C_2T_x/TiO_2$ was formed, with TiO_2 to be anatase nanocrystals [74]. After thermal calcination at 1000 °C, the anatase TiO_2 was converted to rutile TiO_2 phase. It was also observed that both the structure and morphology of the TiO_2 phase in the $Ti_3C_2T_x/TiO_2$ hybrids could be controlled by adjusting the thermal annealing conditions [75, 76].

Structure and electrochemical behavior of the MXene have been studied, after it was annealed in air, Ar, N_2 and N_2/H_2 [76]. Thermal treatment in air altered the structure and morphology the MXene nanosheets, but an enhanced electrochemical performance was interestingly achieved, as compared with that of the as-etched MXene. In comparison, after annealing in Ar, N_2 and N_2/H_2, the morphology of the MXene phase was well retained, while the improvement in electrochemical effect was much more pronounced. For instance, electrodes based on the MXene annealed in N_2/H_2 had a specific capacitance of 51 F/g at 1 A/g, which was the highest among the three samples.

According to XRD results, after annealing at 250 °C in air, the MXene phase was completely converted to anatase TiO_2, together with the presence of graphitic carbon. For those annealing in Ar, N_2 and N_2/H_2 MXene was still the major phase, with a small quantity of anatase TiO_2, due to the partial oxidation. Also, the interlayer spacing was slightly enlarged in these samples. The (002) diffraction peak was shifted from $2\theta = 11.89°$ to 11.82°, 11.84° and 11.79°, corresponding to a decrease in interlayer spacing of 0.743 to 0.748, 0.747 and 0.750 nm, for Ar, N_2 and N_2/H_2, respectively.

Figure 3.33 shows representative SEM images of samples before and after thermal annealing. Evidently, the morphology of the MXene nanosheets annealed in Ar, N_2, or N_2/H_2 was almost unchanged, as seen in Fig. 3.33a, b, while the air annealed sample contained only TiO_2 nanoparticles supported by thin graphitic nanosheets, as demonstrated in Fig. 3.33c. This observation has been further confirmed by TEM/HRTEM results. In addition, the annealing of Ti_2CT_x in controlled atmospheres could result in thinning of the nanolayers, which led to the generation of mesoporous structure that ensured the access of electrolyte ions, when used as electrodes of electrochemical devices.

$Ti_3C_2T_x/TiO_2$ nanocomposites were prepared by coating TiO_2 nanoparticles on $Ti_3C_2T_x$ MXenes, as electrodes of LIBs [77]. At optimal conditions, the nanocomposite-based electrode had a specific capacity of 143 mA h/g, at the current density of 0.5 A/g, after recycling for 200 times. Theoretical studies indicated that structural stability of the nanocomposite during the insertion/de-insertion of Li ions was significantly increased, which was responsible for the outstanding electrochemical performances. To prepare the nanocomposites, $Ti_3C_2T_x$ powder was ultrasonically dispersed in DI water, followed by centrifugation at 4000 rpm for 10 min to obtain the supernatant. Commercial TiO_2 nanoparticles with an average diameter of 4 nm were dispersed in DI water and then mixed with the $Ti_3C_2T_x$ suspension, followed by freeze-drying.

A room temperature surface oxidation process was reported to prepare TiO_2/Ti_2CT_x hybrid, with hydrogen peroxide (H_2O_2) as the oxidation agent [13].

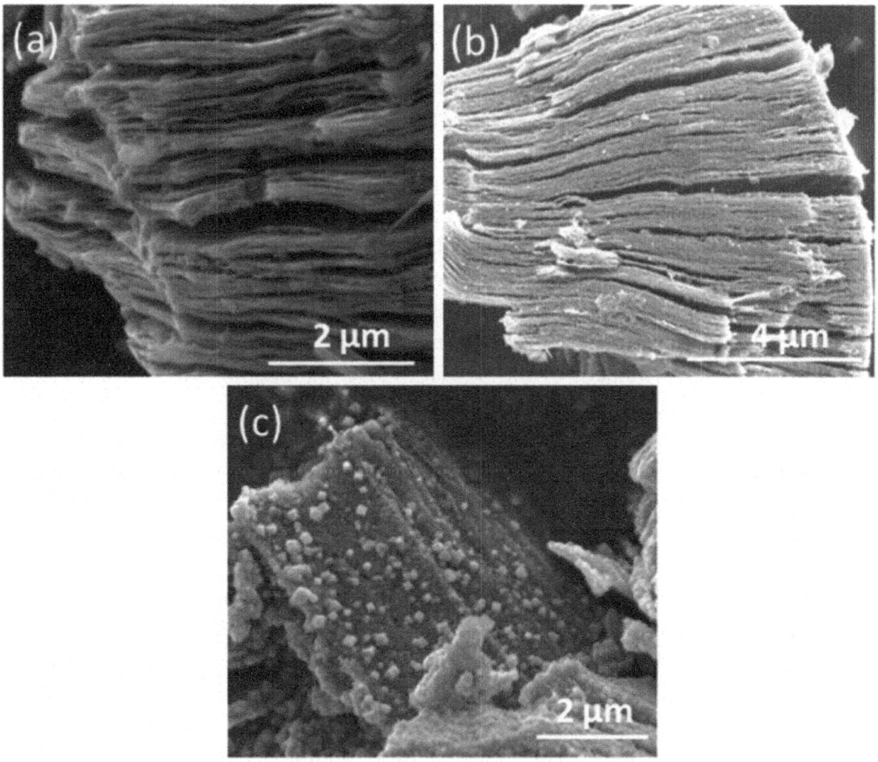

Fig. 3.33 SEM images of the MXene samples: **a** as-etched MXene phase, **b** annealing in N_2/H_2 and **c** annealing in air. Reproduced with permission from [76]. Copyright © 2015, American Chemical Society

TiO_2 nanoparticles were formed on surface of the Ti_2CT_x nanosheets, while the content of TiO_2 could be readily controlled, by adjusting the concentration of H_2O_2 in the reaction systems. During the oxidation process, TiO_2 nanoparticles were formed through the nucleation at the defect sites of the Ti_2CT_x nanosheets. Electrodes based on the H_2O_2 treated MXene exhibited promising electrochemical activity as anode LIBs, as compared was the untreated sample. For example, optimal H_2O_2 treated MXene could deliver specific capacities of 389, 337 and 297 mA h/g at current densities of 100, 500 and 1000 mA/g, respectively.

Similarly, a one-step solvothermal process was employed to produce TiO_2 quantum dots (QDs) on $Ti_3C_2T_x$ nanosheets, with the aid of cetyltrimethylammonium bromide-assisted, as electrodes of Li–S batteries [78]. Due to the presence of the TiO_2 QDs, $Ti_3C_2T_x$ nanosheets were effectively isolates and the restacking behavior was prevented. As a result, a large specific surface and more reactive sites for electrolyte ions to access were available. Density functional theory (DFT) calculation results indicated polysulfides had stronger adsorption energy with TiO_2 than with $Ti_3C_2T_x$, which thus boosted the on-site retention of the polysulfides. Meanwhile the high electrical conductivity of the MXene nanosheets ensured a rapid diffusion of electrolyte ions and electrons, while the strong mechanical properties guaranteed the integrity of the devices. Cathode based on the TiO_2QDs@MXene/S exhibited a specific capacity of 680 mAh/g at 2C after 500 cycles, which was much higher than that achieved by using MXene/S cathode.

A ternary nanocomposite, S@TiO_2/Ti_2C, was developed as cathodes of Li–S batteries [79]. In this special nanostructure, TiO_2 hollow nanospheres with encapsulated sulfur were inserted in between the Ti_2C nanosheets. Electrodes based on the S@TiO_2/Ti_2C nanocomposite with a sulfur content of 78.4 wt% delivered a specific capacity of 1409 mAh/g at 0.2 C. After 200 cycles, the capacity values were still 464 mAh/g and 227 mAh/g, at high current rates of 2 C and 5 C, respectively. The high electrochemical performances of the S@TiO_2/Ti_2C nanocomposites were ascribed to the synergistic effect of all the components.

Figure 3.34 shows a schematic diagram describing fabrication process of the nanocomposites. Firstly, ammonia solution and resorcinol powder were dissolved a mixed solvent of anhydrous ethanol and DI water. Secondly, formaldehyde solution with a concentration of 10.0 mg/ml was added into the solution and kept at 30 °C for one day time prepolymerization. After that, the mixture was subject to hydrothermal treatment at 100 °C for 24 h. Finally, solid resorcinol–formaldehyde (RF) resin spheres were collected through centrifugation.

The RF resin spheres were then used as templates, together with cetyltrimethyl ammonium bromide (CTAB), were dispersed in ethanol to obtain a mixed suspension, into which tetrabutyl titanate ethanol solution at a concentration of 50.0 mg/ml was slowly introduced. Due to the hydrolysis reaction, titanium hydroxide was produced, which was collected and calcined at 550 °C for 3 h to form TiO_2 hollow spheres. The TiO_2 hollow spheres were mixed with the solution of sublimed sulfur in carbon

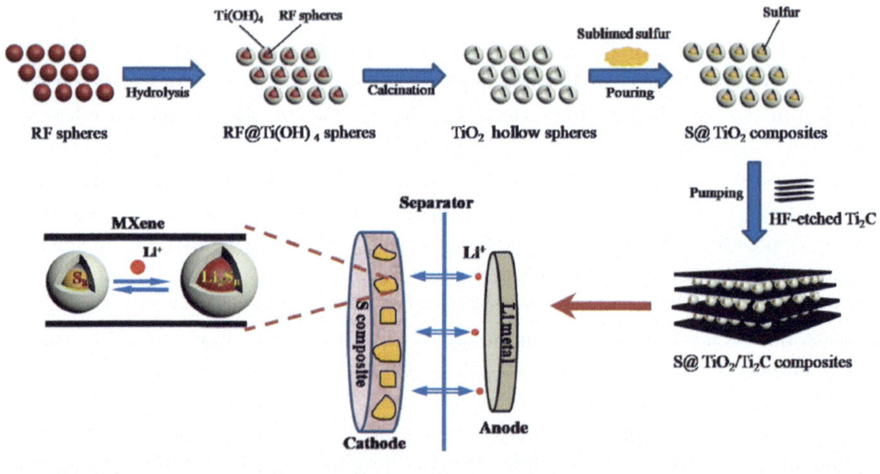

Fig. 3.34 Schematic diagram demonstrating fabrication process of the S@TiO_2/Ti_2C nanocomposite. Reproduced with permission from [79]. Copyright © 2019, Elsevier

disulfide (CS_2). After the CS_2 was completely vaporized, the mixture was hydrothermally reacted at 155 °C for 12 h, resulting in S@TiO_2 nanocomposite samples. S@TiO_2/Ti_2C nanocomposites were prepared by inserting the S@TiO_2 spheres into Ti_2C nanosheets at vacuum.

TiO_2/Ti_3C_2 nanocomposites were obtained by using an in-situ hydrolysis method, followed by heat-treatment process, as electrodes of supercapacitors [16]. In these nanocomposites, the TiO_2 nanoparticles an average diameter of <30 nm, which were uniformly distributed on surface of the Ti_3C_2 nanosheets. Electrodes based on the TiO_2/Ti_3C_2 nanocomposites had a specific capacitance of 143 F/g at 5 mV/s, which was much higher than that of pure Ti_3C_2 based one. In addition, the TiO_2–Ti_3C_2 electrode exhibited high cycling and rate stabilities.

A hydrothermal method was used to synthesize TiO_2/$Ti_3C_2T_x$ nanocomposites with high photocatalytic activities [14]. Ti precursors, heating temperature and reaction time all could be utilized to control the microstructure, composition and photocatalytic performance of the final products. Typically, titanium sulfate ($TiSO_4$) was dissolved in DI water to form solution, into which Ti_3C_2 powder was added. The suspension was homogenized with the aid of strong stirring, followed by hydrothermal treatment at 180 °C for 18 h. After reaction, the TiO_2/Ti_3C_2 nanocomposite samples were collected and dried in vacuum at 80 °C for 10 h.

Figure 3.35 shows SEM images of Ti_3C_2, TiO_2 and the TiO_2/Ti_3C_2 nanocomposites. As observed in Fig. 3.34a, the pristine Ti_3C_2 had a typical multilayered structure, with an average thickness of 40 nm. After hydrothermal treatment at 180 °C, the Ti_3C_2 nanosheets were coated with a layer of nanoparticles with dense packing and irregular morphology, as illustrated in Fig. 3.34b. XRD pattern indicated that the nanoparticles were TiO_2, which were formed because of the partial oxidation of the MXene phase. As demonstrated in Fig. 3.34c, the TiO_2 sample consisted of irregular particles with sizes in the range of 1–4 μm. TiO_2 particles were assembles of nanoparticles with an average diameter of 30 nm.

For the TiO_2/Ti_3C_2 nanocomposites, they were composed of spherical particles, together with a small number of nanoparticles at the edges of the Ti_3C_2 nanolayers, as demonstrated in Fig. 3.35d–g. Comparatively, the morphology of the TiO_2 nanoparticles in the nanocomposites was much more regular, most of them were spherical. Meanwhile, pure TiO_2 sample had agglomeration, which was attributed probably to the high surface interaction in between the nanoparticles, because they had a large specific surface area and thus high surface energy. However, it was not clear whether the aggregation was caused by the drying process or present in the as-obtained products.

TiO_2/$Ti_3C_2T_x$ nanocomposites have also been prepared with tetrabutyl titanate (TBOT) as the precursor that was mixed with $Ti_3C_2T_x$ [15, 16]. TiO_2/Ti_3C_2 nanocomposite was used to immobilize hemoglobin (Hb) to fabricate a mediator-free biosensor [15]. According to spectroscopic results and electrochemical performances, it was demonstrated that TiO_2/Ti_3C_2 nanocomposite was an outstanding immobilization matrix with strong biocompatibility for redox protein, thus leading to high protein bioactivities and stabilities. The special hybrid structure of the TiO_2/Ti_3C_2 ensured the direct electron transfer of Hb, so that the biosensors showed a high sensing

Fig. 3.35 SEM images of the nanocomposite samples: **a** Ti_3C_2, **b** Ti_3C_2-180 °C, **c** TiO_2, **d** 0.001 mol TiO_2/Ti_3C_2, **e** 0.002 mol TiO_2/Ti_3C_2, **f** 0.003 mol TiO_2/Ti_3C_2 and **g** 0.004 mol TiO_2/Ti_3C_2. The insets in the panels are their corresponding high magnification SEM images. Reproduced with permission from [76]. Copyright © 2015, American Chemical Society

performance to H_2O_2, with a wide linear concentration range of 0.1–380 μM, a high sensitivity of 447.3 μA/mM cm^2 and a low limit of detection (LOD) of 14 nM. Moreover, because the TiO_2 nanoparticles on the surface of the nanocomposites were highly biocompatible, they created a protective microenvironment for Hb. As a result, the nanocomposite-based biosensors had long-term stabilities. Specifically, the sensing behavior of the TiO_2/Ti_3C_2 biosensor was retained by about 95% after storage for two months. These TiO_2/Ti_3C_2 nanocomposites could be used to develop mediator-free biosensors for various applications.

A self-assembly method was reported to prepare TiO_2/Ti_3C_2 nanocomposite films, which consisted of TiO_2 opal photonic crystals (IOPCs) and Ti_3C_2 quantum dots (QDs), for biosensor applications [80]. Due to the widened absorption band by the Ti_3C_2 QDs, the IOPCs/QDs nanocomposite film exhibited strong photocurrent effect over 280–900 nm. An internal power conversion efficiency of 26% at 350 nm was achieved. Type-II heterojunction was formed between the Ti_3C_2 QDs and TiO_2, which ensured rapid charge transfer and thus fast electron–hole separation. As a result, photoelectrochemical (PEC) sensor based on TiO_2/Ti_3C_2/Nafion electrode was highly sensitive to glutathione, together with sufficiently high selectivity and long-term stability.

TiO_2/Ti_3C_2 nanocomposites were obtained by using an in-situ hydrolysis method, followed by heat-treatment process, as electrodes of supercapacitors [16]. In these nanocomposites, the TiO_2 nanoparticles an average diameter of <30 nm, which were uniformly distributed on surface of the Ti_3C_2 nanosheets. Electrodes based on the TiO_2/Ti_3C_2 nanocomposites had a specific capacitance of 143 F/g at 5 mV/s, which was much higher than that of pure Ti_3C_2 based one. In addition, the TiO_2–Ti_3C_2 electrode exhibited high cycling and rate stabilities.

$Ti_3C_2T_x$ was incorporated with rutile TiO_2 as catalysts for water splitting to realize visible-light-induced solar hydrogen production [17]. Photocatalytic performance of the $TiO_2/Ti_3C_2T_x$ nanocomposites was highly dependent on the content of $Ti_3C_2T_x$. The optimal content of $Ti_3C_2T_x$ was 5 wt %, with which the photocatalytic hydrogen evolution production was increased by about four times as compared with that of pure rutile TiO_2. The photocatalytic enhancement was also found for other MXenes, such as Nb_2CT_x and Ti_2CT_x.

An in-situ growth method was reported to synthesize Ti_3C_2/TiO_2 nanocomposites with a flower-like morphology, as photocatalysts for water splitting to produce H_2 and O_2 [81]. The nanocomposites were obtained through a synthronic oxidation and alkalization of Ti_3C_2, followed by ion-exchange and thermal treatment. Photocatalytic activity of the Ti_3C_2/TiO_2 could be controlled by controlling the calcination temperature, which was optimized at 500 °C. Photochemical stability was confirmed by the reproducible production of H_2 and O_2 from water splitting, without the requirement of sacrificial agents.

To obtain the Ti_3C_2/TiO_2 nanocomposites, Ti_3C_2 powder was dispersed in a mixed solution of NaOH and H_2O_2, followed by hydrothermal treatment at 140 °C for 12 h. The samples were $Na_2Ti_3O_7/Ti_3C_2$ nanocomposites, which were immersed in 0.1 M HCl solution for one day to substitute the Na^+ ions with H^+ ones, thus leading to $H_2Ti_3O_7/Ti_3C_2$ nanocomposites. Af the $H_2Ti_3O_7/Ti_3C_2$ samples were heated at different temperatures, Ti_3C_2/TiO_2 flower-like nanocomposites were finally formed. The samples heated at 300 °C, 400 °C and 500 °C for 3 h were denoted as Ti_3C_2–TiO_2-300, Ti_3C_2–TiO_2-400 and Ti_3C_2–TiO_2-500, respectively. Figure 3.36 shows a schematic diagram showing synthetic process of the Ti_3C_2/TiO_2 nanocomposites. After hydrothermal treatment, monoclinic $Na_2Ti_3O_7$ phase together with Ti_3C_2 was formed, which was further transferred to $Na_2Ti_3O_7$, thus forming $Ti_3C_2/H_2Ti_3O_7$ nanocomposites. Finally, Ti_3C_2–TiO_2 flower-like nanocomposites were obtained after calcination.

Figure 3.37 shows SEM images of the Ti_3C_2 based nanocomposites prepared at different temperatures. The Ti_3C_2 MXene had a typical multilayered structure with smooth surface, as observed in Fig. 3.37a. After the concurrent treatment of oxidation and alkalization, the sample color was varied from black to light yellow, as seen in Fig. 3.36. The ion exchange and further calcination led to the formation of flower-like

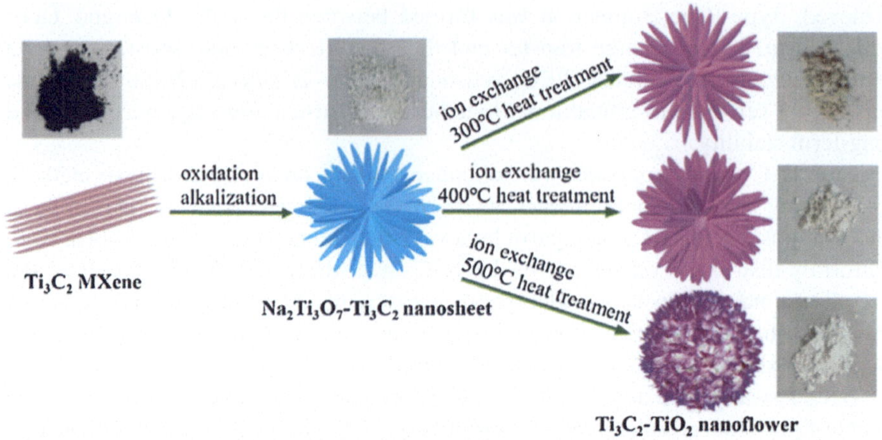

Fig. 3.36 Schematic diagram showing fabrication process of the Ti_3C_2/TiO_2 flower-like nanocomposites. Reproduced with permission from [81]. Copyright © 2018, Elsevier

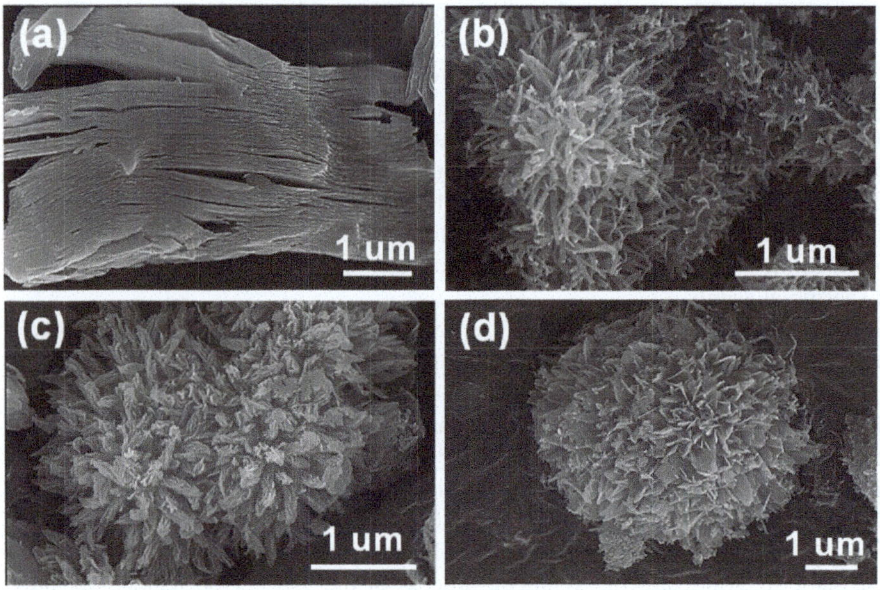

Fig. 3.37 Representative SEM images of the samples: **a** as-etched Ti_3C_2 MXene, **b** annealing at 300 °C, **c** annealing at 400 °C and **d** annealing at 500 °C. Reproduced with permission from [81]. Copyright © 2018, Elsevier

nanostructure, with a special 3D interconnected porous morphology, as illustrated in Fig. 3.37b–d, with the phase composition to be Ti_3C_2/TiO_2. In the Ti_3C_2–TiO_2 nanocomposites, the flower-like particles had an average size of 3 μm.

Meanwhile, the microstructure of the nanocomposites was highly dependent on the calcination temperature. Most interestingly, the building blocks of the Ti_3C_2–TiO_2-300, Ti_3C_2–TiO_2-400 and Ti_3C_2–TiO_2-500 were nano-whiskers, nanoribbons and nanosheets, respectively. Furthermore, as the calcination temperature was increased, the size of the petals was increased, whereas the color of the powders become lighter and lighter. In the Ti_3C_2/TiO_2 nanocomposites, the TiO_2 had a nanobelt morphology, where the belt thickness was fixed at about 200 nm, but the length of the belts was reduced, with increasing calcination temperature. TEM and HRTEM images of the Ti_3C_2–TiO_2-500 sample revealed that the nanoflowers were composed of randomly elongated nanosheets, which were of both the Ti_3C_2 MXene and TiO_2 phases, confirming the formation of the nanocomposites.

A hydrothermal method was used to produce $TiO_2/Ti_3C_2T_x$ nanocomposites, in which the TiO_2 nanosheets with (001) planes to be exposed for photocatalytic applications [18]. The formation mechanism of the nanocomposites was also hydrothermally induced partial oxidation of the Ti_3C_2 MXene. The TiO_2 nanosheets were grown out from the Ti_3C_2 nanosheets, thus suppressing the generation of interfacial defects. Due to the high reactivity of the (001) planes of the TiO_2 nanosheets, the nanocomposites exhibited a high efficiency in the photogeneration of electron–hole pairs. Also, the carrier separation was essentially ensured, because the interfacial Schottky junction with the 2D Ti_3C_2 served as a reservoir of holes, so that the nanocomposites had a strong effect of hole trapping. As a consequence, the nanocomposite displayed high photocatalytic activity for the degradation of methyl orange dye.

A simple calcination process was employed to obtain TiO_2/Ti_3C_2 nanocomposites, with the TiO_2 nanoparticles to be formed in-situ on Ti_3C_2 nanosheets as photocatalyst for CO_2 reduction [82]. With the decoration of TiO_2 nanoparticles on the Ti_3C_2 nanosheets, the nanocomposites had a rice crust-like structure. At optimal conditions, the TiO_2/Ti_3C_2 composites displayed an efficiency of 0.22 μmol/h for the photocatalytic reduction of CO_2 to CH_4, which as higher than that of the commercial TiO_2 (P25) by nearly four times. The nanocomposites contained more surface-active sites, while the high conductivity of Ti_3C_2 ensured photoelectron transfer and prevented the electron–hole recombination.

To prepare the TiO_2/Ti_3C_2 nanocomposites, dried Ti_3C_2 powder samples were heated at different temperature. The samples calcined at 350 °C, 450 °C, 550 °C and 650 °C were denoted as TT350, TT450, TT550 and TT650, respectively. The Ti_3C_2 sample without thermal treatment was named as TT0. After treatment at 350 °C, intensities of the diffraction peaks of Ti_3C_2 were increased, suggesting that the crystallinity of Ti_3C_2 was enhanced. Meanwhile, anatase TiO_2 was present, implying that partial oxidation of Ti_3C_2 started at this temperature. With increasing calcination temperature, the peak intensities of TiO_2 were increased, because of the gradual increase in the content of TiO_2. At the same time, as the calcination temperature was not higher than 550 °C, Ti_3C_2 was still present, which indicated that TiO_2/Ti_3C_2 nanocomposites were formed, whereas the content of Ti_3C_2 was gradually decreased.

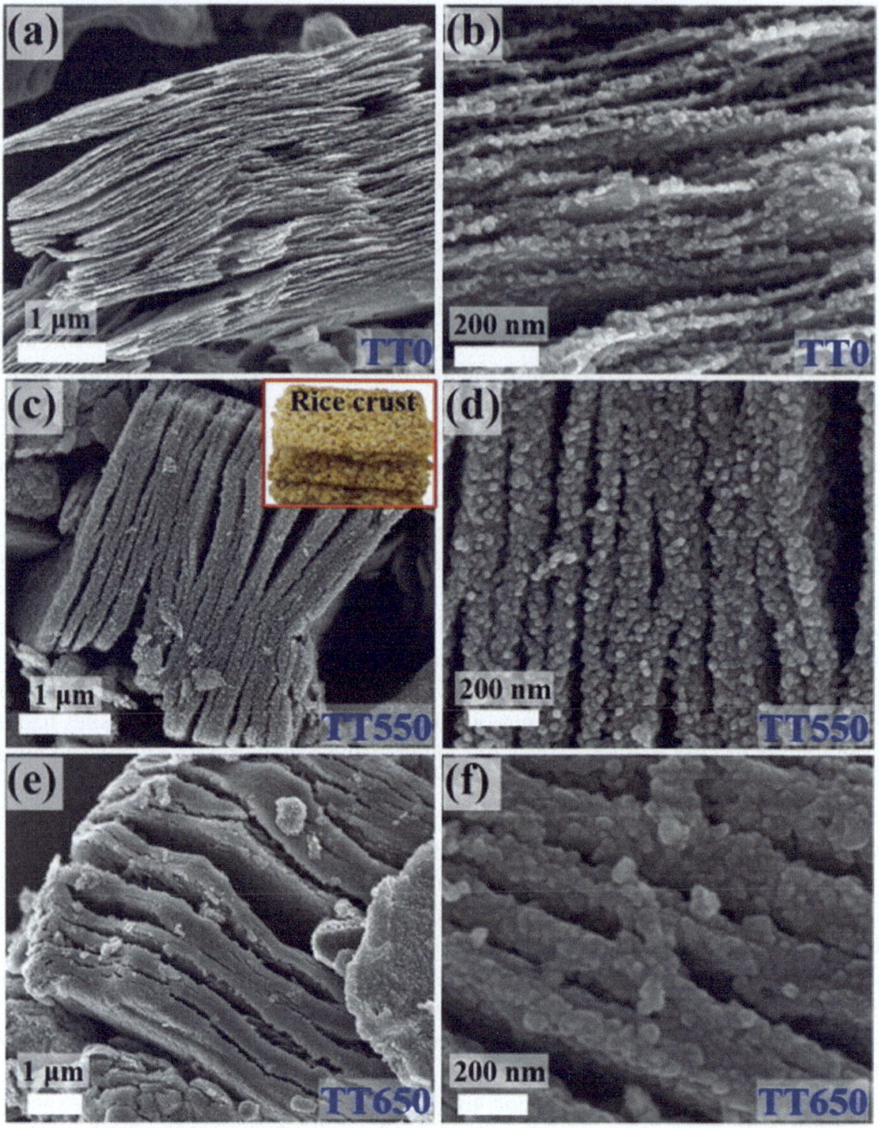

Fig. 3.38 Representative SEM images of the samples before and after calcination: **a**, **b** TT0, **c**, **d** TT550 and **e**, **f** TT650. Reproduced with permission from [82]. Copyright © 2018, Elsevier

This observation implied that the composition of the nanocomposites could be simply controlled by selecting the calcination temperature.

Figure 3.38 shows SEM images of the samples before and after calcination at different temperatures. Although the TT0 sample demonstrated typical accordion-like structure, nanoparticles were present at the edge of the nanosheets. The nanoparticles were amorphous TiO_2 produced during the delamination process of Ti_3C_2, because of the presence of strong oxidation agent. After calcination at 550 °C, surface roughness of the sample was increased, as seen in Fig. 3.38c, d, owing to the surface oxidation of Ti_3C_2 to TiO_2. The TiO_2 nanoparticles had an average size of 25 nm and were uniformly distributed on the Ti_3C_2 nanosheets.

With the in-situ growth nature, the TiO_2 nanoparticles had no problem of agglomeration. Moreover, the presence of the TiO_2 nanoparticles effectively increased the interlayer spacing of the Ti_3C_2 nanosheets. Meanwhile, the TT550 sample had a fluffy morphology, which was similar to that of the rice crust, as shown as the inset in Fig. 3.38c. In addition, since the isolated TiO_2 nanoparticles ensured a large specific surface area of the samples. As the calcination temperature was further increased to 650 °C, the size of the TiO_2 was greatly increases, so that surface area was reduced, as observed in Fig. 3.38e, f. Furthermore, this sample had numerous cracks because of the difference in surface tension of the components.

$Ti_3C_2T_x/TiO_2$ nanocomposites have also been used as catalysts for electrochemical N_2 reduction reaction (NRR) [83]. The nanocomposites were synthesized by using hydrothermal method, with TiO_2 nanoparticles to be decorated on MXene nanosheets. Specifically, in 0.1 M HCl, the $Ti_3C_2T_x/TiO_2$ nanohybrid had a NH_3 yield of 26.32 μg/h $mg_{cat.}$, together with a Faradaic efficiency (FE) of 8.42% at − 0.60 V versus reversible hydrogen electrode (RHE). The performances were much higher than those achieved with TiO_2 or $Ti_3C_2T_x$. In addition, the nanohybrid catalyst also displayed a high selectivity to NH_3 and demonstrated high electrochemical stability.

TiO_2/Ti_3C_2 hybrid mesoporous membranes were developed by assembling Ti_3C_2 nanosheets and TiO_2 nanoparticles on a macroporous support [84]. To develop the mesoporous membranes, Ti_3C_2 nanosheets were dispersed in TiO_2 hydrosols, which were gelated to form the hybrids. In this case, the Ti_3C_2 nanosheets served as floor tiles, which blocked the sols to penetrate the porous support. Therefore, the hydrosols were deposited on macroporous supports to form mesoporous membranes, which exhibited a narrow pore size distribution, a desired rejection property of dextran and a high water permeate flux. The process was simple and scalable for large scale applications.

NiFeCo-layered double hydroxide (LDH) nanoplates were incorporated with $Ti_3C_2T_x$ by using solvothermal process, while the latter was partially oxidized to anatase TiO_2 nanoparticles, thus forming $Ti_3C_2T_x/TiO_2$/NiFeCo-LDH ternary nanocomposite (TTL) [85]. The TTL exhibited much stronger OER effect than pure NiFeCo-LDH and commercial IrO_2-based catalyst. A current density of 10 mA/cm^2 was observed at the potential of 1.55 V versus reversible hydrogen electrode (vs. RHE), in 0.1 M KOH solution. Moreover, the TTL nanocomposite could be coated on the standard $BiVO_4$ film, to assemble TTL/$BiVO_4$ photoanode. The electrode delivered a photocurrent density of 2.25 mA/cm^2 at 1.23 V versus RHE, when illuminated at a power density of 100 mW/cm^2. In this case, the formation of the TiO_2

nanoparticles enhanced the light adsorption, thus generating more electron–hole pairs. At the same time, the LDH nanoplates served as hole scavengers, while the $Ti_3C_2T_x$ nanosheets ensured a smooth charge transport.

An two-step in-situ hydrothermal growth process was used to deposit TiO_2 nanosheets (NSs) are on Ti_3C_2, followed by the growth of MoS_2 NSs on the (101) faceted TiO_2 NSs, thus exposing the high-active (001) facets [86]. As a result, a 2D-2D-2D structured $Ti_3C_2/TiO_2/MoS_2$ nanocomposite was developed. It was found that the optimal content of MoS_2 was 15 wt%, so that the $Ti_3C_2/TiO_2/MoS_2$ nanocomposite exhibited a strong photocatalytic activity for H_2 evolution reaction, as compared with the Ti_3C_2/TiO_2 binary nanocomposite and pure TiO_2 NS. Firstly, the in-situ growth of the TiO_2 NSs ensured their strong contact with the Ti_3C_2, thus allowing effective charge separation. Secondly, surface heterojunction could be formed within an individual TiO_2 NS, promoting charge transfer and separation. Lastly, the MoS_2 NSs suppressed the charge carrier recombination.

V_2O_5/MXene nanocomposites, with the V_2O_5 nanoparticles having different morphologies, were prepared by using hydrothermal method, as electrodes of supercapacitors [87]. In 1.0 M $LiNO_3$ solution, specific capacitances of the electrodes based on the nanocomposites with V_2O_5 nanobelt (NBVM), nanocuboid (NCVM) and nanosphere (NSVM) were 114, 180 and 217 F/g, respectively. These values were higher that of MXene by nearly one order of magnitude. After 100 charge–discharge cycles, specific capacitance retentions of the electrodes based on MXene, NBVM, NCVM and NSVM were 70.6%, 128.8%, 149.2% and 104.3%, respectively. The increased electrochemical properties of the nanocomposites were attributed to the enlarged specific surface area, enhanced charge transfer capability and expanded interlayer spacing of the MXene.

VO_2/MXene nanocomposite with a 3D flower-like structure, was prepared by using hydrothermal process, as anodes of LIBs [88]. A specific capacity of 281 mAh/g was achieved at the current density of 100 mA/g, together with high cycling stability and promising rate capacity. To synthesize the nanocomposite, vanadyl acetylacetonate [$VO(acac)_2$] was dissolved in MXene suspension. Then, a small quantity of HCl was added to the suspension to promoted the dissolution of $VO(acac)_2$. Hydrothermal reaction was conducted at 180 °C for 24 h. The content of VO_2 in the final nanocomposites was simply controlled by the amount of the precursor in the reaction solutions.

A simple chemical synthetic route was utilized to attach ε-MnO_2 nano-whiskers on the surfaces of MXene nanosheets surfaces [89]. The resultant nanohybrids were in the forms of ε-MnO_2/Ti_2CT_x and ε-$MnO_2/Ti_3C_2T_x$, where were suitable as electrode materials of aqueous supercapacitors. The ε-MnO_2 nano-whiskers exhibited significantly enlarged surface area, thus leading to specific capacitance that was increased by about three orders of magnitude, as compared with the those based on the pristine MXenes, in the format of symmetric supercapacitors. Besides the increment in pseudocapacitance, the supercapacitors with the ε-MnO_2/MXene nanohybrids exhibited as the electrodes displayed strong cycling capability, by showing a retention of about 90% of the initial specific capacitance after recycling for 10,000 cycles. In comparison, the supercapacitive devices based on the pristineε-MnO_2 was about

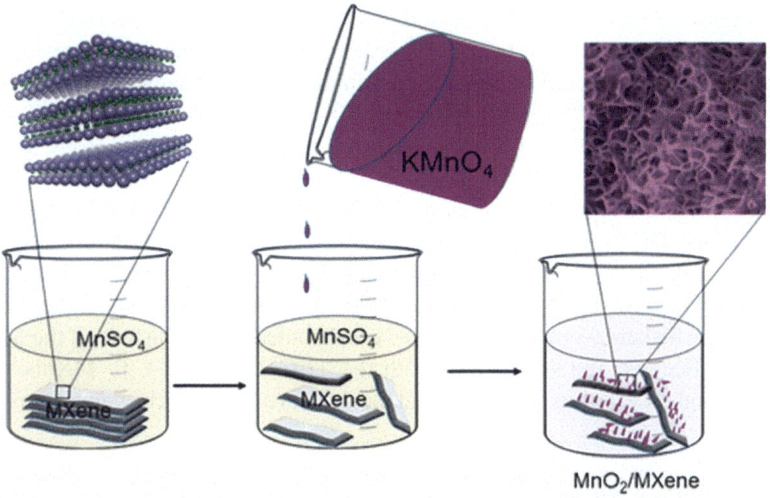

Fig. 3.39 Schematic diagram illustrating synthetic route of the MnO_2/MXene nanohybrid electrode materials. Reproduced with permission from [89]. Copyright © 2016, American Chemical Society

75%. Therefore, the incorporation of the MXene nanosheets enhanced the capacitive effect of MnO_2 and improved the conductivity of the electrode materials.

To prepare the ε-MnO_2/MXene nanohybrids, MXene powder was dispersed in the solution of $MnSO_4$ with a concentration of 0.01 M, with the aid of ultrasonication. To ensure the incorporation of $MnSO_4$ into the interlayer spacing of the MXene nanosheets, the mixture was constantly stirred at 60 °C for 0.5 h. Meanwhile, $KMnO_4$ solution with a concentration of 0.033 M was prepared ant heated to the temperature of 60 °C, which was then slowly dropped into the mixed suspension of MXene and $MnSO_4$. After additional stirring at 60 °C for 15 min, the reactant was collected by filtering and washing, followed by vacuum-drying at 80 °C for 10 h. The sample had a mass ratio of 1:1 MXene to MnO_2 to be nearly 1:1. Figure 3.39 shows a schematic diagram of the synthetic procedure of the MnO_2/MXene nanohybrids.

The formation of MnO_2 and MXenes (Ti_2CT_x and $Ti_3C_2T_x$) was evidenced by the XRD analytical results. The MnO_2 was known as ε-phase, with a polycrystalline orthorhombic crystal structure. The ε-MnO_2 phase attached on the surface of the MXene nanosheets had a relatively low level of crystallinity, as revealed by the broadened diffraction peaks. A trace of TiO_2 (anatase) was detected, which was thought to be generated during the etching reaction of HF and the parent MAX. According to the nitrogen adsorption and desorption isotherms, the Ar-treated HF-exfoliated MXenes possessed specific surface areas of 7.3, 21.1, 125.5 and 183.8 m^2/g, Ti_2CT_x_Ar, $Ti_3C_2T_x$_Ar, ε-MnO_2/Ti_2CT_x_Ar and ε-MnO_2/$Ti_3C_2T_x$_Ar, respectively. A specific surface area of the MXenes was significantly enlarged, due to the presence of the ε-MnO_2 nanophase. This implied that MnO_2 nanoparticles had been well filled into the interlayer spacing of the MXene nanoflakes during the ultrasonication process. In addition, all the samples contained mesopores, with BJH adsorption average pore widths of 38.01, 26.41, 12.79 and 9.80 nm, for Ti_2CT_x_Ar, $Ti_3C_2T_x$_Ar, ε-MnO_2/Ti_2CT_x_Ar and ε-MnO_2/$Ti_3C_2T_x$_Ar, respectively. The corresponding BJH adsorption cumulative pore volumes, with diameters in the range of between 1.7–300 nm, were 0.0280, 0.0540, 0.2650 and 0.4248 cm^3/g, for Ti_2CT_x_Ar, $Ti_3C_2T_x$_Ar, ε-MnO_2/Ti_2CT_x_Ar and ε-MnO_2/$Ti_3C_2T_x$_Ar, respectively. The porous structure became much more pronounced due to the incorporation of the MnO_2 nanosheets.

As seen in Fig. 3.40a, the Ar-calcined MXene had a typical multilayered sheet-like structures. The layer thickness of most $Ti_3C_2T_x$ nanosheets was less than 6 nm, while that of the Ti_2CT_x ones was larger than 50 nm. Representative SEM images of the ε-MnO_2/$Ti_3C_2T_x$_Ar sample are depicted in Fig. 3.40b, indicating that ε-MnO_2 nano-whiskers were homogeneously distributed on the surface of the MXene nanosheets. Because the surfaces of both the $Ti_3C_2T_x$ and Ti_2CT_x MXenes were relatively rough, it was thus suggested that the defects on the MXene nanosheets could

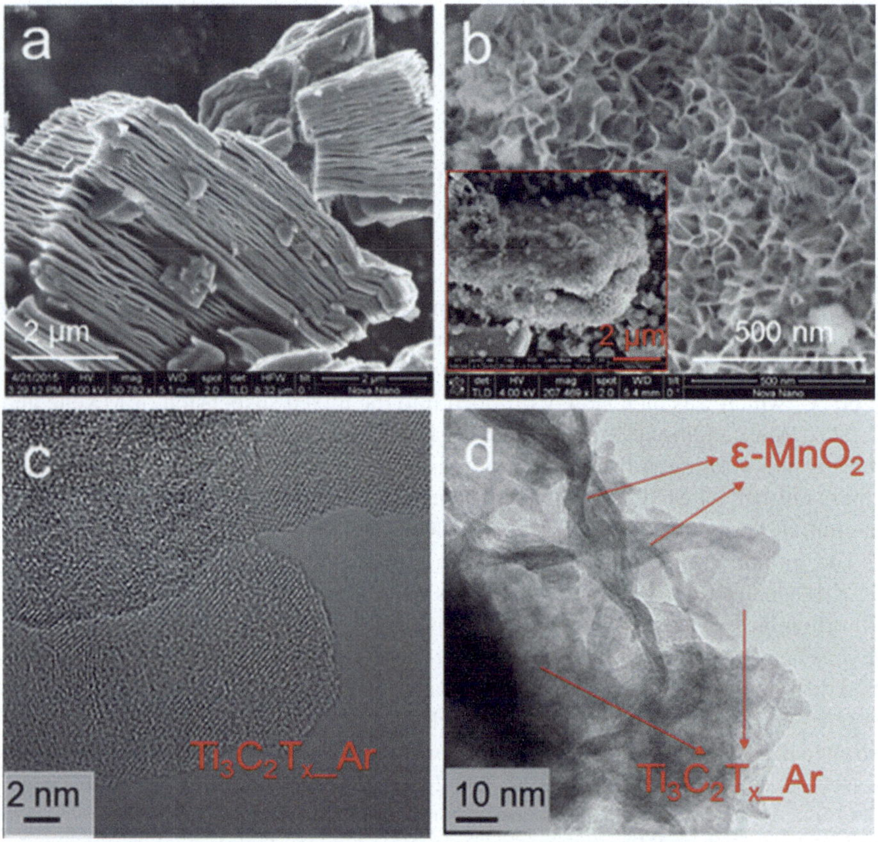

Fig. 3.40 SEM and high-magnification TEM images of the pristine MXene and MnO_2/MXene nanohybrid samples: **a**, **c** $Ti_3C_2T_x$_Ar and **b**, **d** ε-MnO_2/$Ti_3C_2T_x$_Ar. Reproduced with permission from [89]. Copyright © 2016, American Chemical Society

have facilitated the uniform deposition of the MnO_2 nanoparticles. Furthermore, the creation of the surface terminal functional groups, such as –O, –OH and –F, on the surface of the MXene nanosheets during the etching reaction, was also favorite to the attachment of the MnO_2 nanostructures. It was the porous open architecture of the MnO_2 nanoparticles on the surface of the MXene nanosheets that maintained a fast adsorption–desorption of the electrolyte ions. As a result, the reversible redox reactions were strengthened and thus the capacitive effect of the hybrid materials was enhanced when utilized as electrodes of the supercapacitors in aqueous electrolyte solutions.

The MXene nanosheet structure and the formation of the ε-MnO_2/MXene nanohybrid were also demonstrated by the TEM results, as observed in Fig. 3.40c, d. The MXene nanosheets were so thin that they were transparent to the electron beams. In the ε-MnO_2/$Ti_3C_2T_x$_Ar sample, the MnO_2 phase was present as whisker-like structures, in which the MnO_2 nanoparticles were firmly grown on the surface of the MXene nanosheets. Due to their highly porous microstructure, sufficiently large surface area and relatively short diffusion path, the ε-MnO_2 nano-whiskers would promote the diffusion and transportation of protons or alkali cations. Therefore, the ε-MnO_2/MXene nanohybrids exhibited superb electrochemical performances, when used as electrodes for energy storage applications.

MnO_2/$Ti_3C_2T_x$ nanocomposite was prepared by incorporating 1D MnO_2 nanoneedles with 2D $Ti_3C_2T_x$ nanosheets, by using a chemical deposition method, as electrodes of flexible supercapacitors [90]. Both $Ti_3C_2T_x$ powder and $MnCl_2 \cdot 4H_2O$ were mixed in isopropyl alcohol (IPA) to form suspensions. The suspensions boiled at 84 °C and then $KMnO_4$ aqueous solution was poured in quickly. After reaction for 0.5 h, the products black in color were collected after thorough washing and vacuum drying. Figure 3.41 shows a schematic diagram of the nanocomposite fabrication process.

An electrostatic self-assembly process was introduced to obtain MnO_2/$Ti_3C_2T_x$ nanocomposites for supercapacitor applications [91]. Delaminated MnO_2 nanosheets were derived from multilayered H-MnO_2 in the solution of cationic Gemini surfactant (Gem) through intercalation-delamination. The resultant MnO_2 nanosheets possessed surface with positive charges, which could be assembled with the negatively charged $Ti_3C_2T_x$ nanosheets, due to the electrostatic interactions, thus leading to the formation of MnO_2/$Ti_3C_2T_x$ nanocomposite. Electrodes based on the MnO_2/$Ti_3C_2T_x$ nanocomposite delivered a specific capacitance of 340 F/g at the current density of 1 A/g, which was higher that obtained with MXene by two times. A similar MnO_2/$Ti_3C_2T_x$ nanocomposite was reported by Wang et al., which exhibited a specific capacitance of 242 F/g at the current density of 1 A/g [92].

A simple in-situ wet-chemical process was developed to fabricate free-standing flexible MnO_x-Ti_3C_2 hybrid films, which exhibited promising electrochemical performances as electrodes of supercapacitors [93]. The hybrids consisted of few-layered $Ti_3C_2T_x$ nanosheets and MnO_x nanoparticles. The MnO_x phase was a mixture of Mn_2O_3 and MnO. The oxide nanoparticles were attached on the surfaces of the $Ti_3C_2T_x$ nanosheets via electrostatic interactions. The hybrid-based films could be served as electrodes of supercapacitors, possessing high volumetric capacitances of

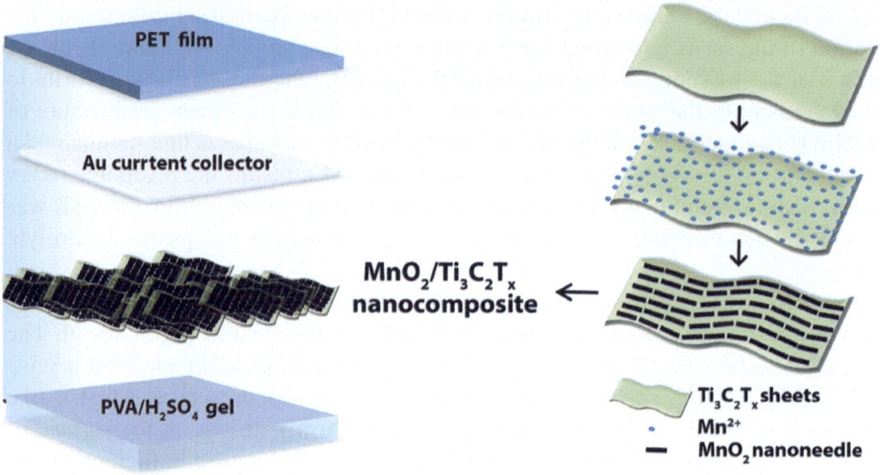

Fig. 3.41 Schematic diagram of fabrication process of the nanocomposite electrodes and flexible supercapacitors. Reproduced with permission from [90]. Copyright © 2018, Elsevier

up to 602.0 F/cm^3 at the scan rate of 2 mV/s. In addition, the MnO_x-$Ti_3C_2T_x$ electrodes were employed to construct symmetric supercapacitors, which displayed a superior energy storage capacity of 13.64 mWh/cm^3. Meanwhile, the devices exhibited a high long-term cycling stability, with a capacitance retention of about 90% after cycling for 10,000 times.

$Fe_2O_3/Ti_3C_2T_x$ nanocomposites, in the Fe_2O_3 nanospheres were encapsulated by the $Ti_3C_2T_x$ nanosheets, were fabricated by using a two-step method, as anodes of lithium-ion capacitors [94]. The $Ti_3C_2T_x$ matrix served as both a conductor to ensure rapid charge transfer and a framework to buffer the volume variation of the active component. Meanwhile, the Fe_2O_3 nanospheres possessed a hollow structure, thus resulting in a large specific capacity. Furthermore, the restacking issue of the $Ti_3C_2T_x$ nanosheets was tackled due to the presence of the Fe_2O_3 nanospheres. Together with 3D nitrogen and sulphur co-doped porous carbon (NS-DPC) derived from biomasses as the capacitor-type cathode, lithium-ion capacitors were constructed. Accordingly, the lithium-ion capacitors delivered an energy density of 216 W h/kg and a power density of 400 W/kg.

To synthesize the Fe_2O_3 hollow nanospheres, w $Na_4Fe(CN)_6$, 0.04 g hexadecyltrimethyl ammonium bromide (CTAB), $(NH_4)_2S_2O_8$ and NaH_2PO_4 were dissolved in distilled water to form a mixed solution, with the aid of magnetic mechanical stirring. The solution was subject to hydrothermal reaction at a temperature of 180 °C for 8 h. After the reaction was finished, the samples were collected and then calcined at 450 °C for 2 h in Ar. The Fe_2O_3 powder was then dispersed in aqueous solution containing cetyltrimethylammonium bromide (CTAB). At the same time, aerogel-like $Ti_3C_2T_x$ powder was dispersed in distilled water. The CTAB-grafted

Fe_2O_3 suspension was slowly dropped into the $Ti_3C_2T_x$ one, followed by ultrasonication for 0.5 h. Finally, freeze-drying was applied to the samples, leading aerogel-like $Fe_2O_3/Ti_3C_2T_x$ nanocomposites, with about 70 wt% Fe_2O_3.

MXene-based hybrid, Ti_3C_2/α-Fe_2O_3, was derived from 2D α-Fe_2O_3 nanosheets and Ti_3C_2 MXene nanolayers, by using a simple ultrasonication aided self-assembly process [95]. The Ti_3C_2/α-Fe_2O_3 hybrid possessed strong photocatalytic performance, according to the degradation efficiency of Rhodamine B (RhB) under the irradiation of visible light. This was because the 2D α-Fe_2O_3 nanosheets were homogeneously distributed on the surface of the Ti_3C_2 MXene nanolayers, thus leading to the formation of a huge number of interfaces. As a consequence, the nanohybrid had high visible light absorbing capabilities and high charge separating efficiencies. Due to the synergistic effect, the Ti_3C_2/α-Fe_2O_3 hybrid exhibited a promising recycling stability, besides the superior photocatalytic performance.

To prepare the hybrids, Ti_3C_2 powder was suspended in alcohol with strong magnetic stirring. Meanwhile, 2D α-Fe_2O_3 nanosheet powder was added into the Ti_3C_2 suspension, together with 5 wt% Nafion solution acting as disperse agent. After that, the mixed suspensions were dispersed by using a strong ultrasonication with an output power of 240 W for 0.5 h. The suspensions were then heated at 80 °C for 12 h evaporate the solvents. After complete drying, the samples were thoroughly washed with ethanol to eliminate the remaining disperse agent, followed by drying at 60 °C for overnight. The samples with α-Fe_2O_3/Ti_3C_2 mass ratios of 1:1, 1:2 and 1:3 were denoted as α-Fe_2O_3/Ti_3C_2-1, α-Fe_2O_3/Ti_3C_2-2 and α-Fe_2O_3/Ti_3C_2-3, respectively. The phase compositions of the hybrids and their components were confirmed by XRD analysis, as depicted in Fig. 3.42.

Figure 3.43 shows SEM images of α-Fe_2O_3, Ti_3C_2 and α-Fe_2O_3/Ti_3C_2-2 hybrid. The α-Fe_2O_3 powder consisted of 2D nanosheets with well-developed morphology and a relatively uniform size distribution in the range of 100–150 nm, as observed in Fig. 3.43a. As seen in Fig. 3.43b, the pristine Ti_3C_2 nanosheets possessed the typical multilayered structure, with layer thicknesses of 5–40 nm. At the same time, the thicknesses of the individual layers were in the range of 5–10 nm, suggesting that they were stacked with few single layers in the Ti_3C_2 MXene powder. In the hybrid, the 2D a-Fe_2O_3 nanosheets were well attached on surface of the Ti_3C_2 MXene nanolayers, as demonstrated in Fig. 3.43c. In addition, the α-Fe_2O_3 nanosheets were also filled into the interlayer spacing of the Ti_3C_2 nanolayers. Figure 3.43d indicated presence of Ti, C, Fe and O elements in the α-Fe_2O_3/Ti_3C_2-2 hybrid, implying the feasibility of the synthetic process in that study.

More recently, $Ti_3C_2T_x$ MXene was combined with Fe_2O_3/CC (carbon cloth), as flexible negative electrodes, for the development of flexible electrodes with sufficiently high areal capacitances, in order to meet the requirements of flexible asymmetric supercapacitors (ASCs) that will be used for power supply of portable and wearable electronic devices [96]. The electrodes were prepared by using a dipping coating process, in which the carbon cloths grown with Fe_2O_3 nanorod arrays were soaked into suspensions of few-layered MXene nanosheets. The process was repeated for several times to ensure that the Fe_2O_3/CC fibers were entirely covered with the MXene nanosheets. The nanocomposite electrodes exhibited high electrical

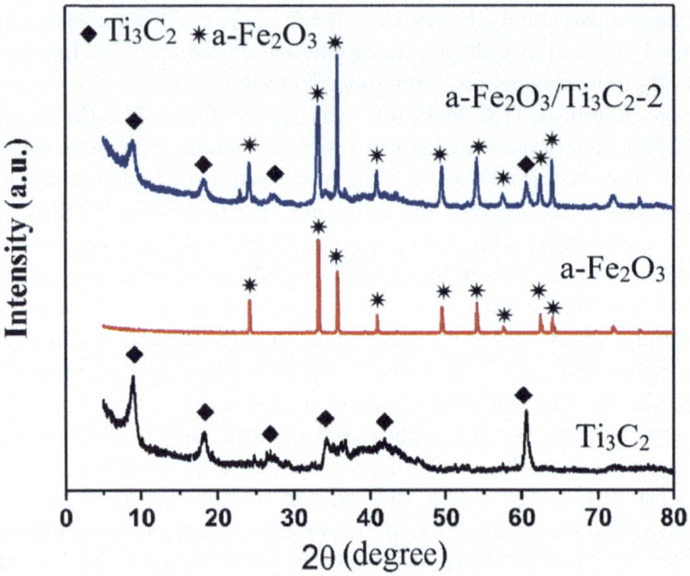

Fig. 3.42 XRD patterns of the nanohybrid and the components. Reproduced with permission from [95]. Copyright © 2018, Elsevier

conductivity, high areal capacitance and strong mechanical flexibility. Specifically, MXene@Fe_2O_3/CC electrodes displayed an areal capacitance of 725 mF/cm^2 at 1 mA/cm^2. In this case, both the MXene nanosheets and the Fe_2O_3 nanorods had contributions to the capacitive effects of the electrodes. In addition, the MXene layers offered unimpeded ion/electron transportation channels. More importantly, ASC devices were assembled with the MXene@Fe_2O_3 electrodes and MnO_2 electrodes, which exhibited an energy density of 1.61 mW h/cm^3, corresponding to a power density of 22.55 mW/cm^3.

Liu et al. synthesized $Ti_3C_2T_x$/Fe_3O_4 nanohybrids by using a simple hydrothermal process for the applications as microwave absorbers [97, 98]. The nanohybrids were present in the form of TiO_2/$Ti_3C_2T_x$/Fe_3O_4, with different ratios of [TiO_2/$Ti_3C_2T_x$]/[Fe_3O_4]. SEM and TEM characterization results indicated that the Fe_3O_4 nanoparticles with an average size of about 5 nm along with TiO_2 nanoparticles were distributed on the surfaces of the MXene nanosheets. The ratio of [TiO_2/$Ti_3C_2T_x$]/[Fe_3O_4] could be tailored to optimize the microwave absorption performance of the nanohybrids, in order to achieve high reflection loss (RL) and small thickness of the absorbers. Specifically, a maximum reflectivity of −57.3 dB was obtained at 10.1 GHz, with a band range below −10 dB over 9.1–11.1 GHz, at a thickness of 1.9 mm.

In the experiment, $FeCl_3 \cdot 6H_2O$ (99%) and $NaHCO_3$ (99.8%) were used to form the reaction solution, which was mechanically stirred for 0.5 h [98]. Meanwhile, aqueous solution of vitamin C was dropped into the Fe^{3+} solution, with a molar

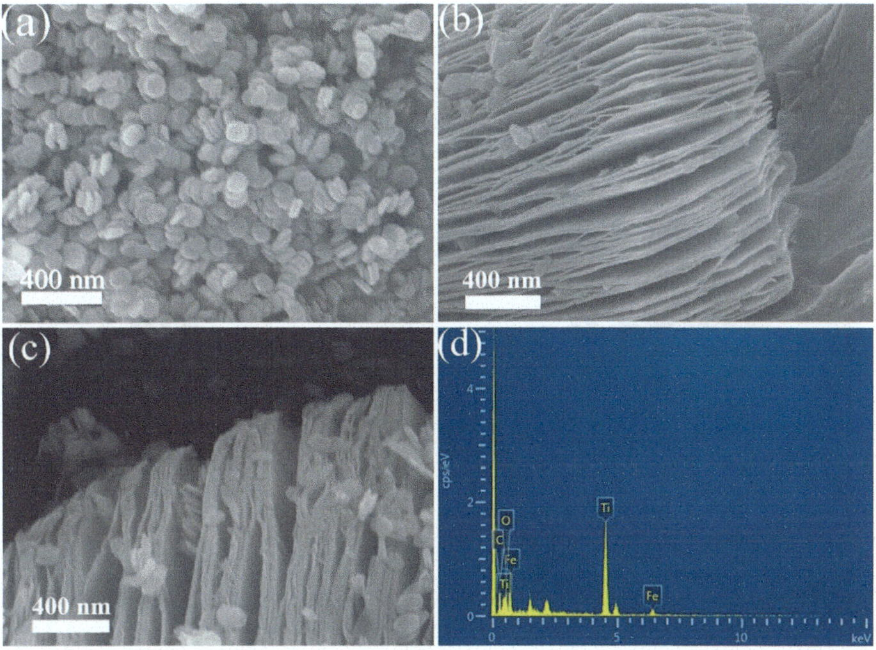

Fig. 3.43 SEM images of the samples: **a** 2D α-Fe_2O_3 nanosheets, **b** Ti_3C_2 MXene nanolayers, **c** α-Fe_2O_3/Ti_3C_2-2 hybrid with α-Fe_2O_3:Ti_3C_2 mass ratio of 1:2 and **d** EDS of the α-Fe_2O_3/Ti_3C_2-2 hybrid. Reproduced with permission from [95]. Copyright © 2018, Elsevier

ratio to Fe^{3+} to be 1:6, under the mechanical stirring for 20 min. Then, MXene was dispersed deionized water to form MXene suspension, which was mixed with the Fe^{3+} solution. The mixed suspension was subject to hydrothermal reaction at 150 °C for 5 h. The products were collected through thorough washing with deionized water, followed by vacuum drying at 60 °C for one whole day. The composition of the hybrid was controlled by adjusting the mass ratio of MXene/Fe_3O_4.

An ultrasonic treatment process was reported to prepare Fe_3O_4@Ti_3C_2 MXene hybrid, with Ti_3C_2 powder and Fe_3O_4 nanoparticles as the precursors, in order to develop electrode materials for LIBs applications [99]. In this case, the MXene nanosheets were employed as a host to include the Fe_3O_4 nanoparticles, by using utilizing the special laminar architecture of the MXene, with lower electronic conduction and Li^+ ion diffusion barriers. At a weight ratio of 5:2, the Fe_3O_4 nanoparticles would be uniformly distributed on the surface of the Ti_3C_2 MXene nanosheets. The Fe_3O_4@Ti_3C_2-2:5 hybrid anode displayed reversible capacities of 747.4 mA h/g and 278.3 mA h/g, at 1C after 1000 cycles and 5C after 800 cycles, respectively. In addition, the hybrid-based anode exhibited a high volumetric capacity of 2038 mA h/cm^3 at 1C, which was attributed to the relatively high density of the hybrid.

To synthesize Fe_3O_4 nanoparticles, ethylene glycol solution was prepared by dissolving 1.35 g $FeCl_3 \cdot 6H_2O$, 0.43 g NaOH and 10 g polyethylene glycol (PEG

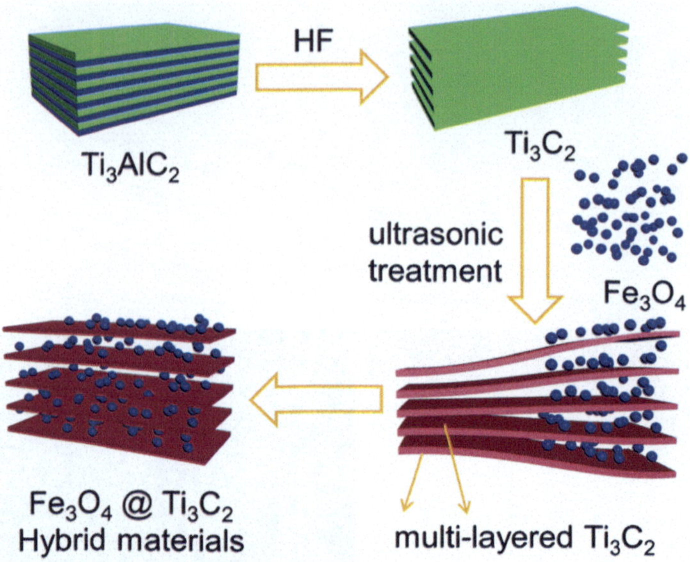

Fig. 3.44 Schematic showing the preparation process of the Fe_3O_4@Ti_3C_2 hybrids. Reproduced with permission from [99]. Copyright © 2018, Royal Society of Chemistry

400), with the aid of magnetic stirring for 0.5 h. The solution was hydrothermally treated at 200 °C for 8 h. After reaction, the product was thoroughly washed with ethanol and distilled water, followed by vacuum drying at 80 °C for 6 h. The as-synthesized Fe_3O_4 nanoparticles and pristine Ti_3C_2 powder were suspended distilled water with the aid of ultrasonication for 0.5 h. Then, the two suspensions were mixed and ultrasonically treated for 6 h. After filtration and washing, Fe_3O_4@Ti_3C_2 hybrid samples were obtained. The mass ratios of Fe_3O_4/Ti_3C_2 were 1:5, 2:5, and 1:1, leading to Fe_3O_4@Ti_3C_2 hybrid samples of Fe_3O_4@Ti_3C_2-1:5, Fe_3O_4@Ti_3C_2-2:5, and Fe_3O_4@Ti_3C_2-1:1, respectively. Figure 3.44 shows a schematic diagram of the synthesis process of the Fe_3O_4/Ti_3C_2 hybrid samples.

Phase compositions and microstructures of the individual components and hybrids have been confirmed by XRD and SEM analysis results.

Figure 3.45 shows TEM images and corresponding analysis results of both the pristine Ti_3C_2 MXene and Fe_3O_4@Ti_3C_2-2:5 hybrid. As revealed in Fig. 3.45a, the Ti_3C_2 nanosheets were of typical multilayered microstructure. The individual layer of the Ti_3C_2 MXene had a thickness of about 1.2 nm. TiO_2 phase was present, according to the *d*-spacing value of 0.236 nm, corresponding to the lattice plane of (001). As discussed earlier, the formation of TiO_2 was ascribed to the oxidation of the MXene phase during the etching reaction process. The Fe_3O_4 nanoparticles were intercalated into the interspacing of the multilayered Ti_3C_2 nanosheets, as seen in Fig. 3.45c. In addition, Fe_3O_4 nanoparticles were also firmed attached on the surfaces of the Ti_3C_2 MXene nanosheets. Therefore, the incorporation of the Fe_3O_4 nanoparticles would have effectively prevented the MXene nanosheets from restacking. The thickness of

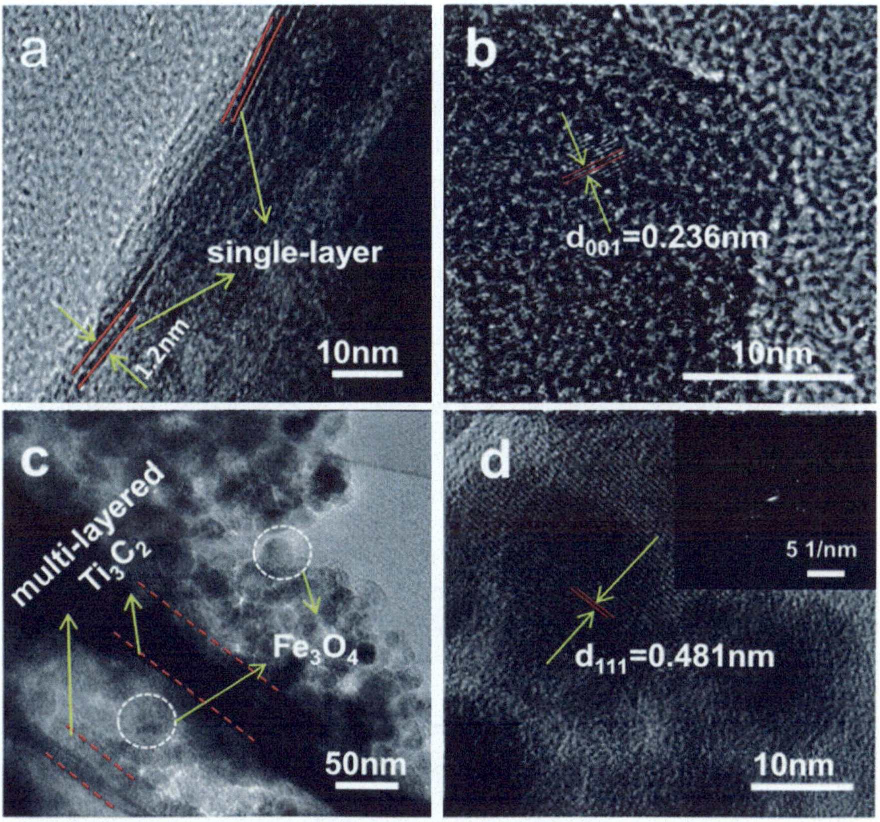

Fig. 3.45 **a** TEM image and **b** high-resolution TEM image of the Ti_3C_2 MXene nanosheets. **c** TEM image and **d** high-resolution TEM image of the Fe_3O_4@Ti_3C_2-2:5 hybrid, with the inset in panel **d** corresponding to the SAED pattern of the cubic spinel Fe_3O_4. Reproduced with permission from [99]. Copyright © 2018, Royal Society of Chemistry

the Ti_3C_2 MXene nanosheets and the size of the Fe_3O_4 nanoparticles were <50 nm and 25 nm, respectively. The phase composition of the Fe_3O_4 nanoparticle was also confirmed by using high-resolution TEM image, as shown in Fig. 3.45d, where the lattice fringe with an interplanar spacing of 0.481 nm corresponded to the (111) plane of cubic spinel Fe_3O_4.

Fe_3O_4/$Ti_3C_2T_x$ nanohybrids have been obtained, through in-situ nucleation and growth of magnetic Fe_3O_4 nanoparticles on the surface and interlayer-space of $Ti_3C_2T_x$ MXene nanosheets, by using a facial solvothermal reaction process [100]. The nanohybrids ensured impedance matching and exhibited multiple loss mechanisms, thus resulting in excellent microwave absorption capabilities. Importantly, during the solvothermal reaction, a reductive environment was generated, so that the $Ti_3C_2T_x$ MXene was well prevented from being oxidized at the temperature of as high as 200 °C. In this case, the $Ti_3C_2T_x$ MXene was more stable when using the solvothermal process than using the normal hydrothermal routes. Specifically, the hybrid with 25 wt% Fe_3O_4 nanoparticles showed optimal electromagnetic wave absorbing performances, with a minimum RL of −57.2 dB at 15.7 GHz, as well as bandwidth of 1.4 GHz for 10 dB absorption, at a thickness of 0.42 cm. The superior microwave absorbing capabilities of the $Fe_3O_4/Ti_3C_2T_x$ hybrid were resulted from the improved impedance matching, enriched interface polarizations and Debye relaxation, which were closely related to the special laminar hetero-architecture of the $Fe_3O_4/Ti_3C_2T_x$ nanohybrids.

To prepare the $Fe_3O_4/Ti_3C_2T_x$ nanohybrids, 1.5 g PEG was used as the dispersant and dissolved in an aqueous solution made of 40 ml EG and 40 ml PG. After that, 0.68 g $FeCl_3 \cdot 6H_2O$, 4.0 g NaAc and $Ti_3C_2T_x$ MXene powder with desired quantities were added into the solution with strong magnetic stirring. The suspensions were solvothermally treated at 200 °C for 8 h, leading to $Fe_3O_4/Ti_3C_2T_x$ nanohybrid powders, which were thoroughly washed with deionized water and ethanol, followed by drying at room temperature for 12 h. The content of Fe_3O_4 was varied in the range of 15–30 wt%, in order to optimize the microwave absorption behaviors of the nanohybrids.

XRD patterns of Ti_3AlC_2, $Ti_3C_2T_x$, Fe_3O_4 and $Fe_3O_4/Ti_3C_2T_x$ hybrid with 30 wt% Fe_3O_4 are depicted in Fig. 3.46a. The original Ti_3AlC_2 sample was phase pure. After etching with HCl + LiF solution, the (002) peak at $2\theta = 9.6°$ was downshifted to 6.6° and was broadened at the same time, suggesting the enlarged *d*-spacing. In the

Fig. 3.46 **a** XRD patterns of Ti_3AlC_2, $Ti_3C_2T_x$, Fe_3O_4 and $Fe_3O_4@Ti_3C_2T_x$ hybrid (with 30 wt% Fe_3O_4). SEM images of the samples: **b** $Ti_3C_2T_x$ MXene nanosheets, **c** pure Fe_3O_4 nanoparticles, **d** $Fe_3O_4/Ti_3C_2T_x$ hybrid (with 30 wt% Fe_3O_4). Low magnification SEM image **e** and EDX elemental mapping profiles **f** of Ti, C, O and Fe in the $Fe_3O_4/Ti_3C_2T_x$ hybrid. Reproduced with permission from [100]. Copyright © 2019, Elsevier

$Fe_3O_4@Ti_3C_2T_x$ nanohybrid, cubic spinel Fe_3O_4 was present as expected. Meanwhile, the (002) peak of the MXene was further shifted to $2\theta = 5.8°$, implying that the MXene nanosheets have been intercalated by the Fe_3O_4 nanoparticles. In addition, neither anatase nor rutile of TiO_2 phase was detected in the XRD patterns, which was attributed to the reductive effect of the ethylene glycol and the 1,2-propylene glycol. The pure Fe_3O_4 nanoparticles and the ones in the $Fe_3O_4/Ti_3C_2T_x$ hybrid had average particle sizes of about 12 nm and 10 nm, respectively. In other words, the grain growth of the Fe_3O_4 nanoparticles was suppressed when incorporating with the MXene nanosheets.

The etching resulted in $Ti_3C_2T_x$ MXene with the unique according microstructure, as seen in Fig. 3.46b. The pure Fe_3O_4 nanoparticles exhibited spontaneous agglomeration, thus forming well-shaped spheres with diameters in the range of 150–200 nm, as demonstrated in Fig. 3.46c. These values were much larger than the crystal size estimated with the Scherrer equation, due to the agglomeration of the Fe_3O_4 nanoparticles, indicating that they were strong magnetic in nature and small in size. The spontaneous aggregation of the Fe_3O_4 nanoparticles was ascribed to the interactions of the magnetic dipoles and the high surface energies. Representative SEM image of the $Fe_3O_4@Ti_3C_2T_x$ hybrid with 30 wt% Fe_3O_4 is shown in Fig. 3.46d. The multilayers structure of the $Ti_3C_2T_x$ MXene was nearly intact, while the Fe_3O_4 nanoparticles were uniformly distributed on the surface and filled into the interlayer spaces of the $Ti_3C_2T_x$ nanosheets. The incorporation with $Ti_3C_2T_x$ MXene nanolayers facilitated the heterogeneous nucleation and growth of the Fe_3O_4 nanoparticles, which were encapsulated by the $Ti_3C_2T_x$ nanosheets. As a result, both the agglomeration and grain growth of the Fe_3O_4 particles were retarded.

Figure 3.47 shows representative HRTEM images of the $Fe_3O_4/Ti_3C_2T_x$ hybrids. Similarly, the images revealed that both the surface and edges of the MXene nanosheet

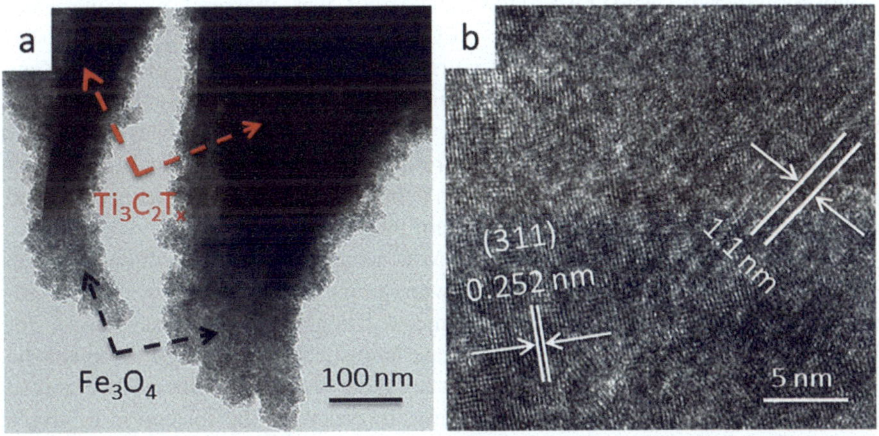

Fig. 3.47 **a** Representative TEM image of the Fe_3O_4/MXene hybrid. **b** HRTEM image at the edge of the hybrid showing multilayered structure of the MXene and the Fe_3O_4 nanoparticles. Reproduced with permission from [100]. Copyright © 2019, Elsevier.

were coated with the Fe_3O_4 nanoparticles, as observed in Fig. 3.47a. The interplanar spacing of 0.252 nm in the lattice fringe corresponded to the (311) plane of the cubic spinel Fe_3O_4, whereas the interlayer spacing along the *c*-axis was about 1.10 nm, corresponding to the (0002) plane of the $Ti_3C_2T_x$ phase, as illustrated in Fig. 3.47b. According to SEM images of the $Fe_3O_4/Ti_3C_2T_x$ hybrid samples with different contents of Fe_3O_4, the agglomeration of the oxide nanoparticles was more and more serious as the content was increased. The agglomerates of the Fe_3O_4 nanoparticles resulted in cluster-like structures and the shell thickness of Fe_3O_4 was also increases, which would pose negative effect on microwave absorption properties of the hybrids. Therefore, uniform dispersion of the Fe_3O_4 nanoparticles in the $Fe_3O_4/Ti_3C_2T_x$ hybrids could be an important factor to influence microwave absorbing behaviors of hybrid materials.

A similar in-situ process was reported to synthesize Fe_3O_4 nanoparticles that were attached on the surface of Ti_3C_2 MXene nanolayers, in order to develop anode materials with enhanced electrochemical performances [101]. The hybrids were further modified by depositing a thin layer of carbon the surfaces. The nanohybrid-based electrodes displayed outstanding electrochemical performances for Li^+ ion storages, due to the synergistic effect of the high capacitive efficiency of Fe_3O_4 and the high electrical conductivity of the MXene nanosheets. Specifically, optimal performances were achieved in the sample Fe_3O_4/Ti_3C_2-2.5, with a weight ratio of 1:1. The reversible capacity was 342.9 mAh/g at 1C, which was higher than the theoretical value of pristine Ti_3C_2 monolayers, with promising rate stability. TEM results indicated that the hybrid particles were uniformly coated by the thin carbon layer, which had a thickness of as small as 1 nm. The Fe_3O_4/Ti_3C_2-2.5 hybrids coated with a thin layer of carbon exhibited specific capacities of 382.9 mAh/g and 236.7 mAh/g at 1C and 5C, respectively.

A nanocomposite film with multiple composition of poly(vinyl alcohol)/poly(acrylic acid)/Fe_3O_4/MXene@AgNP was prepared by using electrospinning process, combined with post-heat treatment and self-reducing step [102]. The self-reducing effect was attributed to the presence of the MXene nanoflakes. The MXene nanosheets were uniformly dispersed in the nanocomposite fibers. Therefore, the restacking issue of the MXene nanolayers was effectively prevented, while the Ag nanoparticles were highly stable in the reducing environment. The nanocomposite fiber-based films exhibited strong catalytic capability and cycling stability 4-nitrophenol and 2-nitroaniline.

The intercalation of the multilayered MXene nanosheets was conducted with DMSO. Colloidal suspensions of MXene were prepared with the aid of ultrasonication in order for the electrospinning experiments. Then, the solutions/suspensions of PVA, PAA, Fe_3O_4 nanoparticles and MXene nanosheets were mixed to obtain the electrospinning precursor solutions. The as-spun nanocomposite fibers were soaked in a $AgNO_3$ solution with moderate stirring, in order for the formation of [Ag^+-DMSO] monomers, through the electrons transferring to generate oxygen lone pair electrons, i.e., [$\ddot{O}S$–$(CH_3)_2$], which resulted in the creation of Ag^+-[DMSO] [71]. As a result, MXene–Ag DMSO dimers were developed, due to the charge transition between the Ag–MXene items. The MXene–Ag dimers were then linked to the –

OH groups, whereas Ag nanosized clusters were obtained. The continuous nucleation and growth of the nanosized clusters on the surface of the Ti_3C_2 nanolayers led to the presence of AgNPs.

Because of their special 2D laminar structure, the MXene nanosheets were seemly inserted into the fibers, so that they were highly dispersed in the nanocomposites, thus avoiding the restacking problem of the MXene nanosheets. Moreover, because no functional groups were involved to modify the MXene nanoflakes, their intrinsic behaviors were well retained. Therefore, in this respect, the formation of the AgNP@MXene hybrids was an in-situ reaction process. It was possible because the MXene nanolayers had sufficiently large specific surface areas and there were numerous reduction and nucleation sites on their surfaces for the formation of the AgNPs. Furthermore, the −OH functional groups attracted the Ag^+ ions. As a result, AgNPs were eventually grown on the surface of the nanocomposite fibers.

Figure 3.48 shows SEM images of the nanocomposite fibers with different compositions and TEM image of the Fe_3O_4 nanoparticles embedded in the fibers. The PVA/PAA fibers were uniform in diameter, with an average thickness of 0.5 μm, as seen in Fig. 3.48a. The PVA/PAA/Fe_3O_4/MXene fibers were slightly different from the PVA/PAA ones, due to the incorporation of the Fe_3O_4 nanoparticles and the MXene nanosheets, although they were prepared with the same processing parameters, as observed in Fig. 3.48b. In addition, the fibers encountered localized fusions

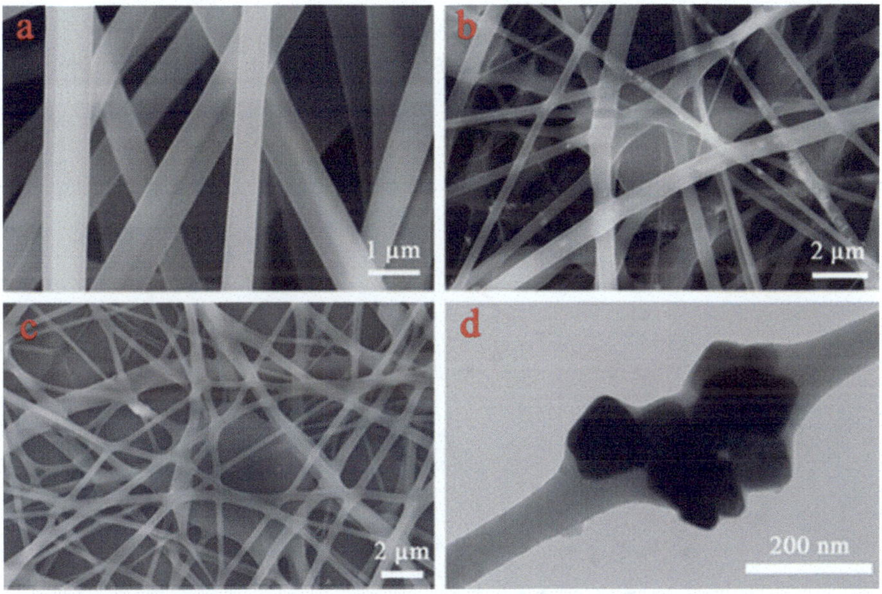

Fig. 3.48 Representative SEM and TEM images of the samples: **a** PVA/PAA nanofibers, **b** PVA/PAA/Fe_3O_4/MXene nanofibers, **c** PVA/PAA/Fe_3O_4/MXene nanofibers after post-heat treatment for 5 h and **d** TEM image of the cube Fe_3O_4 NPs embedded in the nanofibers. Reproduced with permission from [102]. Copyright © 2019, American Chemical Society

at the junctions, which was possibly ascribed to the appearance of some MXene nanosheets with relatively large lateral sizes. As a consequence, the thickness of the fibers was much less uniform. The PVA/PAA/Fe_3O_4/MXene fibers post-heat treated at 120 °C for 5 h were fused at the junctions, as illustrated in Fig. 3.48c. The Fe_3O_4 nanoparticles had a cubic morphology, with an average dimension of 100 nm, which were encapsulated inside the fibers, as demonstrated in Fig. 3.48d. The made the nanocomposite fibers were magnetic in nature and had recycle capability, because of the introduction of the Fe_3O_4 nanoparticles.

EDX results confirmed that the elements of C, O, Fe, Ti, and Ag were homogeneously distributed in the PVA/PAA/Fe_3O_4/MXene@AgNP nanocomposite fibers. The Ag nanoparticles had average sizes of 12 nm, 17 nm and 22 nm, after reaction for 10 min, 20 min and 60 min, respectively. Both the Fe_3O_4 and the Ag nanoparticles exhibited strong XRD peaks, suggesting their high degree of crystallinity. Additionally, the PVA/PAA/Fe_3O_4/MXene@AgNP nanocomposite fibers displayed higher thermal stability as compared with pure PVA and PAA.

A nanohybrid of $Ti_3C_2T_x$/Fe_3O_4/Fe_2O_3, which had magnetic properties and sandwich-like structures, was prepared in order to develop sorbents with strong sorption selectivity and capacity to phosphates [19]. Figure 3.49 shows schematic diagram of the synthetic process of the nanocomposites and their utilization for sorption of

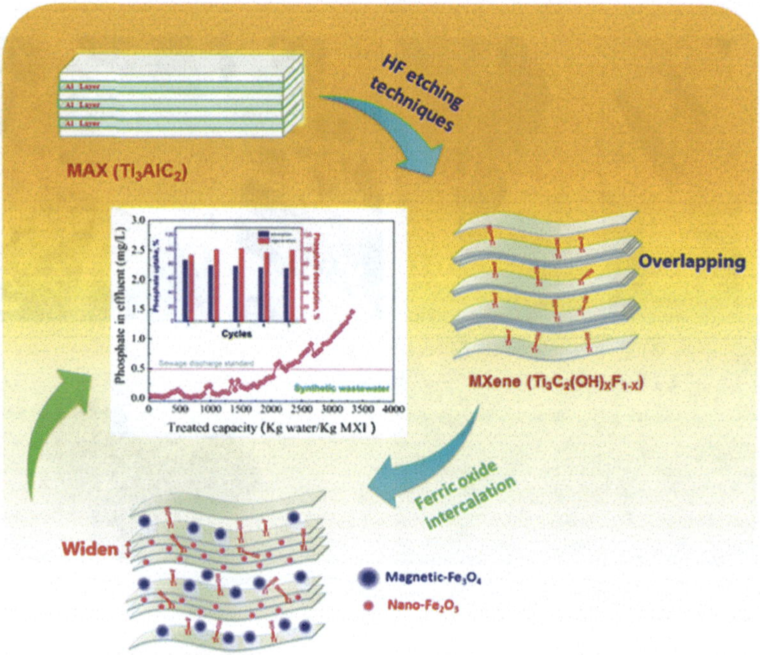

Fig. 3.49 Schematic diagram showing the preparation steps and the application of removing phosphates. Reproduced with permission from [19]. Copyright © 2016, Royal Society of Chemistry

phosphates. Firstly, 1 M $FeSO_4$ and 2 M $FeCl_3$ solutions were mixed, into which $Ti_3C_2T_x$ powder was added. The mixture was heated at 80 °C for 4 h to ensure the ion exchange reaction, where the Ti–H terminal groups were substituted with Ti–Fe ones. Meanwhile, in the Fe^{2+} and Fe^{3+} ions were intercalated into the interlayers of the MXene nanosheets. After that, the mixture was poured into 6 M NaOH solution that was boiled, so that iron hydroxides were formed. After boiling at 100 °C for 5 h, nanosized ferrite phase was derived from the mixed iron hydroxides.

Multi-layered MXene phase, with a composition of $Ti_3C_2OH_{0.8}F_{1.2}$ was derived from the MAX (Ti_3AlC_2) compound, as demonstrated in Fig. 3.50a. TEM images indicated that the MXene nanosheets contained about fifteen layers, with a thickness of about 20, as illustrated in Fig. 3.50b, c. As revealed by the selected area electron diffraction (SAED) pattern, the hexagonally packing structure had been well retained after the etching reaction with HF solution.

Several iron oxide nanoparticles with an average diameter of about 20 nm were present on the surfaces of the MXene nanosheets, as depicted in Fig. 3.50a, b. The lattice constant c of the MXene phase was very close to 2.0 nm, which had been increased, due to the intercalation of the iron oxide nanoparticles. The lattice constant value was increased to about 3.0 nm, as the mass ratio of MXene:Fe_3O_4 was 2:1. It was also found that pure magnetite Fe_3O_4 had an average particle size of about 110 nm. However, after incorporation with MXene phase, the iron oxide phase contained by magnetite Fe_3O_4 and maghemite Fe_2O_3, with average sizes of

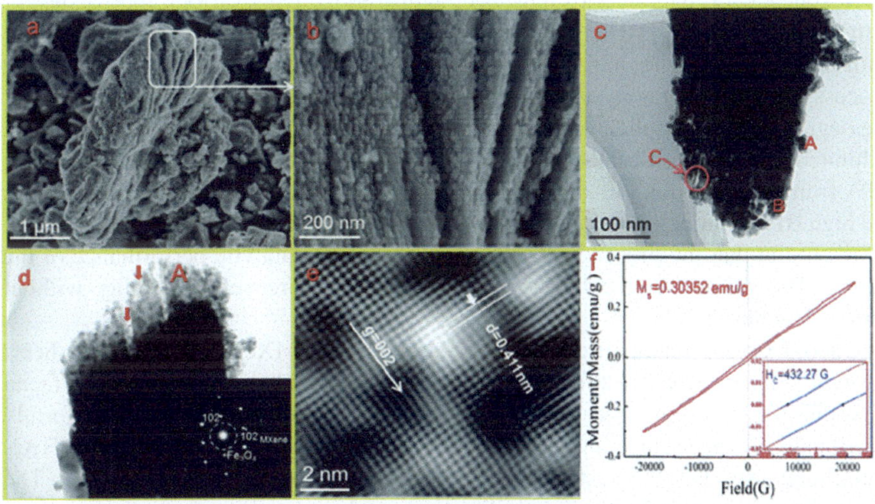

Fig. 3.50 Microstructure and magnetic properties of the MXI samples: **a** representative SEM image, **b** high magnification of the area in **a**, **c** TEM image of an individual MXI nanocomposite with iron oxide nanoparticles (A), anatase TiO_2 particles (B) and nanoscale intercalated iron oxides (C), **d** TEM image of MXI with the inset to be the SAED of the dot A, **e** fast Fourier transform filtered TEM image of the nano-sized Fe_3O_4 phase and **f** magnetic hysteresis loop of the MXI nanocomposite. Reproduced with permission from [19]. Copyright © 2016, Royal Society of Chemistry

13.0 nm and 7.0 nm, respectively. The molar ratio of Fe_3O_4:Fe_2O_3 was estimated to be 4.45–3.95, according to the Rietveld refinement analytic results.

HRTEM images confirmed that the particle size of the magnetite Fe_3O_4 (location A) on the surface of the MXene nanosheet was about 18 nm, as observed in Fig. 3.50d. The plane spacing of d_{001} was about 0.411 nm, as revealed in Fig. 3.50e. Additionally, TiO_2 nanoparticles in the form of anatase were present on the surface of the MXene nanosheets, with a cuboid morphology and an average size of 12.5 nm (location B). The regularly oriented nanolayers had been damaged (location C), while the average plane spacing (d_{0002}) was varied to 1.477 nm. The iron oxide nanoparticles had an average size of about 5 nm.

In summary, the MXI nanohybrids consist of MXene and magnetite Fe_3O_4 as the major phases, with a small quantity of maghemite Fe_2O_3. The magnetite Fe_3O_4 particles with relatively large sizes were either distributed on the surface of the MXene nanosheets or filled in between the multiple-layer spacing. In comparison, the maghemite Fe_2O_3 particles with smaller sizes were mainly intercalated into the MXene. According to the hysteresis loop shown in Fig. 3.50f, the MXI nanohybrid had a saturation magnetization of 0.30 emu/g and a coercive force of 432 G. When pure Fe_3O_4 was used for phosphate sorption, the sample could be quickly picked out by using a commercial magnet in about just 1 s. Comparatively, when using MXene to absorb the phosphate, it took as long as 5 h for the mixture to settle down. As for the MXI nanohybrid, an entire separation could be achieved in 2 min, which is acceptable for practical applications.

MXene/Co_3O_4 nanohybrids have been prepared by using various approaches for different applications [103–106]. For instance, Co_3O_4/MXene (CMs) nanohybrids, with sandwich structure, have been synthesized by using a simple thermal annealing process, as the novel catalysts to boost the activity of peroxymonosulfate (PMS) for the degradation of Bisphenol A (BPA) [103]. The hybrid with optimal composition exhibited a pretty high catalytic property, with an efficiency of 95% in eliminating BPA from the solution with a concentration of 20 mg/L in just 7 min. Importantly, the high BPA elimination efficiency could be retained over a wide range of pH value. It was found that the presence of HCO_3^- reduced the catalytic performance of the hybrid. The hybrid-based catalysts demonstrated a high recycling stability, without serious leaching of Co.

Cuboid Co_3O_4 nanoparticles were used to modify MXene (Ti_3C_2) nanosheets through self-assembly, resulting in MXene/Co_3O_4 nanohybrids, by using a facial solvothermal reaction process [104]. The Co_3O_4 nanoparticles were uniformly distributed on the surface and embedded inside the interlayer spaces of the Ti_3C_2 nanosheets, so that the restacking of the Ti_3C_2 nanosheets was avoided. The Mxene/Co_3O_4 nanohybrids were utilized as catalysts to degrade methylene blue (MB) and Rhodamine B (RhB). Figure 3.51 shows schematic diagram of the fabrication process and catalytic study of the MXene/Co_3O_4 nanohybrids. To prepare the nanohybrids, MXene powder was dispersed in ultrapure water with aid of ultrasonication for 30 min. Then, $Co(Ac)_2$ was dissolved into the suspension at a concentration of 0.2 M, followed by addition of ethanol and continuous stirring for 2 h. The system was solvothermally reacted at 120 °C for 8 h to yield the Mxene/Co_3O_4 nanohybrids.

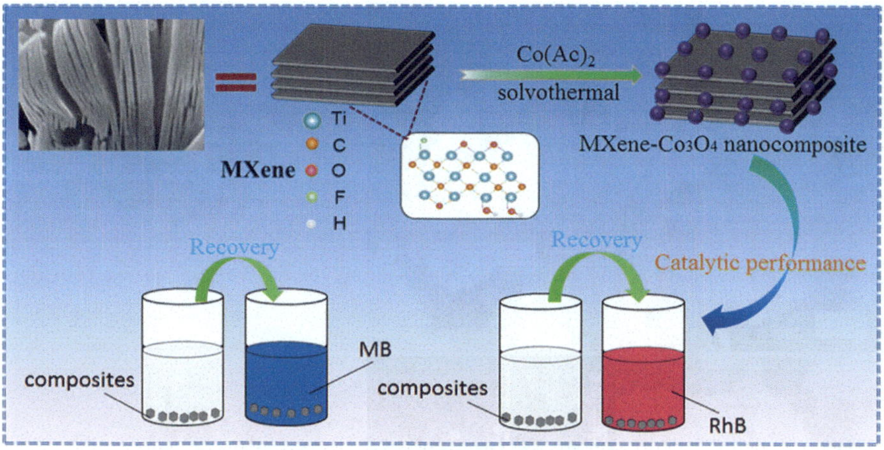

Fig. 3.51 Schematic diagram of the synthetic process and catalytic experiment of the MXene/Co_3O_4 nanohybrids. Reproduced with permission from [104]. Copyright © 2019, American Chemical Society

Figure 3.52 shows microstructural characterization results of MXene/Co_3O_4, Co_3O_4 and MXene nanosheets. The pristine MXene powder showed a typical accordion structure, as seen in Fig. 3.52a, b. The Co_3O_4 nanoparticles possessed a cube-like morphology, with an average size of 20 nm. Representative SEM and TEM images of the MXene/Co_3O_4 nanohybrids are depicted in Fig. 3.52c, d. The surface of the MXene nanosheets was decorated by the Co_3O_4 nanoparticles, while there were also some Co_3O_4 nanoparticles to be filled in the interlayer spacing of the nanolayer. The MXene/Co_3O_4 nanohybrids had a relatively larger specific surface area, they should have strong adsorption to the dyes for the catalytic reactions. The elemental composition of the nanohybrids was confirmed by the XPS analysis results, as demonstrated in the inset of Fig. 3.52c, which was in a good agreement with the element mapping profiles.

Similarly, a hydrothermal method was utilized to prepare $Ti_3C_2T_x$/Co_3O_4 nanohybrids, in concentrated strong base condition and Ar environment, combined with a calcination step [105]. Since the hydrothermal reaction was conducted in concentrated NaOH solution, Na^+ ions were adsorbed at the interface of the $Ti_3C_2T_x$ nanosheets, the decomposition of the MXene was effectively suppressed. The MXene/Co_3O_4 nanohybrid exhibited strong microwave absorption capability, with single-band, multiband and broadband effects, over the frequency range of 2–18 GHz, by tailoring the hybrid composition and selecting different thicknesses. The superb microwave absorption performance was ascribed to the multilayered structures, enriched surface defects, increased conductivity of $Ti_3C_2T_x$ and the capacitive behavior of Co_3O_4.

$Co(NO_3)_2 \cdot 6H_2O$ and $CO(NH_2)_2$ were selected as the precursors, by dissolving in NaOH solution with a concentration of 25 M, while the a molar ratio of the two items was 1:5. The mixture was continuously stirred for 0.5 h, resulting in dark blue sol.

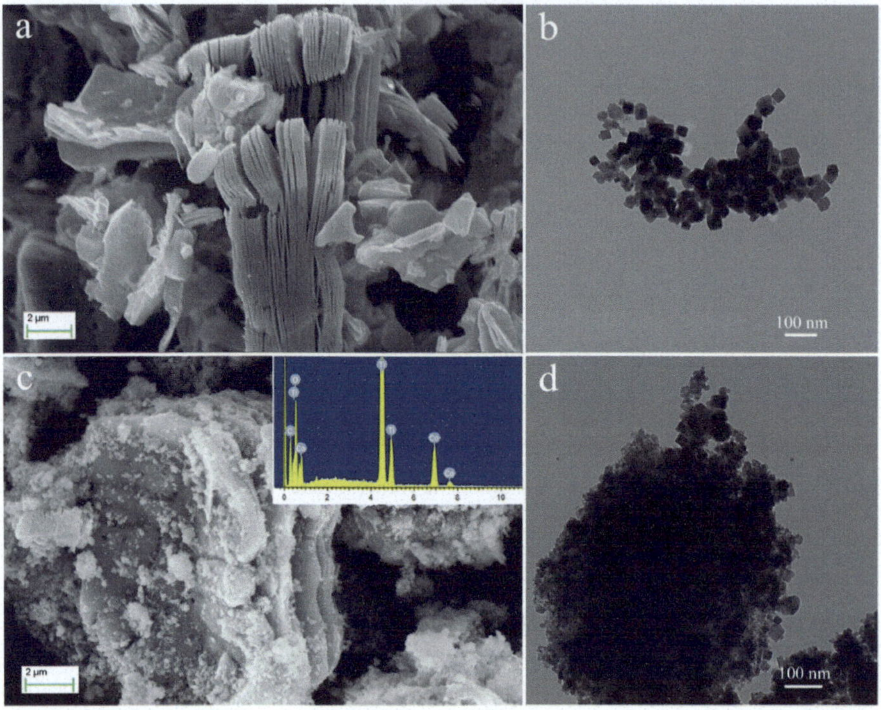

Fig. 3.52 **a** SEM image of the multilayered MXene nanosheet. **b** TEM image of the Co_3O_4 nanoparticles. **c, d** SEM and TEM images of the MXene/Co_3O_4 nanohybrids. Reproduced with permission from [104]. Copyright © 2019, American Chemical Society

After that, $Ti_3C_2T_x$ powder was introduced into the mixture with strong stirring in the environment of Ar. Hydrothermal reaction was performed at 120 °C for one day. The reaction products were collected through filtration and washing, followed by thermal annealing at 280 °C in air for 3 h, leading to final MXene/Co_3O_4 nanohybrid powders. Four samples with molar ratios of $Ti_3C_2T_x$:$Co(NO_3)_2 \cdot 6H_2O$ to be 1:1.5, 1:3, 1:9 and 1:15 were prepared, which were denoted as S1, S2, S3, and S4, respectively.

For comparison, composites with a molar of $Ti_3C_2T_x$ to $Co(NO_3)_2 \cdot 6H_2O$ of 1:9 were hydrothermally treated in pure water, 10 M, 25 M and 30 M NaOH solutions, denoted as SC0, SC1, SC2, and SC3, respectively. Except where noted, the other reaction conditions and process kept unchanged. Figure 3.53 shows SEM images of the pristine MXene and the $Ti_3C_2T_x$/Co_3O_4 nanohybrids. The layered structure was retained after the etching reaction, while the MXene nanosheets had a rough surface, as observed in Fig. 3.53a, b. XRD results indicated that the small particles were anatase TiO_2, due to the in-situ oxidation of the MXene. This oxidizing problem was prevented when using concentrated base solution for the hydrothermal reaction.

The MXene/Co_3O_4 nanohybrids displayed typical laminar structures, as illustrated in Fig. 3.53c–f. Because of the absence of the TiO_2 nanoparticles, the surface

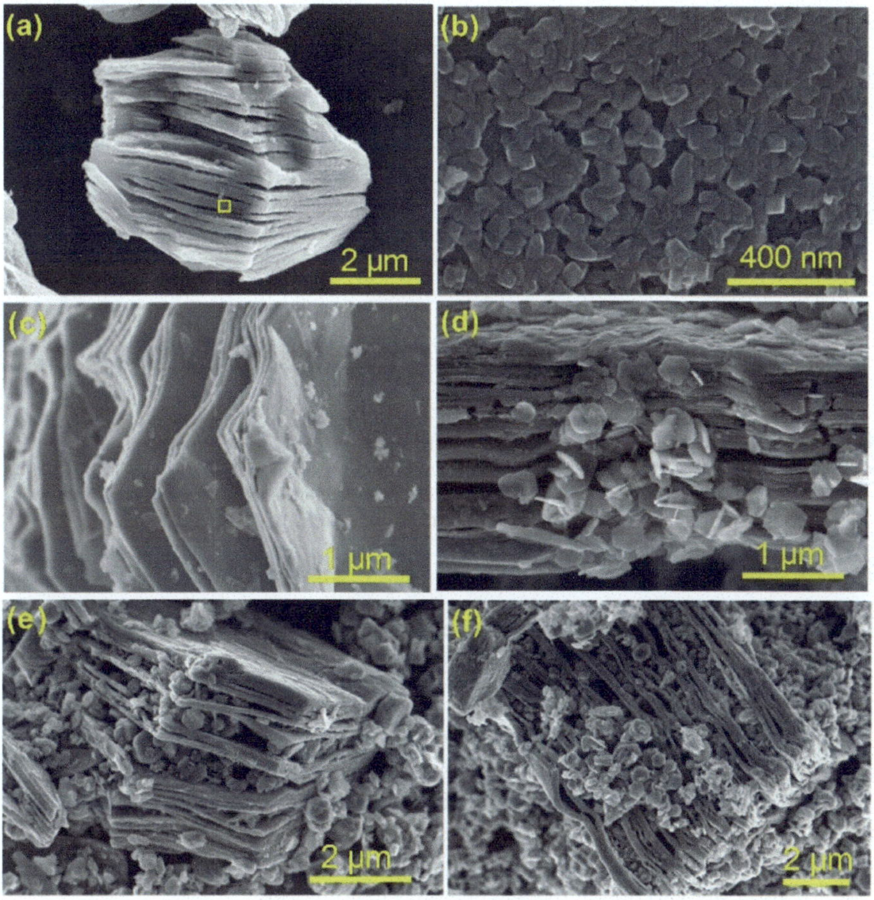

Fig. 3.53 SEM images of **a**, **b** the control sample (SC0) treated with pure water with the panel (**b**) to represent the enlarged image of the zone labelled in panel (**a**). SEM images of the MXene/Co_3O_4 nanohybrids with four compositions: **c** S1, **d** S2, **e** S3 and **f** S4. Reproduced with permission from [105]. Copyright © 2019, Elsevier

of the MXene nanosheets was flat. In the nanohybrids, the Co_3O_4 nanoparticles were distributed not only on the surface of the MXene nanosheets but also encapsulated in the interlayer spaces. The Co_3O_4 nanoparticles possessed a flak-like morphology, with lateral sizes of 400–700 nm. With increasing concentration of the Co precursor, the quantity of the Co_3O_4 nanoparticles was gradually increased.

An in-situ precipitation process was deviced to prepare MXene/NiO nanohybrids, in order to increase the dispersing efficiency of the NiO nanoparticles, thus having enhanced catalytic activity [107]. As a matter of fact, the oxidation of the MXene phase was retarded in the format of MXene/NiO nanohybrid. During the formation process, electrostatic absorption of the Ni^{2+} ions were electrostatically adsorbed on the MXene nanosheets, which then in-situ precipited to $Ni(OH)_2$ at room temperature.

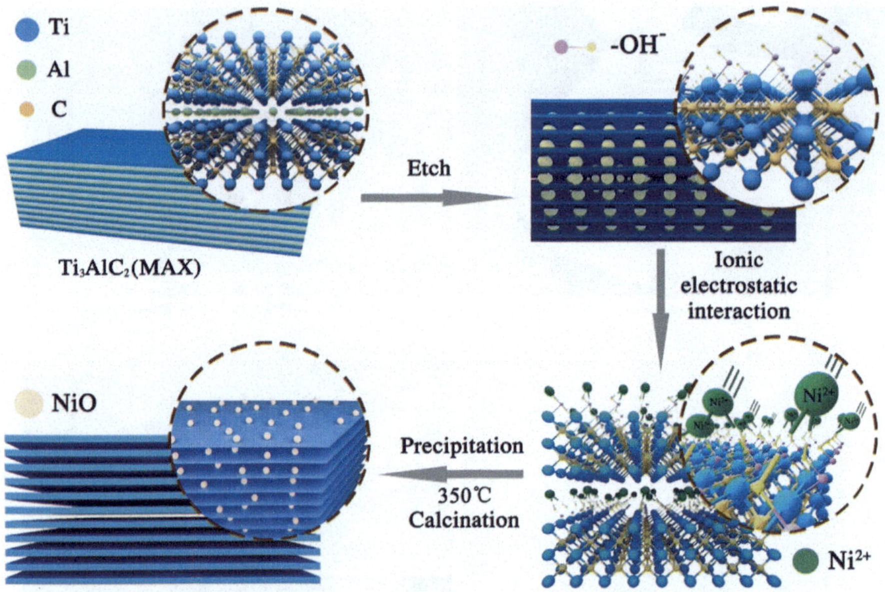

Fig. 3.54 Schematic diagram of the synthesis process to prepare the MXene/NiO nanohybrids. Reproduced with permission from [107]. Copyright © 2019, Elsevier

NiO nanoparticles were obtained as the precipitant was calcined in N_2. IN the final hybrids, the NiO nanoparticles with an average size of 25 nm were evenly distributed on the surface of the MXene nanosheets. Due to the improved dispersing effect, the catalytic activity of NiO for the thermal decomposition of ammonium perchlorate (AP) was significantly boosted. Figure 3.54 shows schematic diagram demonstrating the fabrication steps to obtain the MXene/NiO nanohybrids.

A simple mixing-calcination process was utilized to develop MXene/CuO nanohybrid [108]. Both CuO and MXene powders were added in sufficient volume of ethanol to achieve a thorough mixing. The suspension was heated to evaporate the solvent, which resulted in MXene/CuO nanohybrid powder after calcination in Ar. The CuO nanoparticles had sizes of 60–100 nm, which were evenly attached on the surface and at the edges of the MXene nanolayers. Specifically, when CuO and MXene were hyubridized with 1 wt% each, the catalytic efficiency for the decomposition of ammonium perchlorate (AP) was close to that of the catalyst with 2 wt% CuO.

MXene/Cu_2O hybrids have also been reported in the open literature [109, 110]. For example, a simple wet-chemical route was used to prepare MXene/Cu_2O hyubrid, as well as MXene (Ti_3C_2) and Cu_2O, to be employed as the catalyst for the thermal decomposition of ammonium perchlorate (AP) [109]. In this study, the MXene nanosheets had thicknesses of 25–35 nm, while the Cu_2O nanoparticles were grown uniformly on the surface of the Ti_3C_2 nanosheets. It was found that the Ti_3C_2/Cu_2O

hybrid was much stronger in catalysing the thermal decomposition of AP than both the pristine MXene and pure Cu_2O.

To prepare Cu_2O nanoparticles, reaction solution was obtained by dissolving 0.005 mol $Cu(CH_3COO)_2 \cdot H_2O$ and 0.01 mol $C_6H_{12}O_6$ in 100 ml deionized water, which was heated at 90 °C for 5 h. The reaction product was collected and thoroughly washed with water and ethanol, followed by vacuum drying at 70 °C. To synthesize MXene/Cu_2O nanohybrid, 0.1 g Ti_3C_2 was dispersed in 100 ml deionized water, together with 0.005 mol $Cu(CH_3COO)_2 \cdot H_2O$ and 0.01 mol $C_6H_{12}O_6$, with the aid of continuous stirring for 12 h. The suspension was heated at 90 °C for 5 h, while the product was collected similarly.

A hetero-architecture MXene/Cu_2O nanohybrid, consisting of Ti_3C_2 quantum dots (QDs) and 1D Cu_2O nanowires (NWs), for the application in photo-catalytic transition of carbon dioxide (CO_2) to methanol [110]. In this case, the Cu_2O NWs were coated with Ti_3C_2 QDs by using a self-assembly process. The Cu_2O NWs were tremendously stabilized by the MXene QDs, while the Ti_3C_2/Cu_2O nanohybrid exhibited much strong photo-catalytic activity, due to the enhanced efficiencies in charge transferring charge transportation, carrier densities, light adsorbing capability, along with the reductions in the band bending edges and charge recombining processes.

To prepare Ti_3C_2 nanosheets and Ti_3C_2 QDs, 2.0 g Ti_3AlC_2 power was exfoliated with 20 ml 48 wt% HF solution. Ti_3C_2 nanosheet powder was obtained after centrifugation of the suspensions, followed by thorough washing with deionized (DI) water and vacuum drying at 60 °C for overnight. Then, 1.0 g Ti_3C_2 was dispersed in 20 ml dimethyl sulfoxide (DMSO) for about one day and the resultant suspension was centrifugated at 3500 rpm for 5 min. The powder was added in DI water at a MXene:H_2O weight ratio of 1:300, which was flowed with Ar and ultrasonicated for 3 h. The suspension was further treated with high-rate centrifugation for 20 min to obtain suspendable Ti_3C_2 nanosheets. PEI was put into the centrifugated suspension, followed by heating at 120 °C in N_2 for 5 h. Ti_3C_2 nanosheet powder was formed after dialysis, concentrating and freeze drying.

The Ti_3C_2 nanosheet powder without the addition of PEI was strongly ultrasonicated in the protection of Ar at a power of 120 W for 6 h first and then continued for 10 h at a higher power of 300 W. After that, PEI was introduced to the suspension, followed by heating at 120 °C in the presence of N_2 for about 12 h. After filtration with 0.22 μm membranes, Ti_3C_2 QDs with positive charges were derived through a dialysis treatment. The suspension of the Ti_3C_2 QDs was further heated for the evaporation the solvent and then freeze dried to obtain QDs powder.

Ti_3C_2 QDs/Cu_2O NWs/Cu and Ti_3C_2 sheets/Cu_2O NWs/Cu heteroarchitctures were prepared with Cu_2O NWs/Cu meshes. To do so, the Cu_2O NWs/Cu meshes were soaked in the solution of PSS with a concentration of 1 vol% in ethanol at 90 °C for 120 min. After that, the samples were completely cleaned with ethanol and then dried by using weak flow of N_2. PSS-altered Cu_2O NWs/Cu samples were soaked into the suspensions of the PEI-treated Ti_3C_2 QDs that had positive charges with different concentrations for one day under ambient conditions, followed by washing and drying. The heteroarchitecture-based electrodes were thermally treated at 200 °C for 1 h. Ti_3C_2 sheets/Cu_2O NWs/Cu samples were prepared in a similar way.

$Cu(OH)_2$ NWs/Cu meshes were formed through anodization of Cu meshes in alkali solution. The anodized samples were thermally treated at 550 °C in N_2, leading to Cu_2O NWs/Cu meshes in red color. Representative SEM images of the Cu_2O NWs are depicted in Fig. 3.55a, b, indicating that their surfaces were highly porous and the average diameter was about 0.5 μm. Meanwhile, the Cu_2O NWs were modified with poly(sodium 4-styrenesulfonate) (PSS), so that they would have surface with negative charges. As a result, the Cu_2O NWs and the Ti_3C_2 QDs were assembled to

Fig. 3.55 SEM images of the samples: **a**, **b** Cu_2O NWs/Cu, **c**, **d** Ti_3C_2 QDs/Cu_2O NWs/Cu and **e**, **f** Ti_3C_2 nanosheets/Cu_2O NWs/Cu hetero-architecture. Reproduced with permission from [110]. Copyright © 2019, John Wiley & Sons

Ti_3C_2 QDs/Cu_2O NWs/Cu nanohybrids through electrostatic interactions. After that, both the PSS and PEI were removed by annealing in Ar, while the morphologies of the two components were not varied, as illustrated in Fig. 3.55c, d. The compositions of the nanohybrids were confirmed by the elemental mapping profiles, while the intimate contact of the two components was also evidenced by HRTEM images and other analytic methods. The coating of the Ti_3C_2 nanosheets on the Cu_2O NWs was revealed by SEM images shown in Fig. 3.55e, f.

Ti_3C_2/ZnO hybrids with sandwich architecture were prepared by decorating Ti_3C_2 nanosheets with ZnO nanoparticles, in which the ZnO nanoparticles were obtained by using chemical precipitation process, with $Zn(NO_3)_2 \cdot 6H_2O$ and NaOH as the reactants [111]. The ZnO nanoparticles were evenly attached on both the surface and inside the interlayer spaces of the Ti_3C_2 nanosheets. As a result, the Ti_3C_2 nanosheets were prevented from restacking, due to the formation of the sandwich morphology. Ti_3C_2/ZnO hybrids exhibited promising electrochemical properties as electrodes of supercapacitors. A specific capacitance of 120 F/g was achieved at 2 mV/s, while the electrode displayed a strong cycling behavior by having a capacitance retention of 85% after cycling for 10,000 times at 5 A/g.

A facial chemical precipitation process was developed to obtain ZnO/MXene $Ti_3C_2T_x$ nanohybrid with urchin morphology [112]. The ZnO/MXene nanohybrid exhibited promising microwave absorption capabilities, with an optimized reflection loss (RL) of −26.30 dB, while the pristine $Ti_3C_2T_x$ nanosheets displayed a RL value of only −6.70 dB. The significantly enhanced electromagnetic performance was ascribed to the formation of a special semiconducting porous structure and the presence of enlarged interfaces. The microwave absorbing behaviour could be adjusted over the bands of 14.0–18.0 GHz, by simply altering the reaction time to grow the ZnO nanoparticles.

To grow ZnO nanorods, a thin layer of seeding ZnO was deposited on the surface of the $Ti_3C_2T_x$ nanosheets. The solution of seeding ZnO was made by dissolving 11.0 mg $Zn(CH_3COO)_2 \cdot 2H_2O$ in 40 ml ethanol at 50 °C for 0.5 h with the aid of magnetic stirring. After reaction, 160 ml ethanol was added to dilute the solution. Similarly, NaOH solution was prepared by dissolving 3.2 mg NaOH in 50 ml ethanol, followed by diluting with additional 50 ml ethanol. Then, the $Zn(CH_3COO)_2 \cdot 2H_2O$ and NaOH solutions were mixed for the precipitation of ZnO. Once the reaction was finished, 25 mg $Ti_3C_2T_x$ powder was added, which was continuously stirred at 65 °Cfor 0.5 h. The precipitant was collected through filtration with hydrophobic membranes, followed by vacuum drying at 80 °C for 12 h, leading to ZnO seeding layer on the surface of the $Ti_3C_2T_x$ nanosheets.

The as-obtained $Ti_3C_2T_x$ nanosheet sample coated with a ZnO seeding layer was hydrothermally treated to grow ZnO nanorods. The precursor solution was prepared by dissolving 3.71 g $Zn(NO_3)_2 \cdot 6H_2O$ in 500 ml deionized (DI) water, in which the ZnO seeded $Ti_3C_2T_x$ nanosheet powder was dispersed, with the aid of strong stirring at 70 °C for 0.5 h. Meanwhile, 1.750 g HMTA was introduced into the reaction suspension, followed by stirring for time durations of 0.5, 1.0, 2.0 and 3.0 h. The products were collected and dried at 70 °C in vacuum for one day, thus resulting in ZnO/$Ti_3C_2T_x$ nanohybrid sample, where the surface of the $Ti_3C_2T_x$ nanosheets

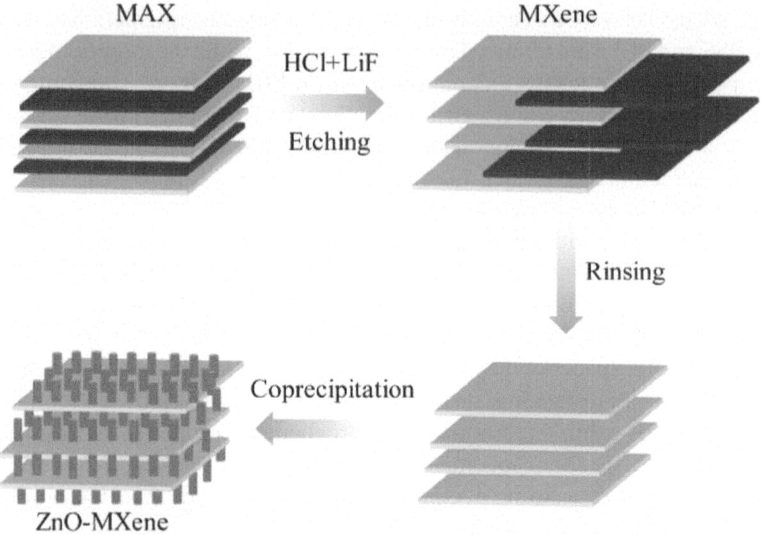

Fig. 3.56 Schematic diagram illustrating the synthetic steps of the $ZnO/Ti_3C_2T_x$ nanohybrids. Reproduced with permission from [112]. Copyright © 2017, Elsevier

were decorated with uniform ZnO nanorods. Figure 3.56 shows a schematic diagram describing the synthetic steps to obtain the $ZnO/Ti_3C_2T_x$ nanohybrids.

Figure 3.57 shows microstructural characterization results of the samples with different compositions. The pristine $Ti_3C_2T_x$ nanosheets exhibited typical laminar structure of MXene, indicating the effective exfoliation of the parent MAX phase, as observed in Fig. 3.57a. During the precipitation step, ZnO nanorods were formed, which were uniformly attached onto the surface and embedded inside the interlayer of the $Ti_3C_2T_x$ nanolayers, as illustrated in Fig. 3.57b, c. The surface structure was detailed by using TEM images, as revealed in Fig. 3.57d. As depicted in Fig. 3.57e, the $ZnO–Ti_3C_2T_x$ nanohybrids contained well-developed ZnO nanorods, in a good agreement with the result of Fig. 3.57b. Phase compositions of various samples have been examined by using XRD. Figure 3.57f shows their XRD patterns, confirming the phase pure parent MAX, effective exfoliation to MXene and the formation of the ZnO/MXene nanohybrids.

The efficacious co-sintering of $ZnO–Ti_3C_2T_x$ hybrid ceramics were prepared by using a newly developed cold sintering process (CSP), with MXene and ZnO as the precursors [113]. The densification process was assisted, due to the presence of a transient aqueous solution during the sintering process, thus requiring a significantly low sintering temperature. The $ZnO–Ti_3C_2T_x$ nanohybrid ceramics could reach relative densities of 92–98% after sintering at a temperature of as low as 300 °C. Because of the very low sintering temperature, the oxidation of the MXene phase and the interdiffusion between ZnO and $Ti_3C_2T_x$ was effectively prevented. Meanwhile, the 2D MXene nanosheets were uniformly distributed at the grain boundaries of the ZnO-based ceramics. It was found that electrical conductivity of the materials was

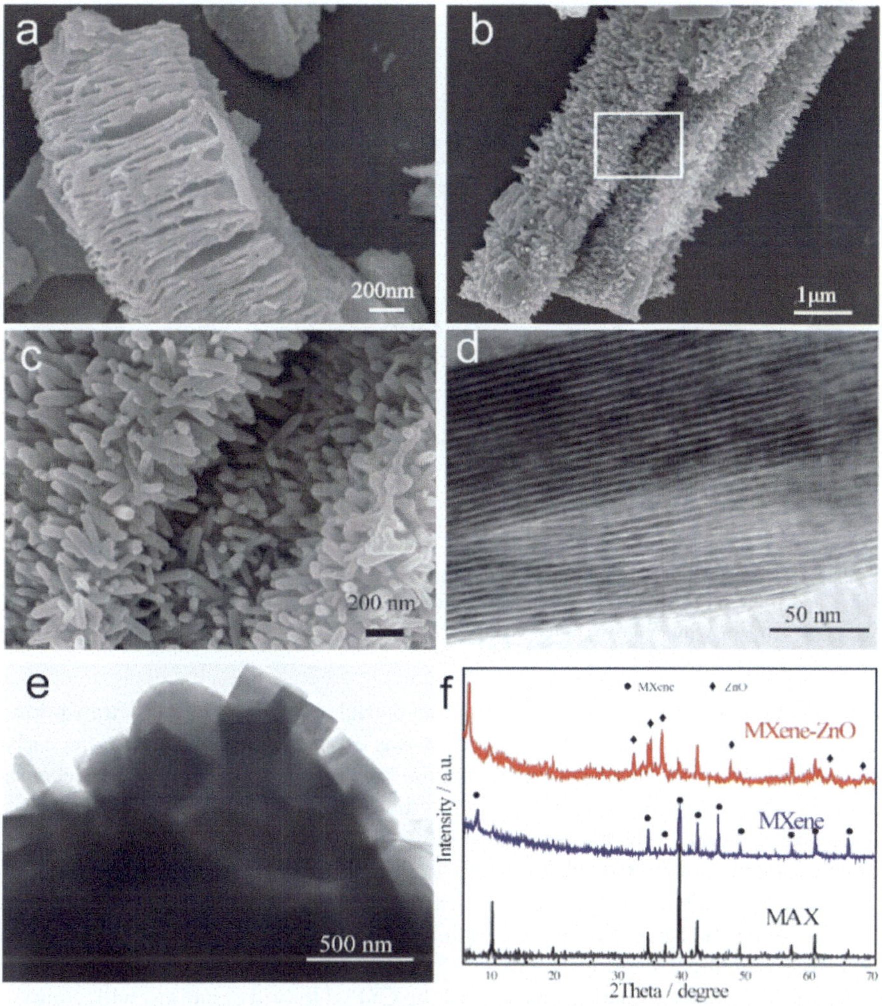

Fig. 3.57 Microstructure characterization results of the $Ti_3C_2T_x$ and $ZnO/Ti_3C_2T_x$. SEM images of different samples: **a** pristine $Ti_3C_2T_x$ nanosheets, **b** $ZnO/Ti_3C_2T_x$ nanohybrids and **c** amplified image of (**b**). TEM images of the samples: **d** pristine $Ti_3C_2T_x$ and **e** $ZnO/Ti_3C_2T_x$ nanohybrid. **f** XRD patterns of Ti_3AlC_2, $Ti_3C_2T_x$ and $ZnO/Ti_3C_2T_x$. Reproduced with permission from [112]. Copyright © 2017, Elsevier

increased by more than order of magnitude, as 5 wt% MXene was included. The hardness and elastic modulus were enhanced by about 50% and 150% for the samples containing 0.5 wt% and 5 wt% MXene, respectively.

Both the densification process and the degree of grain growth of the ZnO-MXene hybrid ceramics could be well controlled by adjusting the content of acetic acid solution in the mixtures. In this case, ZnO was dissolved to a certain degree into the acetic

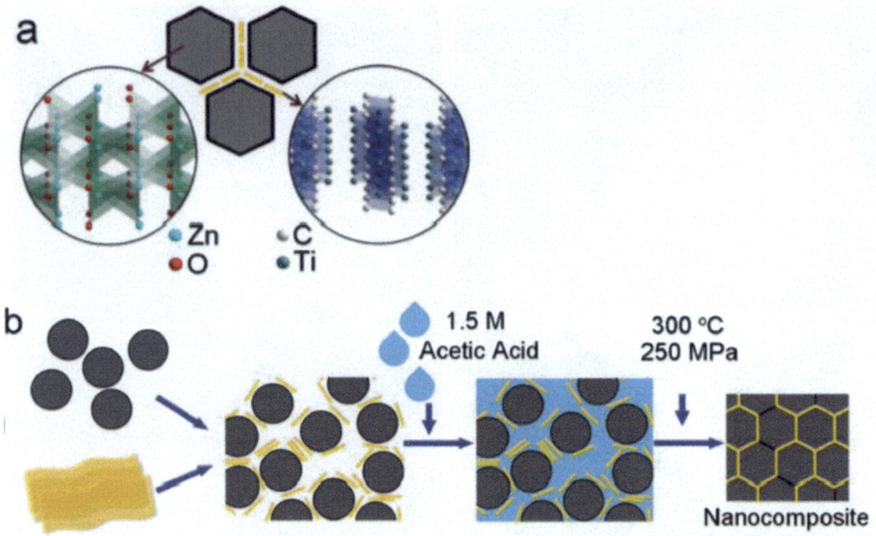

Fig. 3.58 Schematic illustration showing: **a** the grain boundary of (1 − y)ZnO–yTi$_3$C2T$_x$ nanocomposites and **b** the fabrication process via cold sintering. Reproduced with permission from [113]. Copyright © 2018, John Wiley & Sons

acid to form a liquid phase, which wetted the particles to boost the mass transportation during the sintering process. This effect was especially pronounced at the early stage of sintering, i.e., the rearrangement of the ZnO particles was enhanced. Because of the extremely low sintering temperature, the oxidation of the MXene phase was effectively prevented. The distribution of the 2D MXene nanosheets at the ZnO grain boundaries is demonstrated schematically in Fig. 3.58a. The presence of the MXene phase would greatly enhance the electrical conduction behaviour and mechanical strength of the ZnO-based ceramics, with the evolution of the microstructure to be depicted schematically in Fig. 3.58b.

Figure 3.59a shows relative densities of the CSPed hybrid ceramics, with compositions of (1 − y)ZnO–y$Ti_3C_2T_x$, in which y = 0, 0.5, 1, 3, 5 wt%. All samples had a relative density of >90%, confirming the effectiveness and efficiency of the CSP strategy in processing hybrid materials involving MXenes. XRD patterns indicated that the ZnO was of a wurtzite/zincite crystal structure. Meanwhile, the (002) diffraction peak of the MXene was relatively broad, at peaks over 8°–9°. This observation suggested that no chemical reaction occurred between the two components during the CSP sintering process.

Specifically, (1 − y)ZnO–y$Ti_3C_2T_x$ mixtures were blended with 1.5 M acetic acid aqueous solution at concentrations of 17–20 wt%. The semi-wet mixtures were uniaxially pressed at a pressure of 250 MPa. The samples were heated at 300 °C for 60 min, at heating rates of 5–6 °C/min, while the pressure was held throughout the sintering process. After that, the samples were released from the die as the system

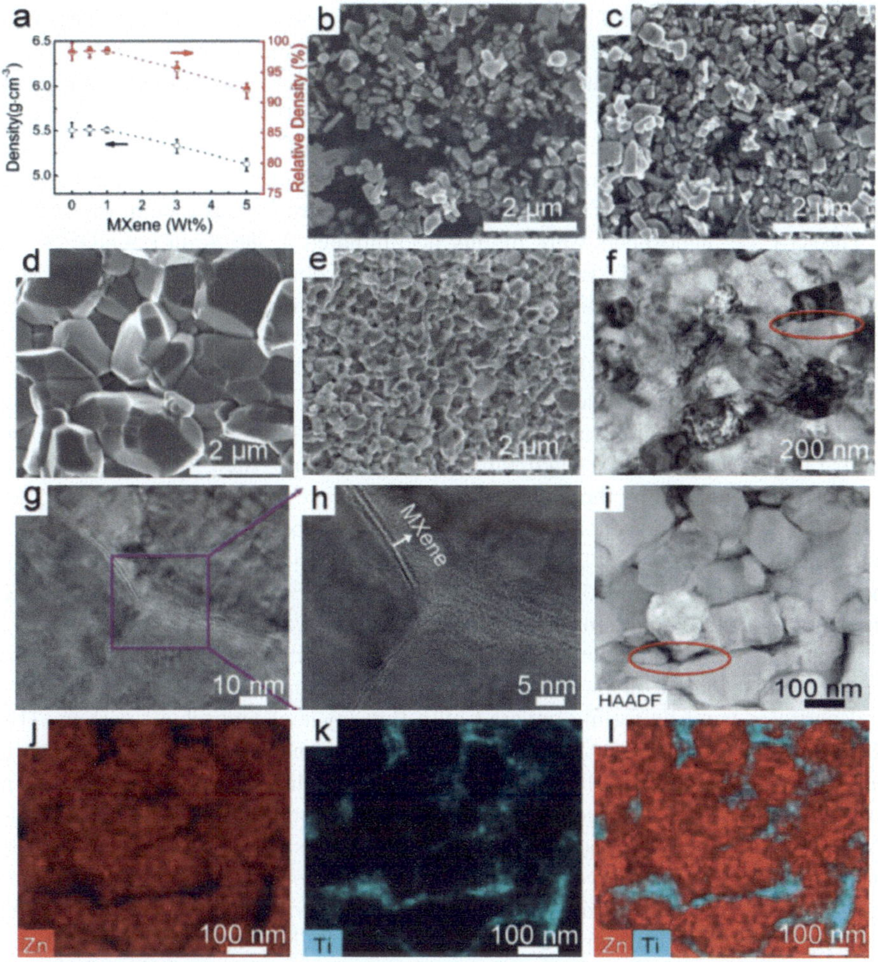

Fig. 3.59 The density and microstructure of ZnO–Ti_3C2T_x nanocomposites cold sintered at 300 °C for 1 h. **a** The densities of cold sintered ZnO–Ti_3C2T_x nanocomposites. SEM images of: **b** ZnO and **c** 99ZnO–1Ti_3C2T_x raw powders, and the cross sections of cold sintered: **d** ZnO and **e** 99ZnO–1Ti_3C2T_x ceramics. **f–h** TEM, (**i**) HAADF-STEM images and (**j–l**) energy dispersive spectroscopy (EDS) elemental mapping of cold sintered 99ZnO–1Ti_3C2T_x nanocomposite. The red circles in TEM and HAADF-STEM images show one example of the Ti_3C2T_x region. In the HAADF image, the bright areas belong to ZnO and the dark areas belong to Ti_3C2T_x. EDS maps, where elemental Zn is shown in red and Ti is shown in cyan, show the presence of Ti_3C2T_x at the ZnO grain boundaries. Reproduced with permission from [113]. Copyright © 2018, John Wiley & Sons

was cooled down below 80 °C. Selected as-CSPed samples were subject to thermal annealing at temperatures of 300 °C and 750 °C for 60 min in Ar.

Representative SEM images of the precursor powders and CSPed samples of both pristine ZnO and the (1 − y)ZnO–y$Ti_3C_2T_x$ (y = 1.0 wt%) hybrid are shown in Fig. 3.59b–e. For pure ZnO, after sintering at 300 °C for 60 min, the grain size was increased from 0.1–0.9 μm to 1–4 μm, as observed in Fig. 3.59b, d. In comparison, the grain size of the CSPed 99ZnO–1$Ti_3C_2T_x$ was much smaller, as seen in Fig. 3.59c, e. The $Ti_3C_2T_x$ nanosheets were mainly accumulated at the grain boundaries, so that grain growth of ZnO was retarded, while the densification process was not influenced. As a result, ZnO–$Ti_3C_2T_x$ nanohybrid ceramics could reach pretty high relative densities.

TEM, HRTEM and STEM images, along with EDS mapping profiles, of the ZnO–$Ti_3C_2T_x$ samples are depicted in Fig. 3.59f–l. As illustrated in Fig. 3.59f, the 99ZnO–1$Ti_3C_2T_x$ sample was characterized by such a microstructure in which the ZnO grains were uniformly coated with $Ti_3C_2T_x$. The core–shell-like microstructure was further evidenced by the high magnification images, as revealed in Fig. 3.59g, h. The $Ti_3C_2T_x$ were present at the ZnO ceramics grain boundaries, with thicknesses at the nanometer-sized scale. The uniform distribution of the $Ti_3C_2T_x$ nanosheets in the hybrid ceramics was also confirmed by the STEM EDS mapping profiles of the elements, as demonstrated in Fig. 3.59i–l.

A multilayered hierarchical hybrid, with orthorhombic Nb_2O_5 nanoparticles to be evenly distributed on the surfaces of $Nb_4C_3T_x$ nanoflakes [12]. The nanohybrid particles were uniformly covered by a thin layer of amorphous carbon. The $Nb_4C_3T_x$ nanoflake powder was obtained through a one-step oxidizing process in the etching solution of HF, followed by thermal annealing at 850 °C for 30 min in carbon dioxide. In these special hierarchical nanostructures, the Nb_2O_5 nanoparticles offered reactive electrochemical sites at the edges of the MXene nanolayers, ensuring high rate-responsive activity, because of the enhanced exposing profile to the electrolyte solution. As a result, the ion diffusion was significantly boosted. At the same time, the thin amorphous carbon layer, together with the remaining MXene nanoflakes, ensured that the whole materials could reach a sufficiently high electrical conductivity. In addition, the amorphous carbon coating served as a binder to provide a strong link between the oxide nanoparticles and the MXene nanosheets.

The effects of thermal oxidation temperature, reaction time and CO_2 partial pressure on the phase evolution, sizes of the Nb_2O_5 nanoparticles, structural characteristics of the hierarchical nanohybrids, and hence the resulting electrochemical activities, have been systematically studied [114]. Specifically, the hybrid-based electrodes with a thickness of about 50 μm offered a high specific capacitance of 330 C/g, corresponding to an areal capacitance of 660 mF/cm^2, at a charging–discharging time duration of 4 min. The electrodes also displayed a promising cycling stability in non-aqueous Li electrolytes. In this case, the kinetics of the charge storing process followed a surface-controlled mechanism. The intrinsically present pseudocapacitive effect and the superb energy storing behaviour of the Nb_2O_5 nanoparticles, along with the smooth charge transportation, owing to the highly conductive 2D Nb_2CT_x

nanosheets and the amorphous carbon layer, were responsible for the promising electrochemical properties of the special Nb_2O_5-MXene nanohybrid architectures.

For the oxidation experiment, 100 mg Nb_2CT_x was first blew with Ar for 1 h, followed by raising the temperature to 800 °C at a heating rate of 5 °C/min, at which the sample heated for 45 min. Then, the sample was thermally oxidized for 60 min in the flow of CO_2 at a flowing rate of 150 sccm. Processing parameters were varied, including annealing temperature (600, 700, 750 and 850 °C), time duration (10, 30 and 60 min) and CO_2 flowing rate (35 sccm, 75 sccm and 150 sccm). The experimental process is illustrated as the schematic diagram shown in Fig. 3.60a.

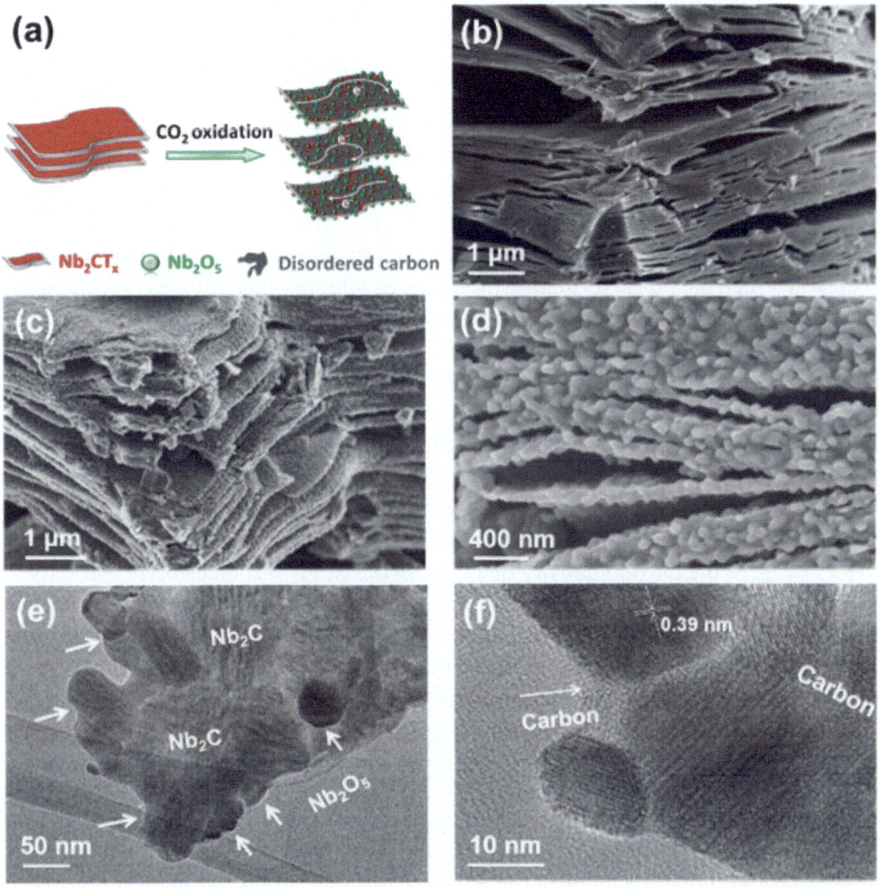

Fig. 3.60 **a** Schematic diagram of the synthetic process for the hierarchical Nb_2O_5/carbon/Nb_2CT_x nanohybrid. SEM images of the samples: **b** as-produced Nb_2CT_x, **c** Nb_2CT_x thermally oxidized at 800 °C for 1 h at CO_2 flow of 150 sccm at low magnification and **d** high magnification image. Low (**e**) and high (**f**) magnification TEM images of the Nb_2CT_x oxidized at 800 °C for 1 h at CO_2 flow of 150 sccm. Reproduced with permission from [114]. Copyright © 2016, American Chemical Society

Theoretical prediction indicated that the oxidation must be conducted at temperatures of <875 °C. As seen in Fig. 3.60b, the Nb_2CT_x powder contained stacked thin layers. After oxidation with CO_2, the roughness of the surfaces of the MXene nanosheets was tremendously increased, due to the generation of the oxide nanoparticles. The nanoparticles were attached both on the surfaces and at the edges, as revealed in Fig. 3.60c, d, with average particle sizes in the range of 20–35 nm, as demonstrated in Fig. 3.60d, e. However, the layered structure of the Nb_2CT_x nanosheets was retained to a certain degree after the oxidation process, as observed in Fig. 3.60e. The Nb_2O_5 nanoparticles had a crystal structure of orthorhombic lattice, as evidenced by the (001) plane with a spacing of 0.39 nm, as depicted in Fig. 3.60f. Furthermore, the Nb_2O_5 nanoparticles were fixed onto the Nb_2CT_x MXene nanosheets, owing to the formation of a thin amorphous carbon layer, as illustrated in Fig. 3.60e, f. In the case, the thin carbon layer acted as a glue so as to enhance the structural integrity of the nanohybrid materials.

Similar to other transition metal oxides (TMOs), WO_3 has also been demonstrated to be a promising candidate as electrode materials for energy storage applications, because of the hollow tunnelling structure of its hexagonal polymorph (i.e., Hexa WO_3) and the highly crystalline monoclinic one (i.e., Mono WO_3). Both Mono WO_3–Ti_3C_2 and Hexa WO_3–Ti_3C_2 hybrids have been prepared by using a hydrothermal process in which the phase formation of WO_3 was intentionally tailored [115]. Experimental data indicated that the electrochemical properties of the hybrids were higher than those of the oxides by about two times, with the Hexa WO_3–Ti_3C_2 hybrid-based electrode to have a specific capacitance of 566 F/g and a cycling stability of 92% after 5000 cycles.

A simple approach was utilized to develop WO_3-MXene hybrid powder through electrostatic interaction between WO_3 nanorods (WNRs) with positive charges and MXene nanosheets with negative charges [116]. Electrode based on the WO_3-MXene hybrid powder had a specific capacitance of 297 F/g, while that made of WNRs was only 121 F/g, at the current density of 1 A/g, in 0.5 M H_2SO_4 electrolyte. Moreover, the hybrid-based electrode had a high rate stability with a retention of 82% at 5 A/g and a strong cyclic stability with retention of 73% after cycling for 5000 times at 4 A/g. The MXene served as an electron conductor, while the electrostatic force enhanced the mechanical stability of the materials.

To hydrothermally synthesize the WO_3 WNRs, 0.825 g $Na_2WO_4 \cdot 2H_2O$ and 0.290 g NaCl were put in 20 mL deionized water, forming precursor solution. Then, the solution was adjusted to pH value of 2.0 with 3 M HCl solution followed by hydrothermal treatment at 180 °C for a whole day. The WNR was collected and dried in vacuum at 40 °C. The as-obtained NWR powder was made into suspension in ethanol, which was then refluxed for 5 h with the addition of cationic surfactant APTES. The sample was finally washed to eliminate the residual APTES. The APTES-WNR powder was dried in vacuum at 40 °C. Due to the opposite signed charges of the two components, WO_3-MXene hybrid could be readily obtained through electrostatic self-assembly.

Figure 3.61 shows microstructural analysis result of the samples, including MXene powder, WNRs and WNRs/MXene hybrid. The MXene possessed a typical

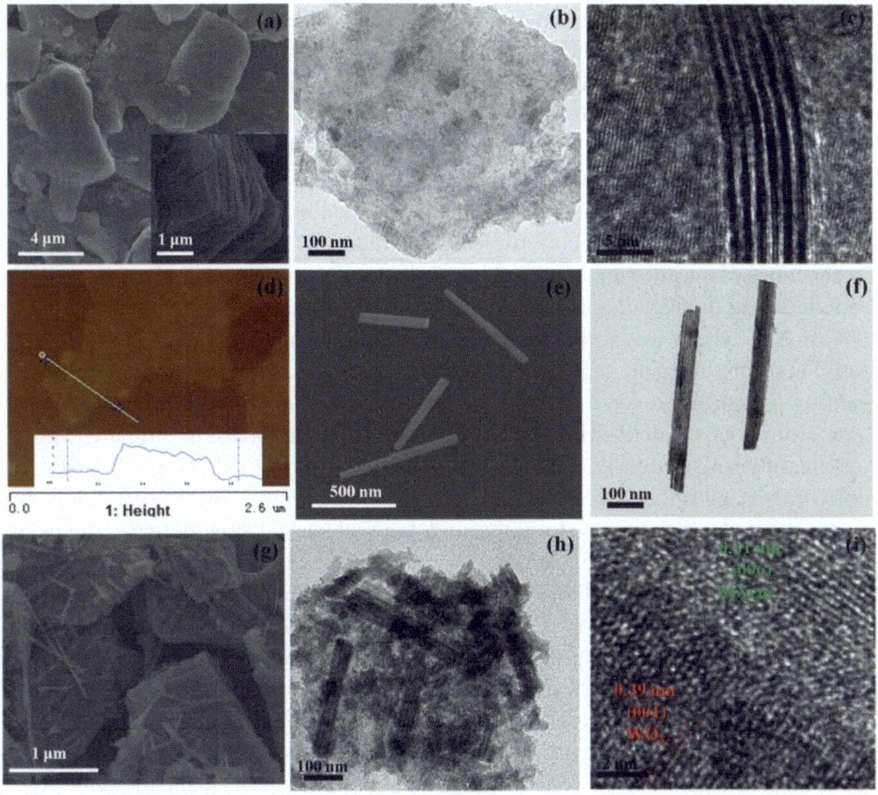

Fig. 3.61 Microstructural characterization results of the samples: **a** SEM, **b** TEM, **c** Cross-sectional TEM, and **d** AFM images of MXene, **e** SEM and **f** TEM of the WNRs **g** SEM, **h** TEM and (i) HRTEM images. Reproduced with permission from [116]. Copyright © 2019, Elsevier

accordion multilayered structure, suggesting effective exfoliation of the etching reaction process, as shown in Fig. 3.61a. The laminar MXene nanosheets had lateral sizes of hundreds of nanometres to micrometer scale, which were thin and transparent, as demonstrated in Fig. 3.61b. The average thickness of the MXene nanosheets was close to 1.33 nm, according to the calculation from the cross-sectional TEM images, as depicted in Fig. 3.61c. This value was well within the thickness range of 1–4 nm, estimated from the AFM images, as revealed in Fig. 3.61d.

SEM and TEM of the WNRs are illustrated in Fig. 3.61e–f. The sample exhibited nanorod-like morphology, with an average diameter of 50 nm. SEM and TEM images of the WNRs/MXene hybrid sample are shown in Fig. 3.61g–h, indicating that the WNRs were homogeneously distributed on the surface of the MXene nanosheets. The WO_3 phase was confirmed by the lattice fringes with a *d*-spacing value of 0.39 nm, corresponding to the (001) plane of the WO_3 crystals, while the lattice fringes with a *d*-spacing of 0.31 nm were ascribed to the (006) plane of the MXene phase, as seen in Fig. 3.61i. Moreover, EDS elemental mapping profiles clarified the close contact between the two components in hybrid made with the electrostatic interaction induced self-assembly.

A facial hydrothermal process was employed to synthesize SnO_2–Ti_3C_2 nanohybrid powder, for the application as electrode of LIBs, in order to enhance the electrochemical storage capabilities [117]. As expected, the d-spacing of the Ti_3C_2 nanolayers was greatly increased, so that the storage capacity of Li^+ ions was enhanced, due to the incorporation of the SnO_2 nanoparticles. At the same time, the presence of the SnO_2 nanoparticles contributed to additional electrochemical effect because of the alloying mechanisms. Specifically, the SnO_2–Ti_3C_2 nanohybrid-based electrode displayed a specific capacity of 1030 mAh/g at the current density of 100 mA/g, which was retained at the level of 360 mAh/g after cycling for 200 times. Therefore, the SnO_2–Ti_3C_2 nanohybrid-based anode exhibited stable cycling capability and strong reversible capacity for the storage of Li^+ ions.

An atomic layer deposition (ALD) process was reported to prepare SnO_2–Ti_3C_2 nanohybrid, similarly for LIBs anode applications [118]. The SnO_2-MXene nanohybrid-based anode not only utilized the high storage capacity of SnO_2 but also maintained mechanical strengths and high electrical conductivity owing to the mechanically strong and electrically conductive MXene frameworks. It was found that the deposition of SnO_2 on MXene nanosheets, which had functional groups of –O, –F and –OH, was crucial to prevent the degrading phenomenon of the MXene nanosheets during the ALD process. The SnO_2-MXene anode demonstrated superb electrochemical properties, because the conductive MXene network effectively buffered the volume variation in the lithiation-delithiation of the SnO_2 nanoparticles. Furthermore, cycling stability was further enhanced by coating a thin layer of HfO_2 on the SnO_2-MXene nanohybrid nanoparticles, by using the same ALD process. Specifically, with the presence of the HfO_2 coating layer, the SnO_2-MXene anode exhibited a specific capacity of 843 mAh/g.

The SnO_2 and HfO_2 thin layers were deposited on the surface of the MXene nanolayers that were coated on a Cu foils via ALD. The MXene coating was deposited with the mixture of the MXene powder, acetylene black and poly(vinylidene fluoride) (PVDF), at a mass ratio of 80:10:10, in NMP as the solvent. The slurry was brushed on the Cu foils by using a doctor blade, with coating thicknesses of up to 100 μm, followed by drying at 80 °C for one in vacuum. The samples with the MXene coatings were subject to ALD process to deposit the SnO_2 and HfO_2 thin layers.

SnO_2 layer was deposited at either 150 °C or 200 °C by using a commercial ALD reactor. The deposition reaction consisted of five steps, i.e., (i) introduction of N_2 at 20 psi, (ii) use of Sn precursor (tetrakis(diethylamido)tin(IV)) for 0.5 s, (iii) reaction of the Sn precursor for 30 s, (iv) introduction of O_3 oxidizing agent for 0.2 s and (v) oxidation reaction for 15 s to achieve a coating rate of 0.1 nm per cycle. The ultrathin layers of HfO_2 were coated on the SnO_2-MXene hybrids at 180 °C in a similar way, where the ALD precursor of HfO_2 was tetrakis (dimethylamino) hafnium ($Hf(NMe_2)_4$)).

The SnO_2 nanoparticles were coated on the MXene nanosheets by using hydrothermal process and sputtering for comparison. Figure 3.62 shows a schematic diagram presenting the three processes to obtain the SnO_2-MXene hybrids. In the

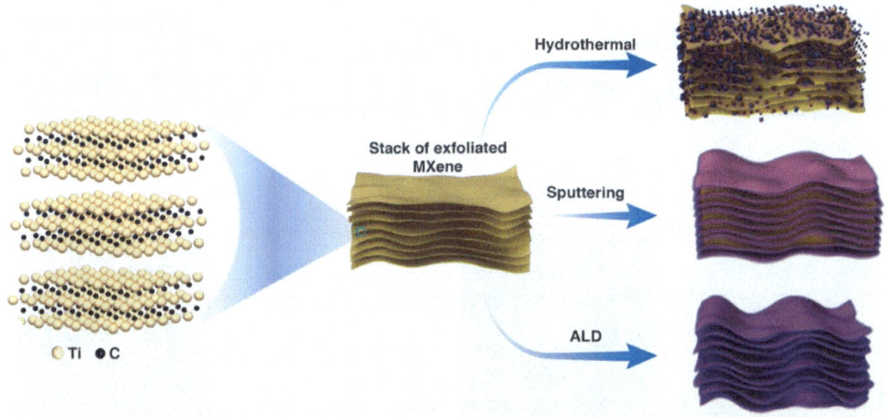

Fig. 3.62 Schematic diagram illustrating the ways to prepare the SnO_2 deposition on Ti_3C_2 MXene nanosheets, with hydrothermal reaction, sputtering and atomic layer deposition (ALD). Atomistic view of the exfoliated MXene nanosheets is demonstrated on the left, the stacked exfoliated MXene nanosheets are in the middle, while the MXene hybrid electrodes with presence of the SnO_2 nanoparticles are on the right. Reproduced with permission from [118]. Copyright © 2017, Elsevier

samples coated with SnO_2 nanoparticles through the hydrothermal method, the MXene nanosheets were distorted, owing to the high reaction temperature, the high pressure and aqueous condition. If sputtering method was used, the SnO_2 thin layer could not entirely cover the MXene nanosheets, because on the outer surfaces of the nanosheets were approachable, while the SnO_2 nanoparticles could not penetrate the MXene interlayer spaces. However, both the structure and morphology of the Ti_3C_2 MXene phase were not altered, simply because the sputtering process was non-conformal. Comparatively, the ALD process resulted in SnO_2 thin layers to be conformal, so that the surfaces of the MXene nanosheets had a nearly full coverage.

To facilitate the ALD process of the SnO_2 nanoparticles on the surface of the MXene nanosheets, the surface should be hydrophilic in nature, by having the functional groups of –O and/or –OH. In this case, the introduction of the oxidizing agent, such as O_3 and water vapor, was to produce the desired functional groups. However, due to their strong oxidation effect, the presence of ozone or water vapor would destroy the MXene nanostructures. To address this issue, the Sn precursor was introduced first before the introduction of water vapor or ozone. XRD results indicated that the MXene layered structures were damaged if the oxidation agent was introduced first due to the strong oxidation reaction. Otherwise, if the Sn precursor was filled in first in the ALD process, the nanostructure of the MXene phase was well retained.

HRTEM images of the MXene-based hybrids with the layer thickness of SnO_2 to be 50 nm coated at 150 °C and 200 °C are shown in Fig. 3.63a, b, respectively. Obviously, the layered and stacked structures of the MXene nanoflakes were well kept after the deposition of the SnO_2 thin layers. Their characteristic hexagonal crystal structure was confirmed by the SAED patterns. In addition, the diffraction of the

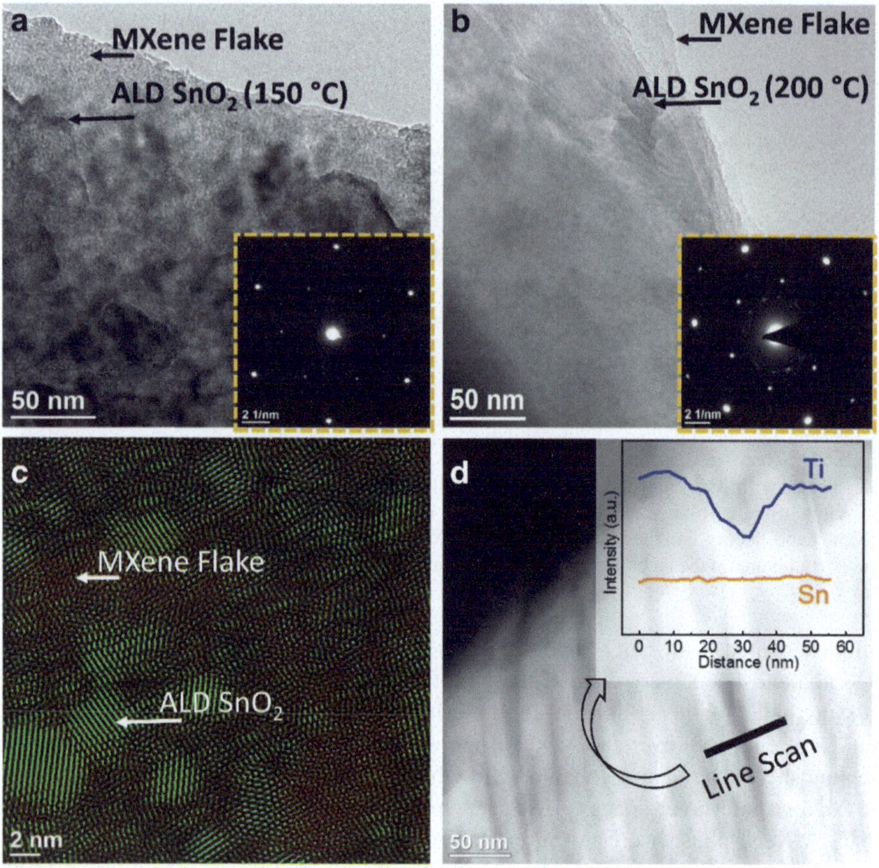

Fig. 3.63 TEM results of the MXene nanosheets deposited with a 50 nm layer of SnO_2 nanoparticles: (**a**) low-magnification TEM image and SAED pattern for ALD at 150 °C, (**b**) TEM image of the sample with ALD at 200 °C, (**c**) Fourier-filtered high-resolution RGB image of the sample with ALD at 200 °C (indicating the presence of two phases) and (**d**) STEM image along with the EDS line-scan for the sample with ALD at 150 °C (demonstrating the conformal SnO_2 layer). Reproduced with permission from [118]. Copyright © 2017, Elsevier

SnO_2 phase was very weak, suggesting that the SnO_2 nanoparticles on the MXene nanosheets after deposition at 150 °C was mainly amorphous, as reflected by the inset in Fig. 3.63a. In comparison, the sample the layer to be deposited at 200 °C, the diffraction pattern related to the SnO_2 phase was quite strong, as evidenced by the inset in Fig. 3.63b, which meant that SnO_2 nanoparticles were crystallized. Meanwhile, the layered structure of the MXene phase was retained at both the atomic and microscopic levels, which was attributed to two factors, i.e., (i) the use of a very low heating rate of 1 °C/min and (ii) the first introduction of the Sn precursor during the first cycle of the ALD process.

Figure 3.63c shows an RGB plot of the Fourier-filtered HRTEM image, demonstrating the presence of the SnO_2 nanoparticles decorated on the surface of the MXene nanosheets. The crystalline state of the SnO_2 nanoparticles was clearly observed in the matrix of the MXene phase. Moreover, the STEM image indicated that layered structure of the MXene nanosheets after the ALD process was well retained, as seen in Fig. 3.63d. The HRTEM image without Fourier filtering was compared with the corresponding images with fast Fourier transform (FFT) and applied filters. EDS line scan results of Sn and Ti are depicted as the inset in Fig. 3.63d, confirming that the SnO_2 thin layers were uniform and the content of Ti in between the nanoflakes was sufficiently low.

An in-situ wet-chemical nucleation process was used to obtain SnO_2–Ti_3C_2 nanohybrid phase through the electrostatic interaction between SnO_2 quantum dots (QDs) and $Ti_3C_2T_x$ nanosheets [119]. Due to the heterogeneous nucleation, the SnO_2 QDs were firmly attached onto the surface of the $Ti_3C_2T_x$ nanolayers. In return, the $Ti_3C_2T_x$ nanolayers thus suppressed the agglomeration of the SnO_2 QDs. Moreover, the specific surface area of the hybrid could be further enlarged, as the colloidal mixtures were freeze-dried. The resultant SnO_2 QDs-$Ti_3C_2T_x$ nanohybrid became highly conductive owing to the formation of the $Ti_3C_2T_x$ porous network. Finally, the nanohybrid-based anode displayed a high specific capacity of 1021 mAh/g at the current density of 100 mA/g, at a loading rate of 2.6 mg/cm. The capacitances were 810 mAh g^{-1} after 200 cycles at 100 mA/g, 697 mAh/g after 520 cycles at 200 mA/g and 500 mAh/g after 700 cycles at 500 mA/g, demonstrating a long-term stability of the anode.

A similar chemical process was reported to prepare SnO_2–Ti_3C_2 nanohybrids, with SnO_2 QDs to be decorated on $Ti_3C_2T_x$ nanosheets, by using self-assembly via electrostatic interactions [120]. The SnO_2-MXene hybrid nanoparticles consisted of SnO_2 QDs with an average diameter of about 5 nm, which were deposited onto the surface of the 2D MXene nanolayers with desirable contents. The SnO_2 QDs acted as both the active component for capacitive effect and the spacer isolate the MXene nanosheets, while the $Ti_3C_2T_x$ MXene nanosheets formed a porous network with high electrical conductivity. At the same time, the volumetric variation related to the lithiation-delithiation of the SnO_2 QDs was mitigated, because they were trapped in between the MXene nanosheets. Specifically, a high specific capacity of 887 mAh/g was observed at the current density of 50 mA/g. A retention of 91% was achieved after cycling for 100 times at 100 mA/g and promising rate stability was present with a specific capacitance of 660 mAh/g at 100 mA/g after 100 cycles.

SnO_2–Ti_3C_2 MXene nanohyrids with various contents of Ti_3C_2 powder, i.e., 0, 0.5, 1.0, 2.0 and 2.5 wt‰, were employed as the electron transport layers (ETLs) of low-temperature processed planar-structured organic–inorganic lead halide perovskite solar cells (PSCs) [121]. It was found that the SnO_2–Ti_3C_2 nanohybrid sample with 1.0 wt‰ Ti_3C_2 could raise the power conversion efficiency (PCE) 18.34% from 17.23% of that without the presence of Ti_3C_2. In comparison, if pure Ti_3C_2 was used as the ETL, the PCE was as low as 5.28%. According to the photoluminescence and electrochemical impedance spectroscopies, the Ti_3C_2 MXene nanolayers ensured smooth charge transfer pathways, due to their high electrical conductivity. As a result, both the electron extraction efficiency and the electron mobility were boosted, while the electron transfer resistance at the ETL/perovskite interfaces was reduced, so that a high photocurrent was achieved.

A chemical coprecipitation process was used to prepare nanohybrids consisting of Gd^{3+}- and Sn^{4+}-doped bismuth ferrite ($Bi_{1-x}Gd_xFe_{1-y}Sn_y$; BGFSO) and 2D MXene nanosheets are synthesized by [122]. The introduction of the 2D nanosheets enlarged the surface area and increased electrical conductivity of the BGFSO nanoparticles, thus leading to longer recombination time and enhanced dye degradation efficiency. Specifically, the BGFO-20Sn-MXene nanohybrid sample displayed a degradation efficiency of 100% to remove Congo dye from the corresponding solution in 2 h, making it promising candidate for large-scale applications.

The same authors prepared nanohybrid powders with La- and Mn-codoped BFO nanoparticles and MXene nanosheets, by using a simple double-solvent sol–gel process, for photocatalysis applications [123]. According to photoluminescence data, the codoped BFO-Ti_3C_2 nanohybrid samples all had lower rate of electron–hole recombination than the pristine BFO powder. The nanohybrids exhibited high generation rate of electron–hole pair, thus resulting in high photocatalytic efficiency for the degradation of organic dyes. The bandgap of the nanohybrids could be tailored to be up to1.73 eV. In addition, the BLFO-Ti_3C_2 and BLFMO-5-Ti_3C_2 showed degradation efficiencies of 92% and 93%, respectively, in removing the organic pollutant in aqueous state.

A simple two-step vacuum filtration approach was used to fabricate $CoFe_2O_4$-$Ti_3C_2T_x$ hybrid free-standing films, as electrode materials [124]. The MXene phase was employed to offer multiple functions, i.e., (i) as electrode to increase electrochemical capacity, (ii) as current collector to raise the electrical and ionic conductivities and (iii) as supports to enhance the mechanical strength of the materials. Figure 3.64 shows schematic diagram of the preparation process of the nanohybrid samples. Suspensions of $CoFe_2O_4$ nanoparticles and $Ti_3C_2T_x$ nanoflakes were mixed and homogenized with the aid of ultrasonication. After filtering with polypropylene membranes, hybrid layers were obtained, followed by drying with an IR light for 10 min, thus leading to free-standing samples.

Representative cross-sectional SEM image of the assembled electrode is depicted in Fig. 3.65a. A bilayer structure was present, with a thickness of about 10 μm, with tan active and a current collector layers. The active layer was made of the $CoFe_2O_4$–$Ti_3C_2T_x$ hybrid, where the $CoFe_2O_4$ nanoparticles were embedded in between the MXene nanoflakes through cross-linking. The current collector layer formed only with MXene nanoflakes. Figure 3.65c shows element maps of the hybrid samples, indicating uniform distributions of the elements in the individual layers. The $CoFe_2O_4$ nanoparticles were covered by the MXene nanoflakes to form the hybrid layer. The porous structure of the hybrid-based electrode ensured the ions of the electrolyte to diffuse into the inner sides of the materials, thus boosting the transportation of the Li^+ ions so as to facilitate the electrochemical reaction at the surface of the $CoFe_2O_4$ nanoparticles. Due to the magnetic characteristic of the $CoFe_2O_4$ nanoparticles, microsized spheres were formed because of their aggregation, if they were dispersed in water.

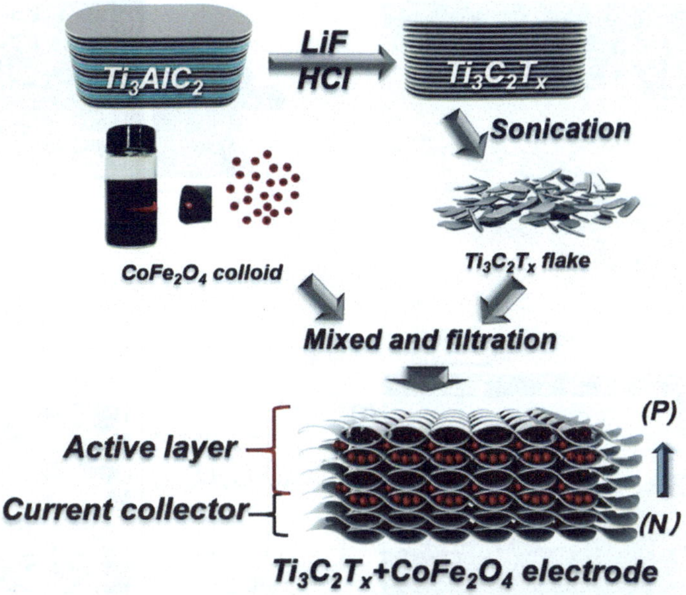

Fig. 3.64 Schematic diagram showing fabrication steps of the MXene-$CoFe_2O_4$ hybrid-based electrodes, with (P) and (N) representing the sides in contact with the electrolyte and the external circuit, respectively. Reproduced with permission from [124]. Copyright © 2019, Royal Society of Chemistry

As observed in Fig. 3.65e, the water drop had a wide angle of 72° on the positive surface of the hybrid electrode due to its hydrophobic behaviour, while the negative surface had a contact angle of only 14° owing to its hydrophobic effect caused by the functional groups on the surface of the $CoFe_2O_4$ nanoparticles. XRD results suggested that the MXene phase and the $CoFe_2O_4$ crystal were detected in the hybrid, in which the $CoFe_2O_4$ nanoparticles had an average crystal size of about 22 nm, according to the Scherrer equation. The hybrid had an electrical conductivity of about 8.2×10^3 S/m, which was much higher than that of pure $CoFe_2O_4$ sample, although it was lower than that of the pristine $Ti_3C_2T_x$ sheets. The unique properties of the $CoFe_2O_4$-$Ti_3C_2T_x$ hybrid-based electrode exhibited promising electrochemical performances.

A conductive 3D Ti_3C_2 porous architecture was reported using by 3D nickel foams (NF) to adsorb Ti_3C_2 MXene nanoflakes, which were used construct nanohybrids with sulfur-doped nickel ferrite nanosheets (S-$NiFe_2O_4$) [125]. The S-$NiFe_2O_4$@Ti_3C_2@NF nanostructures were finally obtained through thiourea-aided electrodeposition step, followed by a low temperature calcination process. Experimental results indicated that the S-$NiFe_2O_4$@Ti_3C_2@NF hybrid-based electrode displayed a high OER activity, with just 1.50 V versus RHE at the current density of 20 mA/cm^2, a Tafel slope of 47 mV/dec and a superb catalytic activity in 1 M KOH aqueous solution. The synergistic effect, of the 3D hierarchical nanostructure, the

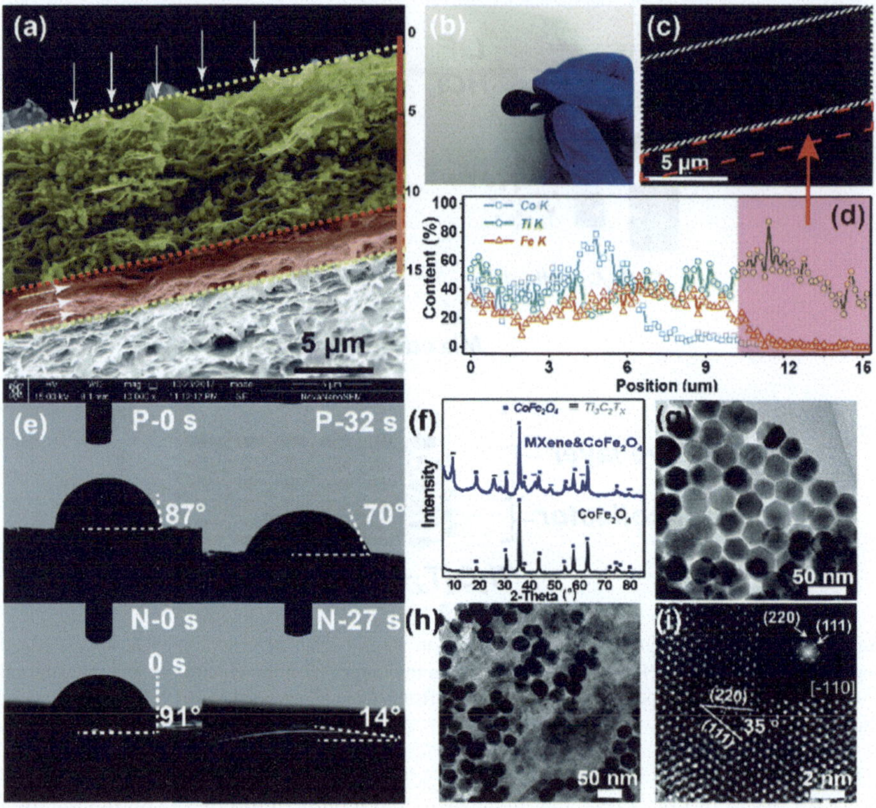

Fig. 3.65 **a** Cross-sectional SEM image, **b** photograph, **c** elemental mapping profiles, **d** line-scan elemental distributions of Ti, Co and Fe and **e** water contact angles of the positive (P) and negative (N) sides of the MXene-$CoFe_2O_4$ hybrid electrode. **f** XRD patterns of the $CoFe_2O_4$ and the MXene-$CoFe_2O_4$ hybrid electrode. TEM images of **g** the $CoFe_2O_4$ and **h** the MXene-$CoFe_2O_4$ hybrid. **i** HRTEM image of the $CoFe_2O_4$ nanoparticles. Reproduced with permission from [124]. Copyright © 2019, Royal Society of Chemistry

high electrical conductivity due to the Ti_3C_2 nanoflake framework, the high catalytic efficiency and the enriched catalytic active sites related to the S-$NiFe_2O_4$ nanosheets, was responsible for the outstanding OER effect of the nanohybrids.

A hydrothermal process was employed to construct $Ti_3C_2T_x$–CNZF, in which the CNZF ($Co_{0.2}Ni_{0.4}Zn_{0.4}Fe_2O_4$) nanoparticles were attached onto the surfaces of MXenes nanosheets [126]. The incorporation of the CNZF nanoparticles made the $Ti_3C_2T_x$–CNZF to have high dielectric loss tangent and multiple scattering effects. As a result, the hybrid possessed enhanced microwave absorbing capacity. Specifically, the promising microwave absorption parameters included a maximum reflectivity of −58.4 dB at the frequency of 6.2 GHz and an effective bandwidth of 3.8–6.0 GHz, at the thickness of 4.2 mm.

To prepare the $Ti_3C_2T_x$–CNZF hybrid, MXene suspensions were mixed with nitrate solutions of Co^{2+}, Ni^{2+}, Zn^{2+} and Fe^{3+}, with a molar ratio according to the ferrite chemical formula mentioned above. The mixture was adjusted to have a pH value of. Hydrothermal reaction of the precursor mixture was conducted at 180 °C for 12 h. At the end of the reaction, the $Ti_3C_2T_x$–CNZF hybrid powder was filtered and dried at 60 °C for 12 h. Similarly, both pristine $Ti_3C_2T_x$ and CNZF samples were prepared for comparison. It was found that the surfaces of MXene nanosheets were densely decorated with the CNZF nanoparticles, as demonstrated in Fig. 3.66a. Meanwhile, CNZF nanoparticles were also entrapped inside the interlayer spaces of the MXene nanosheets, as observed in Fig. 3.66b. Figure 3.66c shows HR-TEM image of the $Ti_3C_2T_x$–CNZF hybrid. The interplanar spacings of the CNZF nanoparticles were 0.479 and 0.248 nm, corresponding to the (111) and (311) crystal planes of spinel CNZF. The average size of ferrite nanoparticles was about 13 nm, similar to that estimated from the XRD patterns.

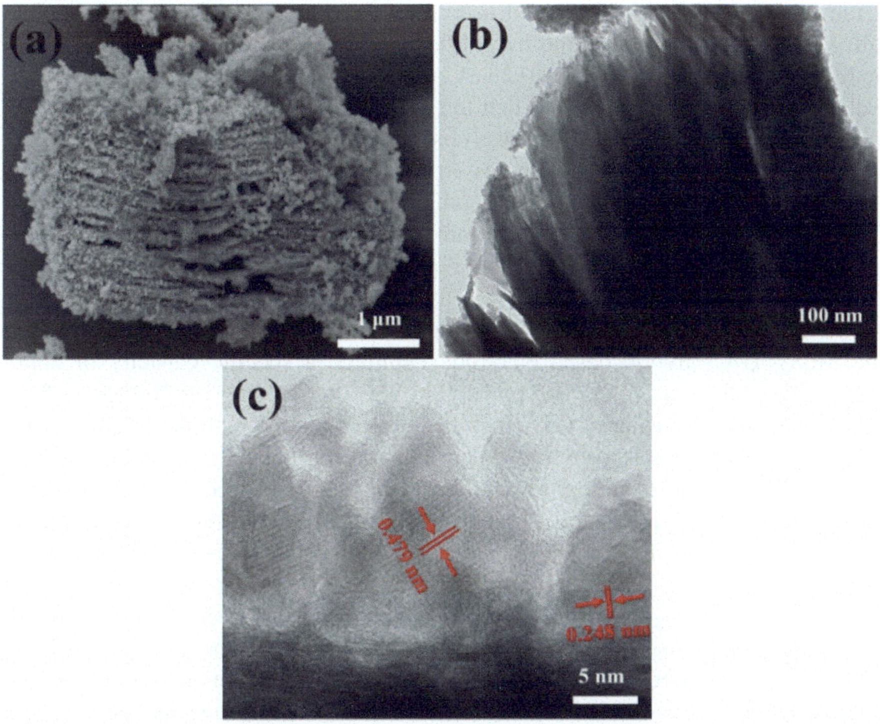

Fig. 3.66 Microstructures of the $Ti_3C_2T_x$–CNZF hybrid sample: **a** SEM, **b** TEM and **c** HR-TEM images. Reproduced with permission from [126]. Copyright © 2019, Elsevier

3.3.4 MXenes-Dichalcogenides

Various transition metal dichalcogenides (TMDC) have been incorporated with MXenes to form 2D-2D nanostructures that exhibited a number of new functionalities for a wide range of applications, especially in energy storages and conversions, which have been reported both experimentally and theoretically in the open literature [127–130]. Selected examples will be presented and discussed in this subsection, without ambitions to have a comprehensive coverage of the special group of materials, due to the space limitation.

It has been demonstrated that $Ti_2C(OH)_xF_y$ could be incorporated with transition metal dichalcogenides, including WSe_2 and MoS_2, to form 2D-2D hybrids, for applications in field effect transistors (FETs) [131]. Work functions of the hybrids were measured by using Kelvin probe force microscopy. The WSe_2-$Ti_2C(OH)_xF_y$ and MoS_2–$Ti_2C(OH)_xF_y$ nanohybrids were made into devices by using mechanical transferring process, in order to study their electronic properties. Schottky barrier heights of the materials were characterized according to temperature-dependent current–voltage transfer behaviors of the hybrid-based devices. The hole barriers between the MXene and the TMDCs were 0.23 eV and 0.19 eV, for WSe_2 and MoS_2, respectively. This observation suggested that the MoS_2–$Ti_2C(OH)_xF_y$ interface displayed pinning effect, while no such effect was observed in the case of WSe_2-$Ti_2C(OH)_xF_y$, simply because WSe_2 and MoS_2 possessed different band structures. With MoS_2 for n-channel and WSe_2 for p-channel, a complementary metal–oxide–semiconductor (CMOS) inverter was constructed, indicating potential applications of $Ti_2C(OH)_xF_y$ in nanoelectronic devices.

A solid state sintering process was used to prepare MoS_2–Ti_3C_2 hybrid, as electrode of LIBs, in order to achieve high electrochemical performances [132]. The MoS_2–Ti_3C_2 hybrid-based electrode exhibited a specific capacity of about 132 mAh/g at the current density of 1000 mA/g, after cycling for 200 times. This electrochemical performance was much better as compared with the pristine Ti_3C_2 and MoS_2. In addition, the hybrid-based electrode displayed outstanding cycling behavior. The enhanced properties of the hybrid-based electrode were ascribed to that fact that the MoS_2 nanoparticles were strong attached on the surface of the MXene nanosheets, thus boosting the charge transport and thus improve the electrochemical reactivity.

An in-situ strategy was employed to develop MoS_2-MXene hybrid through the sulfidation of $Mo_2TiC_2T_x$ MXene, also for the applications as anode of LIBs [133]. Theoretical prediction indicated that the MoS_2-MXene nanohybrid structure possessed metallic conduction behaviors. Meanwhile, the MXene nanolayers helped to promote the adsorbing rate of Li and Li_2S in the intercalating and energy conversion reacting processes, thus offering promising electrochemical performances when using as anode of LIBs.

MXene-1T-2H MoS_2-C hybrid nanoarchitectures were developed for the applications as electrode of Li–S batteries with improved electrochemical performances

[134]. As revealed by the high-resolution TEM and electron paramagnetic resonance (EPR) characterization results, MXene-1T-2H MoS_2-C hybrids contained S-vacancies with positive charges, which offered enriched active sites for adsorption and reaction for the polarized polysulfide intermediates. As a result, redox reactions were effectively accelerated and the dissolution of polysulfides was suppressed. Specifically, the MXene-1T-2H MoS_2-C-S hybrid-based cathode exhibited an initial specific capacity of about 1195 mAh/g at 0.1 C, together with a retained capacity of about 799 mAh/g after cycling for 300 times at 0.5 C.

Fabrication process of the MXene-1T-2H MoS_2-C nanohybrids is schematically illustrated in Fig. 3.67a. MXene powder with micrometer dimension and layered microstructure was used, as shown in Fig. 3.67b. MXene-2H MoS_2-C hybrid nanostructures were prepared by using hydrothermal treatment, starting with molybdate tetrahydrate (AMT) and thiourea, together with the MXene powder and glucose. The as-reacted samples were post-thermally calcined in N_2. Since the Ti_3C_2 MXene nanosheets contained negatively charged functional groups of –OH and –F, so that the cations of AMT were adsorbed through electrostatic interaction, so that MoS_2 was obtained due to the reaction of the cations with thiourea during the hydrothermal process. Meanwhile, a thin layer of carbon was generated on the surfaces of both the MoS_2 nanostructures and the MXene nanosheets, owing to the carbonization of the glucose during the calcination at 200 °C, following by the formation of the MXene-2H MoS_2-C nanoarchitectures after further calcining at 500 °C in N_2. Eventually, the

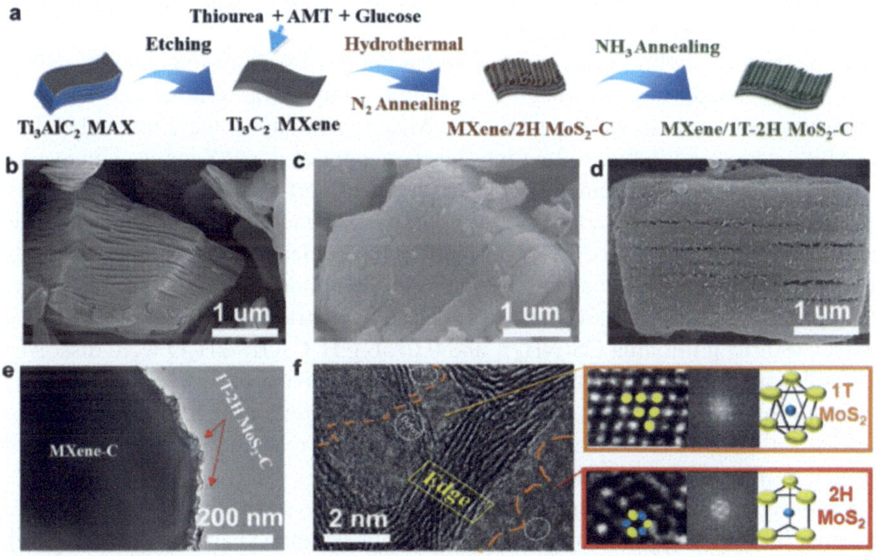

Fig. 3.67 **a** Schematic diagram describing fabrication steps of the MXene-1T-2H MoS_2-C nanohybrid structures. SEM images of **b** MXene, **c** MXene-2H MoS_2-C and **d** MXene-1T-2H MoS_2-C. **e** TEM and **f** HRTEM images of the MXene-1T-2H MoS_2-C. Reproduced with permission from [134]. Copyright © 2018, John Wiley & Sons

MXene-2H MoS_2-C sample was obtained by annealing the MXene-1T-2H MoS_2-C one in the environment of NH_3 at 600 °C for 2 h.

According to SEM and TEM results, the pristine 2H MoS_2 particles were assembled from flower-like nanostructures with sizes of 0.5–0.8 μm. With the addition of MXene, the MoS_2 nanosized items were evenly attached to the surfaces of the MXene nanolayers. In this case, MXene inhibited the agglomeration of the MoS_2 nanosheets. The thin layer of carbon was from the glucose precursor, which wrapped the MXene-2H MoS_2 nanostructures, thus leading to highly porous sandwich3D nanoarchitectures, i.e., MXene-2H MoS_2-C, as depicted in Fig. 3.67c. As observed in Fig. 3.67c, d, show that the final product MXene-1T-2H MoS_2-C had nearly no variation in microstructure, as compared with its precursor after calcination in NH_3.

Representative TEM/HRTEM images of the samples are depicted in Fig. 3.67e–h, revealing that the 1T phase was coexistent with the 2H phase. In addition, the 1T and 2H phases had octahedral and trigonal coordinating configurations, respectively. The coexistence of the two phases was also confirmed by the elemental profiles. The interlayer spacing of the 1T-2H MoS_2 structure was about 0.66 nm. In addition, the MoS_2 item was highly defective, with a large number of edges, as evidenced in the HRTEM images. These edges and defects were expected to serve as reactive sites to result in high catalytic activities because of the decrease in surface energies of the materials. Furthermore, the electrical conductivity of the 1T phase is much higher than that of the 2H phase. Therefore, the generation of the 1T phase would greatly bring out a high conductivity, besides the high catalytic efficiency. Also, it was found that the elements of C, Ti, S and N were all uniformly distributed in the MXene-1T-2H MoS_2-C nanostructures.

MoS_2-$Ti_3C_2T_x$ hybrid was also explored as anode of sodium-ion batteries (SIBs) [135]. In this study, the MoS_2–$Ti_3C_2T_x$ hybrid sample was synthesized by using a hydrothermal process, during which the MXene nanolayers were intercalated by the MoS_2 nanosized items, so that interlayer spacing of the MXene was increased, thus leading to an enhancement in electrochemical performances of the materials. Specifically, electrodes based on the MoS_2–$Ti_3C_2T_x$ nanohybrid exhibited a specific capacity of 251 mAh/g after cycling for 100 times. The electrode also possessed pretty high rate stability, with the specific capacity to be about 163 mAh/g at the current density of 1 A/g. To prepare the MoS_2–$Ti_3C_2T_x$ hybrid, 0.106 g $(NH_4)_6Mo_7O_{24} \cdot 4H_2O$, 0.0913 g thiourea $CS(NH_2)_2$ and 0.0756 g $H_2C_2O_4 \cdot 2H_2O$ were dissolved in 80 ml water, with the aid of strong stirring. After stirring for 0.5 h, 0.1 g $Ti_3C_2T_x$ was dispersed in the solution, followed by ultrasonication for 0.5 h. The mixture was then hydrothermally reacted at 200 °C for one day. The reactant was finally collected through thorough washing and vacuum drying.

Similarly, a simple hydrothermal process was employed to prepare MoS_2-MXene nanohybrid, for the application as cathode of magnesium ion batteries (MIBs) [136]. In the hybrid, few-layered MoS_2 nanolayers were incorporated with highly conductive MXene nanosheets, the hybrid-based electrode displayed a specific capacity of 165 mAh/g at the current density of 50 mA/g, while an excellent rate stability was observed by having a capacity of 93 mAh/g at 200 mA/g. Due to the high electrical conductivity and rich surface functional groups of the MXene, the electrochemical

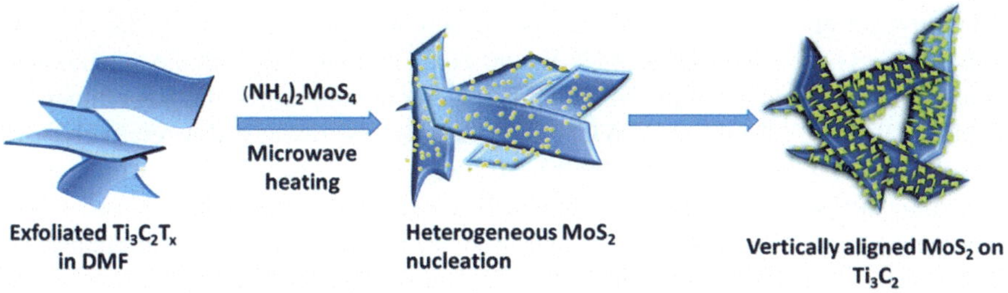

Fig. 3.68 Schematic diagram showing synthetic steps of the vertically aligned MoS_2–Ti_3C_2 nanohybrid. Ti_3C_2 nanoflakes were dispersed in DMF with the aid of ultrasonication, followed by the addition of $(NH_4)_2MoS_4$. The mixture was subject to microwave radiation to trigger the nucleation and vertical orientation of MoS_2 nanosheets on Ti_3C_2. Reproduced with permission from [137]. Copyright © 2018, Royal Society of Chemistry

performance of the hybrid materials was significantly enhanced, which could applicable to the design and fabrication of other electrode materials for energy storage applications.

A microwave-aided route was developed to obtain MoS_2–Ti_3C_2 hybrid, where MoS_2 nanosheets were vertically grown on conductive Ti_3C_2 MXene nanosheets, in order to achieve electrocatalysts with high activity for hydrogen evolution reactions (HER) [137]. In this unique MoS_2–Ti_3C_2 nanohybrid, the high catalytic activity of the edge sites of the MoS_2 phase was entirely utilized, due to the enlarged exposure related to the special arrangement. Meanwhile, the interlayer spacing of the few-layered MoS_2 nanosheets was increased, because of the vertical alignment. The MoS_2–Ti_3C_2 hybrid-based catalyst had a relatively low onset potential of −95 mV versus RHE for HER, together with a low Tafel slope of 40 mV/dec^{-1} and stable catalytic effect for nearly one day. To synthesize the MoS_2–Ti_3C_2 nanohybrid, precursor solutions were prepared by dissolving $(NH_4)_2MoS_4$ (ammonium tetrathio molybdate) in DMF (*N*,*N*-dimethyl formamide). Figure 3.68 shows schematic diagram presenting the formation process of the MoS_2–Ti_3C_2 nanostructure. The Ti_3C_2 nanosheets were employed as the support, owing to their high electrical conductivity and large specific surface area. Moreover, the surface charges well facilitated both the nucleation and growth of the MoS_2 items.

A ternary hybrid system, consisting of multiwall carbon nanotubes (MWCNTs), MoS_2 quantum dots (MoS_2 QDs) and $Ti_3C_2T_x$ QDs, i.e., MoS_2QDs-$Ti_3C_2T_x$QDs-MWCNTs, was developed, for electrocatalytic applications [138]. In the hybrid, MoS_2 QDs and $Ti_3C_2T_x$ QDs were both obtained through the exfoliation of MoS_2 and $Ti_3C_2T_x$ nanosheets by using a hydrothermal reaction process. The as-synthesized MoS_2QDs–$Ti_3C_2T_x$QDs-MMWCNTs sample contained a large number of functional groups and displayed a uniform tube-like nanostructure, thus enabling electrochemical performances.

It was found that the MoS_2QDs–$Ti_3C_2T_x$QDs-MWCNTs nanohybrid were electrocatalytically active in facilitating oxygen reduction reaction (ORR) and methanol oxidation reaction (MOR) in alkaline-based solutions, due to the synergistic electrocatalytic effects of both the QDs of MoS_2 and $Ti_3C_2T_x$, coupling with the outstanding electrochemical behavior of the MWCNTs. Specifically, in terms of ORR efficiency, the MoS_2QDs$Ti_3C_2T_x$QDs-MMWCNTs electrocatalyst displayed a relatively low Tafel slope of 90 mV/dec and a high half potential ($E_{1/2}$) of 0.75 V, which were better than the performances of most non-Pt catalysts and comparable with those of the commercial Pt/C catalysts. When used for electro-oxidation of methanol in alkaline solutions, a maximum methanol oxidation current density of 160 A/g was achieved at the potential of 2.2 V.

A facial route was reported to prepare MoS_2-MXene hybrid, consisting of Co-doped MoS_2 and Mo_2CT_x, which exhibited enhanced HER efficiency in alkaline solutions [139]. The improvement in HER performance was readily attributed to the synergistic effect of the increase in the number of sites for water dissociation and electrical conductivity of the materials. Due to the doping with Co, the electronic structure of MoS_2 was modified, while the intrinsic catalytic activity was enhanced. In the hybrid, the Mo_2CT_x MXene acted substrates that acted as both the active sites for electrochemical reaction and conductive path for electron transport. Specifically, the Co-MoS_2-Mo_2CT_x hybrid had a relatively low overpotential of 112 mV at the current density of 10 mA/cm^2, along with an outstanding long-span stability in alkaline solutions.

There have also been theoretical works to study MXenes that are possible to form hybrids with MoS_2 to shown enhanced performances. For instance, by using first-principles calculations, six MXenes of transition metal carbides or nitrides, including V_2CO_2, Cr_2CO_2, Mo_2CO_2, $V_4C_3O_2$, Cr_2NO_2 and V_2NO_2, could form metal contacts with monolayered MoS_2, which exhibited vanishing p-type Schottky barriers (SB) at the contacting interfaces, so that hole was efficiently injected into the MoS_2 layers [140]. It was found that the SB-free hole contacts were formed at the MoS_2-MXene interfaces, because MXene phases have relatively high work functions and no interfacial gap states were present. As a result, the Fermi level pinning effect in the midgap was avoided. The high efficiency charge injection into the MoS_2 layers of the SB-free contacts were expected to have high HER activity, since the conductivity of the 2H-MoS_2 basal plane was increased and the adsorption efficiency of hydrogen was enhanced.

MoS_2-MXene hybrids, together with other materials, were investigated in terms of primary antimicrobial mode-of-action, toward Gram-positive and Gram-negative bacteria [141]. Specifically, *Bacillus subtilis* and *Escherichia coli* bacteria were treated with different nanomaterials, with a concentration of 100 μg/mL, with either random orientation or vertical alignment configuration. The vertically aligned 2D nanosheets (MnO_2 and MoS_2) were grown on 2D nanosheets of Ti_3C_2 MXene, as well graphene oxide and reduced graphene oxide. Viability of the bacteria in the presence of the 2D nanohybrids was evaluated by adopting two complementary methods, i.e., flow cytometry and fluorescence imaging. It was observed that the antibacterial activities of MoS_2 and MnO_2 nanosheets were different, while the Gram-positive bacteria exhibited higher loss in membrane integrity than the Gram-negative ones. As revealed by SEM images, the 2D nanohybrids were detrimental to the bacteria viabilities, by compromising the cell walls and resulting in alternations in morphology. It was suggested that the peptidoglycan mesh (PM) in the walls of the bacteria could be attached primarily by the nanohybrid materials.

The development of MoS_2-based electronic devices is dependent on finding electrode materials suitable for forming Ohmic contacts with MoS_2. For this purpose, we carried out density functional theory and nonequilibrium Green's function calculations to investigate the possibility of using pristine and modified MXene ($Ta_2C/Ta_2CF_2/Ta_2C(OH)_2$) monolayers as the electrode materials in such devices [142]. These calculations indicated the formation of chemical bonds at the MoS_2/Ta_2C interface, and resulting strong orbital hybridization between the MoS_2 and Ta_2C components, but relatively weak interactions for MoS_2/Ta_2CF_2 and $MoS_2/Ta_2C(OH)_2$. Ohmic contacts were observed in all three cases. Transport properties were further simulated by modeling two-probe field effect transistors (FETs) with $Ta_2C/Ta_2CF_2/Ta_2C(OH)_2$ as electrodes. Interestingly, these simulations indicated the formation of Ohmic contacts between $Ta_2CF_2/Ta_2C(OH)_2$ electrodes and the channel, but an n-type Schottky barrier for the Ta_2C electrode. Furthermore, we found the resistance of the FET based on $MoS_2/Ta_2C(OH)_2$ to be half of that based on MoS_2/Ta_2CF_2. The results of our study not only revealed promising electrode materials for forming full Ohmic contacts with MoS_2 monolayers in FET devices, but also validated the effective role of a small-molecule fragment as a buffer layer in realizing Ohmic contacts between metals and semiconductors.

$MoSe_2$ has also been similarly incorporated with MXene to form nanohybrids for various applications, although it is not as extensively as its counterpart MoS_2. For example, $MoSe_2$/MXene hybrid nanostructures, $MoSe_2$-MXene-C, with a coating of carbon layer, were designed and fabricated for the application as anodes of potassium ion batteries (PIBs) [143]. In this case, the MXene nanosheets interacted with the $MoSe_2$ nanosheets, so that restacking of both items was prevented, while the incorporation of the MXene phase tremendously increased the electrical conductivity of the hybrid materials. Furthermore, the coating of the carbon film not only improved the mechanical integrity but also additionally enhanced electrical conductivity of the nanostructures. At the same time, the $MoSe_2$ nanosheets and the MXene nanoflakes strongly interacted at their interfaces, thus boosting the charge transport capability and enabling structural stability. As a result, the $MoSe_2$-MXene-C nanohybrid-based anode exhibited promising electrochemical performances, with a specific capacity of 355 mAh/g at the current density of 200 mA/g after cycling for 100 times. Also, the anode displayed a superb rate stability, with a specific capacity of 183 mAh/g at the high current density of 10 A/g.

Figure 3.69 shows schematic diagram describing the fabrication process of the $MoSe_2$-MXene-C ternary hybrid nanoarchitectures [143]. MXene nanoflake powder was obtained through the typical etching exfoliation approaches. Sodium molybdate (Na_2MoO_4) was used to offer the Mo component and selenium powder was utilized as the source of Se. $MoSe_2$-MXene nanohybrids were first obtained through hydrothermal reaction at 200 °C for 12 h, during which $MoSe_2$ nanosheets were attached on the surfaces of the MXene nanoflakes as vertical arrays. After that, the $MoSe_2$-MXene nanostructures were covered with a thin layer of polydopamine,

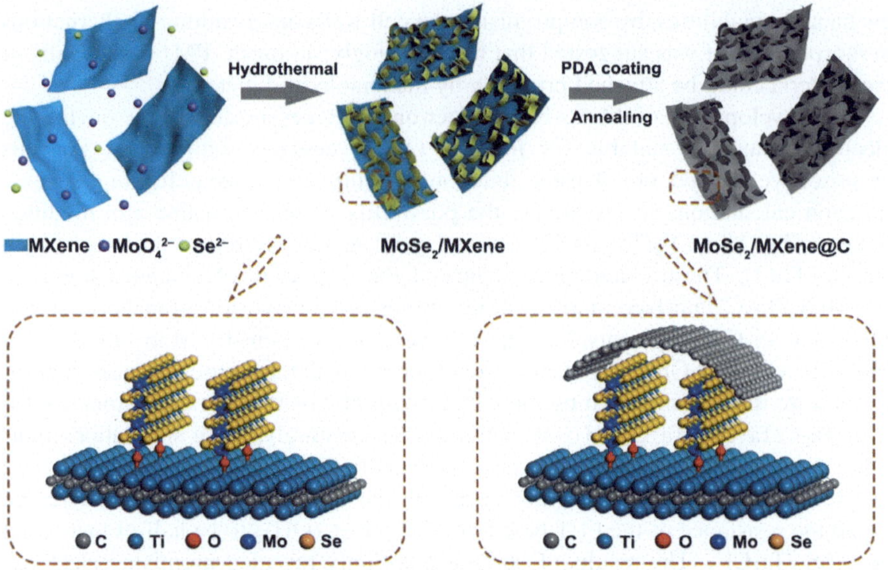

Fig. 3.69 Schematic diagram illustrating preparation process of the $MoSe_2$-MXene-C ternary hybrid nanostructures. Reproduced with permission from [143]. Copyright © 2019, American Chemical Society

leading to $MoSe_2$-MXene-PDA, followed by thermal treatment in H_2/Ar at 600 °C, finally resulting in $MoSe_2$-MXene-C nanoarchitectures. Atomic modelling indicated that the $MoSe_2$-MXene hybrid was formed, due to the covalent bonding between the $MoSe_2$ nanosheets and MXene nanoflakes. As a consequence, the hybrid nanostructure was mechanically stable and electrically conductive. Furthermore, the $MoSe_2$-MXene hybrid nanostructures were further reinforced after a thin layer of carbon was coated, so as to enable additional mechanical strength and electrical conductivity, as mention above.

Figure 3.70 shows representative SEM and TEM images of the nanohybrid samples, as well as elemental distribution profiles. It was found the 2D hierarchical $MoSe_2$-MXene nanostructure was well retained after the formation of the thin layer carbon coating, as revealed in Fig. 3.70a. Obviously, the surface of the MXene nanoflakes was entirely covered by the densely distributed $MoSe_2$ nanosheets, forming a 3D highly porous network structure, as observed in Fig. 3.70b. The presence of the $MoSe_2$-MXene interconnected nanoarchitecture was evidenced by the TEM images, as depicted in Fig. 3.70c. The HRTEM observation indicated that the lattice fringes had an interlayer spacing of 0.72 nm, as seen in Fig. 3.70d, corresponding to the (002) lattice planes of the $MoSe_2$ crystalline phase.

More importantly, the $MoSe_2$ nanosheets were uniformly coated with the thin layer of carbon, which had an average thickness of 2 nm. In addition, the elemental composition of the ternary nanostructure was highly homogeneous, as confirmed by the high-angle annular dark field (HAADF) image and the elemental mapping profiles,

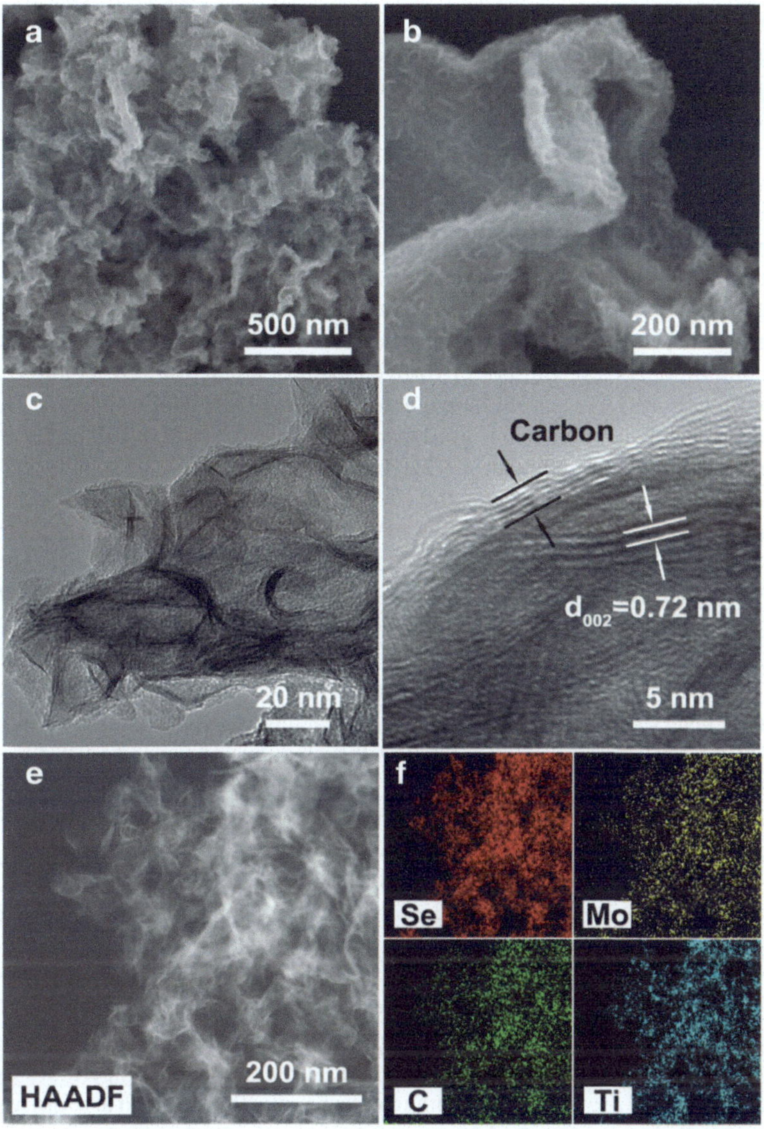

Fig. 3.70 Microstructures and chemical homogeneity of the $MoSe_2$-MXene-C nanohybrids: **a**, **b** SEM images at low and high magnifications, **c** TEM image, **d** HRTEM image, **e** HAADF image and **f** elemental mapping profiles. Reproduced with permission from [143]. Copyright © 2019, American Chemical Society

as illustrated in Fig. 3.70e, f, respectively. According to SEM energy-dispersive spectroscopy analysis results, the contents of $MoSe_2$, MXene (Ti_3C_2), and the carbon thin layer in the hybrid nanostructure were roughly 60.3 wt%, 6.2 wt% and 31.2 wt% respectively. The crystalline phases of $MoSe_2$ and Ti_3C_2 in the $MoSe_2$-MXene-C ternary hybrid were evidenced by the corresponding XRD results.

A simple process was used to prepare 1T/2H $MoSe_2$-MXene hybrid nanoheterostructure, for the application as an electrocatalyst to facilitate overall water splitting [144]. The 1T/2H $MoSe_2$ nanosheets contained rich active sites, which enabled the hybrid materials to have sufficiently high catalytic efficiency. In addition, the MXene nanoflake-based framework was highly conductive, thus boosting the charge transport and tackling the restacking issue of the 1T/2H $MoSe_2$ nanosheets at the same time. As a result, both hydrogen evolution reaction (HER) and oxygen evolution reaction (OER) could be triggered by the 1T/2H $MoSe_2$-MXene hybrid-based electrocatalyst, when it was used in alkaline solution, with much better overall performances as compared with 1T/2H $MoSe_2$, MXene and 2H $MoSe_2$-MXene. Specifically, for HER performance, the 1T/2H $MoSe_2$-MXene displayed an overpotential of 95 mV and a Tafel slope of 91 mV/dec for HER, while the two parameters were 340 mV and 90 mV/dec for OER. Moreover, the bifunctionally electrocatalysed overall water splitting could achieve a current density of 10 mA/cm^2 at just 1.64 V.

A facial vapor transportation (VT) process was employed to prepare WS_2-$Ti_3C_2T_x$ hybrid, by depositing WS_2 nanosheets onto the surfaces of the MXene nanoflakes, which showed enhanced photoluminescence (PL) effect [145]. Interestingly, the number of layers of the WS_2 nanosheets, either multilayer or even monolayer (1 L), could be well controlled by adjusting the chamber pressure and the location of the substrate. It was found that the WS_2 nanosheets were attached at the lateral sides and top surfaces of the MXene nanoflakes. The 1 L WS_2 nanosheets exhibited stronger PL effect than the multi-layered ones, while the PL performance of the 1 L WS_2 nanosheets grown on the lateral sides of the MXene nanoflakes was more pronounced than those deposited on the top surfaces. This kind of nanohybrid could find potential applications in various electronic and electrochemical devices.

A rapid ultrasonication-aided process was adopted to obtain WS_2-$Ti_3C_2T_x$ nanohybrids, by placing 1T WS_2 nanosheets in between the MXene nanoflakes, in order to achieve enhanced charge storage capabilities [146]. Ti_3AlC_2 was exfoliated in 40% HF aqueous solution with magnetic stirring at room temperature for one week to obtain Ti_3C_2-E, followed by intercalation with 40% TBAOH at room temperature for a half-day, leading to delaminated MXene, denoted as Ti_3C_2-D. WS_2 was exfoliated with 1.7 M t-BuLi in pentane in the protection of Ar at room temperature for three days, followed by filtration, high speed centrifugation, washing and drying, resulting in 1T WS_2 powder. The 1T WS_2 was incorporated the MXene samples obtained at different processing stages.

3.4 Conclusions and Outlook

MXenes with different compositions have been explored as components to form composites or hybrids with various other materials, including polymers, nanocarbon materials and inorganic items. Currently, the most widely and extensively studied MXene for the preparation of polymer-based compositions has been mainly $Ti_3C_2T_x$ as the active inclusions, which consisted of single- or multi-layer structures. Also, the MXene is usually obtained through the exfoliation with HF solution and intercalation with DMSO. In comparison, MXenes with other compositions have not been well touched, which are in fact deserved to be paid more attentions. In addition, the polymer-MXene interfacial interactions should be thoroughly examined, so as to reveal more details about the mechanisms. Quantitative characterization of the enhanced mechanical strength of the polymeric-MXene nanocomposites, with different MXene inclusions is still rarely available in the open literature. Homogeneous distribution of the MXene-based fillers in different polymer matrix without the occurrence of restacking variation in processing pathways to manipulate stacking of fillers within polymeric matrices can provide invaluable insights and contribute to the development of future applications.

Polymer-MXene nanocomposites have been demonstrated to show various applications, such as sequestrating heavy metallic pollutants, degrading organic dyes and radionuclides. Being utilized as biocides for the treatment of waste water, pretty high efficiency has been observed. In this case, absorption mechanisms for different metallic ions should be further clarified. Also, characteristics such as adsorption efficiency, reversible capability and long-term stability of the MXene-based nanocomposites should be systematically studied, when they are employed for such applications.

Carbon-based nanomaterials, including carbon blacks (0D), carbon nanotubes and carbon nanofibers (1D) and graphene nanosheets (2D), have been incorporated with MXenes to create a new family of hybrid nanostructures. These hybrid-based materials could be used for applications in sustainable energies, energy storage and conversion. For carbon blacks, the incorporation with MXene nanosheets enabled a high dispersion capability, while the conductive MXene framework led to materials with enhanced electrical conductivity. Other carbon nanomaterials are also capable of improve electrical conductivity of the hybrid materials, while 1D nanotubes and nanofibers facilitated products with more complicated tangling interactions with the MXenes. As for 2D graphene or graphene oxide nanosheets, more interlayer interactions are encountered, so that the restacking issue of the two components could be readily addressed.

For this group of hybrids, systematic and comprehensive data for each form of the nanocarbons should be established as reference for design and fabrication of new materials with more functions and applications. For example, both carbon nanotubes and graphenes can be incorporated with MXenes with either physical or chemical methods. Physical mixing is simple and productively, while the combination of the two components is relatively weak. In comparison, the interactions between

MXene nanosheets and the nanocarbons that are chemically grown on them are much stronger, whereas the products will be more expensive and productivity is relatively low. Most likely, for different applications, the processing methods could be different, because the requirements on the properties of the final materials are different.

Nobel metal nanoparticles have strong catalytic activities. Due to their extremely small size, they strongly tended to agglomerate, thus losing their functionalities. Incorporation with MXenes is an effective way to prevent them from aggregation. Meanwhile, the MXenes acted as a host to support the metallic nanoparticles, without the occurrence of variation in their morphologies during the applications. However, no principles have been established to govern the processing, characterization and applications of the nobel metal-MXene nanohybrids.

The largest group of hybrids based on MXenes and their derivatives have been constructed with various oxides, especially transition metal oxides. The applications of these materials are mainly on electrodes of supercapacitors and various ion batteries, in which the main function of the MXene phase is to offer a porous network structure and enhance mechanical integrity and electrical conductivity, simply because most oxides are electrical insulators. The porous nanoarchitectures are especially important in ensuring the access of ions to facilitate the electrochemical reactions, when using as electrodes of energy storage devices. It seems that all oxides can be combined with the MXene nanosheets. However, it has not been clarified what the advantages and disadvantages of such hybrids.

As a newly emerged 2D materials, transition metal dichalcogenides (TMDC), have been demonstrated to have various special and interesting properties, which could be useful for different applications. It is thus quite natural to combine them with MXenes to further improve their properties and performances and create new materials with new functions. Among various TMDCs, MoS_2 has been the one that is studied most extensively. With the development of the hybrid approach, more and more TMDCs will be adopted to be incorporated with MXenes.

Acknowledgements This work was supported by the National Natural Science Foundation of China (51762023 and 51962013), the Natural Science Foundation of Jiangxi, China (20192ACB20018), and Key R&D Program of Jiangxi Province (20171BBE50006, 20192ACB80007, and 20192ACB80004). Ling Bing Kong would like acknowledge Shenzhen Technology University (SZTU) for financial support through the Start-up Grant (2018) and grant from the Natural Science Foundation of Top Talent of SZTU (grant no. 2019010801002).

References

1. Mashtalir, O., Lukatskaya, M.R., Zhao, M.Q., Barsoum, M.W., Gogotsi, Y.: Amine-assisted delamination of Nb_2C MXene for Li-ion energy storage devices. Adv. Mater. **27**, 3501–3506 (2015)
2. Zhao, M.Q., Ren, C.E., Ling, Z., Lukatskaya, M.R., Zhang, C.F., Van Aken, K.L., et al.: Flexible MXene/carbon nanotube composite paper with high volumetric capacitance. Adv. Mater. **27**, 339–345 (2015)

3. Boota, M., Anasori, B., Voigt, C., Zhao, M.Q., Barsoum, M.W., Gogotsi, Y.: Pseudocapacitive electrodes produced by oxidant-free polymerization of pyrrole between the layers of 2D titanium carbide (MXene). Adv. Mater. **28**, 1517–1522 (2016)
4. Xie, X.Q., Zhao, M.Q., Anasori, B., Maleski, K., Ren, C.E., Li, J.W., et al.: Porous heterostructured MXene/carbon nanotube composite paper with high volumetric capacity for sodium-based energy storage devices. Nano Energy. **26**, 513–523 (2016)
5. Lin, Z.Y., Sun, D.F., Huang, Q., Yang, J., Barsoum, M.W., Yan, X.B.: Carbon nanofiber bridged two-dimensional titanium carbide as a superior anode for lithium-ion batteries. J. Mater. Chem. A **3**, 14096–14100 (2015)
6. Ng, V.M.H., Huang, H., Zhou, K., Lee, P.S., Que, W.X., Xu, J.Z., et al.: Recent progress in layered transition metal carbides and/or nitrides (MXenes) and their composites: synthesis and applications. J. Mater. Chem. A **5**, 3039–3068 (2017)
7. Luo, J.M., Tao, X.Y., Zhang, J., Xia, Y., Huang, H., Zhang, L.Y., et al.: Sn^{4+} ion decorated highly conductive Ti_3C_2 MXene: promising lithium-ion anodes with enhanced volumetric capacity and cyclic performance. ACS Nano **10**, 2491–2499 (2016)
8. Dall'Agnese, Y., Lukatskaya, M.R., Cook, K.M., Taberna, P.L., Gogotsi, Y., Simon, P.: High capacitance of surface-modified 2D titanium carbide in acidic electrolyte. Electrochem. Commun. **48**, 118–122 (2014)
9. Peng, Q., Guo, J., Zhang, Q., Xiang, J., Liu, B., Zhou, A., et al.: Unique lead adsorption behavior of activated hydroxyl group in two-dimensional titanium carbide. J. Am. Chem. Soc. **136**, 4113–4116 (2014)
10. Yang, C.H., Que, W.X., Yin, X.T., Tian, Y.P., Yang, Y.W., Que, M.D.: Improved capacitance of nitrogen-doped delaminated two-dimensional titanium carbide by urea-assisted synthesis. Electrochim. Acta **225**, 416–424 (2017)
11. Naguib, M., Mashtalir, O., Lukatskaya, M.R., Dyatkin, B., Zhang, C., Presser, V., et al.: One-step synthesis of nanocrystalline transition metal oxides on thin sheets of disordered graphitic carbon by oxidation of MXenes. Chem. Commun. **50**, 7420–7423 (2014)
12. Zhang, C.F., Kim, S.J., Ghidiu, M., Zhao, M.Q., Barsoum, M.W., Nicolosi, V., et al.: Layered orthorhombic Nb_2O_5@$Nb_4C_3T_x$ and TiO_2@$Ti_3C_2T_x$ hierarchical composites for high performance Li-ion batteries. Adv. Func. Mater. **26**, 4143–4151 (2016)
13. Ahmed, B., Anjum, D.H., Hedhili, M.N., Gogotsi, Y., Alshareef, H.N.: H_2O_2 assisted room temperature oxidation of Ti_2C MXene for Li-ion battery anodes. Nanoscale **8**, 7580–7587 (2016)
14. Gao, Y.P., Wang, L.B., Zhou, A.G., Li, Z.Y., Chen, J.K., Bala, H., et al.: Hydrothermal synthesis of TiO_2/Ti_3C_2 nanocomposites with enhanced photocatalytic activity. Mater. Lett. **150**, 62–64 (2015)
15. Wang, F., Yang, C.H., Duan, M., Tang, Y., Zhu, J.F.: TiO_2 nanoparticle modified organ-like Ti_3C_2 MXene nanocomposite encapsulating hemoglobin for a mediator-free biosensor with excellent performances. Biosens. Bioelectron. **74**, 1022–1028 (2015)
16. Zhu, J.F., Tang, Y., Yang, C.H., Wang, F., Cao, M.J.: Composites of TiO_2 nanoparticles deposited on Ti_3C_2 MXene nanosheets with enhanced electrochemical performance. J. Electrochem. Soc. **163**, A785–A791 (2016)
17. Wang, H., Peng, R., Hood, Z.D., Naguib, M., Adhikari, S.P., Wu, Z.L.: Titania composites with 2D transition metal carbides as photocatalysts for hydrogen production under visible-light irradiation. ChemSusChem **9**, 1490–1497 (2016)
18. Peng, C., Yang, X.F., Li, Y.H., Yu, H., Wang, H.J., Peng, F.: Hybrids of two-dimensional Ti_3C_2 and TiO_2 exposing 001 facets toward enhanced photocatalytic activity. ACS Appl. Mater. Interfaces **8**, 6051–6060 (2016)
19. Zhang, Q.R., Teng, J., Zou, G.D., Peng, Q.M., Du, Q., Jiao, T.F., et al.: Efficient phosphate sequestration for water purification by unique sandwich-like MXene/magnetic iron oxide nanocomposites. Nanoscale **8**, 7085–7093 (2016)
20. Ling, Z., Ren, C.E., Zhao, M.Q., Yang, J., Giammarco, J.M., Qiu, J.S., et al.: Flexible and conductive MXene films and nanocomposites with high capacitance. Proc. Natl. Acad. Sci. **111**, 16676–16681 (2014)

21. Liu, R.P., Li, W.H.: High-thermal-stability and high-thermal-conductivity $Ti_3C_2T_x$ MXene/poly(vinyl alcohol) (PVA) composites. ACS Omega **3**, 2609–2617 (2018)
22. Sobolciak, P., Ali, A., Hassan, M.K., Helal, M.I., Tanvir, A., Popelka, A., et al.: 2D $Ti_3C_2T_x$ (MXene)-reinforced polyvinyl alcohol (PVA) nanofibers with enhanced mechanical and electrical properties. PLoS ONE **12**, e018370 (2017)
23. Xu, H.L., Yin, X.W., Li, X.L., Li, M.H., Liang, S., Zhang, L.T., et al.: Lightweight Ti_2CT_x MXene/poly(vinyl alcohol) composite foams for electromagnetic wave shielding with absorption-dominated feature. ACS Appl. Mater. Interfaces **11**, 10198–10207 (2019)
24. Naguib, M., Saito, T., Lai, S., Rager, M.S., Aytug, T., Parans Paranthaman, M., et al.: $Ti_3C_2T_x$ (MXene)-polyacrylamide nanocomposite films. RSC Adv. **6**, 72069–72073 (2016)
25. Wu, X.L., Hao, L., Zhang, J.K., Zhang, X., Wang, J.T., Liu, J.D.: Polymer-$Ti_3C_2T_x$ composite membranes to overcome the trade-off in solvent resistant nanofiltration for alcohol-based system. J. Membr. Sci. **515**, 175–188 (2016)
26. Huang, Z.Y., Wang, S.J., Kota, S., Pan, Q.W., Barsoum, M.W., Li, C.Y.: Structure and crystallization behavior of poly(ethylene oxide)/$Ti_3C_2T_x$ MXene nanocomposites. Polymer **102**, 119–126 (2016)
27. Mayerberger, E.A., Urbanek, O., McDaniel, R.M., Street, R.M., Barsoum, M.W., Schauer, C.L.: Preparation and characterization of polymer-$Ti_3C_2T_x$ (MXene) composite nanofibers produced via electrospinning. J. Appl. Polym. Sci. **134**, 45295 (2017)
28. Chen, J., Chen, K., Tong, D.Y., Huang, Y.J., Zhang, J.W., Xue, J.M., et al.: CO_2 and temperature dual responsive "smart" MXene phases. Chem. Commun. **51**, 314–317 (2015)
29. Chen, Z.X., Han, Y.Q., Li, T.X., Zhang, X.W., Wang, T.Q., Zhang, Z.L.: Preparation and electrochemical performances of doped MXene/poly (3,4-ethylenedioxythiophene) composites. Mater. Lett. **220**, 305–308 (2018)
30. Ren, Y.Y., Zhu, J.F., Wang, L., Liu, H., Liu, Y., Wu, W.L., et al.: Synthesis of polyaniline nanoparticles deposited on two-dimensional titanium carbide for high-performance supercapacitors. Mater. Lett. **214**, 84–87 (2018)
31. VahidMohammadi, A., Moncada, J., Chen, H.Z., Kayali, E., Orangi, J., Carrero, C.A., et al.: Thick and freestanding MXene/PANI pseudocapacitive electrodes with ultrahigh specific capacitance. J. Mater. Chem. A **6**, 22123–22133 (2018)
32. Wei, H.W., Dong, J.D., Fang, X.J., Zheng, W.H., Sun, Y.T., Qian, Y., et al.: $Ti_3C_2T_x$ MXene/polyaniline (PANI) sandwich intercalation structure composites constructed for microwave absorption. Compos. Sci. Technol. **169**, 52–59 (2019)
33. Lu, X., Zhu, J.F., Wu, W.L., Zhang, B.: Hierarchical architecture of PANI@TiO_2/$Ti_3C_2T_x$ ternary composite electrode for enhanced electrochemical performance. Electrochim. Acta **228**, 282–289 (2017)
34. Levitt, A.S., Alhabeb, M., Hatter, C.B., Sarycheva, A., Dion, G., Gogotsi, Y.: Electrospun MXene/carbon nanofibers as supercapacitor electrodes. J. Mater. Chem. A **7**, 269–277 (2019)
35. Ghidiu, M., Kota, S., Halim, J., Sherwood, A.W., Nedfors, N., Rosen, J., et al.: Alkylammonium cation intercalation into Ti_3C_2 (MXene): effects on properties and ion-exchange capacity estimation. Chem. Mater. **29**, 1099–1106 (2017)
36. Yan, P.T., Zhang, R.J., Jia, J., Wu, C., Zhou, A.G., Xu, J., et al.: Enhanced supercapacitive performance of delaminated two-dimensional titanium carbide/carbon nanotube composites in alkaline electrolyte. J. Power Sources **284**, 38–43 (2015)
37. Yu, P., Cao, G.J., Yi, S., Zhang, X., Li, C., Sun, X.Z., et al.: Binder-free 2D titanium carbide (MXene)/carbon nanotube composites for high-performance lithium-ion capacitors. Nanoscale **10**, 5906–5913 (2018)
38. Li, X.L., Yin, X.W., Han, M.K., Song, C.Q., Xu, H.L., Hou, Z.X., et al.: Ti_3C_2 MXenes modified with in situ grown carbon nanotubes for enhanced electromagnetic wave absorption properties. J. Mater. Chem. C **5**, 4068–4074 (2017)
39. Li, X.L., Zhu, J.F., Wang, L., Wu, W.L., Fang, Y.: In-situ growth of carbon nanotubes on two-dimensional titanium carbide for enhanced electrochemical performance. Electrochim. Acta **258**, 291–301 (2017)

40. Zheng, W., Zhang, P., Chen, J., Tian, W.B., Zhang, Y.M., Sun, Z.M.: In situ synthesis of CNTs@Ti_3C_2 hybrid structures by microwave irradiation for high-performance anodes in lithium ion batteries. J. Mater. Chem. A **6**, 3543–3551 (2018)
41. Zhang, Y.K., Jiang, H.L., Lin, Y.X., Liu, H.J., He, Q., Wu, C.Q., et al.: In situ growth of cobalt nanoparticles encapsulated nitrogen-doped carbon nanotubes among $Ti_3C_2T_x$ (MXene) matrix for oxygen reduction and evolution. Adv. Mater. Interfaces **5**, 1800392 (2018)
42. Chen, J.N., Yuan, X.L., Lyu, F.L., Zhong, Q.X., Hu, H.C., Pan, Q., et al.: Integrating MXene nanosheets with cobalt-tipped carbon nanotubes for an efficient oxygen reduction reaction. J. Mater. Chem. A **7**, 1281–1286 (2019)
43. Yang, L., Zheng, W., Zhang, P., Chen, J., Tian, W.B., Zhang, Y.M., et al.: MXene/CNTs films prepared by electrophoretic deposition for supercapacitor electrodes. J. Electroanal. Chem. **830**, 1–6 (2018)
44. Lv, L.P., Guo, C.F., Sun, W.W., Wang, Y.: Strong surface-bound sulfur in carbon nanotube bridged hierarchical Mo_2C-based MXene nanosheets for lithium-sulfur batteries. Small **15**, 1804338 (2019)
45. Aïssa, B., Ali, A., Mahmoud, K.A., Haddad, T., Nedil, M.: Transport properties of a highly conductive 2D $Ti_3C_2T_x$ MXene/graphene composite. Appl. Phys. Lett. **109**, 043109 (2016)
46. Li, H.Y., Hou, Y., Wang, F.X., Lohe, M.R., Zhuang, X.D., Niu, L., et al.: Flexible all-solid-state supercapacitors with high volumetric capacitances boosted by solution processable MXene and electrochemically exfoliated graphene. Adv. Energy Mater. **7**, 1601847 (2017)
47. Yan, J., Ren, C.E., Maleski, K., Hatter, C.B., Anasori, B., Urbankowski, P., et al.: Flexible MXene/graphene films for ultrafast supercapacitors with outstanding volumetric capacitance. Adv. Func. Mater. **27**, 1701264 (2017)
48. Xu, S.K., Wei, G.D., Li, J.Z., Han, W., Gogotsi, Y.: Flexible MXene-graphene electrodes with high volumetric capacitance for integrated co-cathode energy conversion/storage devices. J. Mater. Chem. A **5**, 17442–17451 (2017)
49. Yang, Q.Y., Xu, Z., Fang, B., Huang, T.Q., Cai, S.Y., Chen, H., et al.: MXene/graphene hybrid fibers for high performance flexible supercapacitors. J. Mater. Chem. A **5**, 22113–22119 (2017)
50. Ma, Z.Y., Zhou, X.F., Deng, W., Lei, D., Liu, Z.P.: 3D porous MXene (Ti_3C_2)/reduced graphene oxide hybrid films for advanced lithium storage. ACS Appl. Mater. Interfaces **10**, 3634–3643 (2018)
51. Xu, S.K., Dall'Agnese, Y., Li, J.Z., Gogotsi, Y., Han, W.: Thermally reduced graphene/MXene film for enhanced Li-ion storage. Chem. A Eur. J. **24**, 18556–18563 (2018)
52. Ma, Y.N., Yue, Y., Zhang, H., Cheng, F., Zhao, W.Q., Rao, J.Y., et al.: 3D synergistical MXene/reduced graphene oxide aerogel for a piezoresistive sensor. ACS Nano **12**, 3209–3216 (2018)
53. Yue, Y., Liu, N.S., Ma, Y.N., Wang, S.L., Liu, W.J., Luo, C., et al.: Highly self-healable 3D microsupercapacitor with MXene-graphene composite aerogel. ACS Nano **12**, 4224–4232 (2018)
54. Zhao, S., Zhang, H.B., Luo, J.Q., Wang, Q.W., Xu, B., Hong, S., et al.: Highly electrically conductive three-dimensional $Ti_3C_2T_x$ MXene/reduced graphene oxide hybrid aerogels with excellent electromagnetic interference shielding performances. ACS Nano **12**, 11193–11202 (2018)
55. Qu, L.L., Wang, S.P., Yang, X.W., Sun, C.J.: MXene/reduced graphene oxide hydrogel film extraction combined with gas chromatography-tandem mass spectrometry for the determination of 16 polycyclic aromatic hydrocarbons in river and tap water. J. Chromatogr. A **1584**, 24–32 (2019)
56. Ma, T.Y., Cao, J.L., Jaroniec, M., Qiao, S.Z.: Interacting carbon nitride and titanium carbide nanosheets for high-performance oxygen evolution. Angew. Chem. Int. Ed. **55**, 1138–1142 (2016)
57. Shao, M.M., Shao, Y.F., Chai, J.W., Qu, Y.J., Yang, M.Y., Wang, Z.L., et al.: Synergistic effect of 2D Ti_2C and g-C_3N_4 for efficient photocatalytic hydrogen production. J. Mater. Chem. A **5**, 16748–16756 (2017)

58. Su, T.M., Hood, Z.D., Naguib, M., Bai, L., Luo, S., Rouleau, C.M., et al.: 2D/2D heterojunction of $Ti_3C_2/g\text{-}C_3N_4$ nanosheets for enhanced photocatalytic hydrogen evolution. Nanoscale, 30788480 (2019)
59. An, X.Q., Wang, W., Wang, J.P., Duan, H.Z., Shi, J.T., Yu, X.L.: The synergetic effects of Ti_3C_2 MXene and Pt as co-catalysts for highly efficient photocatalytic hydrogen evolution over $g\text{-}C_3N_4$. Phys. Chem. Chem. Phys. **20**, 11405–11411 (2018)
60. Zou, G.D., Zhang, Z.W., Guo, J.X., Liu, B.Z., Zhang, Q.R., Fernandez, C., et al.: Synthesis of MXene/Ag composites for extraordinary long cycle lifetime lithium storage at high rates. ACS Appl. Mater. Interfaces **8**, 22280–22286 (2016)
61. Li, L., Zhang, N., Zhang, M.Y., Wu, L., Zhang, X.T., Zhang, Z.G.: Ag-nanoparticle-decorated 2D titanium carbide (MXene) with superior electrochemical performance for supercapacitors. ACS Sustain. Chem. Eng. **6**, 7442–7450 (2018)
62. Zhang, Z.W., Li, H.N., Zou, G.D., Fernandez, C., Liu, B.Z., Zhang, Q.R., et al.: Self-reduction synthesis of new MXene/Ag composites with unexpected electrocatalytic activity. ACS Sustain. Chem. Eng. **4**, 6763–6771 (2016)
63. Pandey, R.P., Rasool, K., Madhavan, V.E., Aissa, B., Gogotsi, Y., Mahmoud, K.A.: Ultrahigh-flux and fouling-resistant membranes based on layered silver/MXene ($Ti_3C_2T_x$) nanosheets. J. Mater. Chem. A **6**, 3522–3533 (2018)
64. Song, D.D., Jiang, X.Y., Li, Y.S., Lu, X., Luan, S.R., Wang, Y.Z., et al.: Metal-organic frameworks-derived MnO_2/Mn_3O_4 microcuboids with hierarchically ordered nanosheets and Ti_3C_2 MXene/Au NPs composites for electrochemical pesticide detection. J. Hazard. Mater. **373**, 367–376 (2019)
65. Tang, W.T., Dong, Z.L., Zhang, R., Yi, X., Yang, K., Jin, M.L., et al.: Multifunctional two-dimensional core-shell MXene@gold nanocomposites for enhanced photo-radio combined therapy in the second biological window. ACS Nano **13**, 284–294 (2019)
66. Xie, H.H., Li, P.H., Shao, J.D., Huang, H., Chen, Y., Jiang, Z.Y., et al.: Electrostatic self-assembly of $Ti_3C_2T_x$ MXene and gold nanorods as an efficient surface-enhanced Raman scattering platform for reliable and high-densitivity determination of organic pollutants. ACS Sens. **4**, 2303–2310 (2019)
67. Lorencova, L., Bertok, T., Filip, J., Jerigova, M., Velic, D., Kasak, P., et al.: Highly stable $Ti_3C_2T_x$ (MXene)/Pt nanoparticles-modified glassy carbon electrode for H_2O_2 and small molecules sensing applications. Sens. Actuators B Chem. **263**, 360–368 (2018)
68. Filip, J., Zavahir, S., Lorencova, L., Bertok, T., Bin Yousaf, A., Mahmoud, K.A., et al.: Tailoring electrocatalytic properties of Pt nanoparticles grown on $Ti_3C_2T_x$ MXene surface. J. Electrochem. Soc. **166**, H54–H62 (2019)
69. Wang, Y.J., Wang, J.K., Han, G.K., Du, C.Y., Deng, Q.H., Gao, Y.Z., et al.: Pt decorated Ti_3C_2 MXene for enhanced methanol oxidation reaction. Ceram. Int. **45**, 2411–2417 (2019)
70. Yin, J.J., Zhang, L., Jiao, T.F., Zou, G.D., Bai, Z.H., Chen, Y., et al.: Highly efficient catalytic performances of nitro compounds and morin via self-assembled MXene-Pd nanocomposites synthesized through self-reduction strategy. Nanomaterials **9**, 1009 (2019)
71. Satheeshkumar, E., Makaryan, T., Melikyan, A., Minassian, H., Gogotsi, Y., Yoshimura, M.: One-step solution processing of Ag, Au and Pd@MXene hybrids for SERS. Sci. Rep. **6**, 32049 (2016)
72. Zheng, J.S., Wang, B., Ding, A.L., Weng, B., Chen, J.C.: Synthesis of MXene/DNA/Pd/Pt nanocomposite for sensitive detection of dopamine. J. Electroanal. Chem. **816**, 189–194 (2018)
73. Ghassemi, H., Harlow, W., Mashtalir, O., Beidaghi, M., Lukatskaya, M.R., Gogotsi, Y., et al.: In situ environmental transmission electron microscopy study of oxidation of two-dimensional $Ti3C_2$ and formation of carbon-supported TiO_2. J. Mater. Chem. A **2**, 14339–14343 (2014)
74. Li, Z.Y., Wang, L.B., Sun, D.D., Zhang, Y.D., Liu, B.Z., Hu, Q.K., et al.: Synthesis and thermal stability of two-dimensional carbide MXene Ti_3C_2. Mater. Sci. Eng. B **191**, 33–40 (2015)
75. Li, J.X., Du, Y.L., Huo, C.X., Wang, S., Cui, C.: Thermal stability of two-dimensional Ti_2C nanosheets. Ceram. Int. **41**, 2631–2635 (2015)

76. Rakhi, R.B., Ahmed, B., Hedhili, M.N., Anjum, D.H., Alshareef, H.N.: Effect of postetch annealing gas composition on the structural and electrochemical properties of Ti_2CT_x MXene electrodes for supercapacitor applications. Chem. Mater. **27**, 5314–5323 (2015)
77. Xiao, S.H., Zhang, X.Q., Zhang, J.W., Wu, S.M., Wang, J., Chen, J.S., et al.: Enhancing the lithium storage capabilities of TiO_2 nanoparticles using delaminated MXene supports. Ceram. Int. **44**, 17660–17666 (2018)
78. Gao, X.T., Xie, Y., Zhu, X.D., Sun, K.N., Xie, X.M., Liu, Y.T., et al.: Ultrathin MXene nanosheets decorated with TiO_2 quantum dots as an efficient sulfur host toward fast and stable Li-S batteries. Small **14**, 1802443 (2018)
79. Du, C., Wu, J., Yang, P., Li, S.Y., Xu, J.M., Song, K.X.: Embedding S@TiO_2 nanospheres into MXene layers as high rate cyclability cathodes for lithium-sulfur batteries. Electrochim. Acta **295**, 1067–1074 (2019)
80. Chen, X., Lia, J., Pan, G.C., Xu, W., Zhu, J.Y., Zhou, D.L., et al.: Ti_3C_2 MXene quantum dots/TiO_2 inverse opal heterojunction electrode platform for superior photoelectrochemical biosensing. Sens. Actuators B Chem. **289**, 131–137 (2019)
81. Li, Y.J., Deng, X.T., Tian, J., Liang, Z.Q., Cui, H.Z.: Ti_3C_2 MXene-derived Ti_3C_2/TiO_2 nanoflowers for noble-metal-free photocatalytic overall water splitting. Appl. Mater. Today. **13**, 217–227 (2018)
82. Low, J.X., Zhang, L.Y., Tong, T., Shen, B.J., Yu, J.G.: TiO_2/MXene Ti_3C_2 composite with excellent photocatalytic CO_2 reduction activity. J. Catal. **361**, 255–266 (2018)
83. Zhang, J., Yang, L., Wang, H.B., Zhu, G.L., Wen, H., Feng, H., et al.: In situ hydrothermal growth of TiO_2 nanoparticles on a conductive $Ti_3C_2T_x$ MXene nanosheet: a synergistically active Ti-based nanohybrid electrocatalyst for enhanced N_2 reduction to NH_3 at ambient conditions. Inorg. Chem. (2019)
84. Xu, Z., Sun, Y.Q., Zhuang, Y.X., Jing, W.H., Ye, H., Cui, Z.F.: Assembly of 2D MXene nanosheets and TiO_2 nanoparticles for fabricating mesoporous TiO_2-MXene membranes. J. Membr. Sci. **564**, 35–43 (2018)
85. Hao, N.X., Wei, Y., Wang, J.L., Wang, Z.W., Zhu, Z.H., Zhao, S.L., et al.: In situ hybridization of an MXene/TiO_2/NiFeCo-layered double hydroxide composite for electrochemical and photoelectrochemical oxygen evolution. RSC Adv. **8**, 20576–20584 (2018)
86. Li, Y.J., Yin, Z.H., Ji, G.R., Liang, Z.Q., Xue, Y.J., Guo, Y.C., et al.: 2D/2D/2D heterojunction of Ti_3C_2 MXene/MoS_2 nanosheets/TiO_2 nanosheets with exposed (001) facets toward enhanced photocatalytic hydrogen production activity. Appl. Catal. B Environ. **246**, 12–20 (2019)
87. Huang, Y.Y., Zhuo, G., Han, L., Wang, Y.Y., Kang, S.M., Lu, J.J.: Facile synthesis and application of V_2O_5/MXene nanocomposites as electrode materials for supercapacitors. Nanosci. Nanotechnol. Lett. **10**, 1633–1643 (2018)
88. Wu, F., Jiang, Y., Ye, Z.Q., Huang, Y.X., Wang, Z.H., Li, S.J., et al.: A 3D flower-like VO_2/MXene hybrid architecture with superior anode performance for sodium ion batteries. J. Mater. Chem. A **7**, 1315–1322 (2019)
89. Rakhi, R.B., Ahmed, B., Anjum, D.H., Alshareef, H.N.: Direct chemical synthesis of MnO_2 nanowhiskers on transition-metal carbide surfaces for supercapacitor applications. ACS Appl. Mater. Interfaces **8**, 18806–18814 (2016)
90. Jiang, H.M., Wang, Z.G., Yang, Q., Hanif, M., Wang, Z.M., Dong, L.C., et al.: A novel $MnO_2/Ti_3C_2T_x$ MXene nanocomposite as high performance electrode materials for flexible supercapacitors. Electrochim. Acta **290**, 695–703 (2018)
91. Chen, S.G., Xiang, Y.F., Xu, W.J., Peng, C.: A novel MnO_2/MXene composite prepared by electrostatic self-assembly and its use as an electrode for enhanced supercapacitive performance. Inorg. Chem. Front. **6**, 199–208 (2019)
92. Wang, Q.T., Zhang, Z.H., Zhang, Z., Zhou, X.Z., Ma, G.F.: Facile synthesis of MXene/MnO_2 composite with high specific capacitance. J. Solid State Electrochem. **23**, 361–365 (2019)
93. Tian, Y.P., Yang, C.H., Que, W.X., Liu, X.B., Yin, X.T., Kong, L.B.: Flexible and free-standing 2D titanium carbide film decorated with manganese oxide nanoparticles as a high volumetric capacity electrode for supercapacitor. J. Power Sources **359**, 332–339 (2017)

94. Tang, X., Liu, H., Guo, X., Wang, S.J., Wu, W.J., Mondal, A.K., et al.: A novel lithium-ion hybrid capacitor based on an aerogel-like MXene wrapped Fe_2O_3 nanosphere anode and a 3D nitrogen sulphur dual-doped porous carbon cathode. Mater. Chem. Front. **2**, 1811–1821 (2018)
95. Zhang, H.L., Li, M., Cao, J.L., Tang, Q.J., Kang, P., Zhu, C.X., et al.: 2D a-Fe_2O_3 doped Ti_3C_2 MXene composite with enhanced visible light photocatalytic activity for degradation of Rhodamine B. Ceram. Int. **44**, 19958–19962 (2018)
96. Li, F., Liu, Y.L., Wang, G.G., Zhang, H.Y., Zhang, B., Li, G.Z., et al.: Few-layered $Ti_3C_2T_x$ MXenes coupled with Fe_2O_3 nanorod arrays grown on carbon cloth as anodes for flexible asymmetric supercapacitors. J. Mater. Chem. A **7**, 22631–22641 (2019)
97. Liu, P.J., Ng, V.M.H., Yao, Z.J., Zhou, J.T., Kong, L.B.: Ultrasmall Fe_3O_4 nanoparticles on MXenes with high microwave absorption performance. Mater. Lett. **229**, 286–289 (2018)
98. Liu, P.J., Yao, Z.J., Ng, V.M.H., Zhou, J.T., Kong, L.B., Yue, K.: Facile synthesis of ultrasmall Fe_3O_4 nanoparticles on MXenes for high microwave absorption performance. Compos. Part A Appl. Sci. Manuf. **115**, 371–382 (2018)
99. Wang, Y.S., Li, Y.Y., Qiu, Z.P., Wu, X.Z., Zhou, P.F., Zhou, T., et al.: Fe_3O_4@Ti_3C_2 MXene hybrids with ultrahigh volumetric capacity as an anode material for lithium-ion batteries. J. Mater. Chem. A **6**, 11189–11197 (2018)
100. Zhang, X., Wang, H.H., Hu, R., Huang, C.Y., Zhong, W.J., Pan, L.M., et al.: Novel solvothermal preparation and enhanced microwave absorption properties of $Ti_3C_2T_x$ MXene modified by in situ coated Fe_3O_4 nanoparticles. Appl. Surf. Sci. **484**, 383–391 (2019)
101. Adeyemo, A.A., Adeoye, I.O., Bello, O.S.: Metal organic frameworks as adsorbents for dye adsorption: overview, prospects and future challenges. Toxicol. Environ. Chem. **94**, 1846–1863 (2012)
102. Huang, X.X., Wang, R., Jiao, T.F., Zou, G.D., Zhan, F.K., Yin, J.J., et al.: Facile preparation of hierarchical AgNP-loaded MXene/Fe_3O_4/Polymer nanocomposites by electrospinning with enhanced catalytic performance for wastewater treatment. ACS Omega **4**, 1897–1906 (2019)
103. Liu, Y.X., Luo, R., Li, Y., Qi, J.W., Wang, C.H., Li, J.S., et al.: Sandwich-like Co_3O_4/MXene composite with enhanced catalytic performance for Bisphenol A degradation. Chem. Eng. J. **347**, 731–740 (2018)
104. Luo, S.S., Wang, R., Yin, J.J., Jiao, T.F., Chen, K.Y., Zou, G.D., et al.: Preparation and dye degradation performances of self-assembled MXene-Co_3O_4 nanocomposites synthesized via solvothermal approach. ACS Omega **4**, 3946–3953 (2019)
105. Deng, R.X., Chen, B.B., Li, H.G., Zhang, K., Zhang, T., Yu, Y., et al.: MXene/Co_3O_4 composite material: Stable synthesis and its enhanced broadband microwave absorption. Appl. Surf. Sci. **488**, 921–930 (2019)
106. Wang, C., Zhu, X.D., Mao, Y.C., Wang, F., Gao, X.T., Qiu, S.Y., et al.: MXene-supported Co_3O_4 quantum dots for superior lithium storage and oxygen evolution activities. Chem. Commun. **55**, 1237–1240 (2019)
107. Tan, L.H., Lv, J., Xu, X.R., Zhao, H.F., He, C.M., Wang, H., et al.: Construction of MXene/NiO composites through in-situ precipitation strategy for dispersibility improvement of NiO nanoparticles. Ceram. Int. **45**, 6597–6600 (2019)
108. Zhao, H.F., Lv, J., Sang, J.S., Zhu, L., Zheng, P., Andrew, G.L., et al.: A facile method to construct MXene/CuO nanocomposite with enhanced catalytic activity of CuO on thermal decomposition of ammonium perchlorate. Materials **11**, 2457 (2018)
109. Gao, Y.P., Wang, L.B., Li, Z.Y., Zhou, A.G., Hu, Q.K., Cao, X.X.: Preparation of MXene-Cu_2O nanocomposite and effect on thermal decomposition of ammonium perchlorate. Solid State Sci. **35**, 62–65 (2014)
110. Zeng, Z.P., Yan, Y.B., Chen, J., Zan, P., Tian, Q.H., Chen, P.: Boosting the photocatalytic ability of Cu_2O nanowires for CO_2 conversion by MXene quantum dots. Adv. Func. Mater. **29**, 1806500 (2019)
111. Wang, F., Cao, M.J., Qin, Y., Zhu, J.F., Wang, L., Tang, Y.: ZnO nanoparticle-decorated two-dimensional titanium carbide with enhanced supercapacitive performance. RSC Adv. **6**, 88934–88942 (2016)

112. Qian, Y., Wei, H.W., Dong, J.D., Du, Y.Z., Fang, X.J., Zheng, W.H., et al.: Fabrication of urchin-like ZnO-MXene nanocomposites for high-performance electromagnetic absorption. Ceram. Int. **43**, 10757–10762 (2017)
113. Guo, J., Legum, B., Anasori, B., Wang, K., Lelyukh, P., Gogotsi, Y., et al.: Cold sintered ceramic nanocomposites of 2D MXene and zinc oxide. Adv. Mater. **30**, 1801846 (2018)
114. Zhang, C.F., Beidaghi, M., Naguib, M., Lukatskaya, M.R., Zhao, M.Q., Dyatkin, B., et al.: Synthesis and charge storage properties of hierarchical niobium pentoxide/carbon/niobium carbide (MXene) hybrid materials. Chem. Mater. **28**, 3937–3943 (2016)
115. Ambade, S.B., Ambade, R.B., Eom, W., Noh, S.H., Kim, S.H., Han, T.H.: 2D Ti_3C_2 MXene/WO_3 hybrid architectures for high-rate supercapacitors. Adv. Mater. Interfaces **5**, 1801361 (2018)
116. Peng, C., Kuai, Z.Y., Zeng, T.Q., Yu, Y., Li, Z.F., Zuo, J.T., et al.: WO_3 Nanorods/MXene composite as high performance electrode for supercapacitors. J. Alloy. Compd. **810**, 151928 (2019)
117. Wang, F., Wang, Z.J., Zhu, J.F., Yang, H.B., Chen, X.J., Wang, L., et al.: Facile synthesis SnO_2 nanoparticle-modified Ti_3C_2 MXene nanocomposites for enhanced lithium storage application. J. Mater. Sci. **52**, 3556–3565 (2017)
118. Ahmed, B., Anjum, D.H., Gogotsi, Y., Alshareef, H.N.: Atomic layer deposition of SnO_2 on MXene for Li-ion battery anodes. Nano Energy **34**, 249–256 (2017)
119. Xiong, J., Pan, L.M., Wang, H.H., Du, F., Chen, Y.M., Yang, J., et al.: Synergistically enhanced lithium storage performance based on titanium carbide nanosheets (MXene) backbone and SnO_2 quantum dots. Electrochim. Acta **268**, 503–511 (2018)
120. Liu, H., Zhang, X., Zhu, Y.F., Cao, B., Zhu, Q.Z., Zhang, P., et al.: Electrostatic self-assembly of 0D–2D SnO_2 quantum dots/$Ti_3C_2T_x$ MXene hybrids as anode for lithium-ion batteries. Nano-Micro Lett. **11**, 65 (2019)
121. Yang, L., Dall'Agnese, Y., Hantanasirisakul, K., Shuck, C.E., Maleski, K., Alhabeb, M., et al.: SnO_2-Ti_3C_2 MXene electron transport layers for perovskite solar cells. J. Mater. Chem. A **7**, 5635–5642 (2019)
122. Tariq, A., Ali, S.I., Akinwande, D., Rizwan, S.: Efficient visible-light photocatalysis of 2D-MXene nanohybrids with Gd^{3+}- and Sn^{4+}-codoped bismuth ferrite. ACS Omega **3**, 13828–13836 (2018)
123. Iqbal, M.A., Ali, S.I., Amin, F., Tariq, A., Iqbal, M.Z., Rizwan, S.: La- and Mn-codoped bismuth ferrite/Ti_3C_2 MXene composites for efficient photocatalytic degradation of Congo red dye. ACS Omega **4**, 8661–8668 (2019)
124. Lu, M., Li, H.J., Han, W.J., Wang, Y.Z., Shi, W., Wang, J.H., et al.: Integrated MXene&$CoFe_2O_4$ electrodes with multi-level interfacial architectures for synergistic lithium-ion storage. Nanoscale **11**, 15037–15042 (2019)
125. Tang, Y., Yang, C.H., Yang, Y.W., Yin, X.T., Que, W.X., Zhu, J.F.: Three dimensional hierarchical network structure of S-$NiFe_2O_4$ modified few-layer titanium carbides (MXene) flakes on nickel foam as a high efficient electrocatalyst for oxygen evolution. Electrochim. Acta **296**, 762–770 (2019)
126. Liu, P.J., Yao, Z.J., Ng, V.M.H., Zhou, J.T., Kong, L.B.: Novel multilayer-like structure of $Ti_3C_2T_x$/CNZF composites for low-frequency electromagnetic absorption. Mater. Lett. **248**, 214–217 (2019)
127. Li, B.H., Guo, H.R., Wang, Y.R., Zhang, W.X., Zhang, Q.J., Chen, L., et al.: Asymmetric MXene/monolayer transition metal dichalcogenide heterostructures for functional applications. NPJ Comput. Mater. **5**, 16 (2019)
128. Ma, Z.N., Hu, Z.P., Zhao, X.D., Tang, Q., Wu, D.H., Zhou, Z., et al.: Tunable band structures of heterostructured bilayers with transition-metal dichalcogenide and MXene monolayer. J. Phys. Chem. C **118**, 5593–5599 (2014)
129. Wang, C., Zhu, X.D., Wang, K.X., Gu, L.L., Qiu, S.Y., Gao, X.T., et al.: A general way to fabricate transition metal dichalcogenide/oxide-sandwiched MXene nanosheets as flexible film anodes for high-performance lithium storage. Sustain. Energy Fuels **3**, 2577–2582 (2019)

130. Xu, Y., Ang, Y.S., Wu, L., Ang, L.K.: High sensitivity surface plasmon resonance sensor based on two-dimensional MXene and transition metal dichalcogenide: a theoretical study. Nanomaterials. **9**, 165 (2019)
131. Xu, J., Shim, J., Park, J.H., Lee, S.: MXene electrode for the integration of WSe_2 and MoS_2 field effect transistors. Adv. Func. Mater. **26**, 5328–5334 (2016)
132. Shen, C.J., Wang, L.B., Zhou, A.G., Zhang, H., Chen, Z.H., Hu, Q.K., et al.: MoS_2-decorated Ti_3C_2 MXene nanosheet as anode material in lithium-Ion batteries. J. Electrochem. Soc. **164**, A2654–A2659 (2017)
133. Chen, C., Xie, X.Q., Anasori, B., Sarycheva, A., Makaryan, T., Zhao, M.Q., et al.: MoS_2-on-MXene heterostructures as highly reversible anode materials for lithium-ion batteries. Angew. Chem. Int. Ed. **57**, 1846–1850 (2018)
134. Zhang, Y.L., Mu, Z.J., Yang, C., Xu, Z.K., Zhang, S., Zhang, X.Y., et al.: Rational design of MXene/1T-2H MoS_2-C nanohybrids for high-performance lithium-sulfur batteries. Adv. Func. Mater. **28**, 1707578 (2018)
135. Wu, Y.T., Nie, P., Jiang, J.M., Ding, B., Dou, H., Zhang, X.G.: MoS_2-nanosheet-decorated 2D titanium carbide (MXene) as high-performance anodes for sodium-ion batteries. ChemElectroChem **4**, 1560–1565 (2017)
136. Xu, M., Bai, N., Li, H.X., Hu, C., Qi, J., Yan, X.B.: Synthesis of MXene-supported layered MoS_2 with enhanced electrochemical performance for Mg batteries. Chin. Chem. Lett. **29**, 1313–1316 (2018)
137. Attanayake, N.H., Abeyweera, S.C., Thenuwara, A.C., Anasori, B., Gogotsi, Y., Sun, Y.G., et al.: Vertically aligned MoS_2 on Ti_3C_2 (MXene) as an improved HER catalyst. J. Mater. Chem. A **6**, 16882–16889 (2018)
138. Yang, X.L., Jia, Q.J., Duan, F.H., Hu, B., Wang, M.H., He, L.H., et al.: Multiwall carbon nanotubes loaded with MoS_2 quantum dots and MXene quantum dots: non-Pt bifunctional catalyst for the methanol oxidation and oxygen reduction reactions in alkaline solution. Appl. Surf. Sci. **464**, 78–87 (2019)
139. Liang, J.M., Ding, C.Y., Liu, J.P., Chen, T., Peng, W.C., Li, Y., et al.: Heterostructure engineering of Co-doped MoS_2 coupled with Mo_2CT_x MXene for enhanced hydrogen evolution in alkaline media. Nanoscale **11**, 10992–11000 (2019)
140. You, J.X., Si, C., Zhou, J., Sun, Z.M.: Contacting MoS_2 to MXene: vanishing p-type Schottky barrier and enhanced hydrogen evolution catalysis. J. Phys. Chem. C **123**, 3719–3726 (2019)
141. Alimohammadi, F., Sharifian, G.M., Attanayake, N.H., Thenuwara, A.C., Gogotsi, Y., Anasori, B., et al.: Antimicrobial properties of 2D MnO_2 and MoS_2 nanomaterials vertically aligned on graphene materials and Ti_3C_2 MXene. Langmuir **34**, 7192–7200 (2018)
142. Zhao, P., Jin, H., Lv, X.S., Huang, B.B., Ma, Y.D., Dai, Y.: Modified MXene: promising electrode materials for constructing Ohmic contacts with MoS_2 for electronic device applications. Phys. Chem. Chem. Phys. **20**, 16551–16557 (2018)
143. Huang, H.W., Cui, J., Liu, G.X., Bi, R., Zhang, L.: Carbon-coated $MoSe_2$/MXene hybrid nanosheets for superior potassium storage. ACS Nano **13**, 3448–3456 (2019)
144. Li, N., Zhang, Y.F., Jia, M.L., Lv, X.D., Li, X.T., Li, R., et al.: 1T/2H $MoSe_2$-on-MXene heterostructure as bifunctional electrocatalyst for efficient overall water splitting. Electrochim. Acta **326**, 134976 (2019)
145. Su, W.T., Wang, S.G., Fu, L., Chen, F., Song, K.X., Huang, X.W., et al.: Growth of WS_2 flakes on $Ti_3C_2T_x$ Mxene using vapor transportation routine. Coatings **8**, 281 (2018)
146. Vyskocil, J., Mayorga-Martinez, C.C., Szokolova, K., Dash, A., Gonzalez-Julian, J., Sofer, Z., et al.: 2D stacks of MXene Ti_3C_2 and 1T-phase WS_2 with enhanced capacitive behavior. ChemElectroChem **6**, 3982–3986 (2019)

Chapter 4
Energy Related Applications

4.1 Introduction

Due to their special laminated structures, mechanical flexibilities, high electrical/thermal conductivities, unique chemical/physical stabilities, MXenes, together with their nanohybrids and nanocomposites, have been utilized for various applications. These applications include electrodes of energy storage devices, substrates of sensors, absorbents for environmental remediation, electromagnetic interference (EMI) shielding or absorption, to name a few. The applications in energy related fields, such as electrodes of batteries and supercapacitors, hydrogen storage, water splitting, etc., will be discussed in this chapter, while all other applications will be covered in Chap. 5.

4.2 Energy Storage and Conversion

The energy and environmental issues have attracted much attention from all around the world. The key solution to address these problems is to explore renewable energies, such as solar cells, wind energies and biomass derived energies. Most of these renewable energies have encountered a serious challenge in their storage. Among various storage technologies, electrochemical energy storage devices, such as Li-ion batteries (LIBs), Na-ion batteries (SIBs) and supercapacitors, have been widely acknowledged to be the most suitable techniques in terms of energy density and power density. Currently, the commercial batteries are still not sufficiently efficient for many applications. In both batteries and supercapacitors, the performances of the electrodes determine the electrochemical storage capabilities. Since the discovery of MXenes, they have been extensively studied as electrodes of electrochemical storage devices [1–9].

Z. Xiao et al., *MXenes and MXenes-based Composites*, Engineering Materials,
https://doi.org/10.1007/978-3-030-59373-5_4

4.2.1 Ion Batteries

4.2.1.1 Theoretical Considerations

Various batteries, such LIBs, SIBs, ZIBs (zinc ion batteries) and so on, have been developed as electrochemical energy storage systems [9]. As one of most commonly adopted commercial energy storage devices, in rechargeable lithium ion batteries (LIBs), graphite and its composites are being used as anodes. However, due to the low capacity of graphite, graphite anodes cannot meet the requirements for next generation energy storage applications. As a result, it is very urgent to explore new anode materials for LIBs with higher performances in order to replace graphene. As expected, MXenes and their functionalized derivatives have various unique properties, ranging from metallic to narrow band gap semiconducting characteristics, making them potential candidates as the anodes of LIBs, due to their high electronic conductivity.

Naguib et al. were among the first to demonstrate the potential applications of Ti_3C_2 MXene as anodes of LIBs [10]. Since then, the electrochemical properties of $Ti_3C_2T_x$ and $Ti_3C_2T_x$ based materials have been extensively evaluated as anode materials of LIBs both experimentally and theoretically. The exfoliation of Ti_3AlC_2 due to the treatment with HF led to the removal of Al, resulting in $Ti_3C_2T_x$ layered structures with surface terminations. Also, the exfoliated particles exhibited pseudo-ductility of Ti_3AlC_2 and thus could be directly pressed into freestanding paper-like sheet materials. This characteristic makes it potential candidate as electrode materials of LIBs, where volumetric variation should be accommodated.

The $Ti_3C_2T_x$ layered structures were so thin that they were transparent to electrons. Bothe TEM images and simulated results indicated that the two adjacent Ti_3C_2 sheets were terminated with –OH groups. Therefore, after the exfoliation, the Ti–Al bonds were replaced by weaker hydrogen or van der Waals bonds. Besides, the –OH groups, there were also –F groups, whose content could be as high as 12 at%, according to XPS analysis. In this case, the MXene is somehow similar to graphene. Firstly, the exfoliated 2D Ti_3C_2 layers had a multilayered structure, which is analogues to the exfoliated graphite. Secondly, both the Ti_3C_2 and graphene layers are accompanied by the formation of scrolls.

Multilayered structures could serve as host to store Li^+ ions. According to DFT calculations at 0 K in the Li-rich environments, it is possible to form $Ti_3C_2Li_2$ through the intercalation of Li ions into the inter-layer spacing taken originally by the Al atoms, with a reaction given by:

$$Ti_3C_2 + 2Li \rightarrow Ti_3C_2Li_2 \quad (4.1)$$

This reaction has an enthalpy change to be about 0.28 eV. The positive value of the enthalpy change was probably because the radio of Li has an atomic radius of 0.145 nm, which is slightly larger than that of Al (0.125 nm). Theoretical results predicted that a capacity of 320 mAh/g could be reached by the 2D materials, which is very close to value of graphite for LiC_6 (372 mAh/g).

In a separate study, Tang et al. comprehensively clarify the adsorption and migration of Li ions at the surface of bare, F- and OH-terminated Ti_3C_2 MXenes, which further indicated that Ti_3C_2 MXene exhibited strong potential to replace graphite as anodes of LIBs anodes, owing to its relatively low operating voltage and diffusion barrier [11]. Density functional theory (DFT) computational results revealed that bare Ti_3C_2 nanosheet exhibited magnetic metallic behaviors, whereas $Ti_3C_2F_2$ and $Ti_3C_2(OH)_2$ possessed narrow-band gap semiconducting or metallic characteristics, which was dependent on the way of the surface –F and –OH groups to geometrically terminated. At the most stable states, the –F and –OH groups strongly tended to be sitting at the hollow sites formed by the three neighboring C atoms, whereas the resultant I-$Ti_3C_2F_2$ and I-$Ti_3C_2(OH)_2$ derivatives displayed semiconducting properties, with extremely narrow band gaps.

With the special metallic or narrow-band gap semiconducting properties, the Ti_3C_2-based new 2D materials were highly potential as anode materials of LiBs. Namely, the bare Ti_3C_2 monolayer possessed high electrical conductivity, low diffusion barrier, low open circuit voltage and high theoretical Li storage capacity. As for the fluorine and hydroxyl functionalized derivatives, the surface functional groups could bring in degradation of Li diffusion and reduction in Li storage capacity. As a consequence, they should be prevented in the practical synthetic processes.

It is believed that 2D materials could offer high rates of charge and discharge in batteries, because the diffusion of cations observed in a 3D lattice structure is effectively avoided. Er et al. employed first principles density functional calculations to predict the adsorption behaviors of Li^+, as well as Na^+, K^+ and Ca^{2+}, on the surface of Ti_3C_2, one representative MXene [12]. It was found that the alkali atoms demonstrated different adsorption energies, which were highly related to the surface coverage. For example, the adsorption energies of Na^+, K^+ and Ca^{2+} were decreased with increasing surface coverage, whereas that of Li^+ was almost insensitive to coverage profile. The relationship between the adsorption energies and the coverage of the alkali ions on the surface of Ti_3C_2 was closely related to their effective ionic radii. The larger the effective ionic radius, the stronger the interaction between the alkali atoms would, and thus the smaller the surface coverage would be. According to the theoretical calculations, the specific capacities of the bare Ti_3C_2 for Li^+, Na^+, K^+ and Ca^{2+} were 447.8, 351.8, 191.8 and 319.8 mAh/g, respectively.

The theoretical high capacity of the Ti_3C_2 MXene was different from that of graphene nanosheet. Even though pristine single layered graphene nanosheet has very special properties, it is not recommended for the applications as anode of LIBs [13]. The authors calculated the electronic structure of pristine graphene with lithium adatoms by using a $2 \times 2 \times 1$ supercell. According to the total DOS of graphene with a lithium adatom and the DOS projected onto the orbitals of C 2*p* and Li 2*s*, it was demonstrated that the C 2*p* orbital had the largest contribution to the energy states, within the range of 0–16 eV below the Fermi level.

In comparison, below the Fermi level, the Li 2*s* orbital possessed almost no significant peak. The peak of the Li 2*s* orbital was present at 0.815 eV above the Fermi level, while the peak of the C 2*p* orbital was slightly overlapping with the total DOS. In addition, the peak corresponded to an antibonding orbital of *s*–*p*. As a result, there was no hybridization of orbitals, because of the lack of overlapped peaks below the Fermi level, so that it was relatively weak. In contrast, for Ti_3C_2 MXene, the Li 2*s* and Ti 3*d* orbitals were overlapping at 0.70 eV below the Fermi level. Therefore, a distinctive broad peak of Li was overlapping with several peaks of the Ti 3*d* orbital, implying that there would be *s*–*d* hybridization during the adsorption of Li ions.

Shortly after that study, Xie et al. theoretically revealed that O-terminated MXene should process the highest capacity, because the already lithiated O-terminated MXene could allow for an extra Li layers to be adsorbed [14]. The authors combined DFT calculations and experiments, attempting to establish the relation between surface structure and storage capacities of Li ion, for a collection of functionalized 2D MXenes, which included Sc_2C, Ti_2C, Ti_3C_2, V_2C, Cr_2C and Nb_2C. It was found that the storage capacities of Li^+ ion were closely related to the characteristics of the surface functional groups, with O groups to exhibit the highest theoretical storage capacities of Li^+ ion. MXene surfaces were initially covered by –OH groups, which could be removed through high-temperature treatment or reaction with Li^+ ion during the first cycle of lithiation. This conclusion was further confirmed by the samples of f-Nb_2C and f-Ti_3C_2 that were annealed at 673 and 773 K for 40 h in vacuum, in-situ X-ray adsorption spectroscopy (XAS) and the storage capacity of Li^+ ion from the first lithiation/delithiation cycle of f-Ti_3C_2, as illustrated in Fig. 4.1.

The removal of water and OH due to high temperature was evidenced by the results of XRD and inelastic neutron scattering experiments. The surface reactions during the first lithiation cycle were elucidated from the results of voltage profile and X-ray adsorption near edge structure of f-Ti_3C_2. In addition, lithiated surfaces of the O-terminated MXenes could adsorb more Li^+ ions on the monolayer, which was responsible for the enhanced storage capacity. The theoretical diffusion barriers of the Li^+ ion were pretty low, which well supported the experimental high rate performance. The authors also predicted that Cr_2C should exhibit high storage capacity of Li^+ ion, because of the low activation energy of water formation at high temperature. In comparison, a relatively low Li^+ storage capacity was predicted for Sc_2C, owing to the difficulty in removing the –OH groups.

Xu et al. used molecular dynamic simulation to simulate the charging/discharging dynamics in MXene ($Ti_3C_2(OH)_2$) based electrode, by mimicking intrinsic slit nanopores to host ionic liquid at room temperature [15]. Through the modelling of galvanostatic cycling, the dynamic charge storage mechanism was clarified. According to the simulation results, it was found that the $Ti_3C_2(OH)_2$ slit nanosized pores were spontaneously wetted by the electrolyte ions without the application of external potential, with the layer distance of the pores to be as small as 0.7 nm. Inside the nanopores, an electroneutrality breakdown and faster diffusion behaviour took place.

During the charge/discharge process, the charge storage was realized through fast insertion of the counter-ion and exchange of the counter-ion/co-ion. In addition, both the spatial distribution of the ions and the orientation of the cation dipoles inside the nanopores were changed with the cycled charge/discharge processes. As

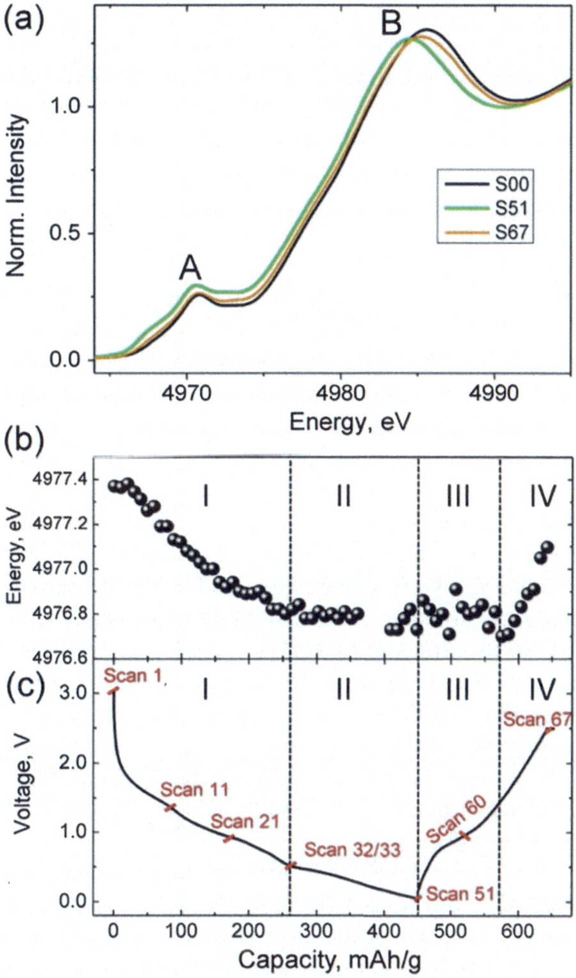

Fig. 4.1 **a** In-situ Ti K-edge XANES analysis during the lithiation and delithiation of f-Ti_3C_2, where S00 is XANES before electrochemical reactions, while S51 and S67 refer to the fully lithiated and fully delithiated states shown in (**c**). **b** Variation of Ti edge energy at half height of normalized XANES spectra versus capacity during lithiation and delithiation. **c** Corresponding voltage profiles. Reproduced with permission from [14]. Copyright © 2014, American Chemical Society

a consequence, the external electric potential was screened. According to the charge mechanism, the effect of the layer distance was examined. The results indicated that the narrower the nanopored, the higher the acceleration and the more ordered the dipole orientation the ions would experience. In other words, $Ti_3C_2(OH)_2$ electrodes with smaller pore sizes would have higher electrochemical performances.

Various other M_2CO_2 MXenes, with M=Ti, V, Cr, Mn, Fe, Co, Ni, Nb and Mo, have been explored for potential applications as electrode materials for LIBs and sodium ion batteries (SIBs). For example, Ashton et al. used dispersion-corrected density functional theory (DFT) calculations to evaluate capacitive properties of O-terminated Ti- and V-based carbide MXenes with bilayered structures as electrodes of ion batteries [16]. The capacity levels of both the V_2CO_2 and Ti_2CO_2 bilayered nanostructures were found to be very close to experimental results, as compared with those obtained previously with single-layers. In addition, the lowest energy pathways and the energy barriers for the diffusion of Li^+ ions in the in-plane [1000] and [0100] directions, crossing the adjacent MXene nanolayers were obtained.

It was found that $V_4C_3O_2$ had a Li^+ ion diffusion barrier of 0.42 eV, which was the lowest among all the items studied, while it delivered a relatively low reversible capacity of 148 mAh/g, due to its high molecular weight. In contrast, the V_2CO_2 MXene could reach a high reversible capacity of 276 mAh/g, where as its diffusion barrier of 0.82 eV was pretty high. In addition, bilayered structures usually have higher diffusion barriers than their monolayer counterparts. Therefore, monolayered MXenes are more suitable for anodes with high electrochemical performances.

Sun et al. used density functional theory (DTF) calculations to study the Li^+ storage capabilities of a series of 2D M_2CO_2, with M=V, Cr, Ta, Sc, Ti, Zr, Nb and Hf, which were terminated with –O functional groups [17]. Depending on the occurrence of structural transformation as the first layer of Li^+ was formed, the Li^+ ions were adsorbed in two ways, i.e., (i) V-group (including V_2CO_2, Cr_2CO_2 and Ta_2CO_2) and (ii) Sc-group (including Sc_2CO_2, Ti_2CO_2, Zr_2CO_2, Nb_2CO_2 and Hf_2CO_2). The V-group members exhibited reversible structural transformation in the process of lithiation-delithiation, according to the ab initio molecular dynamic simulations. The members (except for Nb-MXene) in the V-group tended to adopt sandwich structure of H2H1T-$M_2CO_2Li_4$, whereas those in the Sc-group preferred to have the structure of TH1H2-$M_2CO_2Li_4$, as the second Li^+ layer was adsorbed. Because the –O layer of the H2H1T-$M_2CO_2Li_4$ structure in the V-group members was sandwiched by two Li^+ layers, the formation of Li dendrites was effectively avoided, thus leading to safe battery systems. Also, the –O functional groups tended to bond with Li instead of M for the V-group than for the Sc-group, so that the sandwich structure of H2H1T-$M_2CO_2Li_4$ was more likely for the V-group than for the Sc-group.

A first-principles study was conducted on electronic properties of MXenes as electrode materials of LIBs [18]. The MXenes were monolayers $M_{i+1}N_i$; with M=Ta, Ti and V and $i = 1$ and 2. The MXenes monolayers were of metallic behaviors, while the Ti_2N was antiferromagnetic and Ti_3N_2 was ferromagnetic. The calculation results indicated that Li^+ transportation on the surface of the MXenes monolayers was promoted. Particularly, Ti_2N had a very low barrier of 0.017 eV, due to the pretty small adsorption energies and the long binding lengths. Moreover, the MXenes monolayers maintained to be metallic as the Li^+ was fully intercalated. Also, the open circuit voltage was sufficiently low. Specifically, the Ti_2N monolayer demonstrated the highest electrochemical performances, with a low open circuit voltage of 0.53 V and a high specific capacity of 487 mAh/g. These results were in a good agreement with the previous prediction as discussed above, i.e., MXenes monolayers are preferred as electrode materials of LIBs.

Ti_3CN was theoretically studied as anode materials of LIBs by using DFT and DFT + U simulations [19]. The mechanisms of Li^+ adsorption and diffusion on the surface of Ti_3CN and Ti_3CNT_2, with T=F, O and OH, were elucidated. The Li^+ diffusion barrier of Ti_3CNT_2 was in the range of 0.2–0.3 eV. It was observed that Li atoms were preferentially adsorbed at the N-side of the Ti_3CN nanosheets if the functional groups were not present. However, with the presence of the functional groups of T=O, F, OH, Li^+ adsorption took place at the C-side. The observation could be understood in terms of Bader charge population analysis. In the case of Ti_3CNF_2, a six-membered ring of Li–F could be established, as the concentration of Li^+ was sufficiently high, so that the stability of the system was increased.

MXene/graphene nanostructures were demonstrated to have superb electrochemical properties as anode of LIBs [20]. The mechanisms of the electrochemical performances of such nanohybrids have been elucidated by using first-principle calculations. The MXene hybrids were based on Ti_2CX_2, with X=F, O and OH. It was observed that the incorporation of graphene nanosheets into the MXenes could effectively prevent them from being restacked and also highly increased electrical conductivity of the materials. Meanwhile, Li ion storage capability and mechanical integrity could be maintained. All these characteristics contributed to the outstanding electrochemical properties of the nanohybrid based electrodes.

2D MXenes have also been explored to store Na^+ ions for sodium ion batteries (SIBs). For instance, Yang et al. studied storage capabilities and ion transport characteristics of M_2C-type MXenes, with M=Ti, V, Cr, Mn, Fe, Co, Ni, Nb and Mo, by DFT calculations, which were compared with those of LIBs [21]. According the calculation results on the average voltages and diffusion barriers of the Na^+ ions on surfaces, it was concluded that those with M=Ti, V, Cr, Mn and Mo could be used as anode materials of SIBs. Theoretical specific capacities of the M_2C MXene based electrodes were in the range of 190–288 mAh/g, if two Na^+ ions were taken by one formula unit. With ion diffusion barriers in the range of 0.1–0.2 eV, the M_2C MXenes could be used to develop high power LIBs.

4.2.1.2 Li Ion Batteries

Intercalation of MXenes, including 2D Ti_3C_2, Ti_3CN and TiNbC was studied for the applications as anode of lithium ion batteries [22]. After they were intercalated with hydrazine and hydrazine together with *N*,*N*-dimethylformamide, the *c* lattice constants of the surface functionalized f-Ti_3C_2 were increased from 1.95 to 2.548 and 2.68 nm, respectively. The f-Ti_3C_2 was also intercalated with urea. According to molecular dynamics simulations, hydrazine was present as a monolayer in between the f-Ti_3C_2 nanolayers. Similarly, both f-Ti_3CN and f-TiNbC could also be intercalated with hydrazine. Once the f-Ti_3C_2 was intercalated with dimethyl sulphoxide, stable colloidal suspension of f-Ti_3C_2 was formed after sonication in water, which could be used to make MXene papers through vacuum filtration. The MXene papers exhibited high Li^+ ion storage capacity at high charging rates, with a specific capacity of 410 mAh/g at 1 C rate.

Sun et al. used HF to exfoliate Ti_3AlC_2, obtaining 2D Ti_3C_2, which was intercalated with dimethyl sulfoxide, leading to products as anode materials of Li ion batteries [23]. It was found that the intercalation made the Ti_3C_2 could significantly increase electrochemical performances of the anode materials. A specific capacity of 123.6 mAh/g at 1 C rate was achieved, along with a coulombic efficiency of 47%, which was much higher than that of the 2D Ti_2C and approached the theoretical level of Ti_3C_2 with the termination of –F group. Figure 4.2 shows representative electrochemical characterization results of the intercalated and exfoliated Ti_3C_2 based anode.

Kim et al. reported a method to develop anode materials based on MXenes with sufficiently high mass loadings [24]. Free-standing pellets were derived from the

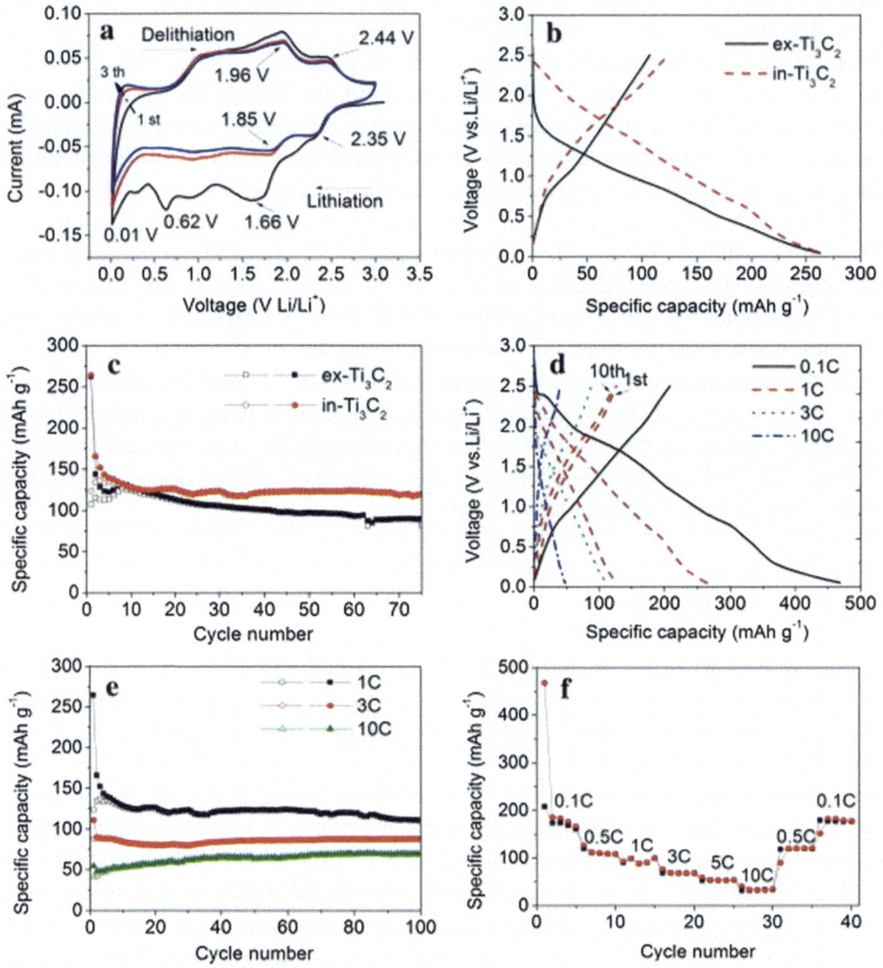

Fig. 4.2 **a** CV curves of the intercalated Ti_3C_2 over 3.0–0.01 V versus Li/Li^+ at the scan rate of 0.2 mV/s. **b** Charge–discharge behaviors of the Ti_3C_2 at 1 C. **c** Cycling performance of the Ti_3C_2 at 1 C. **d** First charge–discharge profiles of the intercalated Ti_3C_2 at 0.1, 1, 3 and 10 C. **e** Cycling properties of the intercalated Ti_3C_2 at different rates. **f** Specific capacities of the exfoliated Ti_3C_2 at 0.1, 0.5, 10, 0.5 and 0.1 C. Reproduced with permission from [23]. Copyright © 2014, Elsevier

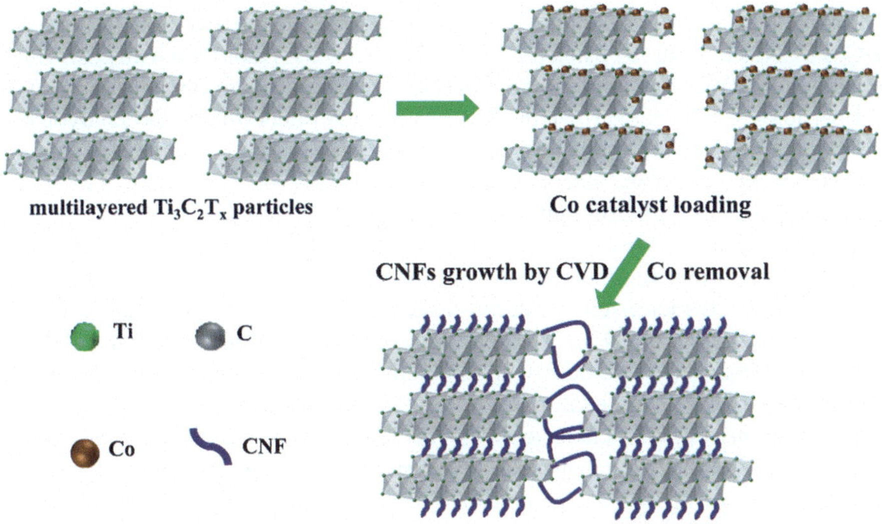

Fig. 4.3 Schematic diagram describing preparation of the MXene/CNFs hybrids. Reproduced with permission from [25]. Copyright © 2015, The Royal Society of Chemistry

powders of the MXenes through cold pressing, without using any binders. A relatively high pressure of 1 GPa was used to obtain samples with thicknesses of up to 300 μm. The Ti_3C_2 based anode possessed an initial reversible areal capacity of about 15 mAh/cm^2, which was declined to about 5.9 mAh/cm^2 after cycling for 50 times, while the value kept quite stable after 20 cycles. In comparison Nb_2C based anode exhibited an initial reversible capacity of about 16 mAh/cm^2. After recycling for 50 time, it was reduced to 6.7 mAh/cm^2. The electrochemical performance was higher than that of Ti_3C_2 by a factor of about 14%.

It was observed that interlayer stacking exists of MXene particles would electrical conductivity of the bulky materials. As a result, the access of electrolyte ions into the interlayers of the MXene nanosheets was hindered, which has been responsible for the low specific capacity and rate stability of multi-stacked MXene particles. To address this problem, Lin et al. developed $Ti_3C_2T_x$/carbon nanofibers (CNFs) hybrid particles as LIB anodes, with enhanced electrochemical properties [25]. The highly CNFs acted as bridges to link the $Ti_3C_2T_x$ nanosheets.

In this case, reversible capacity of the Ti_3C_2/CNFs hybrids was tremendously enhanced, as compared with the pristine $Ti_3C_2T_x$. More remarkably, even, the specific capacity of the Ti_3C_2/CNFs hybrids at a high rate of 100 C was nearly the same as that of the pristine $Ti_3C_2T_x$ at 1 C. Moreover, was no degradation of capacity was observed after 2900 cycles at 100 C, confirming the achievement of high rate capability and long-term stability. The $Ti_3C_2T_x$/CNFs hybrids displayed a high reversible capacity of 320 mAh/g at 320 mA/g, which was increased by about 100% as compared with that of the pristine $Ti_3C_2T_x$.

Figure 4.3 shows a schematic diagram demonstrating preparation step of the $Ti_3C_2T_x$/CNFs hybrids, by using a CVD method. $Co(NO_3)_2$ solution with poly(vinyl pyrrolidone) (PVP) was infilled into the $Ti_3C_2T_x$ nanosheets to obtain catalytic Co nanoparticles. CNFs were grown into the MXene nanosheets due to the catalytic Co nanoparticles. After the growth of CNFs, the Co catalysts were removed, leaving Ti_3C_2/CNFs hybrids. The content of CNFs in the hybrids could be modified by suing precursor solutions with different mass ratios of PVP to Co salt.

Morphological and microstructural properties of the pristine $Ti_3C_2T_x$ and Ti_3C_2/CNFs hybrids are illustrated in Fig. 4.4. The pristine $Ti_3C_2T_x$ accordion-like microstructure, characterized by interlayer spacings at both nanometer and micrometer scales, as seen in Fig. 4.4a. The spacings could facilitate intercalation and deintercalation of various ions, electrical conductivity of the materials was damaged, especially in the vertical direction of the $Ti_3C_2T_x$ nanosheets.

Due to the hydrophilic characteristics, the $Ti_3C_2T_x$ nanosheets could be infilled with Co^{2+} salt, so that Co nanoparticles were produced after reduction, which decorated the surfaces of the $Ti_3C_2T_x$ nanosheets serving as catalysts to grow CNFs.

Fig. 4.4 Microstructures of the $Ti_3C_2T_x$ and Ti_3C_2/CNFs nanoparticles. **a** SEM image of the $Ti_3C_2T_x$ samples after etching. **b** SEM image of the Ti_3C_2/CNFs hybrids, with the inset showing the CNFs that were grown in between the Ti_3C_2 nanosheets. **c** TEM image of the Ti_3C_2/CNFs hybrid. **d** HRTEM image of an individual Ti_3C_2 nanosheet. **e** TEM image of an individual CNFs. **f** TEM image of TiO_2 nanoparticles deposited on the Ti_3C_2 nanosheets. Reproduced with permission from [25]. Copyright © 2015, The Royal Society of Chemistry

As illustrated in Fig. 4.4b, the functional groups were eliminated from the pristine $Ti_3C_2T_x$ after the CVD reaction at 600 °C in Ar/H_2, while the $Ti_3C_2T_x$ nanosheets were decorated by CNFs. More significantly, CNFs were formed inside interlayer spacings in between the $Ti_3C_2T_x$ nanosheets, which linked the individual Ti_3C_2 nanosheets. However, because the presence of the CNFs had no effect on multilayer structure of the materials, the access of electrolyte ions was kept to be the same. This observation was further confirmed by TEM and HRTEM images, as illustrated in Figure 4.4c–e. Meanwhile, TiO_2 nanoparticles of anatase phase, with an average particle size of 20 nm, were formed on surfaces of the Ti_3C_2 nanosheets, due to the oxidation of Ti (Fig. 4.4f).

The $Ti_3C_2T_x$/CNFs hybrids were made into 10 μm thick electrodes for LIB anode testing, including cyclic voltammetry (CV), galvanostatic charge/discharge (GCD) cycling and electrochemical impedance spectroscopy (EIS). Figure 4.5a shows GCD curves of the Ti_3C_2/CNFs based electrode, at a rate of 1 C, with 1 C = charge/discharge for 1 h at 320 mA h/g. It had initial discharge and charge capacities of 848 mAh/g and 407 mAh/g, respectively. The discharge and charge capacities at the second cycle were nearly the same, demonstrating high reversibility

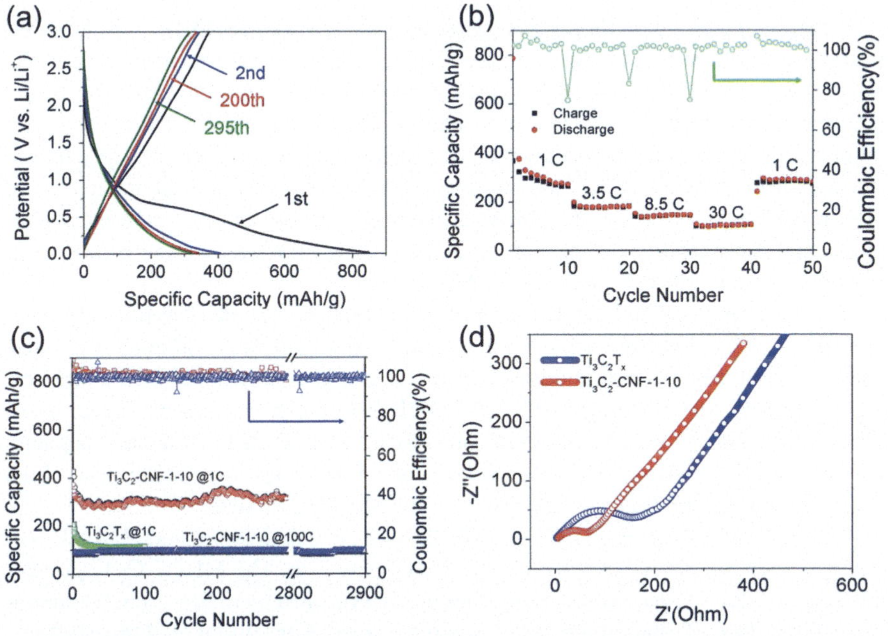

Fig. 4.5 **a** Charging-discharging behaviors of the Ti_3C_2/CNFs hybrids at 1 C. **b** Rate profiles of the Ti_3C_2/CNFs hybrids. **c** Cycling performances of the pristine $Ti_3C_2T_x$ based anode at 1 C and the Ti_3C_2/CNFs hybrids at 1 C and 100 C. **d** EIS curves of the $Ti_3C_2T_x$ and the Ti_3C_2/CNFs hybrids. Reproduced with permission from [25]. Copyright © 2015, The Royal Society of Chemistry

of the electrode. Thereafter, the degradation in capacity became retarded. The presence of the NCFs largely enhanced electrochemical performances of the MXene based electrode.

The hybrid-based electrode had a reversible discharge capacity of 320 mAh/g after 295 cycles, the same as the theoretical value of monolayered Ti_3C_2. In contrast, the reversible capacity of the pristine $Ti_3C_2T_x$ based electrode was only 112 mAh/g after 100 cycles. According to thickness of the electrode, the Ti_3C_2/CNFs hybrid-based electrode had a volumetric capacity of 23 mAh/cm^3. At the same time, the content of CNFs should be optimized. Too low content of CNFs could not sufficiently increase the electrical conductivity, while too much CNFs would dilute electrochemically active sites in the materials.

Specific capacities of Ti_3C_2/CNFs hybrid-based electrode were 320, 180, 145 and 106 mAh/g, at the rates of 1, 3.5, 8.5 and 30 C, respectively, as demonstrated in Fig. 4.5b. The specific capacity was well recovered as the rate was reduced to 1 C. As revealed in Fig. 4.5c, the Ti_3C_2/CNFs based electrode exhibited much higher capacity than the pristine $Ti_3C_2T_x$ at the same rate. According to the EIS results (Fig. 4.5d), the Ti_3C_2/CNFs based electrode had a much smaller semicircle diameter than the pristine $Ti_3C_2T_x$ one, implying that the former possessed much lower charge transfer impedance.

A chemical etching method was reported to prepare porous 2D $Ti_3C_2T_x$ MXenes in aqueous solutions at room temperature [26]. As compared with the pristine sample, the porous $Ti_3C_2T_x$ or p-$Ti_3C_2T_x$ had enlarged specific surface areas (SSA) and more open structure. Moreover, flexible thin sheets could be derived from the p-$Ti_3C_2T_x$, either with or without the presence of CNTs. Specifically, the p-$Ti_3C_2T_x$/CNT hybrid sheets exhibited enhanced Li^+ ion storage capability. An optimum high capacity of 1250 mAh/g was achieved at 32 mA/g, together with outstanding cycling performance and rate stability of 330 mAh/g at 10 C. The effective etching process was also applicable to Nb_2CT_x and V_2CT_x MXenes.

The research group also used $Ti_3C_2T_x$/CNT papers as cathode of hybrid Mg^{2+}/Li^+ batteries [27]. Li^+ ion battery has high capacities, high voltages and quick Li^+ intercalation, while Mg^{2+} metal anodes is of low cost and dendrite-free characteristics. Highly flexible free-standing $Ti_3C_2T_x$/CNTs hybrid papers possessed capacities of 100 mAh/g and 50 mAh/g at 0.1 C and 10 C, respectively. Also, the capacity was retained to be 80 mAh/g after cycling for >500 cycles at 1 C. Similarly, high electrochemical performance was observed for Mo_2CT_x MXene.

Both the multi-layered $Ti_3C_2T_x$ (ML-$Ti_3C_2T_x$) and delaminated $Ti_3C_2T_x$ (d-$Ti_3C_2T_x$) were employed as the cathodes for comparison. The ML-$Ti_3C_2T_x$ powder consisted of $Ti_3C_2T_x$ nanoflakes with stacking profile, as seen in Fig. 4.6a. Comparatively, the d-$Ti_3C_2T_x$ nanoflakes were of single-layered or few-layered morphology, with thicknesses of up to several nanometers and lateral sizes of hundreds from nanometers to micrometers, as demonstrated in Fig. 4.6b. The d-$Ti_3C_2T_x$ nanoflakes could be hybridized with CNTs, thus leading to flexible free-standing papers through vacuum filtration. In the $Ti_3C_2T_x$/CNTs hybrid, the CNTs acted as separators to isolate d-$Ti_3C_2T_x$ nanoflakes, so that they would not be restacked.

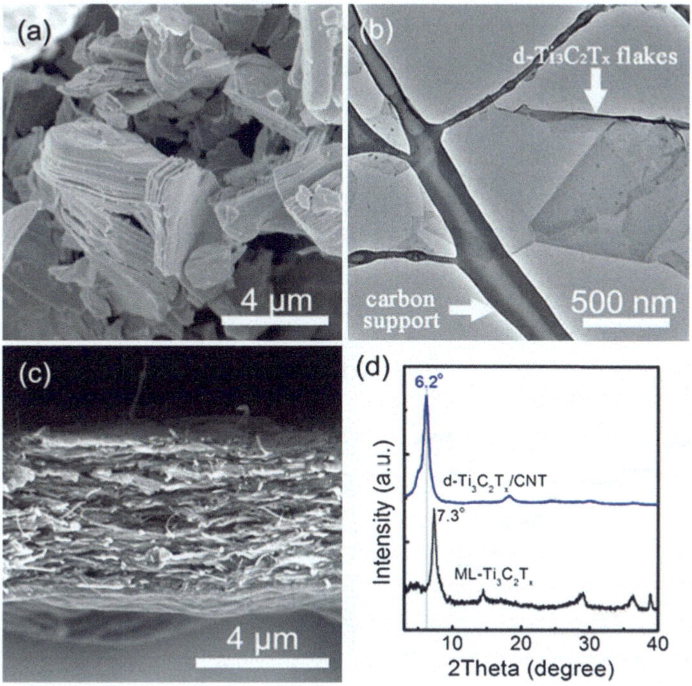

Fig. 4.6 **a** SEM image of the ML-$Ti_3C_2T_x$ nanoparticles. **b** TEM image of the d-$Ti_3C_2T_x$ nanoflakes. **c** Cross-sectional SEM image of the d-$Ti_3C_2T_x$/CNTs hybrid thin sheet. **d** XRD patterns of the ML-$Ti_3C_2T_x$ and the d-$Ti_3C_2T_x$/CNTs hybrid sheet. Reproduced with permission from [27]. Copyright © 2017, American Chemical Society

Figure 4.7a shows CV curves of the ML-$Ti_3C_2T_x$ and d-$Ti_3C_2T_x$/CNTs based electrodes. Both samples had a broadened intercalation/extraction peaks of Li^+ ions. In comparison, the d-$Ti_3C_2T_x$/CNTs hybrid-based electrode exhibited more advanced electrochemical performance. For instance, the ML-$Ti_3C_2T_x$ based cathode possessed a discharge capacity of about 18 mAh/g, whereas the d-$Ti_3C_2T_x$/CNTs hybrid electrode delivered a stable capacity level of 80 mAh/g after 50 cycles, as observed in Fig. 4.7b.

It was found that there the d-$Ti_3C_2T_x$/CNTs based electrode reached high performance without the requirement of pre-cycling at 0.1 C, which could be attributed to the fact that the diffusion rate of Li^+ ions was sufficiently high into the electrodes during each cycle at the low current densities. Meanwhile, the irreversible capacity of first cycle was not present for both the ML-$Ti_3C_2T_x$ and d-$Ti_3C_2T_x$/CNTs cathodes at 0.1 C, as revealed in Fig. 4.7c, mainly because the solid electrolyte interphase (SEI) was not formed over the voltage range, i.e., >0.2 V versus Mg/Mg^{2+} and >0.9 V versus Li/Li^+.

Figure 4.7d shows charge/discharge characteristics of the d-$Ti_3C_2T_x$/CNTs hybrid-based cathode at different current rates. The discharge capacity was associated

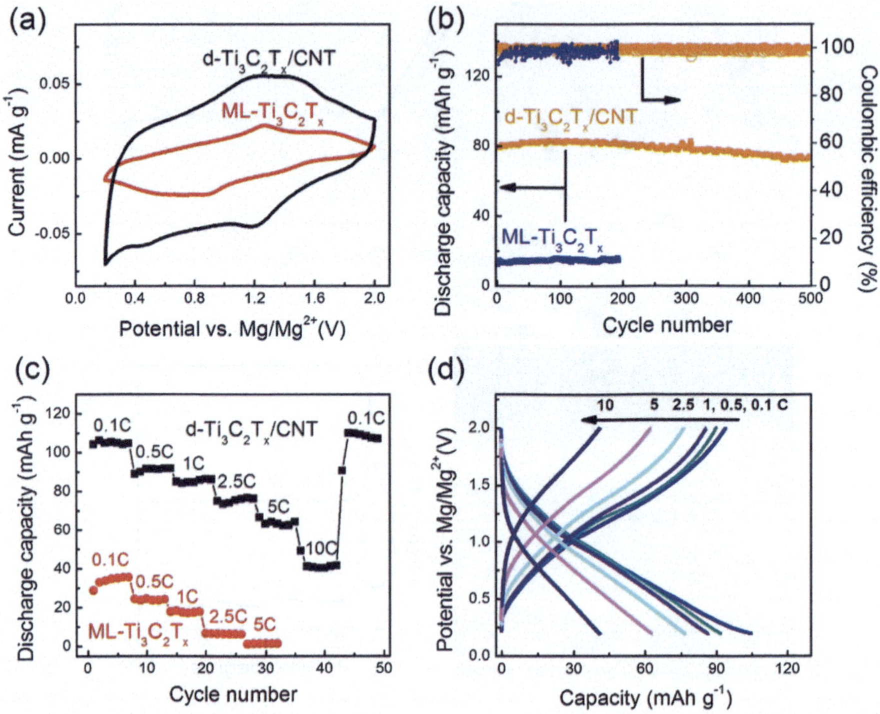

Fig. 4.7 **a** Fifth CV cycles for ML-$Ti_3C_2T_x$ and d-$Ti_3C_2T_x$/CNTs cathodes tested at 0.2 mV/s. **b** Cycle stability of the ML-$Ti_3C_2T_x$ and the d-$Ti_3C_2T_x$/CNTs cathodes pre-cycled for 100 times at 1 C (1 C = 100 mA/g). **c** Rate capability of ML-$Ti_3C_2T_x$ and the d-$Ti_3C_2T_x$/CNTs cathodes. **d** Charge–discharge profiles of a pre-cycled d-$Ti_3C_2T_x$/CNTs cathode at different current rates. Reproduced with permission from [27]. Copyright © 2017, American Chemical Society

with the potentials of <1.7 V versus Mg/Mg^{2+}. Since the 2D d-$Ti_3C_2T_x$ nanoflakes ensured a fast diffusion of Li^+ ions, the charge/discharge curves showed no plateaus characteristics. In addition, the Li^+ ions, rather than the Mg^{2+} ions, were the main contributors to the observed capacities.

Ti_3C_2 based nanocomposites combined with Sn^{4+} ions, or PVP-Sn(IV)/Ti_3C_2, were obtained with enhanced electrochemical performances, by using a liquid phase immersing method, with the aid of polyvinylpyrrolidone (PVP) [28]. The composite based anodes exhibited increased specific capacity due to the pillar effect of Sn^{4+} ions formed in between the interlayers of the intercalated Ti_3C_2 nanosheets (alk-Ti_3C_2) through alkalization. The enhancement in capacities was ascribed to the synergistic effect of both the alk-Ti_3C_2 and Sn^{4+}. An optimum specific capacity of 1375 mAh/cm^3 was achieved at 216.5 mA/cm^3, corresponding to 635 mAh/g at 100 mA/g, along with superb cycling performances and rate stabilities.

According to XRD results, it was found that the Sn(IV) complex attached on the alk-Ti_3C_2 was amorphous. With the loading of the Sn(IV) complex, the layer spacing along the *c*-axis was enlarged from 1.227 to 1.276 nm, suggesting that Sn^{4+}

ions were partially inserted into the interlayers. With the presence of PVP dispersed in $SnCl_4$ solution as the surfactant, the PVP-Sn(IV)/Ti_3C_2 nanocomposite consisted of nanoparticles with a narrow size distribution. The use of PVP had no effect on phase composition of the final product. Both the SEM and TEM results revealed that the Sn(IV) complex was of a spindle-like morphology, with uniform size distribution in the range of 5–10 nm. It was homogeneously distributed in the matrix.

Figure 4.8a shows CV curves of the nanocomposite-based electrode, at the rate of 0.1 mV/s and voltages in the range of 0–3 V. The initial cathodic scanning curve exhibited four reduction peaks. Those at 1.5 V and 1.15 V could be ascribed to the solid electrolyte interphase (SEI) on the working electrode, decomposition of the electrolyte and the formation of Sn irreversibly reduced from the Sn(IV) complex. The one at 0.75 V corresponded to the Li^+ ions intercalated in between the alk-Ti_3C_2 nanosheets. The main peak over 0.5–0 V was due to the presence of Li_xSn.

Further anodic sweeping resulted in the delithiation of Li^+ ions from the previously formed Li_xSn. A broad oxidation peak at about 1.5 V and the peaks at >2.0 V were attributed to the delithiation of Li^+ ions from the alk-Ti_3C_2. After that the PVP-Sn(IV)/Ti_3C_2 nanocomposite based electrode demonstrated very stable cycling performance. Electrochemical interaction between Li^+ ions and the PVP-Sn(IV)/Ti_3C_2 nanocomposites could be described as two stages, i.e., (i) reaction between the Sn(IV) nanocomplex and Li^+ ions and (ii) reaction between the alk-Ti_3C_2 matrix and Li^+ ions.

Figure 4.8b shows charge–discharge characteristics of the PVP-Sn(IV)/Ti_3C_2 nanocomposite based electrode, over 0.01–3 V at 216.5 mA/cm^3 (or 0.1 A/g). The wide slop voltage range of 0.5–1.75 V corresponded to the peak at 1.15 V in the CV curve. For the initial cycle, the discharge and charge capacities were 3219.4 mAh/cm^3 (or 1487 mAh/g), 1842.4 mAh/cm^3 (or 851 mAh/g), respectively, as observed in Fig. 4.8c. The irreversible capacity at the first cycle was caused by the irreversible reduction of the Sn(IV) nanocomplex to Sn and the presence of SEI in the working electrode. Nevertheless, high Coulombic efficiency close to 99% and high cycling stability were maintained after the initial cycling.

Capacity of the PVP-Sn(IV)/Ti_3C_2 nanocomposite based electrode had three contributors. The first one was elemental Sn formed due to the irreversible reduction of the Sn(IV) complex, although the quantity was relatively low owing to the low content. The alk-Ti_3C_2 was the second contributor. However, it was also relatively low in the total value of capacity. The last contributor was the so-called pillar effect [29], which was closely related to the Sn nanoparticles in between the spacings of the alk-Ti_3C_2 nanosheets. Furthermore, the synergistic effect of the alk-Ti_3C_2 nanomatrix and the Sn(IV) nanocomplex played a significant role in determining the capacity level of the nanocomposite electrode. After intercalation, the Sn nanoparticles reacted with Li^+ ions to form alloys, which enlarged the spacings of the Ti_3C_2 interlayers. This in turn facilitate the intercalation of Li^+ ions, thus leading to high capacity.

Cycling behavior of the PVP-Sn(IV)/Ti_3C_2 nanocomposite based electrode at 500 mA/g is illustrated in Fig. 4.8e. It possessed a specific capacity of 544 mAh/g

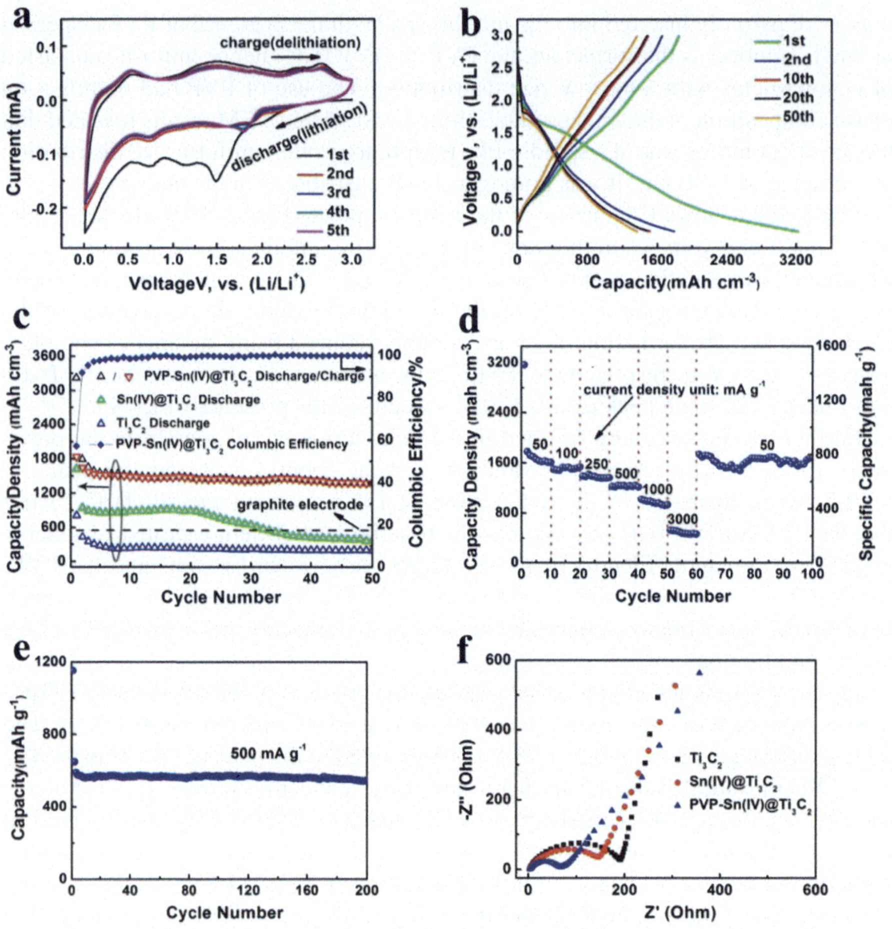

Fig. 4.8 **a** CV curves of the PVP-Sn(IV)/Ti_3C_2 at the scan rate of 0.1 mV/s. **b** Charge/discharge profiles of the nanocomposite based electrode at different cycles at the current density of 216.5 mA/cm^3 (0.1 A/g). **c** Cycling performance and Coulombic efficiency at the current density of 216.5 mA/cm^3 (0.1 A/g). **d** Rate performance of the nanocomposite based electrode. **e** Cycling performance of the PVP-Sn(IV)/Ti_3C_2 at the current density of 500 mA/g. **f** Nyquist plots at frequencies from 100 kHz to 10 MHz for the Ti_3C_2, Sn(IV)/Ti_3C_2 and PVP-Sn(IV)/Ti_3C_2 electrodes before cycling at a perturbation voltage of 10 mV. Reproduced with permission from [28]. Copyright © 2016, American Chemical Society

and capacity retention of 94.3% at 500 mA/g, after cycling for 200 times. In comparison, the capacity of the Sn(IV)/Ti_3C_2 based electrode was quickly decreased to 370 mAh/cm^3 (185 mAh/g) after 50 cycles, while the capacity retention was only 42.5%, at the current density of 216.5 mA/cm^3 (0.1 A/g), although its initial specific capacity was as high as 1626 mAh/cm^3 (813 mAh/g). The rapid decay in capacity of the Sn(IV)/Ti_3C_2 based electrode was caused by the volume expansion of the materials, the increase in particle size of the Sn(IV) nanocomplex enlarges.

Rate performance of the nanocomposite-based electrode is demonstrated in Fig. 4.8d. A steady reduction in discharge capacity was observed with increasing current density, while the high capacity of 1567.9 mAh/cm^3 was readily recovered as the current density was decreased to the low level of 108.3 mA/cm^3. The understanding electrochemical performance of the nanocomposite-based electrode could be explained in terms of the factors as discussed above. The open nanostructures of the 2D Ti_3C_2 MXene matrix offered a relatively short path for the diffusion of the Li^+ ions. The alk-Ti_3C_2 matrix exhibited a low diffusion barrier of 0.07 eV for Li^+ ions, while buffering the volumetric variation during the cycling process. Also, the surface of alk-Ti_3C_2 matrix was rich of negative charges, due to the functional groups of −F and –OH, so that the Sn^{4+} ions were strongly bounded through the chemical adsorption. Moreover, the presence of PVP prevented the growth and aggregation of the nanoparticles.

Figure 4.8f shows Nyquist plots of the electrodes with different compositions. The PVP-Sn(IV)/Ti_3C_2 nanocomposite had a much smaller semicircle than the Sn(IV)/Ti_3C_2 and Ti_3C_2 ones. This implied that the PVP-Sn(IV)/Ti_3C_2 nanocomposite possessed the highest charge transport characteristics, due to its high conductivity. The presence of PVP ensured small particle size of the Sn(IV) nanocomplex and large contact area between the electrode materials and the electrolyte, thus increasing the charge transport efficiency and lowering the resistance to the charge transfer. All these explanations were firmly supported by the microstructural analysis results.

MXene/Ag hybrids were prepared through the reduction of $AgNO_3$ solutions in which MXene ($Ti_3C_2(OH)_{0.8}F_{1.2}$) was dispersed [30]. The MXene/Ag hybrids exhibited promising electrochemical behaviour as anode of lithium-ion batteries, with outstanding cycling stability and high specific capacity at high charge/discharge rates. In this case, the reduction of Ag^+ in the presence of MXene confirmed the low valence state of Ti. The hybrid electrode had a reversible capacity of 310 mAh/g at 1 C, very close to the theoretical value of 320 mAh/g. Meanwhile, high rate capacities of 260 mAh/g and 150 mAh/g were observed at 10 C and 50 C, respectively. Moreover, after cycling for >5000 times, the specific capacity showed almost no decrease at current densities of 1–50 C. The high electrochemical performances were attributed to the significantly lowered interface resistance and the oxidation of Ti(II) to Ti(III) during the charge–discharge cycling process.

Figure 4.9a shows representative SEM image of the MXene/Ag hybrids, revealing that the layered structure of MXene was well retained. There were also Ag nanoparticles on surface of the MXene nanosheets, as seen in Fig. 4.9b. Figure 4.9c shows XRD patterns of the hybrids, demonstrating the formation of face-centered-cubic (fcc) Ag nanocrystals. At the same time, the (002) peak intensity of MXene at near 9° was weakened, implying that the degree of ordering in the *c* direction was reduced. No shift in the (002) peak of MXene was observed, suggesting that the Ag nanoparticles were not intercalated into the MXene nanosheets.

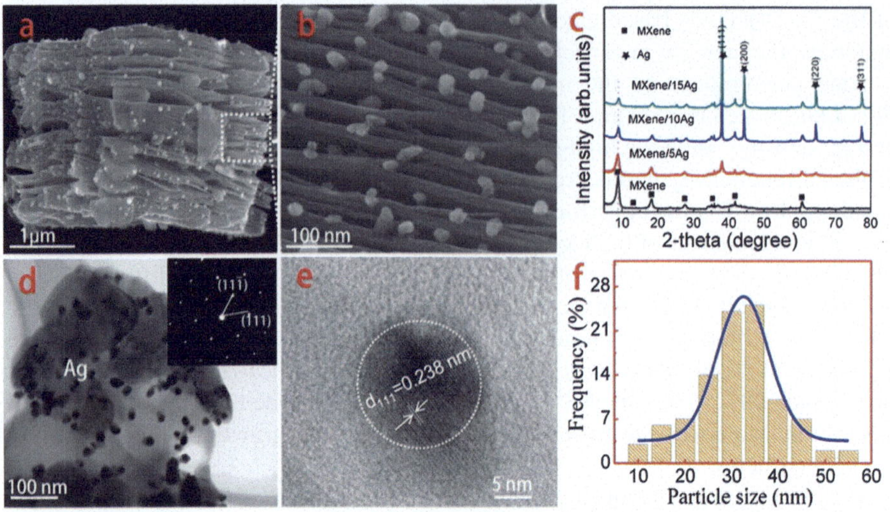

Fig. 4.9 **a** SEM image of the MXene/10Ag hybrid. **b** High-magnification image of the area in panel (**a**). **c** XRD patterns of samples with different contents of Ag nanoparticles. **d** Representative TEM image of the MXene/10Ag hybrid, with the inset showing the SAED along the [1–10] direction. **e** HRTEM image of an individual Ag nanoparticle. **f** Size distribution of the Ag nanoparticles. Reproduced with permission from [30]. Copyright © 2016, American Chemical Society

According to TEM analysis results, the Ag nanoparticles were attached on the surfaces of MXene, when the content of Ag was about 10 wt%, as illustrated in Fig. 4.9d, e. The Ag nanoparticles had an average size of 35.2 nm, as observed in Fig. 4.9f. With increasing level of $AgNO_3$, both the particle size and the content of Ag were increased. As the concentration of Ag was increased from 5 to 15 wt%, the average particle size of the Ag nanoparticles was increased from 10 to >100 nm.

Wang et al. used SnO_2 nanoparticles to modify Ti_3C_2 MXene, forming SnO_2–Ti_3C_2 nanohybrids, by using hydrothermal process, which were utilized as anodes of LIBs [31]. Figure 4.10 shows a schematic diagram describing fabrication steps of the nanohybrids through hydrothermal reaction process. The incorporation of SnO_2 nanosized particles expanded the Ti_3C_2 nanolayers in the thickness direction, so as to increase the Li^+ ion storage capability. In addition, the presence of the SnO_2 nanoparticles enhanced the electrochemical behaviour of the nanohybrids due to the alloying reaction. As a result, the SnO_2–Ti_3C_2 nanohybrids exhibited much higher reversible capacities than the pristine $Ti_3C_2T_x$ nanoparticles. Specifically, the SnO_2–Ti_3C_2 nanohybrids possessed a initial capacity of 1030.1 mAh/g at the current density of 100 mA/g, which was maintained to be 360 mAh/g after cycling for 200 times.

An atomic layer deposition (ALD) process was employed to grow SnO_2 nanolayers on 2D titanium carbide MXenes nanosheets, which led to SnO_2/MXene nanohybrids, with high performances as anodes of Li-ion batteries [32]. Figure 4.11 shows a schematic diagram of the ALD method, together with hydrothermal and sputtering techniques. The SnO_2/MXene hybrid-based electrode possessed high Li^+

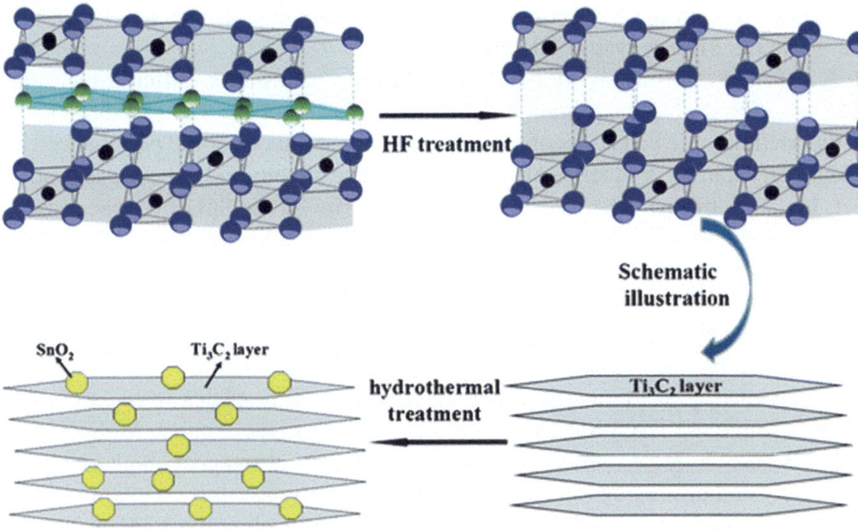

Fig. 4.10 Schematic diagram to show fabrication process of the SnO_2–Ti_3C_2 nanohybrids. Reproduced with permission from [31]. Copyright © 2016, Springer

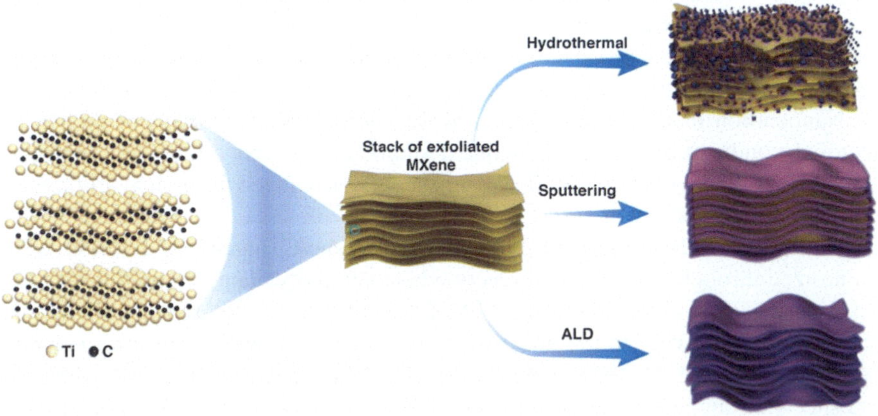

Fig. 4.11 Schematic diagram showing the methods to deposit SnO_2 nanolayers on Ti_3C_2 MXene nanosheets, including hydrothermal reaction process, sputtering deposition and atomic layer deposition (ALD). Atomistic view of exfoliated MXene sheets is shown on the left, the stack of exfoliated MXene sheets is in the middle and the MXene hybrid-based electrodes after the coating of SnO_2 nanolayers are on the right. Reproduced with permission from [32]. Copyright © 2017, Elsevier

ion storage capacity due to the presence of SnO_2, whereas the structural stability and mechanical strength, as well as electrical conductivity, were maintained because of the MXene nanostructures. The ALD parameters could be optimized so as to grow

SnO_2 nanolayers on the MXene nanolayers with functional groups of –O, –F and –OH, while the MXene was not degraded.

It was found that the highly conductive MXene nanosheets buffered the volumetric variation during the lithiation/delithiation process of the SnO_2 nanoparticles. In addition, the cyclic behaviors of the nanohybrid based anode could be enhanced by coating a very thin layer of HfO_2 on the SnO_2/MXene nanohybrid to have passivating effect, by using the ALD process. Structural integrity of the SnO_2/MXene hybrid-based electrode was significantly improved during the cycling process. Importantly, with the coating layer of HfO_2, the SnO_2/MXene hybrid-based anode displayed a high specific capacity of 843 mAh/g after 50 cycles at the current density of 500 mA/g.

Wu et al. developed a simple method to prevent MXenes from structural degradation caused by the spontaneous oxidation [33]. The method was known as carbon nanoplating. Carbon coated few-layered MoS_2 nanoplates were assembled on carbon-stabilized Ti_3C_2 MXene nanosheets, leading to hierarchical MoS_2/Ti_3C_2-MXene@C nanohybrids, which possessed high structural integrity, high electrical conductivity and promising interfacial interactivity, thus ensuring excellent electrochemical activity for Li^+ ion storage. Specifically, after cycling for 3000 cycles, the capacity was maintained at high level of 580 mAh/g, with almost no change at a relatively high rate of 20 A/g.

It was reported that TiO_2 nanosized particles with mixed phase of anatase and rutile could be formed through oxidation of Ti due to the dissolved O_2 in solutions when hydrothermal treatment was conducted at 160 °C. This would damage the structural and electrical properties of MXenes. One of the ways to address this issue was to coat a protection layer as the physical barriers to inhibit the diffusion of O_2. Nanocarbon materials could be used for such purpose, because of their high electrical conductivity, physical/chemical stability and compatibility to MXenes. Carbon nanoplating on Ti_3C_2 MXene nanosheets was realized through hydrothermal carbonization of glucose at 160 °C.

2D hierarchically structured MoS_2/Ti_3C_2-MXene/C nanosized hybrids were synthesized through hydrothermal reaction between molybdate tetrahydrate (AMT) and thiourea, in the dispersion of Ti_3C_2 MXene and glucose, combined with thermal treatment in Ar. Owing to the negative charges of the –F and –O groups, metallic ions in the solution of AMT were adsorbed on the surface of the Ti_3C_2 MXene nanosheets through electrostatic adsorption. The following reaction with thiourea at the hydrothermal environment led to the formation of MoS_2 nano-plates. With the simultaneous carbonization, multiple structured MoS_2/Ti_3C_2-MXene/C nanohybrids were obtained, after the samples were thermally calcined at 500 °C in Ar.

Figure 4.12 shows microstructures and elemental profiles of the nanosized hybrids. According to the SEM and TEM results, the fresh Ti_3C_2 MXene nanosheets were of a flaky morphology. The 2D structure had rich reactive sites to ensure the heterogeneous nucleation and formation of MoS_2 nanosheet coatings, thus leading to 3D hierarchical porous network structure on the MXene nanosheets, as illustrated in Fig. 4.12b. Such special microstructures were confirmed by TEM observations, as revealed in Fig. 4.12c–e. HRTEM image indicated that the MoS_2 nano-plates were

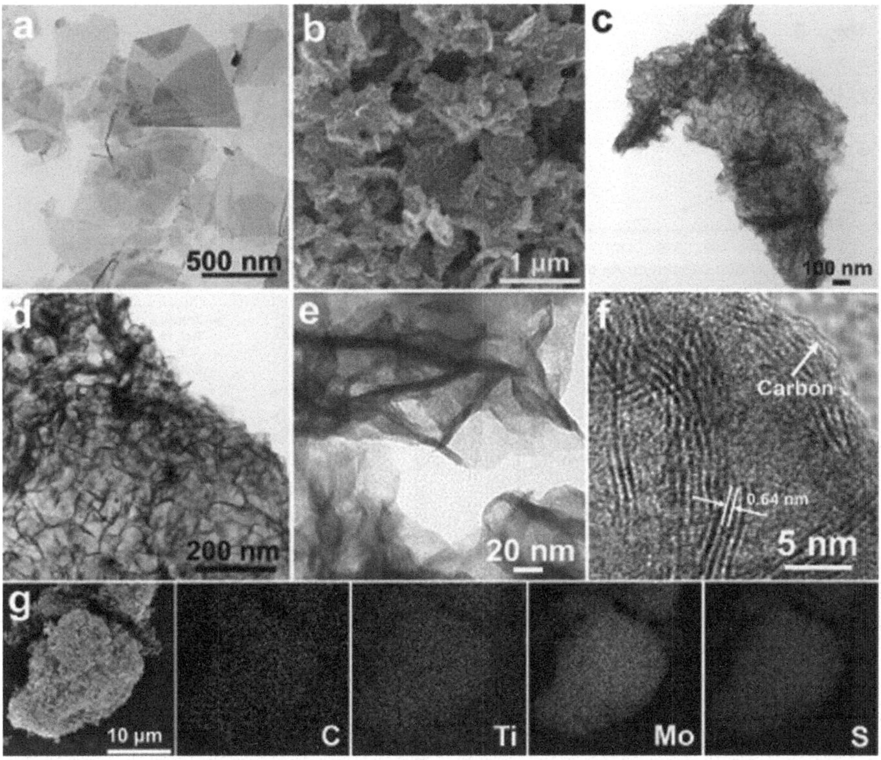

Fig. 4.12 **a** TEM images of the fresh Ti_3C_2 MXene nanosheets. **b** SEM image of the MoS_2/Ti_3C_2-MXene/C nanohybrids. **c–e** TEM images of the MoS_2/Ti_3C_2-MXene/C nanohybrids. **f** Representative HRTEM image indicating carbon nanoplating on the hybrid nanostructure. **g** Elemental mapping profiles of C, Ti, Mo and S elements in the MoS_2/Ti_3C_2-MXene/C nanohybrids. Reproduced with permission from [33]. Copyright © 2017, John Wiley & Sons

highly crystallized, containing a few of layers, with a spacing of 0.64 nm and a thickness of <10 nm, as shown in Fig. 4.12f. The hydrophilic Ti_3C_2 MXene nanosheets were modified to be hydrophobic because of the nanocarbon coatings. C, Ti, Mo and S elements were homogeneously distributed in the MoS_2/Ti_3C_2-MXene/C nanosized hybrids, as observed in Fig. 4.12g. The formation of the MoS_2 phase was also confirmed by XRD and XPS results.

Figure 4.13a shows discharge–charge curves of the MoS_2/Ti_3C_2-MXene/C nanohybrid based electrode at 0.2 A/g, over the voltage window of 0.01–3.0 V, with initial discharge and charge capacities of 1750 and 1210 mAh/g, respectively. Comparatively, the fresh Ti_3C_2 MXene nanosheets had a specific capacity of 270 mAh/g. The MoS_2/Ti_3C_2-MXene/C nanostructure exhibited outstanding high rate capability due to its low resistance, as shown in Fig. 4.13b. This was attributed to the highly active sites on surface of the MXenes nanosheets to host the MoS_2 nanostructures, which offered high conductivity. Also, the MoS_2/Ti_3C_2-MXene/C nanohybrid

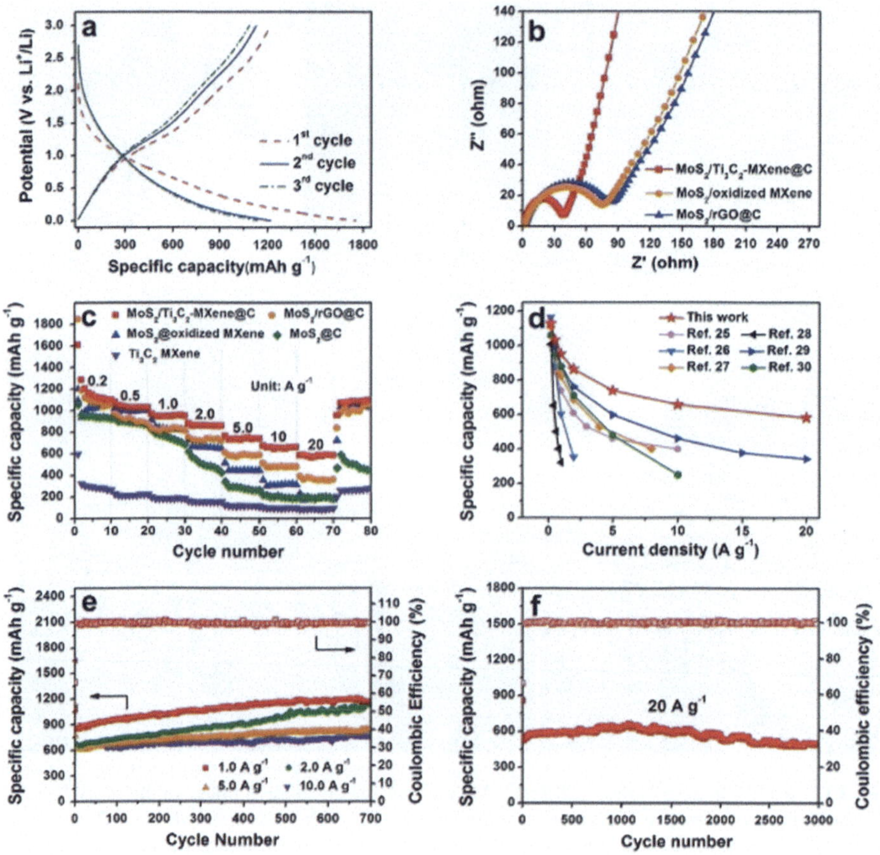

Fig. 4.13 **a** First three cycle discharge-charge curves of the MoS_2/Ti_3C_2-MXene/C nanohybrid based electrodes at 0.2 A/g. **b** EIS spectra of the electrodes based on MoS_2/Ti_3C_2-MXene/C, MoS_2//MXene and MoS_2/rGO/C. **c** High-rate stabilities of different electrodes at current densities of 0.2–20.0 A/g. **d** Li^+ ion storage capabilities of different electrode materials. **e** Cycling stability and Coulombic efficiency of the MoS_2/Ti_3C_2-MXene/C nanohybrid electrodes at current densities of 1.0, 2.0, 5.0 and 10.0 A/g. **f** Cycling up to 3000 cycles of the MoS_2/Ti_3C_2-MXene/C based electrodes at a high current rate of 20.0 A/g. All these tests are conducted between 0.01 and 3.0 V except for EIS. Reproduced with permission from [33]. Copyright © 2017, John Wiley & Sons

based electrode delivered specific capacities of 660–1130 mAh/g at current densities in the range of 0.2–10 A/g, as demonstrated in Fig. 4.13c. Figure 4.13d shows superb high power performance, while outstanding cycling stabilities are evidenced in Fig. 4.13e, f.

Insertion of Li^+ into Ti_2C-based MXene with surfaces that were oxidized was studied [34]. The powders after treatment exhibited slit-type pores according to the N_2 sorption–desorption hysteresis. The specific surface area was 23 m^2/g, which was higher than that of the Ti_2AlC without treatment by about one order of magnitude. The CV curves exhibited peaks of lithiation and delithiation at 1.6 V and 2 V versus

Li^+/Li, respectively. The specific capacity was 225 mAh/g at C/25. It was 110, 80 and 70 mAh/g at 1 C, 3 C and 10 C, after 80, 120 and 200 cycles, respectively. Similarly, MXenes, such as Ti_3C_2, Ta_4C_3, TiNbC and $(V_{0.5}, Cr_{0.5})_3C_2$, could also be potential anode materials of LIBs.

Ti_2C has been treated through surface oxidation at room temperature H_2O_2, which enlarged the interlayer spacing of the MXene nanosheets and triggered the formation of TiO_2 nanoparticles on the surfaces [35]. Both the morphology and composition of the reaction products could be well controlled by adjusting the reaction time and the dosage of H_2O_2 in the experiments. The MXene after the treatment with H_2O_2 showed promising electrochemical performance as anode materials of LIBs. Specifically, after charge–discharge 50 for cycles, the capacities were 389, 337, 297 and 150 mAh/g, at current densities of 100, 500, 1000 and 5000 mA/g, respectively.

Liu et al. obtained V_2C MXene from V_2AlC etched in NaF and HF at 90 °C for 3 days [36]. The V_2C MXene possessed outstanding electrochemical performances as the anode materials of LIBs. It delivered a specific capacity of 260 mAh/g at the current density of 370 mA/g. Considering the difficulty to fully exfoliate V_2AlC to get V_2C MXene Zhou et al. found that partial substituting V with Ti led to $(V_{1-x}Ti_x)_2AlC$ which could be easily exfoliated to form $(V_{1-x}Ti_x)_2C$ [37]. The $(V_{1-x}Ti_x)_2C$ MXene obtained in this way exhibited promising electrochemical properties as anode of LIBs, with a specific capacity of 254 mAh/g at 74.4 mA/g.

Mashtalir et al. employed amine to enlarge the spacings of Nb_2CT_x MXene, in order to form Nb_2CT_x/CNTs with enhanced electrochemical properties [38]. A high specific capacity of 400 mAh/g was observed at 0.5 C, along with high rate stability. In the Nb_2CT_x/CNTs, the content of CNTs was 10 wt% CNTs. To the hybrid obtain electrodes, CNTs were mixed with delaminated Nb_2CT_x in suspensions, which were filtrated through a polymeric membrane.

In the CV curves of the Nb_2CT_x/CNTs paper-based electrodes, pronounced lithiation-delithiation peaks were not present, as seen in Fig. 4.14a. The Nb_2CT_x/CNTs paper-based electrode had an initial capacity of 780 mAh/g and a stable value of 420 mAh/g, At 0.5 C, while its Coulombic efficiency was almost 100%, as illustrated in Fig. 4.14b. In this case, CNTs had neglected contribution to the total capacity, because of the low specific capacity and low content of CNTs. Once the electrode was cycled, the capacity level became stable and had a slight increment. At 2.5 C, the initial capacity was 320 mAh/g, whereas it was raised to 370 mAh/g after 100 cycles and 430 mAh/g after 300 cycles, as observed in Fig. 4.14b. The hybrid electrode also showed high rate capabilities, as demonstrated in Fig. 4.14c, d.

Zhou et al. prepared Hf-based MXene, $Hf_3C_2T_z$, from $Hf_3[Al(Si)]_4C_6$ [39]. According to adhesive energy and atomic charge-related calculations, the etching reaction was facilitated by the Si sublayer. The $Hf_3C_2T_2$ (T=O, F, OH) MXenes had metallic characteristics with high electrical conductivities. The charge storages of the d-$Hf_3C_2T_z$ was attributed to the intercalations of Li^+ and Na^+ ions, instead of conversion reactions. The 2D $Hf_3C_2T_z$ exhibited quite high reversible volumetric capacities of 1567 and 504 mAh/cm^3 as anodes of LIBs and SIBs, respectively, as the current density was 200 mA/g and devices were cycled for 200 cycles.

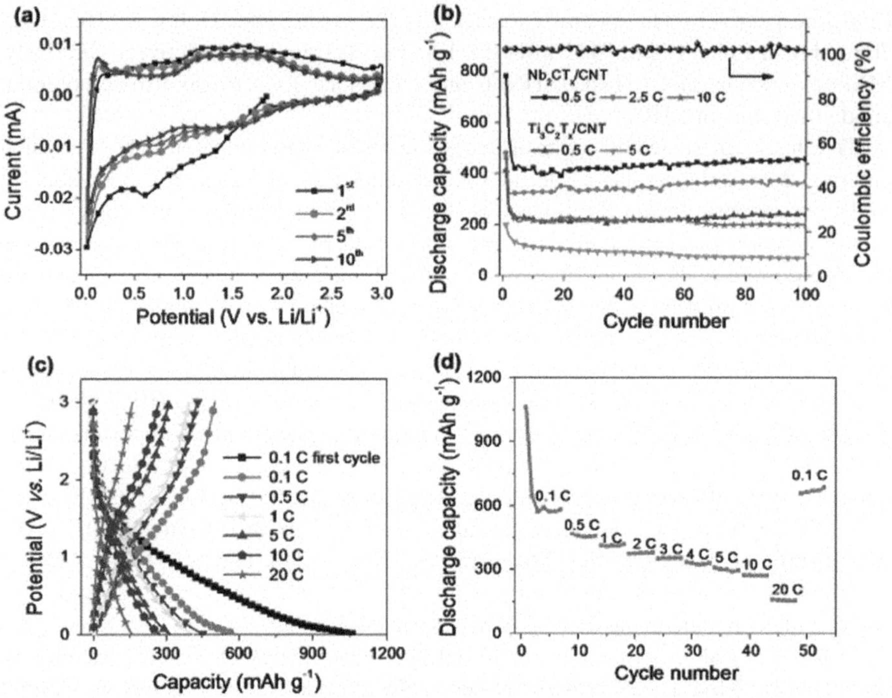

Fig. 4.14 **a** CV curves of the Nb_2CT_x/CNT paper-based electrode at the scanning rate of 0.2 mV/s. **b** Cycling stability of the Nb_2CT_x/CNT and $Ti_3C_2T_x$/CNT paper-based electrodes at different cycling rates. **c** charge/discharge bahaviors of the electrodes. **d** Discharge capacities of the Nb_2CT_x/CNT hybrid electrode at different cycling rates. Reproduced with permission from [38]. Copyright © 2015, John Wiley & Sons

2D $Nb_4C_3T_x$ MXene was also been demonstrated for the applications of LIBs [40]. The Nb-based MXene was derived from Nb_4AlC_3 through etching with HF at room temperature. The $Nb_4C_3T_x$ based electrode showed an increase in charge–discharge capacity with increasing cycles. At the current density of 0.1 A/g, the initial value was 310 mAh/g (194 mAh/cm^3), which was increased to 380 mAh/g (238 mAh/cm^{-3}), after the device was cycled for 100 times. The electrode materials also exhibited high rate capability and stable long-life performance at ultra-high rates.

Fe_2O_3/MXene nanohybrid particles have been synthesized for anodes of LIBs, because of the synergistic effect of high Li + storage capacity of Fe_2O_3 phase and high cycling stability and high conductivity of MXenes [41]. The $Fe_2O_3/Ti_3C_2T_x$ nanohybrids were made with Fe_2O_3 nanosized particles to be embedded inside the spacings of the $Ti_3C_2T_x$ nanolayers with various mass loadings, which were obtained by using the ball milling process. For the samples with 25–50 wt% Fe_2O_3, the Fe_2O_3 nanosized particles were homogeneously distributed on surface of the $Ti_3C_2T_x$ nanolayers.

In this case, the degree of oxidation of the MXene was minimized, as compared with those synthesized by using hydrothermal and wet chemical methods. Furthermore, due to the presence of the Fe_2O_3 nanocrystals, the tendency of restacking of the MXene nanosheets was minimized, so that a high specific surface area was maintained. In the beginning, the capacity of BM50 was lower than that BM25, while the former had high capacity level after cycling for 140 cycles. At 0.1 C, the electrodes of BM25 and BM50 had specific capacities of 650 and 700 mAh/g, respectively. The hybrid sample with 50 wt% Fe_2O_3 had the optimized electrochemical performance, with a specific capacity of 270 mAh/g at 1 C, which was maintained to be 100 mAh/g at a high rate of 10 C.

Figure 4.15a shows CV curves of the $Ti_3C_2T_x/Fe_2O_3$ nanohybrid based electrodes at 0.2 mV/s. Obvious redox peaks at 0.85 and 1.9 V versus Li^+/Li were observed, which corresponded to the reduction of Fe(III) to Fe(0) and the insertion of Li into the Fe_2O_3 nanostructures. The redox peaks in the CV curves were consistent with the voltage plateaus in the charge–discharge profiles, as demonstrated in Fig. 4.15b. The

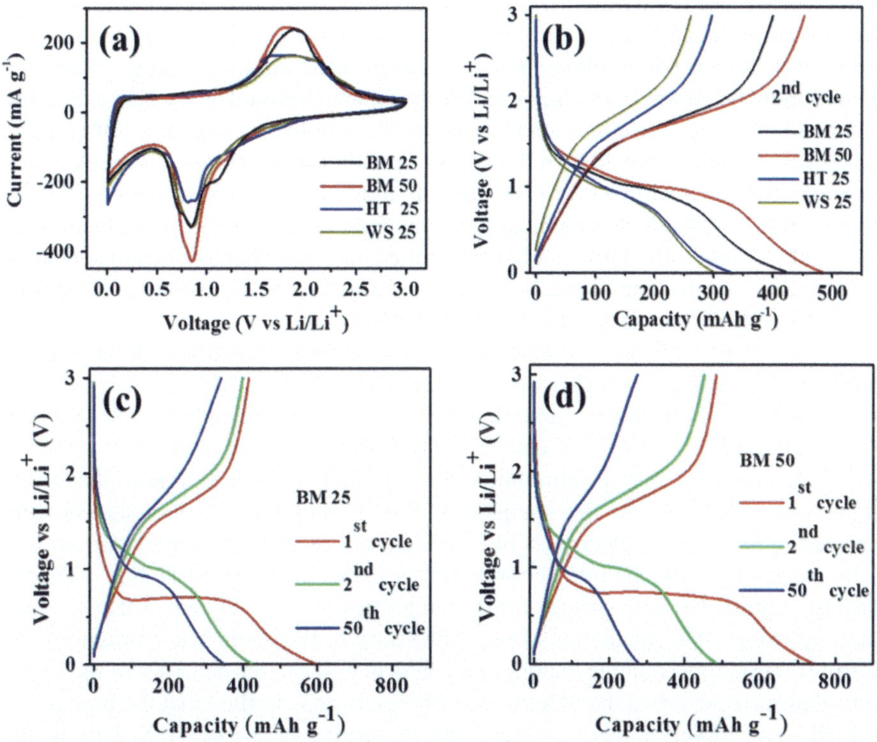

Fig. 4.15 **a** CV curves of the $Ti_3C_2T_x/Fe_2O_3$ nanohybrid based electrode at 0.2 mV/s. **b** Voltage characteristics of the nanohybrid electrode at the second cycle of charge–discharge. **c**, **d** Voltage profiles of the BM25 (25 wt%) and BM50 (50 wt%) samples for the 1st, 2nd and 50th cycles. Reproduced with permission from [41]. Copyright © 2018, American Chemical Society

charge–discharge characteristics of the nanohybrid based electrodes at the earlier cycles were mostly determined by the conversion reactions related to the Fe_2O_3 nanoparticles.

These properties of the BM25 and BM50 electrodes retained after 50 cycles, as illustrated in Fig. 4.15c, d. However, the voltage characteristics were derived from the properties of $Ti_3C_2T_x$. In other words, both the $Ti_3C_2T_x$ MXene and the Fe_2O_3 nanophase had contributions to the capacities of the electrodes. Therefore, the ball milling method is very promising in terms of large scale production of electordes of LIBs. More importantly, synergistic effect of the two components ensured a high capacity of Li ion storage, strong cycling stability and high rate capability.

4.2.1.3 Li–S Batteries

Lithium-sulfur (Li–S) batteries have been extensively studied, because they could offer relatively high energy densities, while there are still critical issues to be addressed for large scale industrial applications [42–46]. On one hand, S and the discharge product Li_2S are both insulating, without being able to offer sufficiently high conductivities. On the other hand, the items formed during the charge–discharge process could be dissolved in organic electrolytes, which would be brought to anodes, causing so-called shuttle effect. As Li_2S was formed, both the Coulombic efficiencies and recycle stabilities would be reduced. As a result, it is necessary to find electrode materials that have sufficiently high surface area and electrical conductivities, in order to improve the electrochemical performances of Li–S batteries. As mentioned before, 2D MXenes are intrinsically of metallic conductive behaviors and have high surface areas, so that they have been acknowledged to be potential candidates as electrodes of Li–S batteries to tackle the problems.

A theoretical study was conducted by using density functional theory (DFT) computations, in order to clarify the potential applications of MXenes as electrodes of Li–S batteries [47]. It was found that the Li^+ ions in Li_2S_n and the –O functional groups at the surfaces of Ti_2CO_2 and $Ti_3C_2O_2$ monolayers were attractive each other, these MXenes were able to immobilize the soluble polysulfides, due to the strong interaction with Li_2S_n. As a consequence, Ti_2CO_2 and $Ti_3C_2O_2$ monolayers were expected to be potential electrodes of Li–S batteries, with high cycling stabilities.

In a separate study, the interaction between Li_2S_m on Ti-based Ti_nX_{n-1} MXenes and Ti_2C terminated with surface functional groups of −F, −O and −OH was evaluated by using DFT calculations [48]. According to the geometric configurations, electronic structures, binding energies and deformation charge densities of the Li_2S_m items that were adsorbed the MXene nanosheets, it was clarified that the interaction of Li_2S_m and MXenes was dominated by the strong bonds of Ti–S. Due to the Coulombic interaction, the S items were strongly confined in the cathode, so that soluble issue of S was addressed. Also, since both the MXenes and Li_2S_m/MXenes were highly conductive, electrochemical performances of the Li–S batteries could be enhanced.

It was demonstrated that the electrochemical performances of Li–S batteries could be improved by depositing a layer of $Ti_3C_2T_x$ MXene nanosheets on commercial membranes of Celgard [49]. Due to the special 2D nanostructures, the coating of $Ti_3C_2T_x$ MXene required a very low loading of 0.1 mg cm^{-2}, thus having an ultra-small thickness of 522 nm. With the presence of the 2D MXene thin layers, electrical conductivity of membranes was significantly increased, while and the polysulfides could be effectively confined. As a result, the lithium–sulfur batteries exhibited high specific capacity and stable cycling capability.

The exfoliated $Ti_3C_2T_x$ MXene had lateral sizes at the micrometer scale, as seen in Fig. 4.16a. Figure 4.16b shows representative TEM image of the MXene nanosheets, evidencing the 2D structures. Morphological characteristics of the MPP (MXene modified polypropylene) separator are illustrated in Fig. 4.16c. Figure 4.16d shows electrochemical impedance spectroscopy EIS curves of the Li–S devices, with PP separator and MPP separator before cycling. The device with the MPP separator had a charge transport resistance of about 45 Ω, which was less than half of the of the

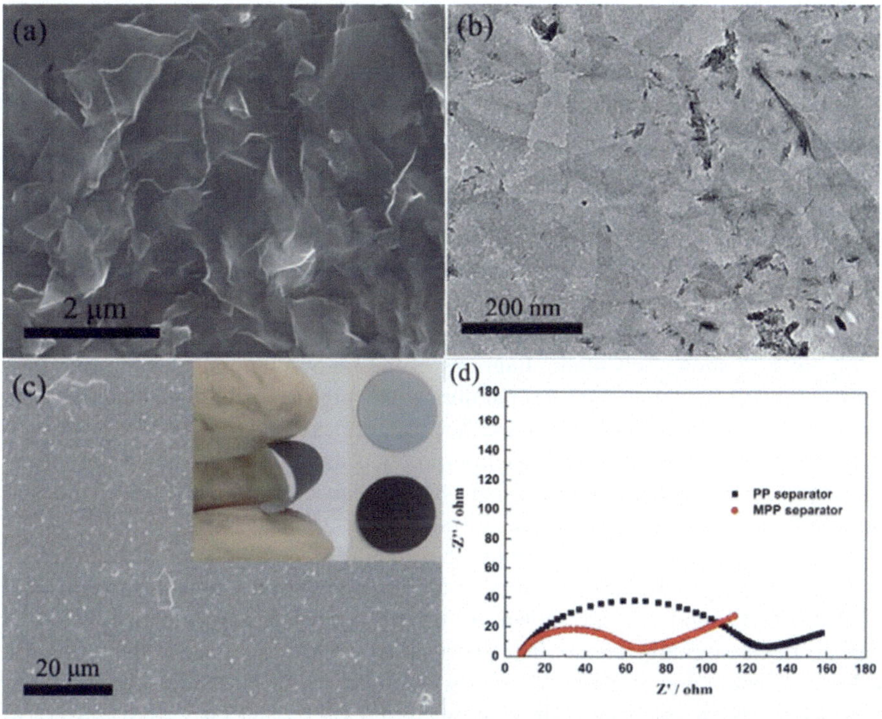

Fig. 4.16 SEM image (**a**) and TEM image (**b**) of the $Ti_3C_2T_x$ MXene nanosheets. **c** Low-magnification surface SEM image and optical photograph (inset) of the MPP separator. **d** EIS curves of the freshly assembled Li–S battery cells with the PP and MPP separators. Reproduced with permission from [49]. Copyright © 2016, American Chemical Society

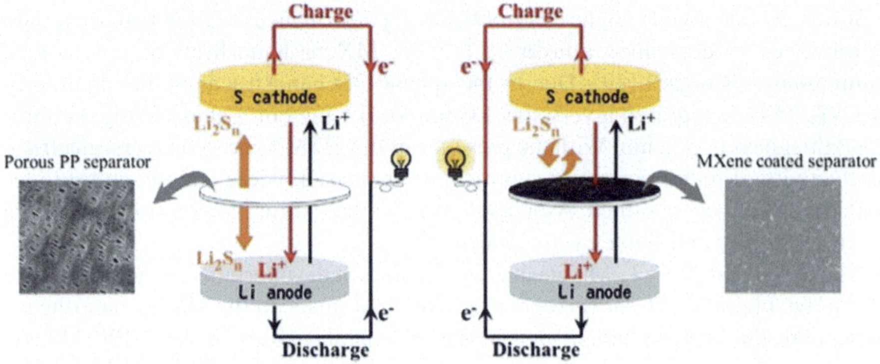

Fig. 4.17 Schematic diagrams of the Li–S cell devices with the PP and MPP separators. Reproduced with permission from [49]. Copyright © 2016, American Chemical Society

value (101 Ω) for that one with the PP separator, demonstrating the high electrical conductivity of the 2D MXene nanosheets.

To identify the effectiveness of the MPP separator in confining the dissolved polysulfide items, the cell devices were dismantled after charge–discharge for 100 cycles at 1 C. After complete cleaning with the solvents of DME/DOL, both the Li piece and the MPP separator were examined to reveal their variations in morphology. SEM analysis results indicated that the surface of fresh metal Li was highly smooth, which was heavily corroded if the separator was PP, because of the accumulation of Li_2S_2/Li_2S formed through the reaction between the Li anode and the polysulfides during the charging-discharging. However, with the application of the MPP separator, the Li anode had a much nice surface profile, confirming that the polysulfides have been captured by the $Ti_3C_2T_x$ nanosheet layers.

Figure 4.17 shows schematic diagrams of the Li–S cell devices with separators of PP and MPP. Although the PP separators with a high porosity that could facilitate rapid transportation of electrolyte ions, the polysulfides could readily diffuse through the pores, resulting in unwanted reaction products and consuming the active materials. After coating with the $Ti_3C_2T_x$ nanosheets, a barrier was formed, which blocked the soluble polysulfides. Instead, the polysulfides were thus confined on the surface of the MPP separator, which in turn participated the electrochemical reactions.

The first charge–discharge profiles of the devices with the separators of PP and MPP at 0.2, 0.5, and 1 C, are shown in Fig. 4.18a, b, respectively. There were two plateaus in the discharge curves. The one at about 2.3 V was related to the formation of Li_2S_{4-8}, while one at about 2.1 V was attributed to the reduction of Li_2S_{4-8} to Li_2S_2/Li_2S. The two plateaus in the charging profiles at 2.2 and 2.4 V were ascribed to the oxidations of Li_2S_2/Li_2S to Li_2S_8/S_8, as the devices were charged. In comparison, the cell device with the MPP separator displayed much more stable and flat charge–discharge plateaus. Also, the difference in potentials of the plateaus was decreased in the charge–discharge profile curves, as illustrated in Fig. 4.18c. This implied that

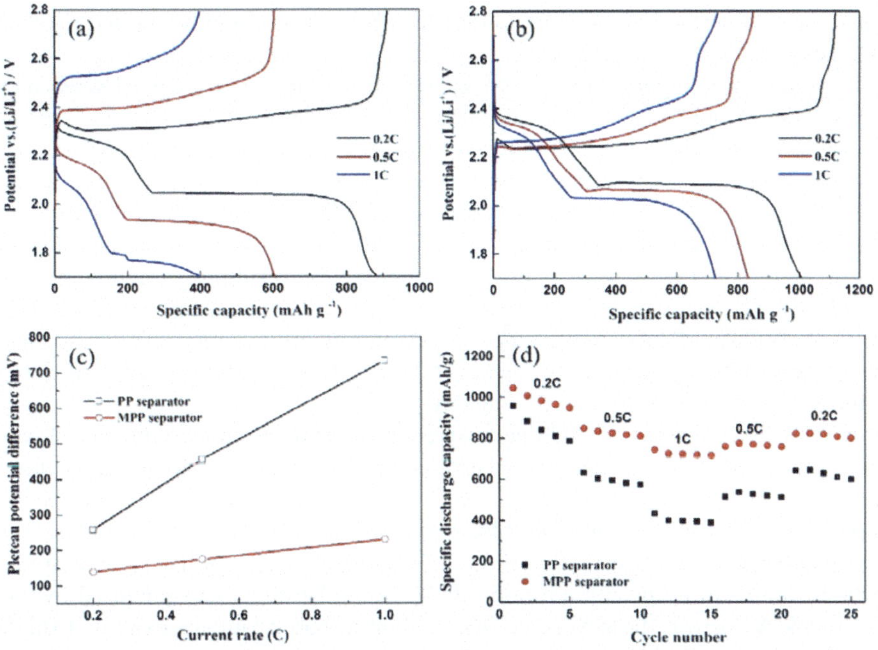

Fig. 4.18 First charge–discharge curves of the Li–S cell devices with the PP separator (**a**) and the MPP separator (**b**) at 0.2, 0.5 and 1 C. **c** Potential differences of the Li–S battery devices with the PP and MPP separators between the charge and discharge plateaus at 0.2, 0.5 and 1 C. **d** Rate characteristics of the Li–S cells with the PP and MPP separators. Reproduced with permission from [49]. Copyright © 2016, American Chemical Society

the use of the MPP separator resulted in more suitable redox reaction kinetics and better reversible capability.

Figure 4.18d shows rate properties of the Li–S battery devices with the PP and MPP separators. With the presence of the MPP separator, reversible capacity of the cell device was largely enhanced, especially when the current density was at the high levels. The capacities were retained by about 91 and 79% at 0.5 and 0.2 C, confirming the stability and reversibility of the devices with the MPP separator. The advantages of the MPP separator were also evidenced by the long-life cycling stability and other electrochemical parameters.

It has been shown that Ti_2C MXenes could be potential candidate as cathode materials of Li–S batteries [50]. Electrodes with a weight percentage of 70 wt% S and Ti_2C were obtained through the reaction between S and the –OH surfaces of the Ti_2C nanolayers. The S/Ti_2C nanocomposites delivered a specific capacity of 960 mAh/g at the current density of 335 mA/g after cycling for 100 times, while 80% of the capacity was retained after cycling for 400 times at a relatively high current density of 838 mA/g. The outstanding performances of the Li–S battery with the MXenes as the hosts was readily attributed to the high electrical conductivity and the functionalized surfaces. It is believed that this observation could be applied to other MXenes.

A strategy was reported to achieve interfacial interaction between MXenes and polysulfides through cleavages of Ti–OH bonds by forming thiosulfates and strong Ti–S bond interactions [51]. The hydroxyl moities of the functionalized MXenes interacted with the polysulfides due to their redox activity, so that the thiosulfate/polythionate groups were attached onto the MXene nanosheets. The process was assembled to the interaction between metallic oxides and graphene nanosheets. The MXene nanolayers were interwoven with CNTs, resulting in highly porous and electrically conductive frameworks that strongly adsorbed the polysulfides. The sulfur hosts ensured outstanding long-term cycling stability, with a low degradation rate of 0.043% per cycle after cycling to 1200 times. In addition, the loading of sulfur in the electrodes was as high as 5.5 mg/cm^2.

A 3D hybrid consisting of MXenes and reduced graphene oxide (rGO) nanosheets was synthesized, with high performances as cathode host materials for lithium–sulfur batteries [52]. The hybrid was developed by using a simple two-step liquid-phase impregnation method. Because the MXenes have a special 3D layered structure and functionalized 2D surfaces, while the rGO nanolayers have similar layered architectures, both sulfur and lithium polysulfides could be effectively captured by the electrodes. As a result, the MXene/rGO/sulfur hybrid-based cathodes exhibited a high specific capacity of 1144.2 mAh/g at 0.5 C, together with a high capacity retention of 878.4 mAh/g after 300 cycles.

A flexible integrated sulfur cathode based on all MXenes, from 3D alkalized Ti_3C_2 MXene nanoribbons, denoted as a-Ti_3C_2 MNR, was developed for Li–S batteries with high electrochemical performances [53]. The 2D delaminated Ti_3C_2 MXenes or d-Ti_3C_2 nanosheets were attached on the polypropylene (PP) separator. The a-Ti_3C_2 MNR frameworks possessed interconnected macropores with sufficiently large surface areas to ensure high mass loading of sulfur and rapid ion diffusion, thus leading to desired lithiation-delithiation kinetics. Due to the 2D d-Ti_3C_2 MXene interlayers, the shuttle effect of the lithium polysulfides was effectively avoided. Therefore, the a-Ti_3C_2–S/d-Ti_3C_2/PP composite could be used as electrode of Li–S batteries, while the metallic current collectors were not necessary.

The Li–S batteries delivered a specific capacity of 1062 mAh/g at 0.2 C, while a high level of 632 mAh/g was retained at 0.5 C after cycling for 50 times, which was higher than those of the a-Ti_3C_2–S/PP electrode (547 mAh/g) and traditional a-Ti_3C_2–S on Al metal current collectors, i.e., a-Ti_3C_2–S/Al (597 mAh/g). In addition, the all MXene cathodes demonstrated excellent rate stability at 10 C and long-life cyclability, with a high specific capacity of 288 mAh/g.

The a-Ti_3C_2 MNRs were derived from m-Ti_3C_2 MXenes that were prepared through HF etching process. The later was shaken at a speed of 250 rpm in 6 M KOH aqueous solution for 72 h in Ar, followed by washing and vacuum drying at 60 °C overnight. The a-Ti_3C_2 MNRs were mixed with commercial sulfur powder at a weight ratio of 3:7, followed by heating at 155 °C for 12 h in Ar, forming a-Ti_3C_2–S hybrids. d-Ti_3C_2 MXenes were formed by intercalating and delaminating MXenes in dimethyl sulfoxide (DMSO).

The d-Ti_3C_2/PP was made by using vacuum filtrating the d-Ti_3C_2 supernatant through PP membrane separator. The dried d-Ti_3C_2/PP had d-Ti_3C_2 mass loading of about 0.4 mg/cm^2. The a-Ti_3C_2–S, carbon black (CB), and PVDF, were mixed with a weight ratio of 7:2:1 in *N*-methylpyrrolidone (NMP) solvent, to form slurries, which were deposited on the d-Ti_3C_2/PP films. After drying at 60 °C for 12 h in vacuum, flexible a-Ti_3C_2–S/d-Ti_3C_2/PP hybrid electrodes were developed. Similarly, m-Ti_3C_2–S/d-Ti_3C_2/PP or S/d-Ti_3C_2/PP hybrid electrodes were also fabricated for the purpose of comparison.

Figure 4.19 shows electrochemical behaviors of the a-Ti_3C_2–S/d-Ti_3C_2/PP integrated electrodes in 1.0 M lithium bistrifluoromethanesulfonylimide dissolved in 1,3-dioxolane/1,2-dimethoxyethane with 1.0 wt% $LiNO_3$ as the electrolyte over 1.6–3.0 V. Properties of the a-Ti_3C_2–S/PP-Al electrodes which were deposited from the a-Ti_3C_2–S slurry on Al foil are illustrated in Fig. 4.19b–d. CV curves of the a-Ti_3C_2–S/d-Ti_3C_2/PP electrode recorded at the scan rate of 0.1 mV/s over 1.6–3.0 V are plotted in Fig. 4.19a. There were two cathodic peaks at about 2.30 V and 2.05 V, corresponding to the phase transitions from the solid S_8 to liquid high ordered Li_2S_x ($4 \leq x \leq 8$) and soluble polysulfide to solid Li_2S/Li_2S_2, respectively.

After cycling for second time, the redox peaks on the CV curves were overlapping, confirming highly reversible characteristics of the electrode materials. The galvanostatic charge–discharge curves of the a-Ti_3C_2–S/d-Ti_3C_2/PP hybrid electrode exhibited two characteristic voltage plateaus at about 2.26 and 2.07 V, as shown in Fig. 4.19b, which were in a good agreement with the CV characteristics. The plateaus were much flatter and more stable, accompanied by a relatively small polarization value of 194 mV at 0.5 C. In comparison, the polarization of the a-Ti_3C_2–S/PP hybrid electrode was 313 mV, which was much higher and displayed a voltage hysteresis.

Moreover, in contrast to the a-Ti_3C_2–S/PP hybrid-based electrodes, the a-Ti_3C_2–S/d-Ti_3C_2/PP electrodes possessed a different impedance characteristics. The latter had a smaller semicircle at frequencies and a more inclined line that was close to the Y axis at low frequencies, implying that its charge transport resistance was much smaller, thus leading to a rapid mass transportation. In this case, the d-Ti_3C_2 nanosheets played a significant role, serving as the internal current collectors. As a result, the internal resistance of the electrodes were largely reduced. Meanwhile, they formed interlayers which effectively prevented the shuttle effect in the a-Ti_3C_2–S/d-Ti_3C_2/PP hybrid electrodes.

The a-Ti_3C_2–S/d-Ti_3C_2/PP hybrid electrodes demonstrated specific capacities of 899 mAh/g and 611 mAh/g at 0.5 C for the first cycle and the 50th cycle, respectively, as illustrated in Fig. 4.19c. Comparatively, these values of the a-Ti_3C_2–S/PP electrode were 626 mAh/g and 539 mAh/g. At the same time, a-Ti_3C_2–S/d-Ti_3C_2/PP hybrid electrodes displayed a significant improvement in cycling stability, whose specific capacity was retained by 68.0%. In contrast, the a-Ti_3C_2–S/Al electrode had specific capacities of 917 and 569 mAh/g, while retention of the capacity was 62.1%. The Coulombic efficiency of the a-Ti_3C_2–S/d-Ti_3C_2/PP was raised rapidly from 94.6 to 99.8% during the first 10 cycles, which was then stabilized at 100%, at the 50th cycle, whereas that of the a-Ti_3C_2–S/Al electrode could not be stabilized.

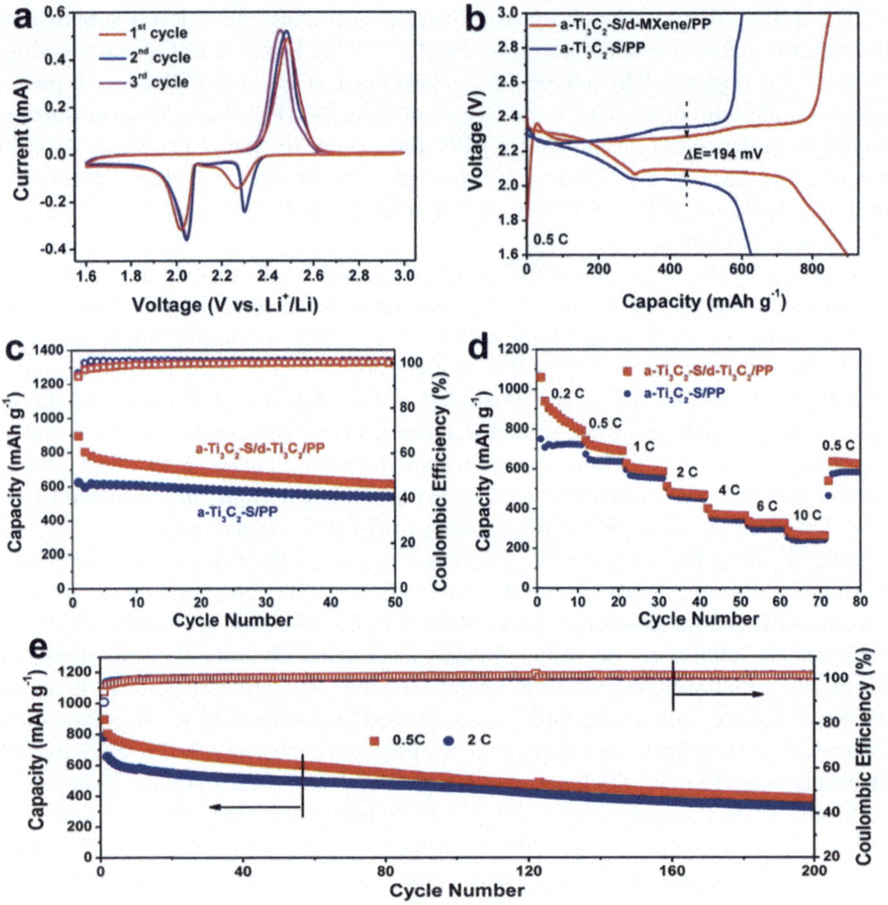

Fig. 4.19 Electrochemical performances of the a-Ti_3C_2–S/d-Ti_3C_2/PP based electrodes. **a** CV curves of the a-Ti_3C_2–S/d-Ti_3C_2/PP electrodes measured at 0.1 mV/s over 1.6–3 V. **b** Galvanostatic charge and discharge curves of the a-Ti_3C_2–S/d-Ti_3C_2/PP and a-Ti_3C_2–S/PP electrodes measured at 0.5 C. **c** Cycling behaviors and **d** rate stability of the a-Ti_3C_2–S/d-Ti_3C_2/PP and a-Ti_3C_2–S/PP electrodes tested at current densities of 0.2–10 C. **e** Long-term cycling profiles of the a-Ti_3C_2–S/d-Ti_3C_2/PP electrodes measured at the current densities of 0.5 and 2 C. Reproduced with permission from [53]. Copyright © 2018, American Chemical Society

Figure 4.19d shows rate stabilities of the a-Ti_3C_2–S/d-Ti_3C_2/PP electrode after evaluated at different densities of 0.2–10 C, with specific capacities of 1062–288 mAh/g. As the current density was bounced back to 0.5 C, the capacity could be recovered to 636 mAh/g. In comparison, the a-Ti_3C_2–S/PP hybrid electrode offered much lower capacities at the same current densities. Furthermore, the a-Ti_3C_2–S/d-Ti_3C_2/PP based electrode demonstrated superior long-term cycling behaviour, with retentions of 47.7 and 50.4% at the 200 and 2 cycles, as seen in Fig. 4.19e, confirming the excellent electrochemical performances.

4.2.1.4 Other Batteries

MXenes have also been used as electrode materials of sodium ion batteries. Mechanisms regarding the electrochemical reactions of Na^+ storages have been deeply studied by using solid-state ^{23}Na magic angle spinning NMR and density functional theory calculations [54]. It was found that Na^+ ions could be intercalated and deintercalated into the interlayer spaces of MXene $Ti_3C_2T_x$, when it was in contact with nonaqueous Na^+ electrolytes. Specifically, the interlayer spaces of the $Ti_3C_2T_x$ were enlarged after the first round Na^+ intercalation. Also, Na^+ ions experienced a reversible intercalation/deintercalation process. The interlayer spacings kept almost unchanged in the intercalation/deintercalation cycling, because of the intercalated Na^+ ions and the inserted solvent molecules in between the $Ti_3C_2T_x$ nanosheets. Because no variation in structure was encountered by the intercalation/deintercalation of the Na^+ ions for the electrochemical reactions, the $Ti_3C_2T_x$ based electrodes exhibited strong long term cycling stabilities and outstanding rate capabilities.

It was believed that that intercalation/extraction process of the Na^+ ions should occur in between the $Ti_3C_2T_x$ nanolayers, while the Na^+ ions could also be adsorbed on surfaces of the $Ti_3C_2T_x$ nanosheets. Figure 4.20 shows typical galvanostatic sodiation/desodiation process of the electrodes. The specific capacity of the first sodiation cycle was 270 mAh/g, corresponding to about two Na^+ intercalation/sorption per chemical unit of $Ti_3C_2T_x$. As a matter of fact, the quantity of Na^+ was only 0.9 Na^+ per unit formula of $Ti_3C_2T_x$, according to the EDX analysis result. Therefore, there must also be electrolyte that decomposed during the first cycle of sodiation. Due to the decomposition of the electrolytes, the current flow of <0.9 V was not reversible in the initial CV curve. In this case, the decomposition of the electrolyte triggered the development of stable solid electrolyte interphases (SEI), which in turn prevented the decomposition process of the electrolyte in the following charge–discharge process. Finally, the $Ti_3C_2T_x$ based electrode reached a pretty stable specific capacity of 100 mAh/g up to 100 cycles, as illustrated in Fig. 4.20b. Correspondingly, the quantity of the Na^+ intercalation/sorption per chemical unit of $Ti_3C_2O_2$ was about 0.75.

It was found that MXene V_2CT_x displayed superior reversible Na^+ charge storage capability, with excellent cycling stabilities and outstanding rate capabilities as the anode materials of Na-ion batteries [55]. The mechanism of charge storages during the intercalation–deintercalation process of the Na^+ ions and the redox reaction of the V species were systematically explored by using a comprehensive method through the combination of synchrotron X-ray diffraction (XRD), hard X-ray absorption near edge spectroscopy (XANES) and soft X-ray absorption spectroscopy (sXAS). The redox reaction involving the V species had a key contribution to the charge storage behaviour and the specific capacity of the V_2CT_x based electrode, which was evidenced by the V *K*-edge XANES and V $L_{2,3}$-edge sXAS experimental analysis.

In addition, the relationship between the level of $CO_3{}^{2-}$ group and the intercalation/deintercalation status of Na^+ ion in the V_2CT_x based electrodes was well established, according to the C and O *K*-edges derived from the sXAS experimental data.

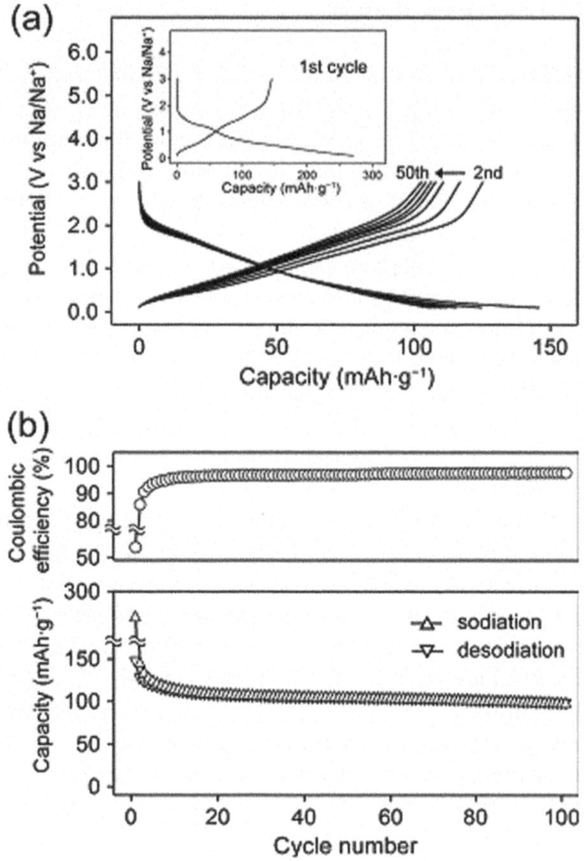

Fig. 4.20 **a** Sodiation/desodiation potential profiles and **b** cycle performance of the $Ti_3C_2T_x$ based electrodes at the current density of 20 mA/g. The first sodiation/desodiation potential profiles are shown in (**a**) as an inset. Reproduced with permission from [54]. Copyright © 2016, American Chemical Society

It was suggested that the Na^+ intercalated V_2CT_x nanostructures could have interacted with the carbonate based nonaqueous electrolytes, thus offering extra reactions to enhance the charge storage capabilities of the electrode materials.

Figure 4.21a shows CV curves of the V_2CT_x base electrode in 1 m $NaPF_6$/ethylene carbonate (EC)–dimethyl carbonate (DMC) electrolyte at the scan rate of 1 mV/s. In the first cathodic scanning, an irreversible current at voltages of <1.0 V (vs. Na/Na^+) was observed, which could be attributed to the generation of the solid electrolyte interphase (SEI) layers. However, the following CV cycles were stabilized, with the presence of quasi-rectangular CV curves. The V_2CT_x electrode had a strong pseudocapacitive charge storage characteristics.

Galvanostatic charge/discharge performances of the V_2CT_x based electrode are plotted in Fig. 4.21b. It was tested over the voltage range of 0.1–3.0 V (vs. Na/Na^+), at a current density of 10 mA/g, in the form half cell with a nonaqueous electrolyte. A voltage plateau at about 1.0 V was observed during the first discharge process, with a total specific capacity of 220 mAh/g. The first charge process demonstrated a lower capacity of 125 mAh/g. As a result, a relatively low Coulombic efficiency

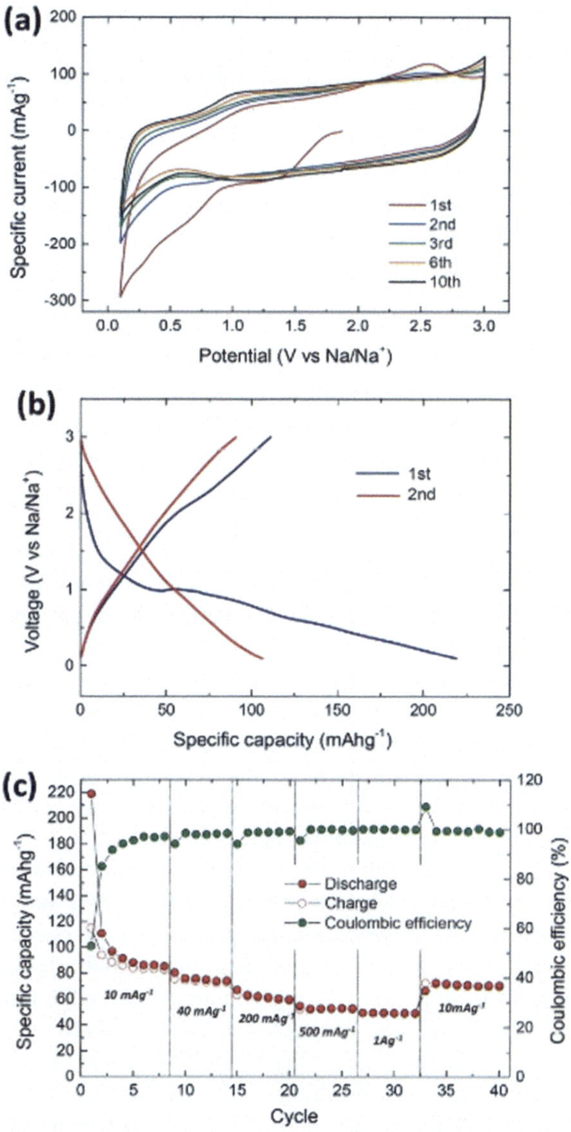

Fig. 4.21 **a** CV curves of the V_2CT_x electrode tested in a 1 m $NaPF_6$/EC–DMC electrolyte at the scan rate of 1 mV/s. **b** Galvanostatic charge/discharge curves over the voltage range of 0.1–3.0 V (vs. Na/Na^+) at the current density of 10 mA/g^{-1}. **c** Variation in capacity during the galvanostatic charge/discharge at different rates. Reproduced with permission from [55]. Copyright © 2017, John Wiley and Sons

was observed for the first cycle, which was 53%. The irreversible loss of capacity was mainly ascribed to the side reactions, like to the presence of SEI layers and the decomposition of the electrolytes. Fortunately, the charge/discharge behaviours were stabilized after several cycles, with the presence of capacitive characteristics over the whole testing voltage range. The corresponding reversible specific capacity was about 90 mAh/g, at the current density of 10 mA/g.

Figure 4.21 shows rate stability characteristics of the V_2CT_x based electrode. The initial capacity of 90 mAh/g at 10 mA/g was retained by 56% at 1 A/g. Finally, as the current density was returned to 10 mA/g, a pretty high specific capacity of 83 mAh/g was obtained, showing a high reversibility. Cycle stability testing indicated that a specific capacity of 78 mAh/g could be well maintained, at a relatively high current density of 20 mA/g, as the charge–discharge process was cycled to 100 times. As claimed by the authors, the electrochemical performance of the V_2CT_x based electrode could be further enhanced, through the optimization of the synthetic process to develop high quality MXenes and proper selection of the electrolytes.

As revealed by the XRD results, after the first sodiation process, the (002) peak shifted to 8.1° from 9.3°, as the electrode was switched from the open circuit voltage (OCV) to a voltage of 0.1 V. It meant that the *c*-lattice constant of the V_2CT_x nanostructure was expanded from 1.9 to 2.18 nm. In this case, the Na^+ ions were mainly adsorbed on surfaces of the V_2CT_x nanosheets, while intercalation of Na^+ ions in between the nanolayers also occurred. Interestingly, after the following deintercalation, the interlayer spacing was reduced by about 1.03 Å, which was much smaller than the initially expanded value of 1.40 Å. This irreversible change in the interlayer spacing was attributed to the fact that a certain number of Na^+ ions were still remained inside the spaces of the V_2CT_x nanolayers, mainly because of the interactions between the Na^+ ions and the functional groups on the surfaces of the MXene nanolayers.

In the V *K*-edge XANES spectra, there was a weak pre-edge peak at 5.47 keV and a main absorption peak at 5.485 keV. The former was due to the transitions of the metallic 3*d* and carbon 2*p* orbitals to hybridized electronic states. For example, 1*s* electron was transferred to the hybridized orbitals of t_{2g} (V 3*d* + C 2*p*) and e_g (V 3*d* + C 2*p*). The main absorption peak was ascribed to the dipole-allowed transition of 1*s* electrons to the unoccupied 4*p* states of V.

The spectra of the V_2CT_x electrode at different discharge–charge stages shared similar profiles, whereas the edge shifted obviously for the discharge and charge processes. As the first sodiation process was proceeded from OCV to 0.1 V, the edge was shifted to lower energy side, implying that V was reduced. Once the desodiation process from 0.1 to 3 V was conducted, the edge returned, i.e., V was oxidized. It was concluded that the redox reaction of vanadium was an important contributor to the charge storage capability of the MXenes in the sodium-ion batteries.

The oxidation state changes of vanadium have been quantified according to the V *K*-edge XANES spectra of the V_2CT_x nanostructure. Based on the linear relationship between the edge energy and the average oxidation state had a rough linear interrelationship, which allowed for estimation of the oxidation state of vanadium in V_2CT_x. In the first desodiation process, the change in the average oxidation state of vanadium was nearly equivalent to 0.2 electron over 0.1–3.0 V.

The formula weight of V_2C was 114 g/mol, which corresponded to a specific capacity of 94 mAh/g. Due to the presence of the functional groups, the chemical weight of V_2CT_x should be used to evaluate the specific capacity. As a result, it was understandable that the real specific capacity was 90 mAh/g, which was slightly smaller than the value should be smaller estimated one based on the molecular weight of V_2C. Nevertheless, the experimental data and the estimated values were in a good agreement.

A template process was reported to obtain hollow spheres of MXenes with 3D architectures from 2D nanosheets, to achieve full exploration of electrochemical energy storage capabilities of MXenes, because they tend to restack [56]. The MXene hollow spheres showed high stabilities in various solvents, including water and alcohol. The 3D macroporous MXene spheres could be developed into free-standing films, with high mechanical flexibility and electrical conductivity, because the spheres had closely connected MXene nanosheets. The 3D MXene sphere films demonstrated superior electrochemical performances to the multilayered MXenes and MXene/CNTs hybrids.

Importantly, the 3D MXene sphere-based films could be employed as the anodes of Na-ion batteries, without the requirement of a current collector and any binders. Specifically, the 3D $Ti_3C_2T_x$, V_2CT_x and Mo_2CT_x film-based electrodes exhibited reversible capacities of 330, 340 and 370 mAh/g at 0.25 C, respectively. Meanwhile, they all displayed excellent rate performances and long-term stabilities. After cycling for 1000 times, their reversible capacities were still retained at 295, 310, and 290 mAh/g at relatively high cycling rate of 2.5 C.

Figure 4.22 shows schematic diagram of fabrication process of the MXene nanosheet sphere-based films. To ensure strong interactions between MXene and PMMA spheres, hollow MXene nanoflakes were prepared first. As the two solutions were mixed, the MXene nanoflakes were randomly adsorbed on the surface of the PMMA spheres, due to the interactions of the surface hydroxyl groups, as shown in Fig. 4.22a–c. After that, the MXene/PMMA spheres collected through centrifugation, as demonstrated in Fig. 4.22c. Then, the PMMA was eliminated by

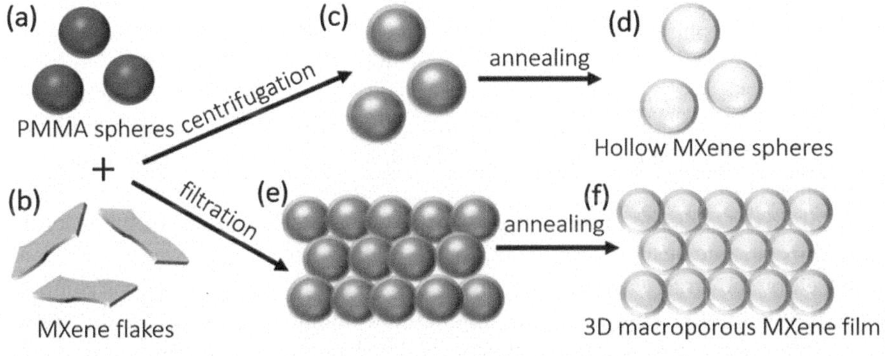

Fig. 4.22 Schematic diagram illustrating formation process of the hollow MXene spheres and the 3D macroporous MXene architectures. **a–c** MXene solutions mixed with PMMA solution to trigger the MXene nanosheets to coat on surface of the PMMA spheres, because of the interactions between the surface HO–groups. **d** Thermal treatment of the MXene/PMMA hybrids resulting in the formation of hollow MXene spheres. **e** Formation of free-standing films through filtration of the MXene/PMMA hybrid spheres. **f** 3D macroporous MXene films after burning the PMMA. Reproduced with permission from [56]. Copyright © 2017, John Wiley and Sons

burning the samples at 450 °C, resulting in hollow spheres of MXenes, as seen in Fig. 4.22d. To get free-standing films, the PMMA/MXene spheres were treated with vacuum-filtration, as illustrated in Fig. 1e. Similarly, the PMMA was removed through thermal treatment. The 3D macroporous MXene sphere based free-standing films were highly flexible, as observed in Fig. 4.22f.

Representative SEM and TEM images of the PMMA spheres and $Ti_3C_2T_x$ MXene nanosheets are shown in Fig. 4.23a, b. The PMMA spheres had diameters of 2–3 μm and their surface was relatively rough. The $Ti_3C_2T_x$ nanosheets possessed dimensions at the scale of micrometers and thickness of about 1 nm. Figure 4.23c shows SEM image of the $Ti_3C_2T_x$/PMMA hybrid spheres. Comparatively, the hybrid spheres had smoother surfaces than their PMMA counterparts, because of the deposition of the $Ti_3C_2T_x$ nanosheets. Accordingly, $Ti_3C_2T_x$ hollow spheres were developed after the PMMA templates were removed during the thermal treatment, as seen in Fig. 4.23d. Since the $Ti_3C_2T_x$ nanosheets consisted of only very less layers, they were transparent to the electron beams, as revealed in Fig. 4.23e, f.

Figure 4.24a shows a photograph of the free-standing macroporous $Ti_3C_2T_x$ films derived after vacuum filtration and calcination, demonstrating their high flexibility. Figure 4.24b shows thermogravimetric analysis (TGA) curves of PMMA and the $Ti_3C_2T_x$/PMMA hybrid samples. Clearly, the PMMA phase was entirely eliminated above 400 °C. The $Ti_3C_2T_x$/PMMA hybrid sample experienced a weight loss of about 80%, corresponding to a PMMA/$Ti_3C_2T_x$ mass ratio of about 4/1. In addition, the weak weight increase at about 400 °C of the hybrid sample was related to either the oxidation of $Ti_3C_2T_x$ due to the O–functional groups or the decomposition of PMMA.

Figure 4.24c plots cross-sectional SEM image of the macroporous $Ti_3C_2T_x$ film, revealing the presence of the $Ti_3C_2T_x$ hollow spheres, which formed highly ordered 3D networks. The $Ti_3C_2T_x$ film had a thickness of about 13 μm. Because of the hollow structure of the spheres, the macroporous $Ti_3C_2T_x$ film exhibited a low packing density of 0.4 g/cm^3, which was much lower than that (3.8 g/cm^3) of its compact counterpart. The macroporous $Ti_3C_2T_x$ film displayed an electrical conductivity of about 200 S/cm. Although the level of conductivity was slightly lower than that of the compact $Ti_3C_2T_x$ films, it was higher than that (12 S/cm) of similarly structured 3D graphene films [57].

Owing to the stacking of the $Ti_3C_2T_x$ hollow spheres, the surface of the macroporous films became be more roughness, so that high hydrophobicity was observed, as evidenced by the large water contact angle, as seen in Fig. 4.24e. In comparison, the dense films exhibited a much smaller water contact angle of about 35°, as illustrated in Fig. 4.24f. Figure 4.24g shows interfacial TEM image of two adjacent $Ti_3C_2T_x$ hollow spheres. The two spheres were closely contacted each other, which could be attributed to van der Waals interactions and hydrogen bonds between the two $Ti_3C_2T_x$ nanosheets. It was this close contact that ensured the formation of the highly flexible and conductive 3D network architectures.

Furthermore, both diameter and wall-thickness of the hollow MXene spheres can be intentionally tailored to have different properties. In this case, the wall-thickness of the $Ti_3C_2T_x$ hollow spheres could be readily modified by controlling

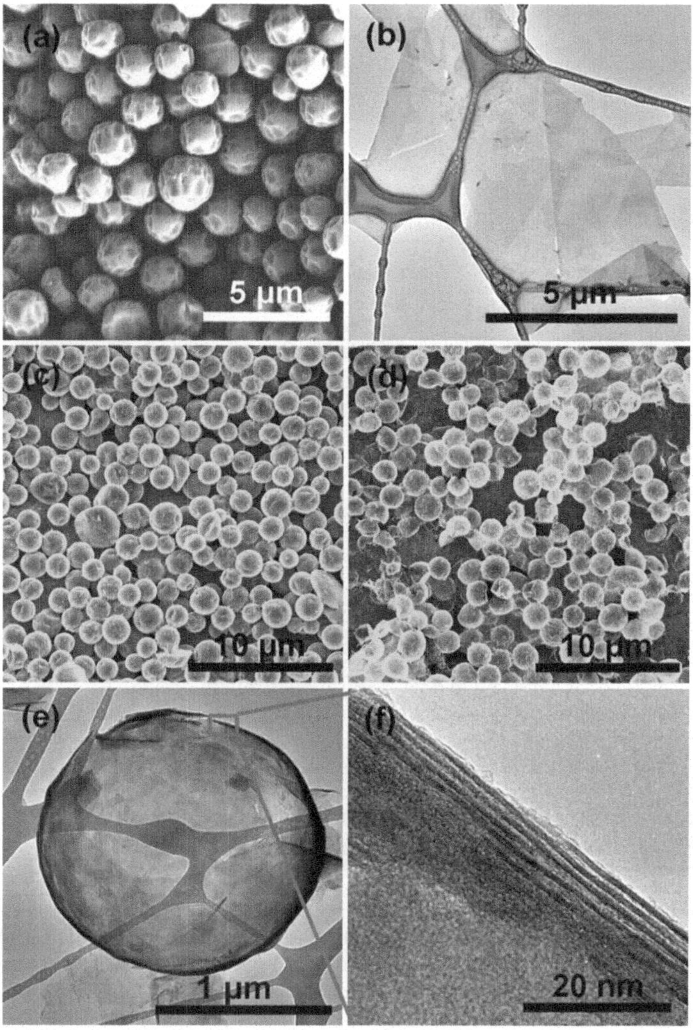

Fig. 4.23 **a** SEM image the PMMA spheres. **b** TEM image of the $Ti_3C_2T_x$ nanosheets. **c** SEM image of the $Ti_3C_2T_x$/PMMA hybrids. **d** SEM image of the hollow $Ti_3C_2T_x$ spheres after the elimination of PMMA. **e** Representative TEM image the hollow $Ti_3C_2T_x$ spheres **f** TEM image of wall structure of the MXene nanolayers. Reproduced with permission from [56]. Copyright © 2017, John Wiley and Sons

the PMMA/$Ti_3C_2T_x$ weight ratio. As a consequence, the flexibility of the 3D macroporous films was tailored accordingly. Generally, the lower the content of PMMA, the thicker the walls and the stronger the $Ti_3C_2T_x$ films would be. It was found that the PMMA/$Ti_3C_2T_x$ weight ratio could be as high as 10:1, while the $Ti_3C_2T_x$ films still possessed sufficiently high flexibility and electrical conductivity. In addition, the size of the PMMA spheres determined the size of the $Ti_3C_2T_x$ hollow spheres.

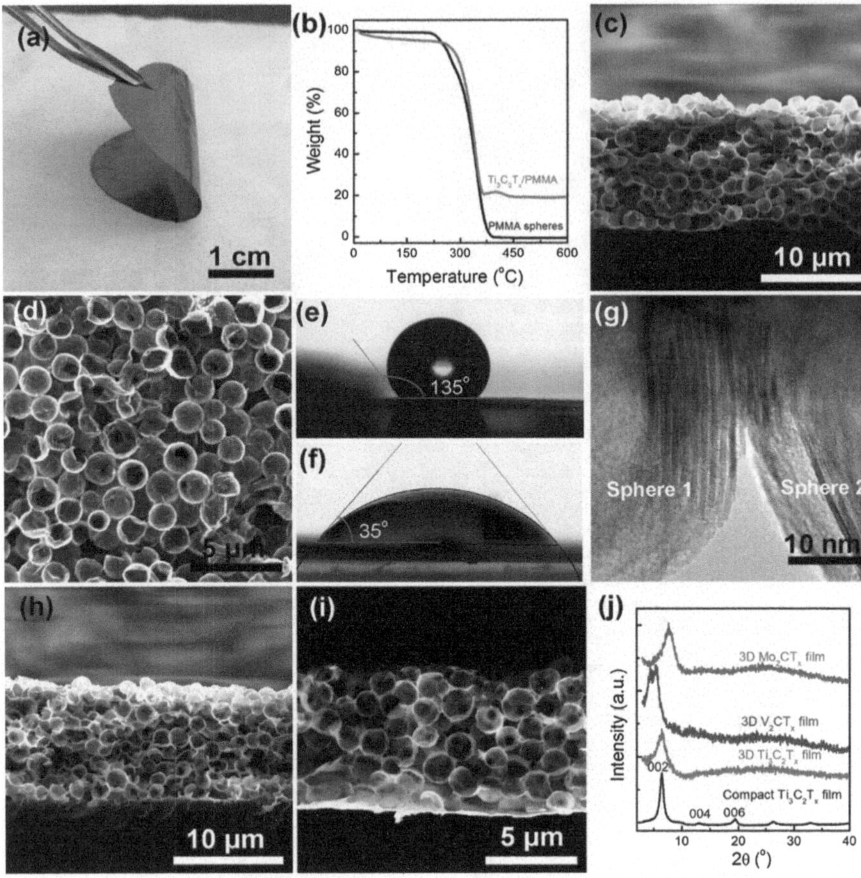

Fig. 4.24 **a** Photograph of the flexible 3D macroporous $Ti_3C_2T_x$ films. **b** TGA curves of pure PMMA spheres and the $Ti_3C_2T_x$/PMMA hybrid films in Ar. **c** Cross-sectional and **d** top-view SEM images of the 3D macroporous $Ti_3C_2T_x$ films. Water contact angles of the **e** 3D macroporous $Ti_3C_2T_x$ film and **f** the dense $Ti_3C_2T_x$ film. **g** Interfacial TEM image of two adjacent $Ti_3C_2T_x$ hollow spheres in the 3D films. Cross-sectional SEM images of the 3D macroporous **h** V_2CT_x and **i** Mo_2CT_x films. **j** XRD patterns of dense $Ti_3C_2T_x$ film and the 3D macroporous MXene films. Reproduced with permission from [56]. Copyright © 2017, John Wiley and Sons

The versatility of the synthetic process of the MXene network nanostructures was demonstrated by V_2CT_x and Mo_2CT_x. As shown in Fig. 4.24h, j, the two samples both possessed a stacking structure, with hollow characteristics and high flexibility. Figure 4.24j shows XRD patterns of a dense $Ti_3C_2T_x$ film and the 3D macroporous films of $Ti_3C_2T_x$, V_2CT_x and Mo_2CT_x. All samples had the typical (002) peak of MXenes. The *c*-lattice constant of the samples was about 2.73 nm, corresponding to the (002) peak at about 6.47°, no matter whether they were dense or macroporous films. Comparatively, the microporous films displayed a wider full width at half maximum of the diffraction peak, while higher order of (00*l*) peaks, e.g., (004) and (006), were absent, mainly because the nanosheets were bent and randomly oriented in the 3D films. No peaks of the corresponding transition metal oxides were observed in the XRD patterns of the macroporous films, suggesting the robustness of the MXenes as the thermally induced oxidation.

Similarly, the 3D Mo_2CT_x sample had a *c*-lattice constant of 2.28 nm. The (002) peak of the 3D V_2CT_x film were split into two, with *c*-lattice constants of 4.02 nm and 3.24 nm. However, the presence of the larger interlayer spacing and the two (002) peaks of the 3D V_2CT_x film was not fully understood. The possible reason was the adsorption of water or carbon species derived from the organic items in the films, because the delamination process utilized for the V_2CT_x sample was different.

A $MoS_2/Ti_3C_2T_x$ hybrid has been prepared to with significant enhancement in ion accessibility in the MXene nnaolayers [58]. The specific capacity was largely increased because of the enlargement in the interlayer spacing. In this case, the MoS_2 nanosheets were intercalated into the $Ti_3C_2T_x$ nanolayered structures by using a hydrothermal treatment. The $MoS_2/Ti_3C_2T_x$ hybrid based electrode had a specific capacity of 251 mAh/g up to 100 cycles. Moreover, the $MoS_2/Ti_3C_2T_x$ hybrid based electrode exhibited excellent rate capability with a specific capacity of about 163 mAh/g at a relatively high current density of 1 A/g. To prepare the $MoS_2/Ti_3C_2T_x$ hybrid, ammonium molybdate tetrahydrate ($(NH_4)_6Mo_7O_{24} \cdot 4H_2O$), thiourea ($CS(NH_2)_2$) and oxalic acid ($H_2C_2O_4 \cdot 2H_2O$) were dissolved in H_2O. $Ti_3C_2T_x$ powder was added into the mixture with the aid of ultrasonication. Hydrothermal treatment of the mixture was conducted at 200 °C for one day. The samples with 200 and 50 mg MXene were denoted as $MoS_2/Ti_3C_2T_x-200$ and $MoS_2/Ti_3C_2T_x-50$.

SEM and TEM images of the $MoS_2/Ti_3C_2T_x$ hybrids indicated that the spacing of the MXene nanolayers was effectively enlarged, where the ultra-thin MoS_2 nanoflakes had an average size of 50 nm. The MoS_2 nanosheets were uniformly distributed on surface of the MXene nanolayers. The ultra-small dimension and the homogeneous distribution of the MoS_2 nanosheets promoted the diffusion of Na^+ ions during the charge–discharge processes and boosted the electrochemical reactions between the active materials and electrolytes. HRTEM image revealed lattice structure of the MoS_2 nanosheets, with a spacing of 0.62 nm for the (002) planes. The (110) lattice planes of $Ti_3C_2T_x$ had *d* parameter of 0.31 nm. According to TGA profile of the $MoS_2/Ti_3C_2T_x$ hybrid in N_2, the MoS_2 phase had a mass content of 14%.

Figure 4.25a shows discharge–charge characteristics of the $MoS_2/Ti_3C_2T_x$ hybrid-based electrode, over the potential range of 0.01–3.0 V (vs. Na/Na^+), at the current density of 100 mA/g. The first cycle discharge and charge capacities were 490 mAh/g and 331 mAh/g, respectively. The capacities in the subsequent charge–discharge curves were all highly reversible, confirming high rate stability and reversibility of the $MoS_2/Ti_3C_2T_x$ hybrid electrode. The charge–discharge curves were also characterized by an intercalation-type slope over 0.4–2.3 V, together with promising cycle stability. A weak discharge plateau at about 0.4 V was present at the subsequent

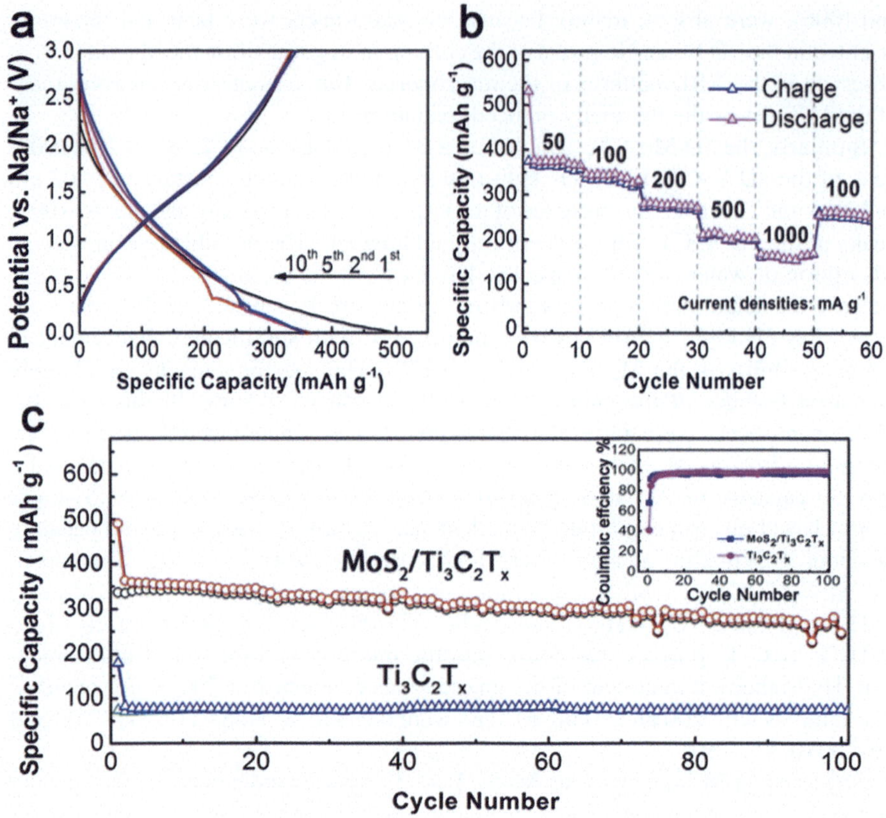

Fig. 4.25 Electrochemical performances of the $MoS_2/Ti_3C_2T_x$ hybrid in a two-electrode sodium-ion battery cell with Na foil as the counter and reference electrode: **a** charge/discharge curves at the current density of 100 mA/g over the voltage window of 0.01–3.0 V, **b** rate characteristics, **c** cycling behaviour of pure $Ti_3C_2T_x$ and the $MoS_2/Ti_3C_2T_x$ hybrid at the current density of 100 mA/g (with the coulombic efficiencies to be shown as insets). Reproduced with permission from [58]. Copyright © 2017, John Wiley and Sons

cycles, which was attributed to the reduction of Mo^{4+} to Mo because of the reaction of Na^+ and MoS_2. In other words, MoS_2 also participated the electrochemical reactions, thus having contribution to the total capacities.

Based on CV curves of the hybrid electrode, the first cathodic scan displayed an irreversible current at voltages of <1 V, due mainly to the formation of the SEI films. Another possible reason of the irreversible current could be the electrochemical activation related to the desolvation of Na^+ ions that were intercalated during the early reduction process. In nonaqueous Na^+ electrolytes, as the Na^+ ions were intercalated into the spacing of the MXene nanolayers during the first cathodic process, there were also Na^+ ions were adsorbed as pillars on the surface of the nanolayers throughout the charge–discharge processes. The broad oxidation and reduction peaks at about 1.5 V were attributed to the extraction of the Na^+ ions from the $Ti_3C_2T_x$ matrix.

The CV curves of the second and fifth cycles were overlapped, suggesting that the $Ti_3C_2T_x/MoS_2$ hybrid electrode possessed high electrochemical reversibility during the charge-discharged processes. Figure 4.25c shows cycling behaviours of the $MoS_2/Ti_3C_2T_x$ hybrid and bare $Ti_3C_2T_x$ electrodes. The $MoS_2/Ti_3C_2T_x$ hybrid electrode had a specific capacity of 251 mAh/g at the current density of 100 mA/g current density after cycling for 100 cycles, corresponding to a low capacity loss of about 12%. In this case, the $MoS_2/Ti_3C_2T_x$ hybrid electrode exhibited higher cycling stability and Coulombic efficiency than the bare $Ti_3C_2T_x$ electrode. The promising electrochemical performances of the hybrid electrode was ascribed to the hierarchical microstructure together with the enlarged spacing of the $Ti_3C_2T_x$ nanolayers and the ultra-small MoS_2 nanoflakes.

The bare MoS_2 electrode encountered a quick degradation after 100 cycles, with discharge capacity to be about 115 mAh/g. As a consequence, cycling behaviour of the $MoS_2/Ti_3C_2T_x$ hybrid electrodes was decreased with increasing content of MoS_2, though an unexpected high capacity was observed in the early cycling. Therefore, the presence of the MoS_2 nanosheets was not to increase the specific capacity. Instead, they generated more active sites and offered more diffusion paths across the MXene nanolayers that would possess enlarged spacing to host the electrolyte reactions. In addition, the high conductivity of both the $Ti_3C_2T_x$ and MoS_2 nanolayers boosted electron transport and the diffusion of ions.

Figure 4.25b shows rate behaviours of the $MoS_2/Ti_3C_2T_x$ hybrid electrode at current densities in range of 50–1000 mA/g. The specific capacity could be retained at high current densities, demonstrating highly stable charge–discharge characteristics of the $MoS_2/Ti_3C_2T_x$ hybrid electrodes. Generally, the stored charge of electrodes has three contributions: (i) diffusion-related faradaic component due to the intercalating reactions, (ii) faradaic component because of the charge transport of the atoms on surface and (iii) non-faradaic item owing to the electrical double-layer effects.

The CV curves of the $MoS_2/Ti_3C_2T_x$ hybrid electrode recorded at different scan rates are shown in Fig. 4.26a. With those data, reaction kinetics of the $MoS_2/Ti_3C_2T_x$ based sodium-ion batteries could be clarified. As the scan rate was increased from 0.1 to 0.9 mV/s, the shape of the CVs was kept almost unchanged. Figure 4.26b indicated that the Na^+ ion storage had a non-diffusion limited mechanism at the surface of the $MoS_2/Ti_3C_2T_x$ hybrid. Because the MoS_2 nanosheets were attached on the MXene nanolayers, high surface area was achieved, which enhanced the diffusion behaviour of both the Na^+ ions and the electrolytes. The $MoS_2/Ti_3C_2T_x$ hybrid also had strong non-diffusion controlled contribution, according to the quantitative capacitive analytic results, as shown in Fig. 4.26c. At the scan rate of 0.9 mV/s, the non-diffusion effective current offered a contribution of 89.4% to the total charge storage. Moreover, the non-diffusion limited charge storage contribution was increased with increasing scan rate, as observed in Fig. 4.26d.

A wet-chemical process was devised to prepare Sb_2O_3/MXene ($Ti_3C_2T_x$) hybrid used as electrode of sodium-ion batteries, which exhibited greatly improved electrochemical performances [59]. In the hierarchical structures of the $Sb_2O_3/Ti_3C_2T_x$ hybrids, the Sb_2O_3 nanosized particles with an average size of about 50 nm were

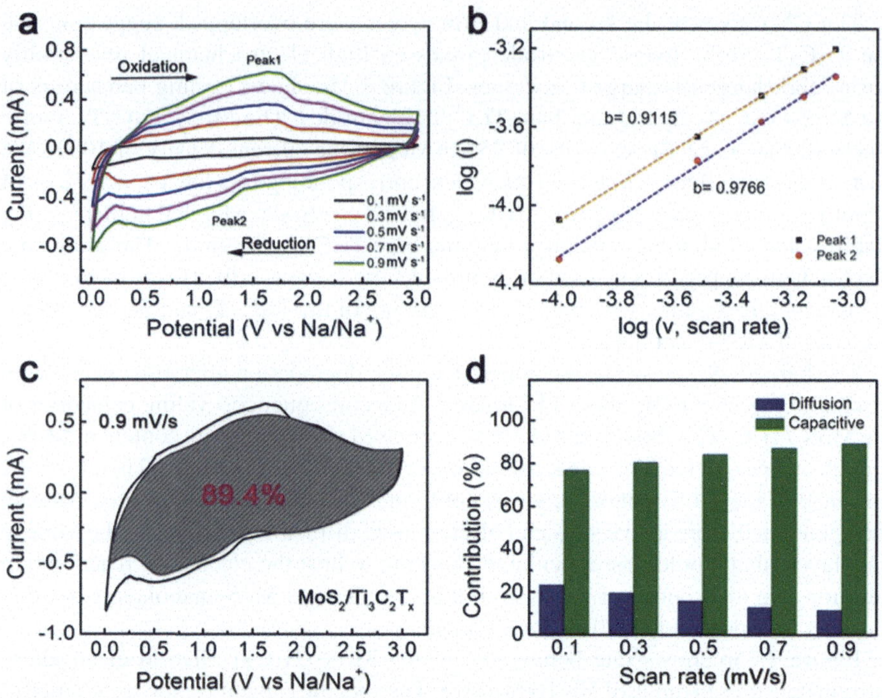

Fig. 4.26 Quantitative capacities of the $MoS_2/Ti_3C_2T_x$ hybrid sodium storage behaviours: **a** CV curves at different scan rates and **b** relationship between logarithm cathode peak current and logarithm scan rates. **c** Capacitive contribution and diffusion contribution at 0.9 mV/s. **d** Normalized contributions of capacitive capacities at different scan rates. Reproduced with permission from [58]. Copyright © 2017, John Wiley and Sons

homogeneously distributed inside the 3D $Ti_3C_2T_x$ networks. The Sb_2O_3 nanoparticles acted as a reservoir of Na^+ ions, while the $Ti_3C_2T_x$ networks ensured pathways of electrons and Na^+ ions. Since the Sb_2O_3 nanoparticles were trapped in between the $Ti_3C_2T_x$ nanolayers, their volume expansions due to the sodiation/desodiation could be readily accommodated. Therefore, the $Sb_2O_3/Ti_3C_2T_x$ hybrid anodes were mechanically robust and electrochemically active. Specifically, a capacity of 295 mAh/g was achieved at the current of 2 A/g. After cycling for 100 times, the capacity at 100 mA/g was 472 mAh/g was retained.

To prepare the $Sb_2O_3/Ti_3C_2T_x$ hybrid, $Ti_3C_2T_x$ suspension was mixed with $SbCl_3$ solution, in the presence of polyvinylpyrrolidone (PVP) that served as the surfactant. The Sb^{3+} ions were adsorbed onto the $Ti_3C_2T_x$ nanosheets with negative charges, due to ion exchanges and electrostatic interactions. Eventually, Sb_2O_3 nanoparticles were formed by using a hydrolysis process with excessive alkali. Microstructural analysis results revealed that the nanoflakes had dimensions at the scales from submicron to microns. The $Ti_3C_2T_x$ was single layer structure with a thickness of about 1 nm. In the $Sb_2O_3/Ti_3C_2T_x$ hybrids, the Sb_2O_3 nanoparticles were uniformly distributed on the $Ti_3C_2T_x$ nanolayers, while the $Ti_3C_2T_x$ nanolayers were inter-linked one another, resulting in 3D network architectures. There were spacing in between the Sb_2O_3 nanoparticles and the $Ti_3C_2T_x$ nanoflakes, so that volumetric variation of the $Sb_2O_3/Ti_3C_2T_x$ hybrid electrodes during the sodiation/desodiation process could be readily buffered.

XRD patterns of the $Sb_2O_3/Ti_3C_2T_x$ hybrids indicated that the Sb_2O_3 nanoparticles had an orthorhombic structure. Due to its low crystallinity, no diffraction peaks of $Ti_3C_2T_x$ were were present. Raman spectra revealed that the peak intensity of Sb_2O_3 was largely reduced, as compared with those of the bare Sb_2O_3. This was because the Sb_2O_3 nanoparticles were isolated by the $Ti_3C_2T_x$ nanosheets, so that the Raman signal of Sb_2O_3 was decreased. In addition, TiO_2 was not detected, which implied that oxidation of the MXene was avoided. According to XPS results, bare $Ti_3C_2T_x$ had functional groups of –O, –OH and –F, while the $Sb_2O_3/Ti_3C_2T_x$ hybrid possessed almost no peaks of –OH and –F, which might be consumed due to their reactions with the Sb^{3+} ions or eliminated from the MXene nanosheets during the formation of the $Sb_2O_3/Ti_3C_2T_x$ hybrids.

Figure 4.27a shows 1st, 2nd and 50th discharge–charge curves of the $Sb_2O_3/Ti_3C_2T_x$ hybrid-based electrode, at the current density of 50 mA/g, over the voltage window of 0.01–2.5 V. The $Sb_2O_3/Ti_3C_2T_x$ electrode exhibited initial discharge and charge capacities of 796 and 471 mAh/g, respectively, which led to a Coulombic efficiency of about 59%. The presence of the SEI film was responsible for the initial irreversibility in capacity of the electrode. Comparatively, the Sb_2O_3 electrode had a reversible capacity of 441 mAh/g at the current density of 50 mA/g, with a Coulombic efficiency of 60%, while the bare $Ti_3C_2T_x$ electrode had a reversible capacity of 89 mAh/g at 50 mA/g with a Coulombic efficiency of 35%.

At the early state, the Na^+ irreversibly reacted with the surface functional groups of –F, –OH and –O and there were water molecules trapped in between the MXene nanolayers, which let to low initial Coulombic efficiencies of the electrodes. In comparison, the functional groups of the MXene in the $Sb_2O_3/Ti_3C_2T_x$ hybrids reacted with Sb^{3+} ions instead. Also, the groups could have been eliminated in the synthesis of the materials. The absence of the functional groups was confirmed by XPS results. As a result, the $Sb_2O_3/Ti_3C_2T_x$ hybrid-based electrode had a first-cycle Coulombic efficiency to be similar to that of the bare Sb_2O_3 electrode, which was much higher than that of the bare $Ti_3C_2T_x$ electrode. In other words, the $Ti_3C_2T_x/Sb_2O_3$ hybrid was more promising than $Ti_3C_2T_x$ as electrodes of Na-ion batteries.

Figure 4.27b shows cycling behaviours of the $Sb_2O_3/Ti_3C_2T_x$ hybrid and bare Sb_2O_3 electrodes at the current density of 50 mA/g. The $Sb_2O_3/Ti_3C_2T_x$ hybrid electrode displayed much higher cycling stability, which corresponded to a specific capacity of 470 mAh/g up to 50 cycles. In comparison, the bare Sb_2O_3 electrode had a relatively low capacity of 190 mAh/g after cycling for 50 times. Meanwhile, only 43% of the initial capacity was retained. The poor performance of the bare Sb_2O_3 electrode was suffered from the electrical disconnection of the active materials from the conductive additives or current collectors. In this case, the Sb_2O_3 nanoparticles were pulverized because of the volumetric expansions.

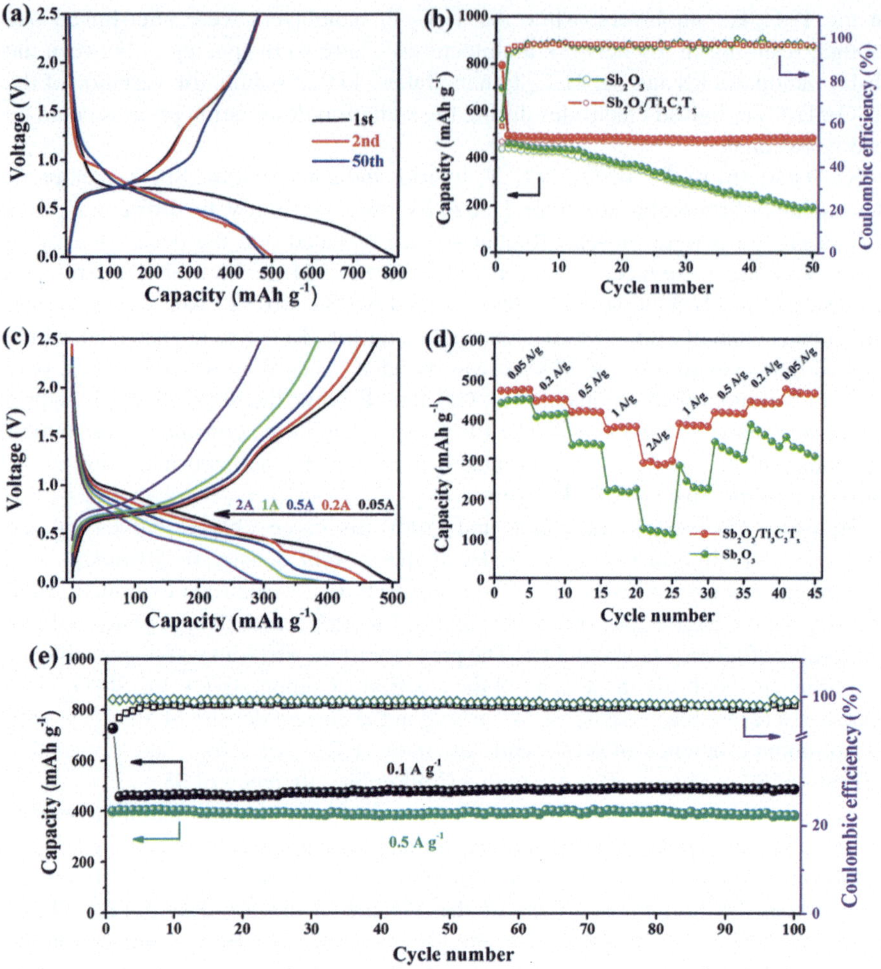

Fig. 4.27 **a** Galvanostatic charge–discharge curves of the $Sb_2O_3/Ti_3C_2T_x$ hybrid electrode for the 1st, 2nd and 50th cycles at 50 mA/g. **b** Cycling behaviour of the $Sb_2O_3/Ti_3C_2T_x$ hybrid and the bare Sb_2O_3 electrodes at the current density of 50 mA/g. **c** Charge–discharge curves of the $Sb_2O_3/Ti_3C_2T_x$ hybrid electrode at different current densities. **d** Rate performances of the $Sb_2O_3/Ti_3C_2T_x$ hybrid and the bare Sb_2O_3 electrodes of Na-ion batteries. **e** Reversible capacity and Coulombic efficiencies of the $Sb_2O_3/Ti_3C_2T_x$ hybrid electrodes for the first 100 cycles at the current densities of 0.1 and 0.5 A/g. Reproduced with permission from [59]. Copyright © 2017, The Royal Society of Chemistry

Figure 4.27c depicts discharge–charge curves of the $Sb_2O_3/Ti_3C_2T_x$ hybrid electrode at various current densities. Specifically, at current densities of 0.5 A/g, 1 A/g and 2 A/g, the $Sb_2O_3/Ti_3C_2T_x$ hybrid electrode possessed capacities of 422 mAh/g, 383 mAh/g and 295 mAh/g, respectively. In comparison, the values of the bare Sb_2O_3 were 331, 216 and 111 mAh/g. Because of the high electrical conductivity, the $Sb_2O_3/Ti_3C_2T_x$ hybrid electrode had lower polarization than the bare Sb_2O_3 electrode. Figure 4.27d demonstrates rate performances of the $Sb_2O_3/Ti_3C_2T_x$ hybrid and bare Sb_2O_3 electrodes. Specific capacity of the $Sb_2O_3/Ti_3C_2T_x$ hybrid electrode was well recovered to the initial levels, confirming its high stabilities and excellent rate capabilities. The outstanding cycling behaviors of the $Sb_2O_3/Ti_3C_2T_x$ hybrid electrode at high rates illustrated in Fig. 4.27e. Nyquist plots indicated that the $Sb_2O_3/Ti_3C_2T_x$ hybrid and bare Sb_2O_3 electrodes had similar impedance characteristics.

Ultrathin nanoribbons of sodium titanate, $NaTi_{1.5}O_{8.3}$/M-NTO and potassium titanate $K_2Ti_4O_9$/M-KTO were prepared through an oxidation/alkalization process from Ti_3C_2 MXene [60]. The M-NTO and M-KTO had inter-layer spacings of 0.90 and 0.93 nm, respectively, while both samples had thicknesses of $\leq$11 nm, widths of nanoribbons of $\leq$60 nm. Together with the presence of the macroporous nanostructures, ion insertion/extraction kinetics were greatly enhanced, the M-NTO based electrode delivered a specific capacity of 191 mAh/g at the current density of 200 mA/g for sodium-ion batteries, which was higher than those of the electrodes based on Ti_3C_2 and TiC. Similarly, the M-KTO based electrode had specific capacities of 151 mAh/g and 88 mAh/g at 50 mA/g and 300 mA/g, respectively. Also, long-term stability of the two electrodes was demonstrated by cycling up to 900 times.

To obtain M-NTO and M-KTO nanoribbons, Ti_3C_2 MXene powder was dispersed in solution with NaOH/KOH and H_2O_2. Hydrothermal process of the mixtures was conducted at 140 °C for 12 h. SEM images of the Ti_3C_2 powder are shown Fig. 4.28a, b, revealing typical accordion-like multilayered nanostructures of MXenes. The exfoliated multi-layered nanostructures had thicknesses of 30–50 nm, as seen in Fig. 4.28b. After the Ti_3C_2 was hydrothermally reacted in the solution NaOH and H_2O_2, the color of the powder was varied from dark red to white, as observed in

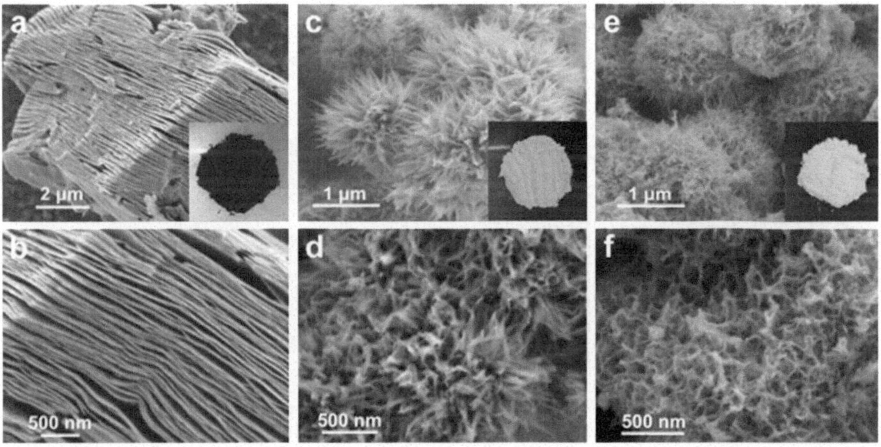

Fig. 4.28 SEM images of the samples: **a, b** Ti_3C_2 MXene, **c, d** M-NTO and **e, f** M-KTO. The insets in **a, c, e** are their corresponding photographs in the form of powder. Reproduced with permission from [60]. Copyright © 2017, American Chemical Society

Fig. 4.28b, c, respectively, implying the formation of M-NTO from MXene. The M-NTO powder consisted of nanoribbons with an urchin-like morphology and curved microstructures, as illustrated in Fig. 4.28d. Similarly, M-KTO nanoribbons were obtained after the hydrothermal treatment of the Ti_3C_2 MXene in KOH and H_2O_2, with a light yellow color, as observed in Fig. 4.28e, f.

Figure 4.29 shows representative TEM and HRTEM images of the Ti_3C_2 MXene and the M-NTO/M-KTO nanoribbons. As revealed in Fig. 4.29a–c, the Ti_3C_2 MXene displayed typical multilayered structures. HRTEM image demonstrated that the interlayer spacing was about 0.98 nm, while crystalline characteristic of the powder was evidenced by the SAED pattern. The hydrothermal reaction treatment resulted in agglomerates with nanosized sphere-like morphologies, which diameters of <1 μm. The nanoshperes were assembled from nanoribbons, as observed in Fig. 4.29e. Widths and thicknesses of the M-NTO nanorribons 15–50 nm and 4–11 nm, respectively, as illustrated in Fig. 4.29f. The M-KTO sample exhibited similar microstructure and lattice characteristics, as demonstrated in Fig. 4.29g–i.

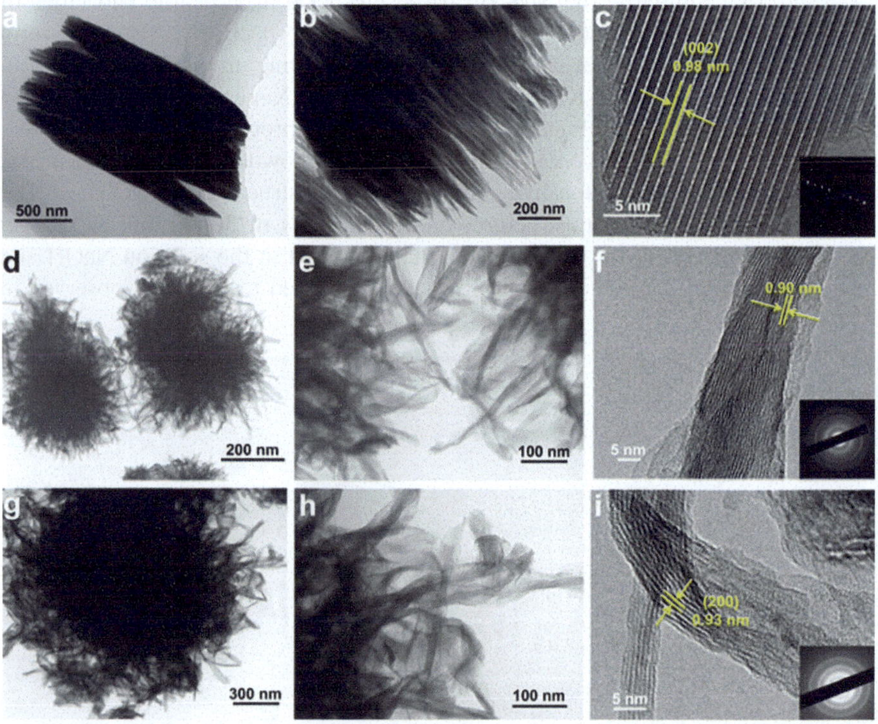

Fig. 4.29 **a**, **b** TEM and **c** HRTEM images of the Ti_3C_2 MXene. **c**, **d** TEM and **f** HRTEM images of the M-NTO. **g**, **h** TEM and **i** HRTEM images of the M-KTO. The insets in **c**, **f** and **i** are their corresponding SAED patterns. Reproduced with permission from [60]. Copyright © 2017, American Chemical Society

A hydrothermal process was recently reported to synthesize $CoNiO_2$/MXene hybrids as electrode materials of sodium-ion batteries [61]. The $CoNiO_2$ hybrid nanoparticles were attached onto the MXene nanosheets uniformly, which increased roughness of the multilayered structures. As a result, the transport of Na^+ ions was greatly enhanced, while the volumetric expansion of the materials during the charge/discharge process could be effectively buffered. The $CoNiO_2$/MXene hybrid-based electrodes offered a discharge capacity of 223 mAh/g at the current of 0.1 A/g, with an outstanding rate capability of 188 mAh/g at 0.3 A/g.

To prepare the hybrid, $Ti_3C_2T_x$ was added to the solution of ethanol and deionized water with a volumetric ratio of 1:1. After that, $Ni(NO_3)_2 \cdot 6H_2O$ and $Co(NO_3)_2 \cdot 6H_2O$ were dissolved into the suspension with the aid of ultrasonication. Then, $C_6H_{12}N_4$ was mixed with the solution, followed by hydrothermal treatment at 85 °C for 8 h. The products were collected and then thermally treated at 450 °C for 4 h in Ar to form $CoNiO_2/Ti_3C_2T_x$ hybrid powders. Bare $CoNiO_2$ powder was prepared similarly for the purpose of comparison.

Nanoflake hybrid of cobalt sulfide and Ti_3C_2 MXene, CoS/MXene, with special hetero-layered architectures, was developed as electrode of sodium-ion batteries [62]. In such hybrids, ultrafined CoS nanoparticles with an average size of 6 nm were attached onto MXene nanoflakes, in which the MXene acted similarly as the printed circuit board to maintain highly conductive networks. With such characteristics, the CoS/MXene hybrid-based electrode retained a reversible capacity of 267 mAh/g after cycling for 1700 cycles at the current density of 2 A/g, corresponding to a capacity degradation rate of as low as 0.007% per cycle. Also, an excellent rate performance with 272 mAh/g was maintained at a relatively high current density of 5 A/g.

Ti_3C_2 suspension was mixed with deionized water and ethylene glycol. Then, $Co(NO_3)_2 \cdot 6H_2O$ and hexamethylenetetramine were introduced into the mixture, followed by ultrasonication for 20 min. Hydrothermal reaction was carried out at 130 °C for 6 h. The products retracted and thoroughly washed with ethyl alcohol. The dried powder was dispersed in ethanol, with the presence of thioacetamide. Then, the samples were treated hydrothermally at 160 °C for 3 h, leading to the formation of CoS/Ti_3C_2 hybrids. Bare CoS was also prepared similarly for comparison.

According to CV curves of the first three cycles over 0.4–2.9 V at 0.1 mV/s, the CoS/Ti_3C_2 hybrid had two main pairs of redox peaks at 0.73/1.77 V and 1.03/2.14 V, respectively, corresponding to the initial insertion of Na^+ and the displacement reactions between Na_2S and Co. The cathodic peaks were found to shift to high potentials in the following cycles, which could be attributed to the electrochemical activations and structural reorganization of the CoS nanoparticles [63, 64]. In addition, the cathodic peak at 0.73 V was suddenly dropped when further cycling was conducted, due to the presence of the electrolyte interphase (SEI). Also, the relatively low anodic peak at near 1.5 V could be possibly related to the extraction of Na^+ ions from the MXene phase, implying that the MXene also had contributed to the total capacities.

It was observed that CV curves of the CoS/MXene hybrid electrode was similar to that of bare CoS electrode, suggesting that the CoS nanosized particles contributed mainly to the total specific capacities. Also, the first-cycle discharge capacity was 315 mAh/g, which was slightly lower that for the second cycle. This was mainly

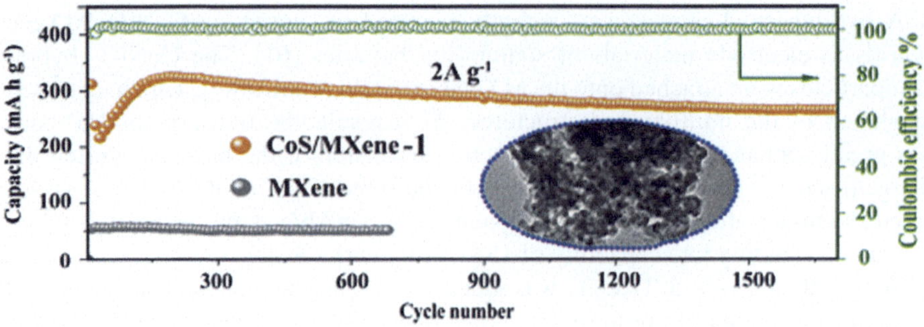

Fig. 4.30 Cycling performances at the current density of 2 A/g with the inset showing TEM image of the CoS/MXene hybrid electrode after 1000 cycles. Reproduced with permission from [62]. Copyright © 2018, Elsevier

because the initial voltage was relatively low. In addition, the plateaus of charge and discharge were consistent with those observed in the CV curves.

According to the charge–discharge curves, the CoS/MXene hybrid electrode exhibited discharge capacities of 508, 454, 405, 365, 323 and 273 mAh/g, at current densities of 0.1, 0.2, 0.5, 1, 2 and 5 A/g, separately. Most importantly, a high specific capacity of 505 mAh/m was retained, as the current density returned to the low level of 0.1 A/g. Therefore, it could be concluded that the CoS/MXene hybrid electrode possessed a strong mechanical strength at high current densities. The weak difference in the charge–discharge plateaus at different current densities implied that the CoS/MXene hybrid electrode experienced a low polarization, while the reaction kinetics was sufficiently fast.

Figure 4.30 shows long-term cycling behaviours of the CoS/MXene hybrid and bare MXene electrodes, at the current density of 2 A/g. The Coulombic efficiency of the CoS/MXene electrode was almost 100% throughout the cycling process, which strongly suggested that the CoS/MXene hybrid had as highly stable structural integrity. TEM image of the CoS/MXene hybrid after 1000 cycles indicated that both morphology and size of the CoS nanoparticles were well retained. In contrast, the CoS electrode suffered serious degradation in capacity, which was mainly caused by the structural instability of the CoS nanoparticles, as evidenced by the variation in their morphology before and after the cycling.

Ti_3CNT_z MXene was used as electrodes of potassium-ion batteries with promising electrochemical performances [65].

Figure 4.31a shows Galvanostatic voltage characteristics of the electrode for several cycles at the current density of 20 mA/g over 5.0 mV–3.0 V. At 20 mA/g, the Ti_3CNT_z electrode delivered a first-cycle charge capacity of 710 mAh/g and a first-cycle discharge capacity of 202 mAh/g, corresponding to 1.5 K^+ per unit of Ti_3CNOF, with a first-cycle Coulombic efficiency to be about 28%. The second-cycle Coulombic efficiency was 56%, with capacities of 275 mAh/g and 154 mAh/g. As seen in Fig. 4.31c, the capacity was degraded in the cycling process, approaching 75 mAh/g at the 100th cycle, which corresponded to 0.6 K^+ per unit of Ti_3CNOF.

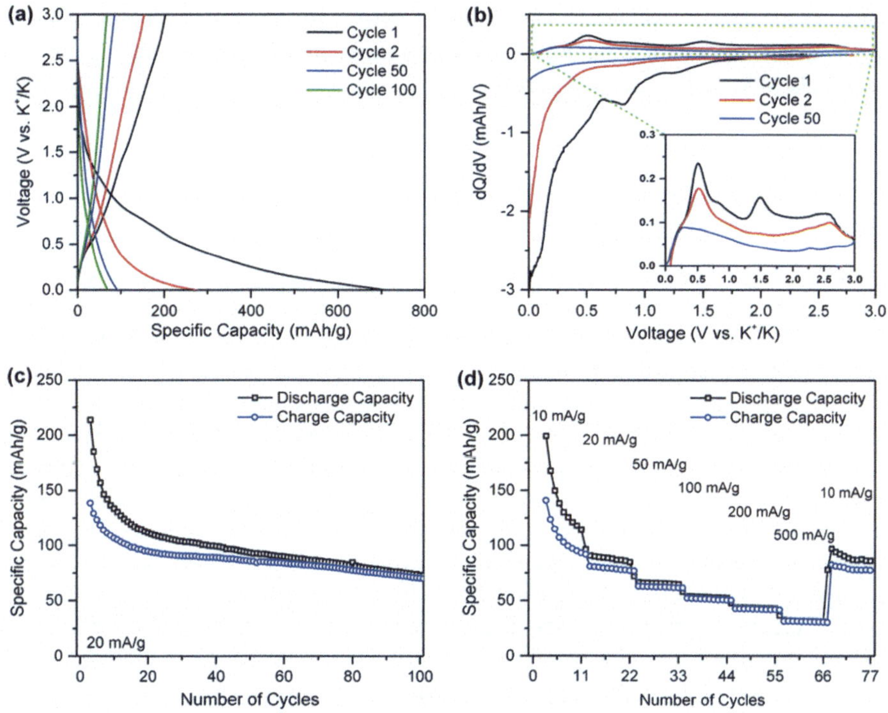

Fig. 4.31 Electrochemical performances of the Ti_3CNT_z based electrode: **a** Galvanostatic voltage characteristics of the 1st, 2nd, 50th and 100th cycles at the current density of 20 mA/g over 5.0 mV–3.0 V, **b** d*Q*/d*V* of the results plotted in (**a**), with the inset showing a higher magnification of the potassiation part of the plot in **b**, **c** cycling stabilities of the damples at 20 mA/g, **d** rate stabilities over the current densities of 10–500 mA/g. Reproduced with permission from [65]. Copyright © 2017, The Royal Society of Chemistry

Figure 4.31b shows d*Q*/d*V* plot of the electrode, where there were three peaks at 1.25, 0.81 and 0 V, in the first potassiation cycle, corresponding to depotassiation peaks at 1.50, 0.80 and 0.50 V, respectively, as illustrated as the inset in Fig. 4.31b. The two potassiation peaks at 1.25 and 0.80 V were badly decreased in the second cycle and almost disappeared up to 50 cycles. Meanwhile, the potassiation peak at about 0 V and its depotassiation counterpart at 0.50 V kept to be pronounced in the second cycle, whereas the depotassiation peak was broadened and shifted to about 0.25 V after cycling for 50 times.

Similar to Li^+ and Na^+, K^+ could be confined onto the surface of MXene nanosheets, due to the presence of defects and passivation sites, which was an irreversible process, thus leading to relatively high irreversible capacity decay. Decomposition of the electrolytes and formation of solid electrolyte interphases (SEI) could also result in irreversible degradation of capacity, as evidenced by the XPS data. Figure 4.31d shows rate capability of the electrode at current densities of 10–500 mA/g. Although the electrode exhibited relatively low specific capacities, the materials displayed promising rate capability with very weak degradation at high current densities, which could be related to the intercalation mechanisms due to the special 2D MXene layered structures.

4.2.2 Supercapacitors

Supercapacitors have various advantages, such as high power densities, quick charging-discharging rates and long-life stabilities, making them to be promising energy storage devices. With high power outputs, supercapacitors could find applications as start-stop controlling systems in hybrid electric vehicles and fuel cell vehicles. Supercapacitors also have shortcomings, one of which is their relatively low energy densities, as compared with most of the batteries. Depending on the mechanisms of energy storage, supercapacitors can be classified into two groups, i.e., (i) electrochemical double-layer capacitors (EDLCs) and (ii) pseudocapacitors. In EDLCs, ions are reversibly accumulated at the interfaces between the electrodes and electrolytes. As a result, the electrochemical performances of EDLCs are determined by the capability to host the ions of electrolytes by the electrodes. In this case, surface areas of electrodes that allow the ions to access are important factor for energy storage capabilities. Pseudocapacitors are based on the reversible Faradic processes. For more detailed description on the concepts of supercapacitors, the readers are referred to Ref. [66].

As mentioned earlier, MXenes have special 2D structures, with high electrical conductivities and large specific surface areas (SSA), thus having a strong potential to be electrode materials for supercapacitor applications. It was found that $Ti_3C_2T_x$ MXene could be intercalated spontaneously by using cations, such as Na^+, K^+, NH_4^+, Mg^{2+} and Al^{3+}, in their salt solutions [67]. In KOH solution, $Ti_3C_2T_x$ paper-based electrode could deliver a volumetric capacitance of 340 F/cm^3 at the scan rate of 2 mV/s. Furthermore, the capacity value kept almost unchanged at cycling of up to 10,000 times at a relatively high current density of 1 A/g.

The electrochemical behaviour of $Ti_3C_2T_x$ has been deeply analysed to reveal the factors that contributed to the capacitive effect [68]. The safe voltage window was optimized, which could be well extended to various other materials, especially in understanding decomposition behaviours of the electrolytes. The variation in the oxidation state of Ti due to the electrochemical reaction during eh charge–discharge cycling process, by using in-situ XAS measurement techniques. It was found that values obtained were very close to those calculated according to the gravimetric capacitance of the samples by assuming that there was merely redox-related capacitive bahviours. It was thus concluded that the electrochemical properties of $Ti_3C_2T_x$ in sulfuric acids were highly pseudocapacitive.

As a result, $Ti_3C_2T_x$ MXenes exhibited real pseudocapacitive effect, which was ascribed to the continuous change in the oxidation state of Ti in the charge–discharge process, thus leading to CVs with a well-developed rectangular shape. The pseudocapatitive property was simply attributed to the 2D characteristic of the $Ti_3C_2T_x$

MXenes. In this case, the ions were spontaneously intercalated into the nanolayers and attached onto surfaces of the transition metal oxides with high electrochemical activities, whereas the highly conductive carbide layers promoted the transport of the charges. More importantly, it implied that the average oxidation state of Ti in the $Ti_3C_2T_x$ MXene was is very close to +2, instead of +4.

The mechanism of capacitive effect of $Ti_3C_2T_x$ MXene has been well studied [69]. The charge/discharge processes of $Ti_3C_2T_x$ based electrodes in the electrolytes of sulfate ions were evaluated with three cations, combined with in-situ Raman spectroscopic technique. During the discharging and charging processes, hydronium in the H_2SO_4 electrolytes was bonded and debonded with the –O functional groups, respectively. The bonding/debonding processes were reversible, which resulted in reversible variation in the valence state of Ti in the MXene. Therefore, in acidic electrolyte, it was pseudocapacitive effect.

However, in $(NH_4)_2SO_4$ or $MgSO_4$, the capacitance was only contributed by electric double layer capacitive effect. In terms of charge storage characteristics, ion exchange was dominant in H_2SO_4, whereas counter-ion adsorption was more pronounced in other electrolytes. Comparatively, hydronium possessed the highest mobility, due to its smaller hydration radius and less amount of charge. As a consequence, it was the most favorable to penetrate the electrode and reach the active sites in between the MXene nanolayers. Therefore, the high capacity of the $Ti_3C_2T_x$ MXene in H_2SO_4 was readily ascribed to two aspects, (i) pseudocapacitive effect due to the bonding/debonding process related to the surface functional groups and (ii) ion-exchange enabled charge storage mechanism, which concurrently contributed to the total capacity. Figure 4.32 shows a schematic diagram of the two different charge storage mechanisms in the two electrolytes.

As one of the early examples, Ghidiu et al. demonstrated supercapacitive effect of $Ti_3C_2T_x$ MXene as the electrodes with H_2SO_4 solution as the electrolyte [70]. The electrodes exhibited volumetric capacitances of as high as 900 F/cm^3, corresponding to specific capacitances of 245 F/g. At a relatively high current density of 10 A/g, the specific capacitance was almost unchanged after cycling for 10,000 times. Due to its smaller radius, H^+ ion displayed much higher electrochemical performances, as compared with other ions. In addition, after the etching reaction with LiF and HCl, the interlayer spacings in between the MXene nanosheets were greatly enlarged to ensure the high accessibility of electrolyte ions. The extra contribution of the increase surface redox processes was also responsible for the enhanced electrochemical activities of the 2D MXene materials.

A modified electrophoretic deposition (MEPD) process was reported to obtain binder-free $Ti_3C_2T_x$ MXenes as electrodes of supercapacitors, which had outstanding electrochemical properties [71]. The MEPD strategy enabled the deposition of binder-free films of MXenes, with controllable mass loadings and large deagglomerations, on both Ni foams and flexible fabrics, with organic suspensions of $Ti_3C_2T_x$ nanosheets. The electrodes developed in this way possessed a specific capacitance of 140 F/g in alkaline based electrolytes. The special electrodes showed a high cycling stability, with nearly unchanged specific capacity loss after 10,000 cycles and good rate performance, because of the decreased degree of aggregation of the $Ti_3C_2T_x$

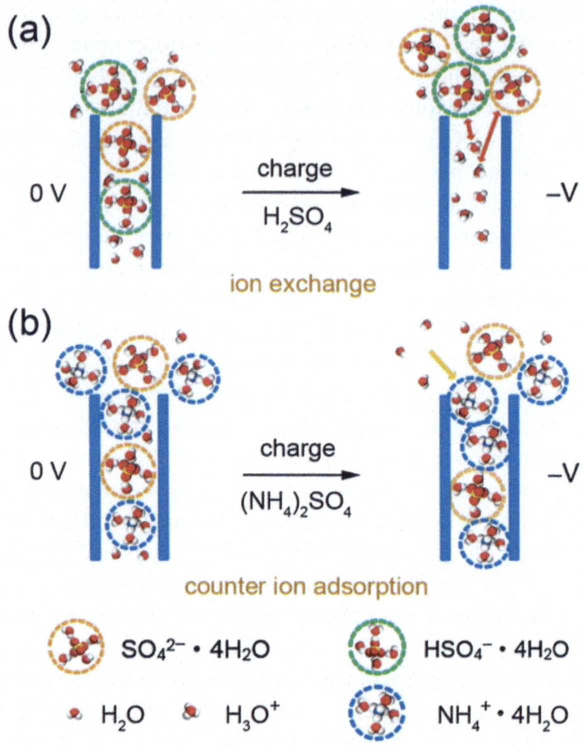

Fig. 4.32 Schematic diagrams illustrating charge storage mechanisms of the $Ti_3C_2T_x$ based electrodes, for different electrolytes: **a** ion-exchange in H_2SO_4 solution and **b** counter-ion adsorption in $(NH_4)_2SO_4$ solution. Reproduced with permission from [69]. Copyright © 2016, American Chemical Society

nanolayers, in the form of self-assembled structures, leading to appropriate pore size distribution, with an average dimension of 4 nm. Also, high electrical conductivity was maintained, together with high accessibility to the electrolyte ions and full utilization of surfaces of the MXene nanoflakes.

Four electrodes were made after deposition for 10, 30, 40 and 50 s, denoted as MEPD10, MEPD30, MEPD40 and MEPD50. Figure 4.33 shows electrochemical properties of the samples in 1 M KOH electrolyte. CV curves of the sample MEPD10 were measured at different scan rates of 5–150 mV/s within the potential window from −0.75 to −0.25 V. At different scan rates, the CV curves were similarly rectangular-shaped, suggesting their typical capacitive effect. No redox peaks were observed, indicating that d-$Ti_3C_2T_x$ film-based electrodes had a constant rate for electrochemical redox reactions in the cycling range, which was well attributed to the intercalation pseudocapacitive mechanism. GCD curves of the MEPD10 electrode are shown in Fig. 4.33b, which were recorded at the current densities of 0.5, 1 and 1.5 A/g. The charge–discharge processes all demonstrated a linear profile, while a symmetrical equicrural triangle shape was observed, implying that the redox reactions on the surface of the electrodes, were highly reversible, in a good agreement with the CV curves.

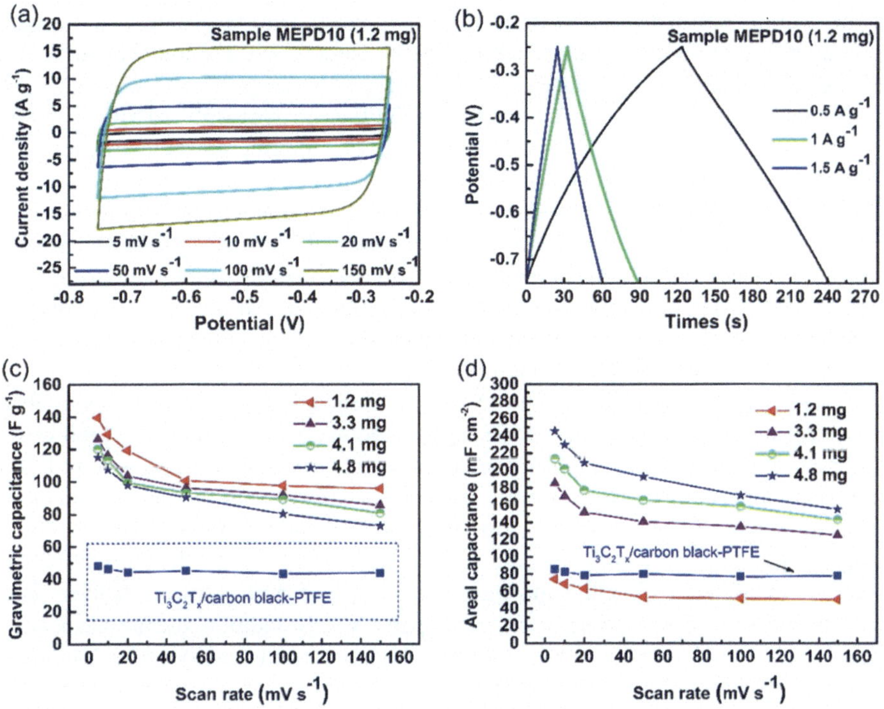

Fig. 4.33 CV curves of the sample MEPD10 at different scan rates measured with the three-electrode configuration. **b** GCD curves of the sample MEPD10 at different current densities. **c** Gravimetric capacitances of the four d-$Ti_3C_2T_x$ films and the $Ti_3C_2T_x$/carbon black-PTFE electrodes as a function of scan rate. **d** Areal capacitances of the d-$Ti_3C_2T_x$ film and the $Ti_3C_2T_x$/carbon black-PTFE electrodes as a function of scan rate. Reproduced with permission from [69]. Copyright © 2017, Elsevier

Figure 4.33c shows specific capacitances of the film electrodes as a function of scan rate. At 5 mV/s, the specific capacitances of MEPD10, MEPD30, MEPD40 and MEPD50 were 140, 127, 120 and 115 F/g, respectively. As the scan rate was increased from 5 to 50 mV/s, their capacity retentions were 78.5%, 77.6%, 76.2% and 72.2%, respectively, evidencing high rate stability of the d-$Ti_3C_2T_x$ film-based electrodes. Obviously, the specific capacitance was decreased gradually with increasing mass loading. This was because the thickness of the $Ti_3C_2T_x$ films was increased with increasing mass loading. The thicker the film, the longer the transport path of the electrolyte ions and the less the utilization rate of the active materials would be. As a result, the electrochemical performance was decreased with mass loading.

Meanwhile, the lower the mass loading, the lower the areal capacitance would be, which should be paid attention to in terms of real application of the materials for supercapacitors. As seen in Fig. 4.33d, the areal capacitances were 74, 186, 214 and 246 mF/cm^2 for the samples MEPD10, MEPD30, MEPD40 and MEPD50, respectively. Definitely, real applications always require sufficiently high areal capacitance.

In this aspect, optimal mass loading of the d-$Ti_3C_2T_x$ film electrode was 4.1 mg, corresponding to optimal electrochemical parameters of 120 F/g and 214 mF/cm^2, for specific and areal capacitances.

Comparatively, the $Ti_3C_2T_x$/carbon black-PTFE electrode exhibited a much lower specific capacitance of 48 F/g at 5 mV/s. The significantly improved electrochemical properties of the MEPD based electrodes could be well ascribed to their sufficiently high conductivity due to the absence of binders and the retained stacking nanostructure. At the same time, the sizes of the pores were all very close to 4 nm, which was appropriate for ion transport of the electrolytes. Also, the functional groups on surface of the MXene nanosheets were entirely utilized. All these factors concurrently contributed to the entire capacitance of the $Ti_3C_2T_x$ film-based electrodes made with the promising MEPD method.

A 3D $Ti_3C_2T_x$ aerogel architecture was proposed and fabricated, which was able to prevent the restacking of the 2D $Ti_3C_2T_x$ nanolayers, so as to ensure high electrochemical performances when it was used as electrode of supercapacitors [72]. The enhanced electrochemical properties were benefited from the enhancement in diffusion capability of electrolyte ions into the electrode materials. The $Ti_3C_2T_x$ aerogel had a relatively large specific surface area of 108 m^2/g. The aerogel-based electrode exhibited a specific capacitance of 349 F/g, 3 M H_2SO_4 electrolyte. The high specific capacitance could be retained up to 2 V/s. In addition, the initial specific capacitance was retained by 90%, at an even higher current density of 20 A/g, after charging–discharging for 20,000 cycles.

A simple method was reported to develop low cost, flexible, tough and scalable electrodes of supercapacitors based on silk textiles combined with $Ti_3C_2T_x$ MXene nanoflakes [73]. In this case, the silk textile was first carbonized to obtain highly conductive flexible free-standing substrates. Then, $Ti_3C_2T_x$ nanosheet suspensions were loaded onto the carbonized silk cloth (CSC) substrates, leading to nanometer-thick $Ti_3C_2T_x$ layers, with high uniformity and strong adhesion. Due to the high electrical conductivity and mechanical integrity, MXene-coated flexible textile electrodes possessed a high areal capacitance of 362 mF/cm^2, with outstanding cycling stability and mechanical flexibility. Furthermore, the electrodes could withstand very large bending deformations, making them promising candidates for flexible supercapacitor applications.

To prepare the CSC and CSC@$Ti_3C_2T_x$ electrodes, pristine silk textiles were soaked in ethanol to clean their fabric surfaces. After the silk textiles were carbonized, the cloths could have carbon contents of 30% and 65%, respectively. According to SEM images, there were impurities on surface of the fibers, which could be resulted from the inorganic compounds brought during the sample fabrication process. The impurities could be eliminated readily through treatment in acids, so that fibers were cleaned and smoothened again, as shown in Fig. 4.34a, b.

It is well known that silks were mechanically strong. More importantly, the mechanical strengths could be retained after the carbonization process. To obtain CSC@$Ti_3C_2T_x$ hybrid-based electrodes, the as-obtained CSCs were treated with oxygen plasma, in order to introduce more –O functional groups, so that the $Ti_3C_2T_x$ suspensions could be easily attached onto the surface of fabrics. In this case, the

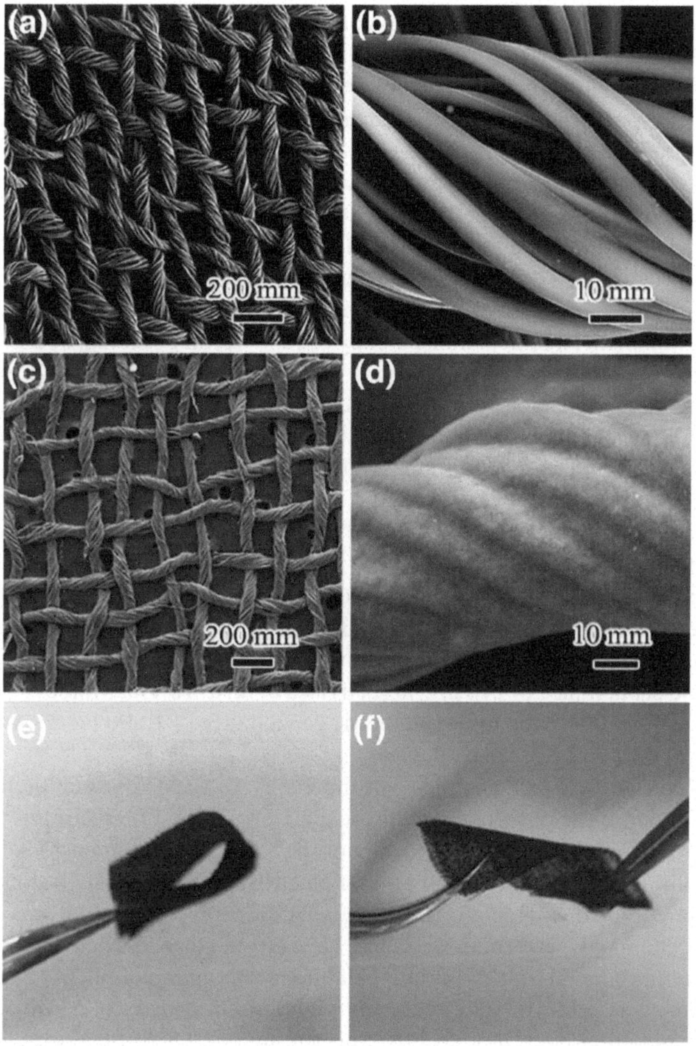

Fig. 4.34 SEM images of the samples: **a**, **b** carbonized silk cloth (CSC) and **c**, **d** CSC@$Ti_3C_2T_x$. The $Ti_3C_2T_x$ films were homogeneously coated onto the carbonized silk fabric. Photographs of the CSC@$Ti_3C_2T_x$ fabric electrodes: **e** bended and **f** twisted states. Reproduced with permission from [73]. Copyright © 2018, Elsevier

$Ti_3C_2T_x$ suspension was coated onto the CSCs in the manner of drop by drop on a hot plate. After the water molecules were all evaporated, MXene nanosheets were coated uniformly on the fibers, as shown in Fig. 4.34c. The presence of the $Ti_3C_2T_x$ nanosheets made the fiber bundles in the textiles to be more compacted, without the formation of any defects. Moreover, the MXenes-covered CSCs were still highly flexible, which could be bended and twisted very significantly, as seen in Fig. 4.34e,

f. Therefore, such special fiber hybrids could be potential candidates as electrodes of flexible supercapacitors.

$Ti_3C_2T_x$ films with high optical transparency and electrical conductivity were utilized as transparent solid-state supercapacitors [74]. The transparent MXene films were deposited by using spin-casting method, from $Ti_3C_2T_x$ nanosheet colloidal suspensions, followed by heat treatment at 200 °C in vacuum. The samples had transparencies of 93% and 29%, together with DC conductivities of 5736 and 9880 S/cm, for thicknesses of 4 nm and 88 nm, respectively. The highly transparent and electrically conductive $Ti_3C_2T_x$ films demonstrated a volumetric capacitance of 676 F/cm^3, with rapid electrochemical responses. Meanwhile, transparent solid-state asymmetric supercapacitors with an optical transparency of 72% were assembled, by using hybrid films of $Ti_3C_2T_x$ and single-walled carbon nanotubes (SWCNTs). These electrodes displayed an areal capacitance of 1.6 mF/cm^2, corresponding to an energy density of 0.05 μWh/cm^2, along with a long term stability of above 20,000 cycles. The electrochemical performances were much higher than those of the similar transparent devices based on graphene or SWCNTs.

It has been found that MXenes could be operated at rates much higher than those of the traditional EDLCs, while both the volumetric and areal capacitances were maintained to be sufficiently high [75]. Meanwhile, ion accessibilities to the active sites of redox reactions could be enhanced through special designs of the MXenes based electrodes. When made into macroporous film, the $Ti_3C_2T_x$ MXenes offered a high specific capacitance of 210 F/g, at scan rates of up to 10 V/s. Moreover, MXenes hydrogels exhibited an even high volumetric capacitance of 1,500 F/cm^3, which was very close to that of RuO_2 based devices.

The influence of surface chemistry of the 2D $Ti_3C_2T_x$ MXenes on their electrochemical properties as electrodes of supercapacitors in H_2SO_4 electrolyte has been clarified [76]. The electrochemical behavior of Ti_3C_2, a two-dimensional titanium carbide from the MXene family, in H_2SO_4 electrolyte is reported. To do so, Ti_3C_2 MXenes were treated with delamination and intercalation processes. It was observed that electrochemical performances could be well enhanced, when –O functional groups were introduced. In case of intercalation, a very high volumetric capacitance of 415 F/cm^3 was achieved, at a relatively high current density of 5 A/g, while the materials had a relatively low specific surface area of 98 m^2/g. For delaminated MXenes, the volumetric capacitance could be up to 520 F/cm^3, at a scan rate of 2 mV/s.

Figure 4.35 shows schematic diagrams demonstrating the two chemical modification processes for the 2D $Ti_3C_2T_x$ MXenes. The delaminated Ti_3C_2 was denoted as d-Ti_3C_2, while the K^+ ion intercalated sample was named as K^+–$Ti_3C_2T_x$. Both KOH and KOAc were employed to conduction the intercalation experiments. All the $Ti_3C_2T_x$, KOH–Ti_3C_2, KOAc–Ti_3C_2 and d-Ti_3C_2 samples were characterized by using XRD. After the intercalation processes, the interlayer spacings were enlarged, evidenced by the increase in *c*-lattice parameter of the $Ti_3C_2T_x$ MXenes. For KOH and KOAc, the difference in their *c*-lattice parameter was also dependent on other factors, like pH value. For instance, although acetate ion was larger than the –OH ion, the *c*-lattice parameter of KOH–Ti_3C_2 was larger than that of KOAc–Ti_3C_2. However,

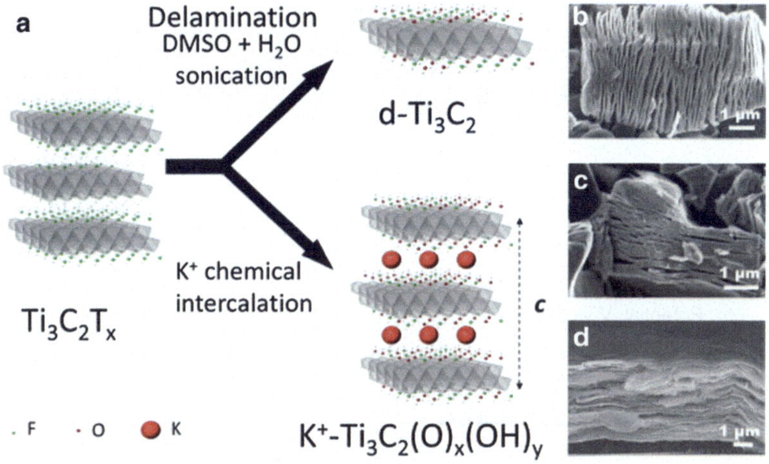

Fig. 4.35 **a** Schematic diagrams of the two modifications processes for the $Ti_3C_2T_x$ MXenes: delamination and K^+ intercalation. Representative SEM images of the samples: **b** $Ti_3C_2T_x$, **c** KOH–Ti_3C_2 and **d** d-Ti_3C_2. Reproduced with permission from [76]. Copyright © 2017, Elsevier

the treatments with KOH and KOAc resulted in different surface characteristics, thus leading to different *c*-lattice parameters.

CV curves of the samples recorded at the scan rate of 10 mV/s in 1 M H_2SO_4 solution are illustrated in Fig. 4.36a. According to XRD characterization results, *c*-lattice parameters of all the samples had almost no variation, which could be attributed to the fact that the spacings in between the adjacent Ti_3C_2 nanolayers were sufficiently large to host the intercalated ions. The d-Ti_3C_2 electrodes exhibited the highest electrochemical performance, with a volumetric capacitance of 520 F/cm^3 and a specific capacitance of 325 F/g at the scan rate of 2 mV/s, as revealed in Fig. 4.36b. This could be understood in terms of several factors. Specifically, the d-Ti_3C_2 electrode had much smaller thickness, thus allowing a rapid charge transport. Also, it had relatively high specific surface area and a high packing density, due to the aligned MXene nanoflakes. However, the performance of the d-Ti_3C_2 based electrode was highly dependent on the scan rate, which could be attributed to its special microstructure, i.e., the MXene nanoflakes were aligned parallelly to the current collector. In addition, the transport path for electrolyte ions was increased, as the thickness of the film was increased.

Since specific surface area of the $Ti_3C_2T_x$, KOH–Ti_3C_2 and KOAc–Ti_3C_2 samples was same, the difference in electrochemical properties was just ascribed to the difference in their surface characteristics. The $Ti_3C_2T_x$ delivered relatively low specific capacitance, because of the absence of the –F functional groups. The KOAc–Ti_3C_2 and KOH–Ti_3C_2 electrodes possessed almost the same electrochemical performance, as shown in Fig. 4.36b. The surface groups included –OOH, =O and –OH, all with pseudocapacitive effect in acidic electrolytes. During the electrochemical reactions, the oxidation state of Ti component in the MXenes could be varied between +3 and

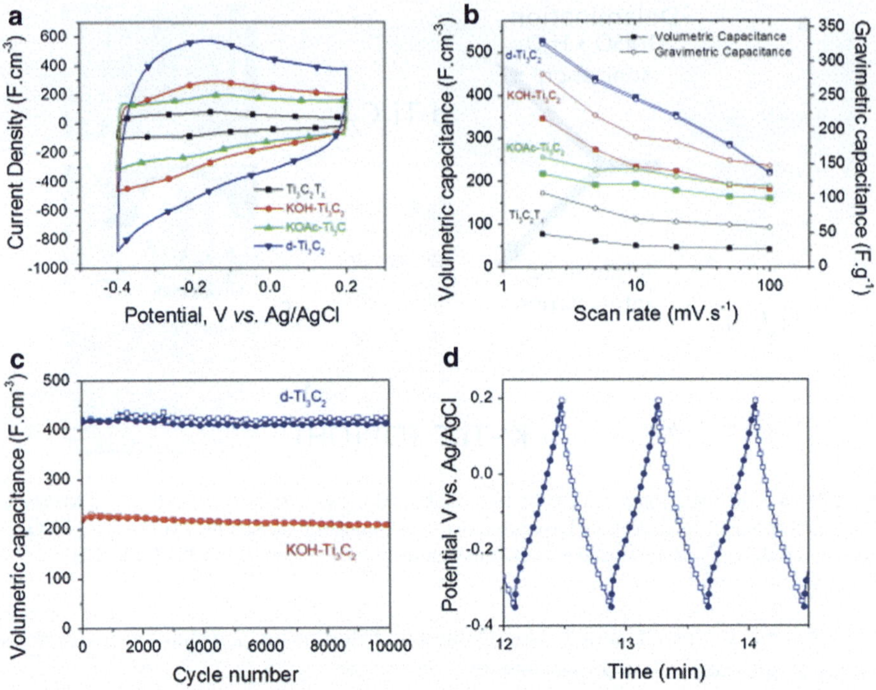

Fig. 4.36 Electrochemical behaviors of the Ti_3C_2-based electrodes in 1 M H_2SO_4 electrolyte: **a** CV curves at 10 mV/s and **b** rate performances. **c** Charge and discharge volumetric capacitances of the KOH–Ti_3C_2 and d-Ti_3C_2 electrodes derived from their Galvanostatic cyclings at a current density of 5 A/g. **d** Galvanostatic charge–discharge profiles of the d-Ti_3C_2 based electrodes. Reproduced with permission from [76]. Copyright © 2017, Elsevier

+4. Figure 4.36c shows capacitances of the samples, which were derived from the Galvanostatic charge–discharge data illustrated in Fig. 4.36d. The capacitances of the d-Ti_3C_2 and KOH–Ti_3C_2 electrodes were stable at 415 and 215 F/cm^3, at the current density of 5 A/g. Also, they all demonstrated a sufficiently high cycling stability.

A similar approach was developed to enhance the capacitive effect of $Ti_3C_2T_x$ MXenes through cation intercalations combined with surface modifications [77]. With the intercalation of K^+ ions and the removal of –OH/–F groups, the pseudocapacitive effect was enhanced by about three times, as compared with that of the pristine MXenes. A high specific capacitance of 517 F/g was achieved at a high current density of 1 A/g. In addition, the initial specific capacitance could be fully retained after cycling for 10,000 times. The significant enhancement in electrochemical properties was ascribed to the enlarged interlayer spacings in between the Ti_3C_2 nanosheets and the reduced number of the surface groups.

Another example of intercalation was the utilization of hydrazine, so as to alter surface chemistries of $Ti_3C_2T_x$ MXenes, resulting in a decrease in the contents of –F/–OH groups and water molecules [78]. At the same time, the intercalation of hydrazine also physically enlarged the spacings of the MXene nanolayers, thus enhance the accessibility of electrolyte ions. As a consequence, electrodes based on the hydrazine intercalated MXenes delivered a specific capacitance of 215 F/g at a high current density of 5 A/g in 1 M H_2SO_4 electrolyte, while the capacitance was nearly not degraded after a long cycling up to 10,000 cycles.

A simple treatment with LiCl solution was employed to increase electrochemical performances of 2D $Ti_3C_2T_x$ as electrode of supercapacitors [79]. The 2D $Ti_3C_2T_x$ was made into flexible free-standing papers, which exhibited a pretty high volumetric capacitance of 892 F/cm^3. Meanwhile, the $Ti_3C_2T_x$ paper-based electrode displayed a strong cycling stability. Four reasons were responsible for the outstanding electrochemical properties of the $Ti_3C_2T_x$-Li paper-based supercapacitor electrode.

Firstly, the synthetic approach and the delamination process were so appropriated that the resultant MXenes met the requirement of electrochemical reactions. Secondly, $Ti_3C_2T_x$-Li paper had sufficiently high electrical conductivity, thus offering a shortened diffusion pathway for the H^+ ions and promoting the necessary electronic conduction. Meanwhile, the $Ti_3C_2T_x$-Li paper was able to release the volume expansion/contraction during the repeated charging-discharging processes. Thirdly, the intercalation of Li^+ ions further opened up the 2D layered structure to promote the electrochemical reactions. Finally, the pseudocapacitive effect of the 2D MXenes was excessively demonstrated in the acidic electrolytes.

The intercalation of Li^+ ions into the $Ti_3C_2T_x$ MXene interlayers was also demonstrated by using ion-exchange method, with trimethylalkylammonium (AA) cations having different alkyl chain lengths [80]. The expansion of the MXene interlayer spacing was discontinuous, because there was critical chain length for the complete packing of the nanosheets. The number of cations per Ti_3C_2 formula unit could be estimated according to the critical chain length. The results were highly consistent with the observations based on other analytic techniques. Density functional theory (DFT) and molecular dynamics were used to predict the cation concentrations, which were at the comparable levels. Furthermore, electrical conductivity of the MXenes could be controlled by using different AA cations.

Figure 4.37 shows CV curves of the samples recorded at the scan rate of 5 mV/s. The area of the CV loops was increased initially from the control sample C_0 to C_1, while it was then decreased gradually to C_{10}. The authors couldn't figure out the underlying reasons. In addition, the stable potential window of C_{10} was narrower than those of the control, C_1 and C_6 samples by about 50 mV, as shown in Fig. 4.37b. The specific capacitance of the C_1 sample higher than that of the control sample C_0 by about 12%, at the lowest scan rate of 2 mV/s. At the same time, the capacitance level of the C_{10} sample was lower that of the control sample by about 30%. Figure 4.37c shows specific capacitances as a function of n_c, at different scan rates. The C_1 sample was the only one that had an increased capacitance value at the scan rates of below 20 mV/s. Above that scan rate, the specific capacitance was nearly not dependent on n_c. Because the capacitance values of all the samples were converged at high scan rates, the electrical double-layer on surface of the particles or shallow sites in the interlayers were responsible for the charge storage, as the scan rates were $\geq$50 mV/s.

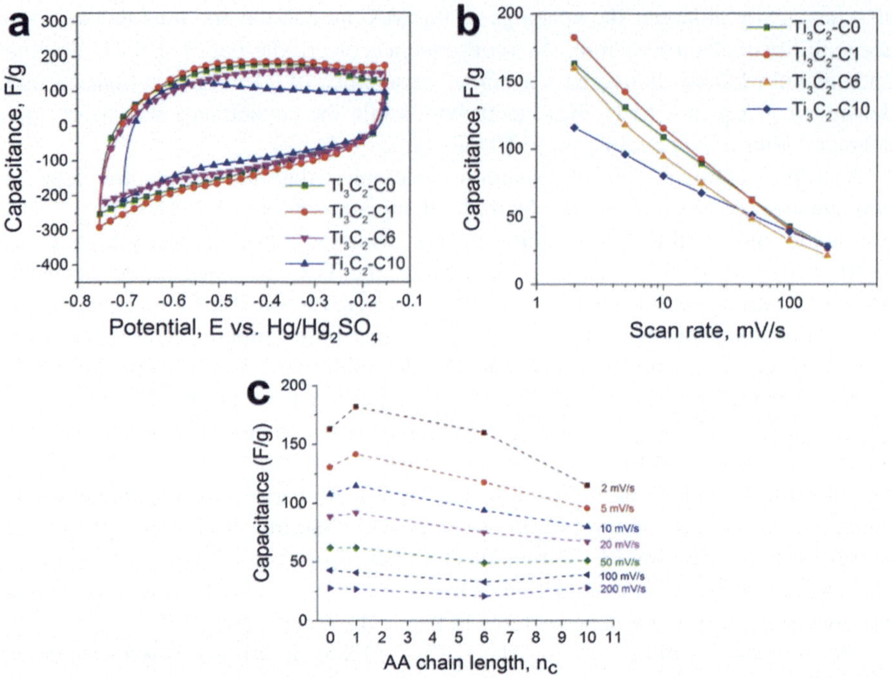

Fig. 4.37 Electrochemical properties of the AA-intercalated $Ti_3C_2T_x$ MXenes: **a** CV curves of the nonintercalated (C_0) and intercalated C_1, C_6, and C_{10} samples, **b** gravimetric capacitances as a function of scan rate and **c** gravimetric capacitance at different scan rates as a function of alkylammonium chain length. Reproduced with permission from [80]. Copyright © 2017, American Chemical Society

The increase in specific capacitance of the electrodes derived from the MXenes intercalated with smaller AA chains was due to the positive pillaring effect, which resulted in interlayer spacings that enhanced the accessibility of the electrolyte ions. Once the AA chains were too long, electrical conductivity of the materials was reduced, thus leading to the decrease in specific capacitance. If it was assumed that the AA cations had no electrochemical activity over the potential range and the number of molecules of each AA that preintercalated into the electrodes was the same, the larger the AA cations, the lower the specific capacitance would be. This was because longer AA chains posted more dead weight/volume in the electrodes.

A nitrogen-doped 2D MXene, N-$Ti_3C_2T_x$, was reported as electrode of supercapacitor with enhanced electrochemical performances [81]. The N-$Ti_3C_2T_x$ was obtained by annealing the as-etched $Ti_3C_2T_x$ in NH_3 environment. The content of nitrogen could be controlled in the range of 1.7–20.7 at%, by controlling the annealing temperature in the range of 200–700 °C. The incorporation of nitrogen atoms into the $Ti_3C_2T_x$ nanosheets largely increased the *c*-lattice parameter of MXenes, which could be tuned from 1.92 nm of pristine $Ti_3C_2T_x$ to 2.46 nm of N-doped sample treated at 200 °C. Optical specific capacitances were 192 F/g in 1 M H_2SO_4 and

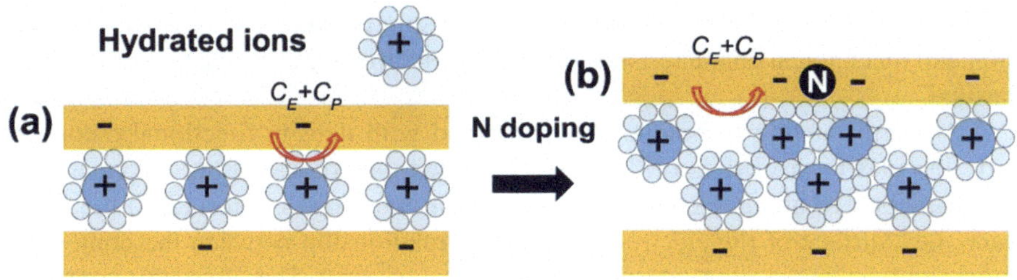

Fig. 4.38 Schematic diagrams showing charge storage process of the hydrated electrolyte ions: **a** MXenes and **b** N-doped MXenes. The concurrent effect of the enlarged interlayer spacing and the N-based functional groups is present in the N-doped MXenes. In the pristine $Ti_3C_2T_x$, the capacitive effect consists of both electrical double-layer capacitance (C_E) and pseudocapacitance (C_P) due to the variable oxidation state of Ti. In the case of the N-doped $Ti_3C_2T_x$, the enlarged interlayer spacing had a significant contribution to the capacitance enhancement. Reproduced with permission from [81]. Copyright © 2017, Elsevier

82 F/g in 1 M $MgSO_4$, which were much higher than those of the pristine $Ti_3C_2T_x$. Figure 4.38 shows schematic diagrams of the $Ti_3C_2T_x$ MXenes with and without N-doping.

A similar process was reported to develop nitrogen-doped delaminated Ti_3C_2 MXene, which was aided with urea [82]. In this case, urea played a crucial role in obtaining N-d-Ti_3C_2, where it acted as an intercalant for delamination and a source of nitrogen for functionalization. In a very similar way, parameters, such as layered structure, specific surface area, distribution of pores, N-doping concentration, could be optimized. Optical specific capacitance was 266.5 F/g at the scan rate of 5 mV/s. Also, the electrode showed promising rate capability, with 79% capacitance retention at scan rates of up to 200 mV/s, together with outstanding cycling stability, i.e., 86.4% specific capacitance retention at the current density of 2 A/g in 6 M KOH solution.

It was found that etching time could influence the morphology, structure, surface chemistry and hence electrochemical performances of Ti_3C_2 MXenes [83]. The Ti_3C_2 MXene nanosheets were obtained by selective etching of Al layer from Ti_3AlC_2 MAX with HF solution for different etching times. The electrochemical performance of the Ti_3C_2 MXene was optimized after etching reaction form 216 h (Ti_3C_2-216), while no etching time effect was observed as the reaction was proceeded from 2 to 168 h. An optimal specific capacitance of 118 F/g at the scan rate of 5 mV/s, along with promising rate characteristics and cycling stabilities. The enhanced capacitive effect was readily ascribed to the sufficiently high content of carbon to ensure a high electrical conductivity and rapid ion and electron transport. Meanwhile, the enlarged specific surface area promoted the accessibility of the electrolyte to the interior of the electrodes.

It is important to understand the fundamental properties of the electrodes at the nanometer scale for the design of high performance supercapacitors, such as ions insertion and interactions with active materials [84]. In most cases, the insertion and desertion of ions are accompanied by large variation in volume of the electrode materials, which leads to mechanical strain and stress that are harmful to the cycling stability of the electrodes. Therefore, the relationship between the dimensional variation and mechanical characteristics during the electrochemical reactions should be clarified.

An in-situ imaging technique was combined with density functional theory to study the elastic properties of 2D $Ti_3C_2T_x$ based electrodes during the intercalation/extraction of alkaline cations [84]. It was in the direction normal to the basal plane, i.e., surface of the electrodes. A strong relationship between the content of cations and the out-of-plane elastic modulus was observed. Therefore, this approach could be used to identify the preferential pathways of intercalation within individual particles of the active materials, so as to reveal ionic transport characteristics.

The authors further demonstrated that the intrinsic properties of MXene based electrodes could be studied by using an electrochemical quartz-crystal microbalance with dissipation (EQCM-D), which was highly sensitive to the intercalation-related Gravimetric and viscoelastic changes [85]. In this study, $Ti_3C_2(OH)_2$ MXene was used as an example. It was found that, for every Li^+ ion inserted into the thin MXene electrodes, there was one water molecule to be inserted at the same time. Corresponding to the reversible intercalation/deintercalation of the Li^+ ion into the MXene electrodes, there were viscoelastic changes and periodic stiffening/softening behaviours, which matched well with the long-term cycling stabilities.

Composites based on MXene with various polymers have also been explored as electrodes of supercapacitors. For example, $Ti_3C_2T_x$ nanoflakes could be blended with PVA to form multifunctional $Ti_3C_2T_x$/PVA composite films with promising electrochemical performances and mechanical strength [86]. Especially, mechanical properties of the $Ti_3C_2T_x$/PVA composite films were much higher than those of pristine $Ti_3C_2T_x$ and PVA. In KOH electrolyte, the composite based electrodes exhibited high volumetric capacitances of 528 F/cm^3 and 306 F/cm^3, at 2 mV/s and 100 mV/s, respectively, together with outstanding cycling stability.

An in-situ polymerization method was used to confine pyrrole in between $Ti_3C_2T_x$ nanolayers [87]. According to XRD and TEM characterization results, the pyrrole was intercalated into the $Ti_3C_2T_x$ nanolayers, with the polymer chains to have a high degree of alignment after polymerization, as shown schematically in Fig. 4.39. The $Ti_3C_2T_x$/PPy composites could be employed as electrodes of supercapacitors,

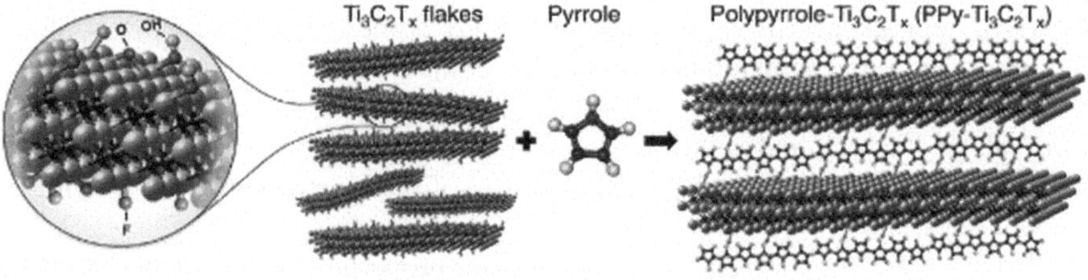

Fig. 4.39 Schematic diagram demonstrating the in-situ polymerization of pyrrole in between the MXene nanolayers. The terminating groups on the latter contributed to the polymerization process. Reproduced with permission from [87]. Copyright © 2015, John Wiley & Sons

with a volumetric capacitance of 1000 F/cm^3 and a capacitance retention of 92% after cycling for 25,000 times. The outstanding electrochemical properties of the composites were attributed to the self-assembled highly ordered multilayer structures, in which the PPy chains were strongly oriented, thus enabling high electrical conductivity, rapid reversible redox reactions and smooth ion transport. Meanwhile, the introduction of the aligned PPy enlarged the interlayer spacings. Moreover, the PPy also facilitated surface redox reaction.

A similar approach was reported to make Ti_3C_2/PPy composites, in order to enhance the capacitive effect of PPy [88]. Freestanding conductive composite films were developed by intercalating PPy into the multilayered Ti_3C_2 MXene. With the incorporation of MXene, the areal capacitance was increased from 150 to 203 mF/cm^2. More importantly, the capacitance was retained by nearly 100%, after charging/discharging for 20,000 cycles. In this case, the stacking of PPy was prevented, due to the presence of the layered MXene nanosheets. As a result, the infiltration of the electrolyte ions into the electrode materials was enhanced. In addition, the PPy backbones and the surfaces of the Ti_3C_2 nanosheets were strongly bounded, which ensured high conductivity and enhanced the structural stability of the PPy backbones. The freestanding Ti_3C_2/PPy films could be employed to assemble ultra-thin all-solid-state supercapacitors, which displayed an areal capacitance of 35 mF/cm^2, with the electrochemical performance to stable against both mechanical bending and charging/discharging cycling.

$Ti_3C_2T_x$ was also incorporated with conjugated polymers [polyfluorene derivatives (PFDs)] [89]. Three derivatives were studied, which were derived through the Suzuki polycondensation reaction process, with nonpolar, polar and charged N-containing functionalities. The $Ti_3C_2T_x$/PFD composites were thoroughly characterized by using various techniques to clarify the underlying interaction mechanisms. Among the three samples, polar polymers with charged N-containing ends interacted with the $Ti_3C_2T_x$ nanolayers in the strongest way, resulting in enlarged interlayer spacing. Furthermore, the interaction between the polymers and the $Ti_3C_2T_x$ nanolayers influenced pseudocapacitive effect of the composites, showing potential applications in energy storages.

Polyaniline (PANI) modified 2D Ti_3C_2 composites exhibited high electrochemical performances as electrode materials of supercapacitors [90]. The Ti_3C_2/PANI composites were prepared by using in-situ polymerization technology. The incorporation of PANI into the Ti_3C_2 nanolayers resulted in composites with significantly improved electrochemical properties, due to the amino groups in PANI which led to high electrical conductivity of the materials and enhanced surface wettability of the Ti_3C_2 nanosheets to create more active sites. As a result, additional capacitive effect was obtained because of the Faradaic reactions. The Ti_3C_2/PANI composite based electrodes exhibited a specific capacitance of 164 F/g at 2 mV/s, which was higher than that of pristine Ti_3C_2 by about 30%. In addition, the composite electrodes possessed promising cycling stability, with a 96% capacity retention after cycling for 3000 times.

PANI@$TiO_2/Ti_3C_2T_x$ ternary composites with hierarchical structures were developed, by using a hydrothermal method combined with in-situ polymerization technique [91]. The composites were used as electrodes of supercapacitors, demonstrating promising electrochemical performances. In this case, the $Ti_3C_2T_x$ nanosheets acted as both the active materials with large number of reactive sites and the framework to ensure smooth insertion/extraction of the electrolyte ions. The presence of the PANI nanoflakes and the TiO_2 nanosized particles resulted in an enhanced electrochemical performance, because of their pronounced pseudocapacitive effect and the enlarged surface area of the active materials.

Specifically, the hierarchical PANI@$TiO_2/Ti_3C_2T_x$ ternary composites-based electrode displayed a specific capacitance of 188.3 F/g at the scan rate of 10 mV/s, which was higher than that of $TiO_2/Ti_3C_2T_x$ base one by nearly two times. The areal capacitance was about 435.4 F/cm^2, at the current density of 0.5 A/g, in KOH solution. Moreover, the composite electrode exhibited excellent cycling stability, where the initial specific capacitance was retained by as high as 94%, after cycling for 8000 times, at the current density of 1 A/g. The outstanding electrochemical properties of the composite electrodes were well ascribed to the hierarchical architectures and the synergistic effects of the PANI nanoflakes and $TiO_2/Ti_3C_2T_x$ hybrids.

Figure 4.40 shows CV curves of the electrodes based on $Ti_3C_2T_x$, $TiO_2/Ti_3C_2T_x$ and PANI@$TiO_2/Ti_3C_2T_x$, at the scan rate of 100 mV/s. All the samples had rectangular-like shaped CV curves, suggesting that they all exhibited double layer capacitive effect. The PANI@$TiO_2/Ti_3C_2T_x$ composite based electrode possessed the largest area, implying its strongest capacitive effect. In the composite based electrodes, the layered $Ti_3C_2T_x$ nanosheets facilitated electronic conductive pathways and served as a support t hierarchical structure. At the same time, charge transfer and ion diffusion were enhanced due to the high electrical conductivity and the large surface areas. Also, the intimate contact between the $Ti_3C_2T_x$ matrix and the in-situ oxidation of TiO_2 prevented the active materials from dissolving into the electrolyte solutions. In addition, the introduction of the PANI nanoflakes further enlarged the

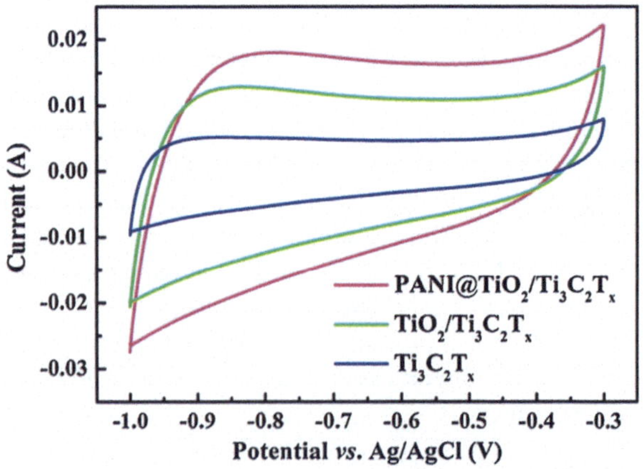

Fig. 4.40 CV curves of the electrodes based on $Ti_3C_2T_x$, $TiO_2/Ti_3C_2T_x$ and PANI@$TiO_2/Ti_3C_2T_x$. Reproduced with permission from [91]. Copyright © 2017, Elsevier

surface area and increased the electrical conductivity of the active materials, thus promoted the charge transport and electrolyte ion diffusion.

Interestingly, Li et al. reported an ultrafast polyaniline@MXene cathode of supercapacitors, which was fabricated by using a simple casting method [92]. The electrode was made with 3D porous $Ti_3C_2T_x$ MXene coated with a homogenous polyaniline layer. The composite exhibited a stable operation for the MXene at positive potentials, owing to the increased work function with the incorporation of polyaniline, as predicted by the first-principle calculations. Flexible positive electrode based on the polyaniline@MXene composite possessed a volumetric capacitance of 1632 F/cm^3 and an ultrahigh rate capability of 827 F/cm^3 at a high scanning rate of 5 V/s. These performances were much higher than those of the similar positive electrodes reported in the open literature. An asymmetric capacitor was assembled with MXene as the anode and the polyaniline@MXene as the cathode, delivering an energy density of 50.6 Wh/L and a power density of 127 kW/L.

A quasi-core–shell structure, made of few layer Ti_3C_2 nanosheets (FL-Ti_3C_2) with negative charges and polyethyleneimine with positive charges as building blocks, were assembled by using an electrostatic LbL process, in which Ni foam was utilized as the supporting skeleton [93]. In the quasi-core–shell structures, the ultrathin layers of the Ti_3C_2 nanosheets offered a smooth electron migration and ion diffusion, as well as enlarged effective contact areas, which resulted in a significant enhancement in electrical conductivity and thus electrochemical performance of the materials as electrodes of supercapacitors. Without the use of binders, the FL-Ti_3C_2@Ni foam-based electrodes exhibited a specific capacitance of 370 F/g at the scan rate of 2 mV/s. Importantly, the specific capacitance was retained to be 117 F/g at a high scan rate of 1000 mV/s, in Li_2SO_4 solution electrolyte. Furthermore, the electrode displayed high long-term cyclic stability, with a retention of about 86% after cycling for 10,000 times.

Besides polymer-based composites, various nanocarbon materials, such as CNTs and graphene, have also been explored as the interlayer spacing enlarge agents to prevent MXene nanoflakes from restacking, so as to maximize the effective utilization of the surfaces. Electrodes based on these hybrids demonstrated excellent electrochemical performances as supercapacitors.

A facial filtration process was used to obtain sandwich-like multilayered papers consisting of MXenes and CNTs, from their aqueous suspensions [94]. The free-standing MXenes/CNTs papers exhibited strong flexibility. The papers displayed much higher electrochemical performances and more excellent rate stabilities, as compared with the bare MXenes and randomly mixed MXene/CNT papers. At optimized conditions, the electrodes delivered a volumetric capacitance of 390 F/cm^3 at the scan rate of 2 mV/s. At the scan rate was increased to 5 A/g, the volumetric capacitance was retained to be 350 F/cm^3, without obvious degradation after cycling for 10,000 times. It was reasonable to expect that the electrochemical properties of the MXene/CNT papers could be further enhanced, the processing parameters are thoroughly optimized.

A layer-by-layer (LbL) assembly aided with spray-coating technique was developed to fabricate free-standing highly flexible MXene/MWCNTs hybrid films,

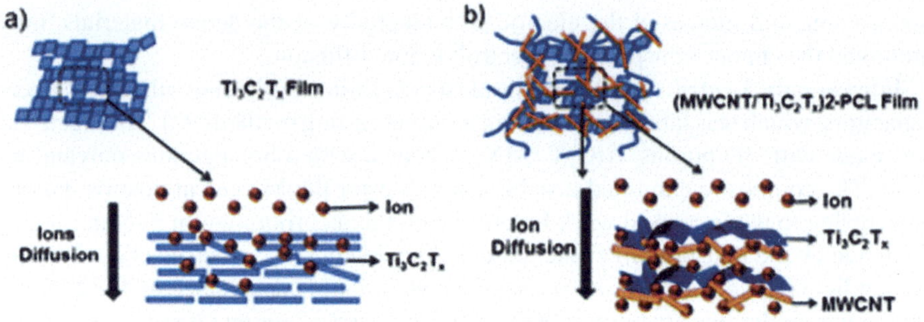

Fig. 4.41 Schematic diagrams of ion diffusion pathways in films: **a** $Ti_3C_2T_x$ and **b** ($MWCNT/Ti_3C_2T_x$)2-PCL. Reproduced with permission from [95]. Copyright © 2018, Royal Society of Chemistry

as electrodes of flexible supercapacitors [95]. $Ti_3C_2T_x$ (MXene) nanosheets and MWCNTs were spray-coated on electrospun polycaprolactone (PCL) fiber network substrates. The PCL networks and the MWCNTs served as interlayer spacing enlargement agents to restrict the restacking of the $Ti_3C_2T_x$ nanosheets. Meanwhile, they also increased the accessible surface of the active materials, ensuring rapid diffusion of the electrolyte ions inside the electrodes, as schematically illustrated in Fig. 4.41. The hybrid-based electrodes delivered areal capacitances in the range of 30–50 mF/cm^2, with promising rate stabilities. More importantly, the hybrid electrodes had high mechanical flexibilities, with potential applications in flexible electronic devices.

Titanium disulfide (TiS_2) was coated on vertically aligned carbon nanotubes (VACNTs) to form hybrid based electrodes of supercapacitors [96]. The electrodes had a stable working range up to 3 V, with a high specific capacitance of 195 F/g in an electrolyte with rich lithium. Symmetric cells based on the electrodes possessed an energy density of 60.9 Wh/kg, which was among the highest for the symmetric pseudocapacitors with transitional metal oxides, conducting polymers, 2D transition metal carbides (MXene) and other transition metal dichalcogenides. Since the nanostructures were constructed by using the atomic layer deposition (ALD) technique combined with sulfurization process, ion transport and surface reactions were effectively promoted, thus leading to a high power density of 1250 W/kg, together with a high cycling stability up to 10,000 cycles.

Figure 4.42a shows structure of the TiS_2-VACNT hybrid electrodes. TiN was coated onto the VACNTs, followed by sulfurization process through heat treatment in the presence of sulfur powder at 300 °C, in order not to damage the configuration of the VACNTs. The outstanding electrochemical performances of the hybrid-based electrode are shown in Fig. 4.42b. The TiS_2–VACNT hybrids had a specific energy in ultra-highly concentrated LiTFSI electrolyte that was higher than that of the pristine VACNTs in 0.5 m H_2SO_4 electrolyte by about 900 times. Moreover, the hybrid-based electrodes also displayed extraordinary energy and power densities, as compared with other electrodes of similar structures. In addition, the special TiS_2-VACNTs hybrids offered a high capacitance retention of about 95%, after cycling for 10,000 times,

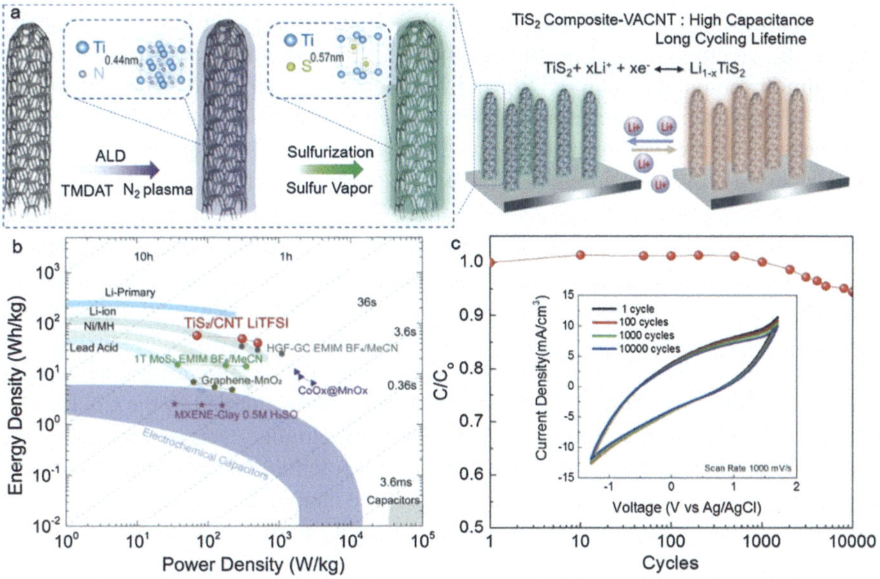

Fig. 4.42 Fabrication steps and advantages of the TiS_2–VACNTs hybrids as electrodes of pseudocapacitors. **a** Two-step process to synthesize the TiS_2–VACNT hybrid electrodes: (i) coating of TiN onto the VACNTs with atomic layer deposition and (ii) conversion of TiN to TiS_2 in sulfur environment. **b** Ragone plot of the state-of-the-art energy storage systems showing the TiS_2–VACNT hybrid to have the highest energy density. **c** Long-term cycling stability of the TiS_2–VACNT hybrid electrode in 21 m LiTFSI electrolyte. Reproduced with permission from [95]. Copyright © 2017, John Wiley & Sons

as demonstrated in Fig. 4.42c. The VACNT forest integrated the TiS_2 layers, which buffered the stress and deformation of the materials during the charging–discharging cycling processes, thus leading to long-term cycling stabilities.

An ultrasonic process was used to prepare Ti_3C_2/CNTs hybrids for the application of supercapacitors [97]. According to FTIR characterization results, the ultrasonic agitation did not damage the –O functional groups of the MXenes. The hybrid-based electrodes possessed pretty high volumetric capacitances and large capacitance retention up to scan rate of 100 mv/s, which were much better than those of the electrodes based on pristine Ti_3C_2 MXene nanosheets. Specifically, a volumetric capacitance of 393 F/cm^3 was achieved at 5 mv/s in KOH electrolyte, as the Ti_3C_2/CNT weight ratio optimally to be 2:1. After cycling for 10,000 times, almost no degradation was observed in the capacitance level. Similarly, the presence of the CNTs reduced the restacking of the Ti_3C_2 nanolayers, thus ensuring the sufficiently large interlayer spacing and high electrical conductivities.

Electrochemical properties of Ti_3C_2 MXene in 1 M solution of 1-ethly-3-methylimidazolium bis- (trifluoromethylsulfonyl)-imide (EMITFSI) in acetonitrile and two other general organic electrolytes were studied, by using MXene/CNTs hybrid electrodes [98]. The focus was on the effect of architectures and compositions on the electrochemical profiles of the electrodes of supercapacitors. Specific and volumetric capacitances of 85 F/g and 245 F/cm^3 were achieved at the scan rate of 2 mV/s, with both high rate capabilities and cycling stabilities. In-situ XRD results confirmed the intercalation of the large EMI^+ cations into the spacing of MXene nanolayers, which were responsible for the outstanding electrochemical properties.

Figure 4.43 shows CV curves of the three Ti_3C_2-based electrodes at different scan rates in 1 M EMITFSI in acetonitrile. They displayed similar electrochemical profiles, consisting of capacitive envelopes and redox peaks at about −0.2 V and −0.4 V versus Ag. However, their potential differences between the oxidation and the reduction peaks were not same, implying that they experienced different kinetics and Ohmic limitations. The CNT-$Ti_3C_2T_x$ based electrode had smaller difference, which could be attributed to the rapid diffusion of electrolyte ions enabled by the CNTs. The $Ti_3C_2T_x$ one had a similar electrochemical behavior a higher overpotential, whereas the d-$Ti_3C_2T_x$ electrode possessed the largest overpotential and was more resistive, which could be ascribed to the restacking of MXene nanolayers caused by the device assembly process. The potential range was relatively narrow, owing mainly to the

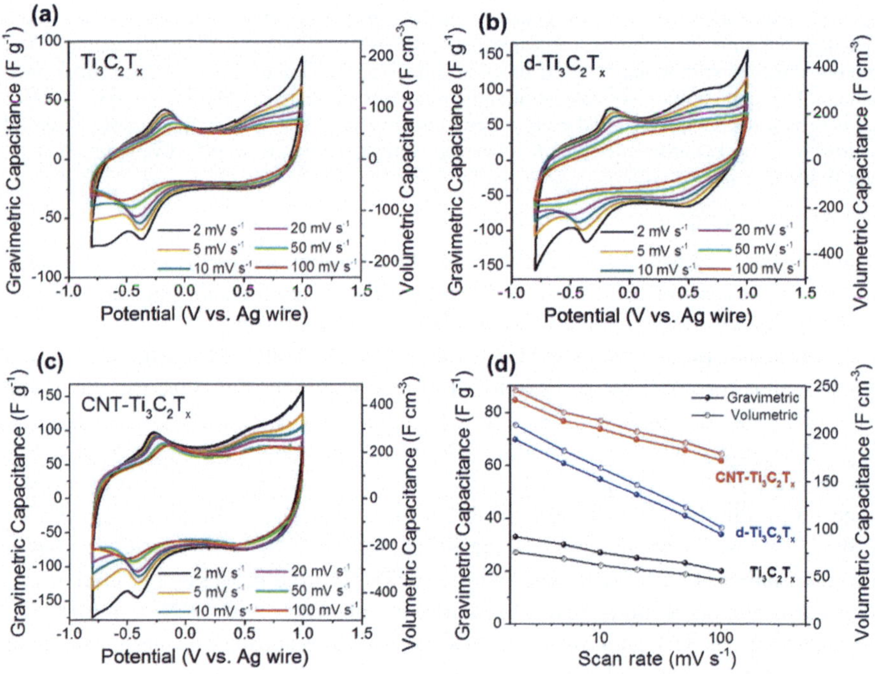

Fig. 4.43 CV curves of the samples in 1 M EMITFSI in acetonitrile electrolyte at different scan rates: **a** $Ti_3C_2T_x$, **b** d-$Ti_3C_2T_x$ and **c** CNT-$Ti_3C_2T_x$. **d** Specific capacitances of the samples as a function of the scan rate. All the data were recorded by using the three-electrode Swagelok cells. Reproduced with permission from [98]. Copyright © 2016, Elsevier

presence of water molecules in between the MXene nanolayers. Specific capacitance values of the samples derived from CV curves, as a function of scan rate, are illustrated in Fig. 4.43d.

MXene was explored to fabricate yarn-shaped supercapacitors (YSCs), which could be integrated with fabrics to develop energy storage devices for the applications in flexible, wearable and portable electronics [99]. The MXene-based YSCs were prepared by using a biscrolling method. The yarns made in this way were highly flexible, while the 2D MXene offered pseudocapacitive effect. The biscrolling MXene yarns with CNTs were note as BMX yarns. In the biscrolling method, MXene dispersion was drop-cast on stacked CNT sheets, which were then twisted to stack into yarns with ≈2000 turns/m of inserted twist, as shown in Fig. 4.44. The BMX yarns could have MXene loading of above 50 wt%. The amount of MXene in the yarns was controlled by tuning the concentration of the MXene dispersions. In this case, the MXene loading could be up to 98 wt%.

At the highest loading of MXene, the electrodes delivered record specific volumetric, aerial, gravimetric and linear capacitances of 1083 F/cm^3, 3188 mF/cm^2, 428 F/g and 118 mF/cm, respectively, at the current density of 2 mA/cm^2. The BMX yarns could be used assemble free-standing asymmetric YSC devices, by pairing with biscrolled RuO_2 yarns. Consequently, the devices exhibited maximum

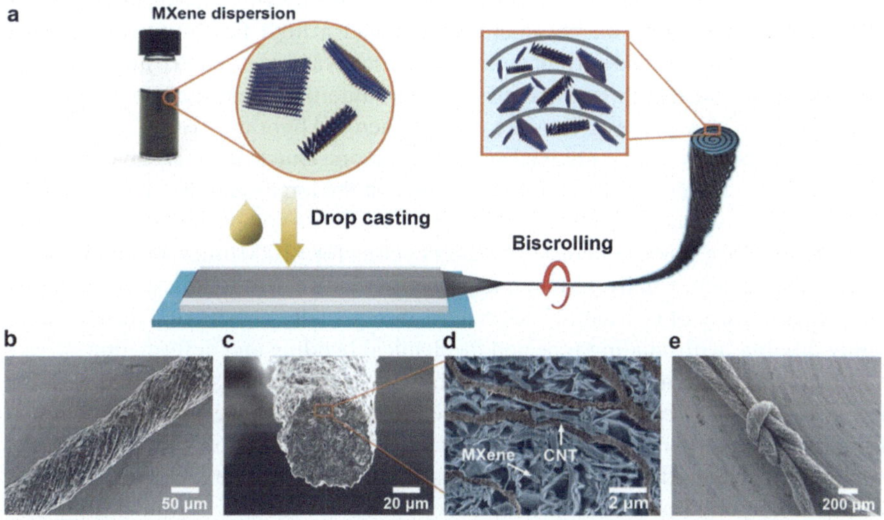

Fig. 4.44 **a** Schematic diagram of fabrication process of the BMX yarns. MXene dispersion was coated on CNT sheets and biscrolled to form a composite BMX yarn, in which the MXene was evenly distributed within the CNT frameworks. SEM images showing the **b** surface and **c** cross-section morphologies of the BMX yarn containing 97.4 wt% MXene. **d** High-magnification colored SEM image of the marked section in (**c**) with enhanced contrast to shown how the MXene sheets are trapped within the CNT yarn corridors. **e** SEM image the BMX yarn with 90 wt% MXene to be tied into a reef knot. Reproduced with permission from [99]. Copyright © 2018, John Wiley & Sons

energy and power densities of 61.6 mWh/cm^3 (168 μWh/cm^2 and 8.4 μWh/cm) and 5428 mW/cm^3 (14.8 mW/cm^2 and 741 μW/cm), respectively. The high capacitances could be well retained against repeated bending testing.

Reduced graphene oxide (rGO) has been incorporated with MXenes to form flexible conductive MXene/rGO hybrids as electrodes of supercapacitors [100]. The hybrids were obtained by using an electrostatic self-assembly process, with positively charged rGO modified with poly(diallyldimethylammonium chloride) and negatively charged $Ti_3C_2T_x$ nanoflakes. The rGO nanolayers were confined inside the spacings of the MXene nanosheets, so that the restacking of the MXenes was prohibited and interlayer spacings were enlarged. Accordingly, the diffusion of the electrolyte ions were significantly enhanced.

The flexible MXene/rGO hybrid electrode with the content of rGO to be 5 wt% displayed a volumetric capacitance of 1040 F/cm^3, at the scan rate of 2 mV/s. Also, the electrode demonstrated a sufficiently high rate capability, with the specific capacitance to be retained by about 61% at the scan rate of 1 V/s. Furthermore, the electrodes could be used to assemble binder-free symmetric supercapacitors, which delivered a volumetric energy density of 32.6 Wh/L. This value was a record among those of carbon and MXene based electrodes in aqueous electrolytes. The high performances made the MXene/rGO hybrids to be promising candidates for flexible energy storage device applications.

The same authors applied MXene/rGO hybrid to asymmetric micro-supercapators [101]. The MXene/rGO micro-supercapacitor had a high flexibility, without the requirement of binders and current collectors, which was constructed by using lithography technique on flexible transparent substrates. Interdigitated electrodes were used with a spray coating process. To make the MXene/rGO hybrids, $Ti_3C_2T_x$ nanolayers and rGO nanosheets were employed as the precursors. Due to their 2D layered structures, the hybrids were formed without sacrifice of the ion diffusion behaviour of the MXenes. The asymmetric device could be operated over a voltage window of 1 V. The initial capacitance was retained by 97% after cycling for >1000 times. Meanwhile, the micro-supercapacitor delivered an energy density of 8.6 mWh/cm^3, corresponding to a power density of 0.2 W/cm^3. Moreover, the specially constructed devices were highly flexible, withstanding repeated mechanical bending and deformation.

A facial process was reported to get flexible free-standing films consisting of modified MXene nanosheets and holey graphene [102]. The films were prepared through filtration of the dispersions containing alkalized MXene and holey graphene oxide, combined with a moderate heating process. In order to increase the relative content of Ti to enlarge the pseudocapacitive effect of MXene, surface terminal groups of –F/–OH were eliminated. In addition, because of their holey structures, the presence of the graphene nanosheets restrained the self-restacking of the MXene nanoflakes. As a result, a highly nanoporous connected network was formed, thus enabling rapid ion transport and shortening the transfer pathways of both the electrolyte ions and mobile electrons inside the materials. The hybrid-based electrode delivered a volumetric capacitance of as high as 1445 F/cm^3, at the scan rate of 2 mV/s, along with high rate stability and high mass loadings. Accordingly, a high volumetric energy density of 38.6 Wh/L would be readily achieved, which was a record at the time.

A ternary composite consisting of graphene-encapsulated Ti_2CT_x and polyaniline (GMP) was synthesized, which could be used to electrodes of supercapacitors [103]. The GMP based electrode demonstrated promising electrochemical performances, due to the incorporation of the mechanically robust and chemically inert graphene nanosheets. Also, the polyaniline intercalated into the multilayered MXenes was highly conductive, which had a positive effect on charge transport during the electrochemical reactions.

A high specific capacitance of 635 F/g was observed at the current density of 1 A/g, corresponding to a volumetric capacitance of 1143 F/cm^3, along with outstanding cycling stability of 97.5% retention up to 10,000 cycles. Additionally, the asymmetric pouch type supercapacitors, constructed with the GMP as the positive electrode and graphene as the negative electrode, delivered an energy density of 42.3 Wh/kg, at a power density of as high as 950 W/kg, together with promising cycling stability of 94.3% retention at the current density of 10 A/g after cycling for 10,000 times.

Nickel-aluminum layered double hydroxide (LDH) was combined with MXene to form MXene/LDH hybrids to be used as electrodes of supercapacitors, by using an in-situ growth method to deposit LDH on MXene nanosheets [104]. The LDH platelets were uniformly attached onto surface of the MXene nanosheets, leading 3D porous network structures that had increased number of active sites to facilitate electrolyte ions to access. Meanwhile, the network structures could buffer the volumetric variation of the materials related to charge/discharge processes. In addition, the high conductivity of the MXene matrix was well retained, ensuring the charge transport of the active materials. At optical conditions, the MXene/LDH hybrid based electrode displayed a specific capacitance of 1061 F/g, at the current density of 1 A/g, together with an outstanding cycling stability. The initial specific capacitance was retained by 70% after cycling for 4000 times, at the current density of 4 A/g, while a rate capability of 556 F/g to be achieved at the high current density of 10 A/g.

A simple two-step approach was conceived to prepare flexible coplanar asymmetric microsupercapacitors, by using screen-printing, with 2D $Ti_3C_2T_x$ nanolayers as negative electrode and Co-Al LDH nanosheets as positive electrode [105]. The microscale hybrid device (MHD) delivered higher energy density than the carbon-based device devices. It was fabricated in a coplanar form and all-solid-state, demonstrating an energy density of 8.84 μWh/cm^2, with a high cycling stability, characterized by a retention of 92% after cycling for 10,000 times. In addition, the MHD could serve as portable power source units, as well as force sensing resistors, thus allowing to design lightweight integrated force sensors to detect the variation of pressure.

FeOOH has a typical channel-type nanoporous structure and high theoretical specific capacitance, while its relatively low electrical conductivity restricts its real applications in supercapacitors. To address this issue, amorphous FeOOH was incorporated with 2D $Ti_3C_2T_x$ MXene nanosheets to form hybrid, by using a one-step method at room temperature, with $FeCl_3 \cdot 6H_2O$, NH_4HCO_3 and $Ti_3C_2T_x$ as the precursors [106]. It was found that the hybrid exhibited superb electrochemical properties, due to the synergistic effect of the highly conductive $Ti_3C_2T_x$ nanosheets and the strongly capactive amorphous FeOOH. The hybrid-based electrode delivered a specific capacitance of 217 F/g at the current density of 1 A/g, which was much higher than that of the amorphous FeOOH base one. The enhanced electrochemical performances of the hybrid-based electrode were originated from the open layered structures of the 2D MXene nanosheets, which were intercalated with the amorphous FeOOH nanosized particles, thus leading to enlarged electrode–electrolyte interface area, shortened ion diffusion pathways and enhanced electrical conductivity.

It is expected that combination of transitional oxides with MXenes could be an effective strategy to develop supercapacitors with promising electrochemical performances, given that both have strong capacitive effects. Representative examples are discussed as follows. For example, TiO_2 nanoparticles were employed decorate Ti_3C_2 MXene by using an in-situ hydrolysis combined with a post-thermal treatment, which can be used to construct high performance supercapacitors [107]. The TiO_2 nanoparticles had an average diameter of 30 nm, which were uniformly distributed on surfaces of the Ti_3C_2 MXene nanosheets. The hybrid-based electrode delivered a specific capacitance of 143 F/g, at the scan rate of 5 mV/s, which was higher than that the pristine Ti_3C_2 by about 50%. Additionally, the Ti_3C_2/TiO_2 hybrid electrode brought out outstanding cycling stability, whereas the initial capacitance was retained by 92% after cycling for 6000 times.

A simple in-situ wet-chemistry route was developed to fabricate flexible free-standing Ti_3C_2/MnO_x hybrid films, from Ti_3C_2 nanosheets and MnO_x nanoparticles, which demonstrated outstanding electrochemical performances as electrodes of supercapacitors [108]. The MnO_x nanoparticles were composed of Mn_2O_3 and MnO, which were uniformly grown on surfaces of the Ti_3C_2 nanosheets through an electrostatic interaction process. The hybrid film-based electrodes displayed a volumetric capacitance of 602 F/cm^3, at the scan rate of 2 mV/s. In addition, symmetric supercapacitors were assembled with the Ti_3C_2/MnO_x hybrid film electrodes, which demonstrated an energy storage capacity of 13.6 mWh/cm^3. A promising cycle stability with 89.8% retention of the initial capacitance was achieved after cycling up to 10,000 times.

A $MnO_2/Ti_3C_2T_x$ nanohybrid was synthesized by incorporating 1D MnO_2 nanoneedles with 2D $Ti_3C_2T_x$ MXene nanosheets, as electrodes of flexible supercapacitors [109]. According to XPS results, there was charge transfer from $Ti_3C_2T_x$ nanosheets to MnO_2 nanoneedles, in which the $Ti_3C_2T_x$ nanosheets acted as 2D conductive substrate to support the transport of electrons in the nanohybrids. Due to the coupling effect of $Ti_3C_2T_x$ and MnO_2, the electrochemical reactions were tremendously enhanced, leading to high electrical conductivity. As a consequence, the $MnO_2/Ti_3C_2T_x$ nanohybrid based electrodes exhibited outstanding electrochemical performances, including high specific capacitance, high rate stability and strong structural integration. The hybrid was used to construct symmetrical flexible supercapacitors, which delivered an energy density of 0.7 μWh/cm^2 at the power density of 80 μW/cm^2, with nearly 100% capacitance retention after 1000 cycles at 0.2 mA/cm^2.

Similar Ti_3C_2/MnO_2 hybrid nanostructures were prepared by using a simple hydrothermal method [110]. In this case, the 2D-Ti_3C_2 nanosheets were taken as a conductive that enhanced the electrical conductivity of the whole electrode materials and maintained structural integrity of the active MnO_2. Furthermore, due to the 2D layered nanostructures, the presence of Ti_3C_2 created more active sites on surface of the MnO_2 particles. Therefore, the Ti_3C_2/MnO_2 hybrid nanostructure-based electrode exhibited a high specific capacitance of 254 F/g, at the current density of 0.5 A/g, with a capacitance retention of 95.5% after cycling for 5000 times.

$Ti_3C_2T_x/\alpha\text{-}Fe_2O_3$ nanohybrids were synthesized by using a self-assembly process through the electrostatic attraction between negatively charged $Ti_3C_2T_x$ nanosheets sand positively charged $\alpha\text{-}Fe_2O_3$ nanosized particles at room temperature, in order to achieve high performance electrode materials of supercapacitors [111]. The hybrid-based electrodes exhibited promising electrochemical performances, with an operating potential of 1.2 V (from −1.2 to 0 V). In 5 M LiCl solution, specific capacitances of 405 F/g and 198 g/F were observed, at the current densities of 2 A/g and 20 A/g, respectively. In terms of cycling stability, 97.7% capacitance retention was achieved for the cycling up to 2000 times.

NiO nanosheets were incorporated with $Ti_3C_2T_x$ MXene nanolayers to form hybrid electrodes by using a hydrothermal reaction route [112]. The NiO nanosheets were deposited on the carbon-supported TiO_2 layers, which were formed out of the $Ti_3C_2T_x$ matrix, during the post-thermal treatment process. At optimal conditions, the hybrid electrode possessed capacities of 92 mAh/cm^3 and 54 mAh/cm^3, at the current densities of 1 A/g and 10 A/g, respectively. With the hybrid electrodes, an asymmetric supercapacitor (ASC) was assembled, with $Ti_3C_2T_x$/NiO as the positive electrode and $Ti_3C_2T_x$ MXene as the negative electrode, which delivered an even higher energy density of 1.04×10^{-2} Wh/cm^3, at a power density of 0.22 W/cm^3, together with a high cycling performance. The initial capacity was retained by 72% after 5000 cycles.

Similar to $Ti_3C_2T_x$, Ti_2CT_x was also utilized as the electrodes of supercapacitors [113]. In this study, the effects of the post-etching treatment conditions (Ar, N_2, N_2/H_2, and air) on structural and electrochemical properties of the MXene nanosheets were systematically evaluated. Air annealing resulted in changes in morphology, structure and electrochemical behaviors of the MXenes, while annealing in Ar, N_2 and N_2/H_2 had no effect. More interestingly, capacitive effect was largely enhanced after the MXenes were thermally treated in Ar, N_2 and N_2/H_2. The electrodes based on the MXene treated in N_2/H_2 had the highest specific capacitance of 51 F/g at the current density of 1 A/g, with a high rate stability of 86% retention. The enhanced capacitive behavior of the electrodes after being treated in Ar, N_2 and N_2/H_2 was readily ascribed to high level of carbon and the minimized content of fluorine, while 2D layered structures was well retained, ensuring the accessibility of the electrodes into the electrodes.

Ti_2CT_x MXene was used to fabricate wire-type supercapacitors (WSCs) [114]. The formation of –O and –F functional groups on surfaces of the MXene nanolayers was confirmed by XPS results. The WSCs exhibited outstanding capacitive properties, with a specific length capacitance of 3.09 mF/cm, corresponding to a gravimetric capacitance of 4.64 F/g. The specific energy densities were 210 nWh/cm and 315 mWh/kg, in length and gravimetric, respectively. The devices also displayed a superior cycling stability.

A mild strategy was reported to develop flexible $Ti_3C_2T_x$ MXene based electrodes of supercapacitors, by using self-assembly method with few-layered $Ti_3C_2T_x$ as the precursor, combined with alkalinization and thermal treatment [115]. The alkalinized and thermally treated $Ti_3C_2T_x$ films demonstrated a volumetric capacitance of as high as 1805 F/cm^3 at the current density of 1 A/g, with a high cycling stability characterized by a capacitance retention of 98% after cycling for 8000 times. The superb electrochemical performances of the flexible $Ti_3C_2T_x$ MXene based electrodes were ascribed to the increased number of redox-active sites, due to the elimination of the surface functional groups during the alkalinization and thermal treatment processes, as well as the enhanced electrical conductivity of the films because of the increased degree of crystallinity of the materials caused by the thermal treatment. A volumetric energy density of 45.2 Wh/L was achieved.

Ag nanoparticles were used to decorate $Ti_3C_2T_x$ nanosheets through the filtration method, from the dispersion mixture of $Ti_3C_2T_x$ nanosheets and Ag in water [116]. The $Ti_3C_2T_x$/Ag hybrid film had a high specific surface area of 107 m^2/g. The hybrid based electrode delivered a high areal capacitance of 332 mF/cm^2 at the scan rate of 2 mV/s, together with promising rate stability, i.e., 63.2% retention of the initial capacitance, as the scan rate was increased from 2 mV/s to 100 mV/s. The device also exhibited a long-term cycling stability, with 87% capacitance retention after 10,000 charging-discharging cycles. An even high areal capacitance of 1173 mF/cm^2 could be achieved by increasing the mass loading.

The hybrids could be used to construct asymmetric supercapacitors (ASC), with $Ti_3C_2T_x$/Ag as negative electrode and MnO_2/ESCNF as positive electrode, in 1 M Na_2SO_4 solution as electrolyte. The storage device offered an areal capacitance of 246.2 mF/cm^2, at the current density of 2 mA/cm^2, with capacitance retentions of 69.4% up to the current density of 20 mA/cm^2 and 82% after cycling for 10,000 tiems. Moreover, a maximum energy density of 121.4 μWh/cm^2 and a maximum power density of 17 395 μW/cm^2 were achieved. The stored energy could light commercial red light-emitting diode (LED), demonstrating the capability in practical applications.

Besides aqueous-based electrolytes, other electrolytes, such as organic electrolytes and ionic liquid electrolytes, have also been explored when using MXenes derived electrodes, because of the wider voltage windows and higher energy densities. For example, Lin et al. prepared $Ti_3C_2T_x$ MXene ionogel films by using vacuum filtration method and used them as supercapacitor electrodes with 1-ethyl-3-methylimidazolium bis(trifluoromethylsulfonyl)imide (EMI-TFSI) neat ionic liquid as the working electrolyte [117]. Since the $Ti_3C_2T_x$ hydrogel films possessed a disordered structure and stable spacings were present in the vacuum dried samples due to the immersion of the $Ti_3C_2T_x$ hydrogel films in the ionic liquid electrolyte. As a result, the surface of the $Ti_3C_2T_x$ MXene nanosheets readily allowed the EMI$^+$ and TFSI$^-$ ions to access. In such neat EMI-TFSI electrolytes, the MXenes based electrode delivered a high capacitance of 70 F/g, with a wide voltage window of 3 V, at the scan rate of 20 mV/s.

Wang et al. demonstrated that the pseudocapacitive charge storage capacity of Ti_3C_2 MXene based electrode could be tremendously enhanced by simply varying the solvent for the electrolyte systems [118]. Based on the experimental results on the charge storage behaviors of the Ti_3C_2 in Li-containing electrolytes, in various solvents, including nitrile-, carbonate- and sulfoxide-based ones, it was found that the charge storage capacity when using carbonate solvent was higher than those using other solvents by about two times. The authors claimed that the chemical properties of the electrolyte solvents could influence on the arrangement of molecules/ions in the Ti_3C_2 MXene, so as to determine the total charges to be stored in the materials. Because Li^+ ions could be almost entirely desolvated in Ti_3C_2 when using carbonate-based electrolyte, thus leading to high volumetric capacitances at high charging/discharging rates. Therefore, all aspects of an electrochemical system should be taken into account for practical applications.

Li et al. developed a redox-controlled A-site etching process to etch MAX phases in Lewis acidic melts [119]. They demonstrated the synthesis of various MXenes from the unconventional MAX-phase precursors, with A site to be occupied by Si, Zn and Ga. The Ti_3C_2 MXene based anode exhibited a high lithium ion storage capacity of 738 C/g, corresponding to a value of 205 mAh/g in 1 M LiPF6 carbonate-based electrolyte. The electrode had high charge–discharge rate capacity. Pseudocapacitive-like electrochemical nature was observed in the MXenes, which could be potential candidates as negative electrodes for high-rate electrochemical energy storage applications.

With the rapid development of next-generation portable and wearable flexible electronic devices, which are expected to have miniaturized dimension and enhanced performances, microsupercapacitors (MSCs) have been considered to be an important type of energy storage systems in this field. Usually, MSCs are fabricated with interdigital electrodes on planar substrates by using lithography technology. Owing to their special structure and electrochemical properties, MXenes and MXene-based composites have also been employed to fabricate MSCs devices.

For instance, Li et al. proposed a simple, fast and almost zero-cost "scratch" method to pattern few-layered $Ti_3C_2T_x$ nanoflakes on various substrates, in order to construct high performance MSCs [120]. The all-solid-state MSCs based on the patterned $Ti_3C_2T_x$ MXene could deliver a high areal capacitance of 25.5 mF/cm^2. The high performance was attributed to the unique layered structure and high electrical conductivity of the MXene-based electrode. The planar MSCs also exhibited promising cycling stability, together with strong mechanical flexibility. Furthermore, the process could be highly applicable to MXenes with other compositions for the design of MSCs.

MSCs with 3D structure could have enhanced energy storage capabilities, while 3D structured electrodes are usually suffered from damage due to mechanical deformation in real applications. To address this problem, Yue et al. developed a self-healable 3D MSC, with MXene ($Ti_3C_2T_x$)–graphene (rGO) composite aerogels as the electrodes. The $Ti_3C_2T_x$–rGO composite aerogels were encapsulated with a self-healing polyurethane in the form of outer shells [121]. In these MXene–rGO composite aerogels, the synergistic effect of large specific surface area offered by the rGO and high conductivity related to the MXene suppressed the self-restacking of the lamella structures of the two components and minimized the oxidization of MXene. The MSCs exhibited an area specific capacitance of 34.6 mF/cm^2, at the scan rate of 1 mV/s. The micro-devices also possessed an excellent cycling stability, with more than 90% capacitance retention after cycling for 15,000 times. In addition, the 3D MSCs displayed a strong self-healing capability, with about 82% specific capacitance retention after five round of healing experiment.

Similarly, Couly et al. reported an asymmetric MXene-based MSC, with mechanical flexibility but without binder and current-collector [122]. The interdigitated structure was developed using a scalable spray deposition method onto a flexible transparent substrate through a custom-made mask. The electrode consisted of $Ti_3C_2T_x$ MXene and rGO. This MXene-based asymmetric MSC device had a voltage window of 1 V. The initial capacitance was retained by 97% after cycling for 10,000 times. The energy density was 8.6 mW h/cm^3 while the power density was 0.2 W/cm^3. Furthermore, the MSCs were highly flexible with strong resistance to mechanical deformation. This achievement offered a useful approach to design MSCs with both high energy and high power densities.

Representative photograph of the as-assembled 14-finger MSC device is shown in Figure 4.45a. Different electrode materials could be identified according to the different colors, as seen in Fig. 4.45b. Because the density of $Ti_3C_2T_x$ MXene is higher than that of rGO, $Ti_3C_2T_x$ fingers are brighter than the rGO ones. Moreover, well-patterned electrodes had a uniform spacing, confirming effectiveness of the fabrication route. Figure 4.45d depicts top-view SEM image of the 14-finger MSC, from which the interfinger spacing and the width of the fingers could be measured.

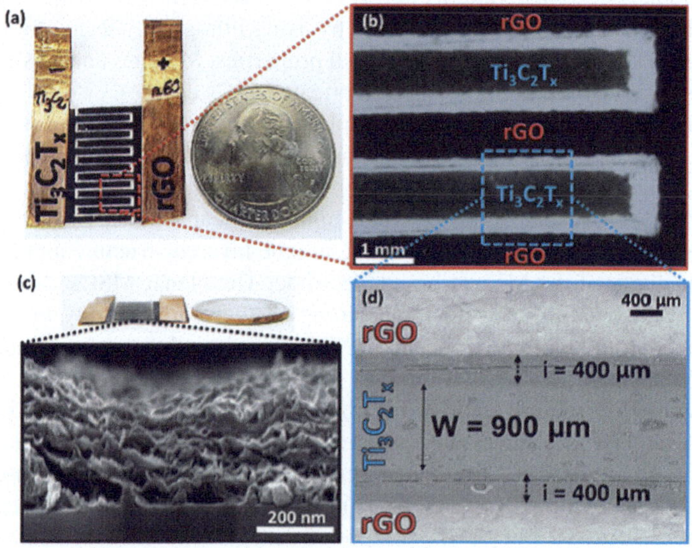

Fig. 4.45 **a** Photograph of top-view of the asymmetric MSC device together with a quarter US dollar coin. **b** zoomed-in image of $Ti_3C_2T_x$ and rGO interdigitated electrode configurations and the interspace. **c** Side view of the device with a US quarter dollar coin and cross-sectional SEM image of the $Ti_3C_2T_x$ electrode. **d** Top-view SEM image of the interdigitated fingers, showing one $Ti_3C_2T_x$ electrode between two rGO electrodes. Reproduced with permission from [122]. Copyright © 2018, John Wiley & Sons

The interfinger spacing was about 400 μm whereas the finger width was about 900 μm. The thickness of the layers was about 300 nm, as illustrated in Fig. 4.45c. Because the electrodes were sufficiently thin, they also served as current collectors, without the requirement of additional current collectors either on top or at bottom.

In a separate study, Chen et al. developed a MSC with free-standing MXene-molybdenum disulfide (MoS_2) films, by using a laser etching technology [123]. By incorporating MoS_2 into MXene, the electrochemical performances could be highly enhanced, with the specific capacitance to be increased by about 60%. Specifically, the MSC exhibited a maximum specific capacitance of 173.6 F/cm^3 at 1 mV/s, estimated with the total volume of positive and negative electrodes. Meanwhile, the device possessed maximum energy density and power density of 15.5 mWh/cm^3 and 0.97 W/cm^3, respectively. Additionally, the MSC device displayed excellent cycling stability and strong mechanical flexibility. After 6000 charge–discharge cycles and bending to 150°, the MSC still had 98% and 89% capacitance retentions, respectively. Th laser-etched MXene-MoS_2 composite based MSC could be a potential candidate as high-performance micro-scale energy storage systems.

More recently, Wu et al. reported an in-situ synthetic process, in which $Ti_3C_2T_x$ MXenes were capped with sodium ascorbate (SA) to for homogeneous dispersion of SA-MXene [124]. The SA-MXene dispersion was highly resistant against oxidation and stable in ambient environment for more than two months. Due to the special in-situ synthesis process, the interlayer spacing of the SA-MXene nanosheets was enlarged, thus leading to enhanced energy storage capability, while the electrical conductivity was not compromised. The SA-MXene could be made into printable ink with Triton X-100 and propylene glycol as modifiers. The ink could be used to print interdigitated electrodes for MSCs by using an inkjet printer. The solid-state MSC device worked without the requirement of additional current collectors, showing areal and volumetric capacitances of 108.1 mF/cm^2 and 720.7 F/cm^3, respectively. The development of ligand-capped stable MXenes served as an example of water-based inks for printing micro-electronic and supercapacitor devices.

4.2.3 Water Splitting

An ab initio calculation method has been employed to study 48 2D MXenes to reveal their photocatalytic properties for water splitting to produce H_2 [125]. Among the MXenes, 2D Zr_2CO_2 and Hf_2CO_2 were highly potential as single photocatalysts, with possibly high efficiency for photocatalytic water splitting. It was found that Zr_2CO_2 and Hf_2CO_2 had a high anisotropic carrier mobility, which could be favourite to the migration and separation of photogenerated electron–hole pairs. At the same time, they also showed high optical absorption coefficient over the wavelength range of 300–500 nm. According to the ab initio molecular dynamics simulations, Zr_2CO_2 and Hf_2CO_2 were especially stable in water. The adsorption and decomposition of H_2O molecules on the surface of 2D Zr_2CO_2 were evaluated, while the formation process of H_2 was clarified, thus elucidating the micro-mechanisms of the photocatalytic H_2 production on MXenes.

To evaluate the electrocatalystic effect of MXenes for the hydrogen evolution reaction (HER), computational screening study on 2D M_2XT_x [126]. For both Mo_2CT_x and Ti_2CT_x MXenes, the experimental results were in a good agreement with the theoretical predictions, whereas Mo_2CT_x had a much higher HER activity than Ti_2CT_x. According to the theoretical predictions, the basal planes of Mo_2CT_x were of catalytic activity for HER, while MoS_2 had active sites only at edges of the 2H phase.

Theoretical studies on transition-metal carbonitrides M_3CN MXene were conducted by using density functional theory (DFT) calculations for potential applications in water splitting [127]. Pristine, surface-terminated and transitional metal (TM)-modified M_3CNO_2 were comprehensively investigated in terms of structural configurations, H-adsorption free energy (ΔG_H) and charge transfer. It was revealed that all pristine transitional metal carbonitrides showed promising interactions between H atoms and the photocatalysts. However, only Ti_3CNO_2 and Nb_3CNO_2 could be used as HER active catalysts according to the calculation results of ΔG_H. In addition, through incorporation of –O and –OH mixed functional groups and modification with TMs on the surface of Ti_3CNO_2, the HER activity could be significantly enhanced. This work further confirmed that carbonitride MXenes could be potential candidates as noble-metal-free co-catalysts for highly efficient and low-cost water splitting applications.

A capillary-forced assembling method was developed to make MXenes into hierarchical 3D architectures with high resistance to aggregation of the nanosheets [128]. Due to the aggregate-resistant properties of the 3D MXene, the surface area was not lost and the kinetics-favorable framework was maintained. Meanwhile, the high robustness and outstanding processability were retained in solutions and solid state. Such 3D MXene nanostructures allowed incorporation of other active components to realize specific functionalities. In this study, the CoP-3D MXene hybrids were synthesized, which exhibited high electrocatalytic activity for evolutions of O_2 and H_2 in alkaline electrolyte. Their cell voltage and durability were higher than those of the combined RuO_2/Pt catalysts for overall water splitting in basic solution.

MXenes have been combined with rutile TiO_2 as co-catalysts for visible-light-induced solar H_2 production through water splitting [129]. The optimal content of $Ti_3C_2T_x$ was 5 wt% in the $TiO_2/Ti_3C_2T_x$ hybrids, which showed an enhancement in the photocatalytic H_2 evolution reaction by 400%, as compared with pure rutile TiO_2. Figure 4.46 shows water splitting performances of the samples with different contents of MXene. This approach could be readily extended to other MXenes, such as Nb_2CT_x and Ti_2CT_x, as co-catalysts with TiO_2.

The photocatalytic hydrogen evolution reaction was conducted with an air-tight quartz cell under the irradiation of visible-light at $\lambda > 400$ nm. Figure 4.45 shows hydrogen production rates of different samples. The pure $Ti_3C_2T_x$ sample had a very low hydrogen production rate of about 1 μmol/h $g_{catalyst}$. Because $Ti_3C_2T_x$ was of metallic characteristics, the hydrogen production behavior was only attributed to the TiO_x items on the surface. According to O 1 s XPS spectrum of the $Ti_3C_2T_x$ sample, a clear peak was present, which was related to the Ti–O structures, confirming the

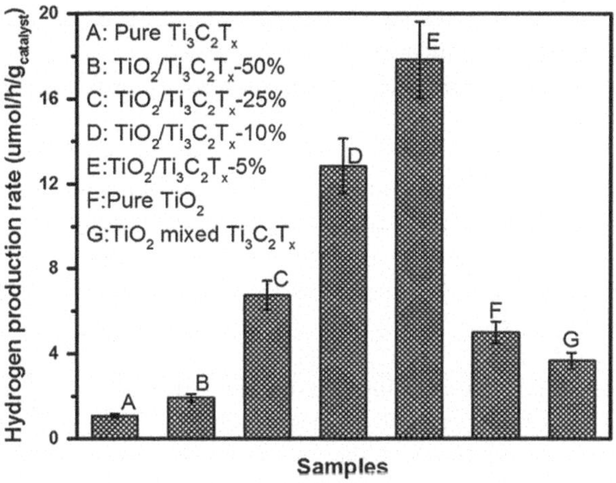

Fig. 4.46 Photocatalytic H_2 production rates of pure TiO_2, $TiO_2/Ti_3C_2T_x$ (5 wt%), $TiO_2/Ti_3C_2T_x$ (10 wt%), $TiO_2/Ti_3C_2T_x$ (25 wt%), $TiO_2/Ti_3C_2T_x$ (50 wt%), pure $Ti_3C_2T_x$ and a physically mixed $TiO_2/Ti_3C_2T_x$ sample with 5 wt% $Ti_3C_2T_x$ loading. Reproduced with permission from [129]. Copyright © 2016, John Wiley & Sons

presence of TiO_x items in the sample. Once the sample was irradiated with visible light, excited $[Ti^{3+}-O^-]^*$ items were generated, thus triggering the photoredox reaction to produce H_2. Since the number of the TiO_x items on the surface was relatively small, pure $Ti_3C_2T_x$ exhibited a weak hydrogen production activity.

The $TiO_2/Ti_3C_2T_x$ sample with 5 wt% $Ti_3C_2T_x$ offered a hydrogen evolution rate of 17.8 μmol/h $g_{catalyst}$, which the highest among all the samples. As the concentration of TiO_2 in the samples was increased, the amount of visible light captured was increased, thus generating photoexcited electron–hole pairs. In other word, the more the TiO_2 contained in the sample, the more the photoinduced charge carriers would be. Methanol and $Ti_3C_2T_x$ acted as the hole-scavenger and electron sink, respectively, which separate the charges and hence raised the solar hydrogen production rate. A high content of $Ti_3C_2T_x$ implied a low content of TiO_2.

Additionally, the physically mixed sample had an even lower hydrogen production activity than that pure TiO_2. This was simply because the presence of $Ti_3C_2T_x$ negatively affected the light absorption of TiO_2. Also, the loose contact between TiO_2 and $Ti_3C_2T_x$ could not effectively facilitate an efficient charge transfer from TiO_2 to $Ti_3C_2T_x$. However, in the chemically connected $TiO_2/Ti_3C_2T_x$ hybrids, the $Ti_3C_2T_x$ nanosheets were covered by the TiO_2 nanoparticles. The chemical interaction ensured sufficient light absorption of TiO_2 and efficient charge transfer from TiO_2 to $Ti_3C_2T_x$. Therefore, a close contact between TiO_2 and $Ti_3C_2T_x$ was crucial to support the co-catalytic effect.

An in-situ process was developed to prepare Ti_3C_2–TiO_2 nanoflowers through simultaneous oxidation and alkalization, combined with ion exchange and calcination of Ti_3C_2 MXene, which were employed as photocatalysts to produce H_2 and O_2 through water splitting [130]. It was found that calcination temperature had a strong effect on photocatalytic behaviors of the Ti_3C_2–TiO_2 hybrid nanoflowers. The optimal calcination temperature was 500 °C, at which the Ti_3C_2–TiO_2 nanoflowers had much higher photocatalytic effect than pure TiO_2. As the sacrificial reagent was removed, overall water splitting was achieved with the Ti_3C_2–TiO_2 nanoflowers, without the application of noble metals.

Moreover, the Ti_3C_2–TiO_2 nanoflowers were highly stable and reproducible. The outstanding catalytic performances were resulted from the special 3D porous structures, which offered increased number of reactive sites, enhanced reflection and scattering of light, and shortened diffusion length of photogenerated holes and electrons. In addition, the close contact between the Ti_3C_2 MXene nanosheets and the TiO_2 particles led to synergetic effect and Schottky junction, which was favorite to the charge separations and effectively prevented the recombination. As a result, more electrons would participate in the photoreduction of H_2 evolution, while more holes would participate the photooxidation of O_2 evolution.

Ternary NiFeCo-layered double hydroxide (LDH) nanoplates were incorporated with $Ti_3C_2T_x$ through an in-situ hybridization by using a simple solvothermal reaction method, in which the $Ti_3C_2T_x$ phase was controllably oxidized to anatase TiO_2 nanoparticles [131]. The $Ti_3C_2T_x$/TiO_2/NiFeCo-LDH nanohybrids (TTL) exhibited high performance oxygen revolution reaction (OER), when compared with the pristine NiFeCo-LDH and commercial IrO_2 catalysts. A current density of 10 mA/cm^2 was observed, at the potential of 1.55 V versus a reversible H_2 electrode (vs. RHE) in 0.1 M KOH. Moreover, the hybrid was combined with a standard $BiVO_4$ film to form TTL/$BiVO_4$ photoanode, thus leading to greatly enhanced photocurrent density of 2.25 mA/cm^2 at 1.23 V versus RHE under illumination of 100 mW/cm^2. The outstanding PEC-OER behaviour could be ascribed to the presence of the TiO_2 nanoparticles, which broadened the light adsorption range and hence promoted the generation of electron/hole pairs. Meanwhile, the ternary LDH nanoplates served as hole scavengers, while the metallic $Ti_3C_2T_x$ nanosheets ensured smooth transport of electrons and/or ions.

Harnessing the transfer and flow of photogenerated electrons and holes, in order to prolong the life-times of charge carriers and increase the activity of semiconductor photocatalysts, is always a challenge [132]. For this purpose, the photocatalysts are usually incorporated with proper co-catalysts that have suitable electronic properties. MXenes could serve as a hole mediator to increase the photocatalytic activity of TiO_2. TiO_2 sheets with exposed (001) surfaces were grown on $Ti_3C_2T_x$ nanosheets, followed by the photodeposition of Cu_2O nanodots on TiO_2. The $Ti_3C_2T_x$ MXene showed a low work function of $\phi = 3.4$ eV. Therefore, the photogenerated electrons on the TiO_2 sheets hybridized with $Ti_3C_2T_x$ were accumulated and tunneled to Cu_2O, so that Cu was formed to serve as a co-catalyst. The ternary Cu/TiO_2@$Ti_3C_2T_x$ photocatalysts exhibited strong effect of water splitting, producing H_2 at a production rate of 860 μmol/g h.

Density functional theory calculations were conducted to enable atomic-level exploration, design and fabrication of Ti_3C_2 nanoparticles, as highly efficient co-catalysts for water splitting [133]. The Ti_3C_2 nanoparticles were hybridized with cadmium sulfide (CdS) by using a hydrothermal reaction, which led to superb high visible-light photocatalytic hydrogen production activity of 14 342 mmol/h g, with an apparent quantum efficiency of 40% at the wavelength of 420 nm. The high performance was attributed to the favourable Fermi level position, electrical conductivity and hydrogen evolution capacity of the Ti_3C_2 nanoparticles. Moreover, the Ti_3C_2 nanoparticles also served as an efficient co-catalyst with ZnS or $Zn_xCd_{1-x}S$. Therefore, MXene materials are potential candidates as high performance and low cost photocatalysts or photoelectrodes.

Photocatalysts based on 2D MXenes (Ti_2C) and graphitic carbon nitride (g-C_3N_4 were developed to promote water splitting [134]. It was observed that the enhanced water splitting effect was readily ascribed to the synergistic interaction between the 2D Ti_2C and the g-C_3N_4. Optimal content of Ti_2C was 0.4 wt%, with a hydrogen production rate of 47.5 μmol/h, which was higher than that of pure g-C_3N_4 by about 13 times and slightly better than that of Pt-loaded g-C_3N_4. Also, the Ti_2C/g-C_3N_4 hybrid demonstrated high stability and outstanding reproducibility.

0D–2D nanohybrids based on a nickel-based bimetal phosphorus trisulfide ($Ni_{1-x}Fe_xPS_3$) nanomosaic particles deposited on the surface of MXene nanosheets (NFPS-MXene) were prepared water splitting [135]. The nanohybrids were synthesized by using a simple self-assemble method, with transition metal layered double hydroxide (TMLDH) to be grown on the surface of the MXenes, combined with an in-situ solid-state reaction process at relatively low temperatures. With an optimal ratio of Ni:Fe, the NFPS-MXene nanohybrids displayed outstanding electrocatalytic activities for overall water splitting. Specifically, the $Ni_{0.7}Fe_{0.3}PS_3$-MXene nanohybrid had a low overpotential of 282 mV and a Tafel slope of 36.5 mV/dec, for oxygen evolution reaction (OER) in the solution of 1 m KOH. At the same time, the $Ni_{0.9}Fe_{0.1}PS_3$-MXene sample possessed a low overpotential of 196 mV for the hydrogen evolution reaction (HER) in 1 m KOH. As the $Ni_{0.7}Fe_{0.3}PS_3$-MXene and the $Ni_{0.9}Fe_{0.1}PS_3$-MXene were coupled for overall water splitting, a relatively low onset potential of 1.42 V was observed, while a high current density of 10 mA/cm^2 could be readily approached at 1.65 V. The performance was high than the all noble metallic IrO_2||Pt/C electrocatalyst.

Figure 4.47a shows HER polarization curves of the $Ti_3C_2T_x$ nanosheets, $NiPS_3$-MXene, $Ni_{0.9}Fe_{0.1}PS_3$-MXene and the bare $Ni_{0.9}Fe_{0.1}PS_3$ samples. HER polarization curves of the NFPS-MXene nanohybrids with different ratios of Ni:Fe and the NFPS compounds without MXene were also recorded. The $Ni_{0.9}Fe_{0.1}PS_3$-MXene nanohybrids possessed an overpotential of 196 mV at the current density of 10 mA/cm^2, which was lower than those of the $NiPS_3$-MXene nanohybrids (364 mV) and bare $Ni_{0.9}Fe_{0.1}PS_3$ (318 mV). The pristine $Ti_3C_2T_x$ nanosheets displayed almost no HER activity. As the content of Fe was increased, the overpotential was gradually increased, with the values of the $Ni_{0.8}Fe_{0.2}PS_3$-MXene, $Ni_{0.7}Fe_{0.3}PS_3$-MXene, and $Ni_{0.6}Fe_{0.4}PS_3$-MXene to 297 mV, 359 mV and 475 mV, respectively.

Tafel slopes of the $Ni_{0.9}Fe_{0.1}PS_3$-MXene, $Ni_{0.8}Fe_{0.2}PS_3$-MXene, $Ni_{0.7}Fe_{0.3}PS_3$-MXene and $Ni_{0.6}Fe_{0.4}PS_3$-MXene were 114, 137, 140, and 173 mV per decade, respectively. Without the presence of MXenes, the bare $Ni_{0.9}Fe_{0.1}PS_3$, $Ni_{0.8}Fe_{0.2}PS_3$ and $Ni_{0.7}Fe_{0.3}PS_3$ samples demonstrated Tafel slopes of 142, 148, and 153 mV per decade, respectively, which were higher than those of their nanohybrid counterparts. Among all the samples, the $Ni_{0.9}Fe_{0.1}PS_3$-MXene exhibited the highest HER activity. Chronoamperometric curve of the $Ni_{0.9}Fe_{0.1}PS_3$-MXene nanohybrid for HER was measured at the constant overpotential of 196 mV for 50 h. The current density

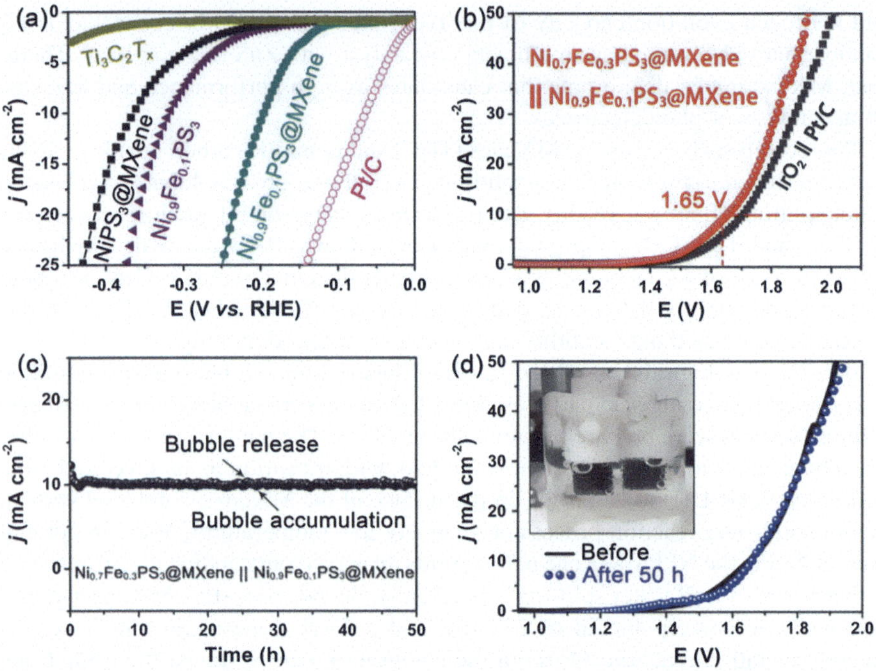

Fig. 4.47 **a** HER polarization curves of the $NiPS_3$-MXene, $Ni_{0.9}Fe_{0.1}PS_3$-MXene, $Ni_{0.9}Fe_{0.1}PS_3$ and $Ti_3C_2T_x$ in 1 M KOH at the scan rate of 5 mV/s. **b** LSV curves of the $Ni_{0.7}Fe_{0.3}PS_3$-MXene || $Ni_{0.9}Fe_{0.1}PS_3$-MXene and the all noble metallic IrO_2 || Pt/C electrocatalysts in 1 m KOH at the scan rate of 5 mV/s. **c** Chronoamperometric curves of the $Ni_{0.7}Fe_{0.3}PS_3$-MXene || $Ni_{0.9}Fe_{0.1}PS_3$-MXene couple at the static voltage of 1.65 V for 50 h. **d** LSV curves of the $Ni_{0.7}Fe_{0.3}PS_3$-MXene || $Ni_{0.9}Fe_{0.1}PS_3$-MXene couple before and after the testing for 50 h. The inset shows photographs of the water splitting at the applied potential of 1.65 V. Reproduced with permission from [135]. Copyright © 2018, John Wiley & Sons

was nearly unchanged during the testing of 50 h, while the LSV curves before and after the HER test were almost the same, suggesting that the HER activity of the $Ni_{0.9}Fe_{0.1}PS_3$-MXene nanohybrid was highly stable.

To verify that the currents were generated from water splitting, the theoretical and experimental H_2 productions of the $Ni_{0.9}Fe_{0.1}PS_3$-MXene electrode, at the constant current of −10 mA, were compared. The experimental production of H_2 was evaluated by using a gas chromatography, demonstrating Faradaic efficiency of about 100% as the electrolysis reaction was lasted for 1 h. In addition, the $Ni_{0.7}Fe_{0.3}PS_3$-MXene electrode also had a nearly 100% Faradaic efficiency at the constant current of 10 mA, for O_2 evolution for 1 h.

Accordingly, overall water splitting was examined by using $Ni_{0.7}Fe_{0.3}PS_3$-MXene || $Ni_{0.9}Fe_{0.1}PS_3$-MXene as the representative electrocatalyst couple. As seen in Fig. 4.47b, the $Ni_{0.7}Fe_{0.3}PS_3$-MXene || $Ni_{0.9}Fe_{0.1}PS_3$-MXene couple displayed a relatively low onset potential of 1.42 V. The current density could approach 10 mA/cm^2

as the potential was 1.65 V. The performance was much higher than that of the all-noble-metal IrO_2 || Pt/C electrocatalysts (1.71 mV@10 mA/cm^2) and comparable with the literature values of advanced electrocatalysts. The $Ni_{0.7}Fe_{0.3}PS_3$@MXene || $Ni_{0.9}Fe_{0.1}PS_3$@MXene couple was also evaluated at a constant voltage of 1.65 V for water splitting reaction over 50 h, as shown in Fig. 4.47c. Clearly, the chronoamperometric curve possessed a very stable current density during the 50 h testing process.

Figure 4.47d shows LSV curves before and after 50 h overall water splitting reaction testing. The two curves were highly overlapped, implying that the $Ni_{0.7}Fe_{0.3}PS_3$-MXene || $Ni_{0.9}Fe_{0.1}PS_3$-MXene couple had a sufficiently high stability. Large number of bubbles could be observed on surface of the electrodes, as illustrated as inset of Fig. 4.47d. Also, peaks of current density surge were present in the serrate-like chronoamperometric curve, as revealed in Fig. 4.47c, which was closely related to the alternate bubble accumulation and production of H_2 and O_2 on surface of the electrode during the reaction. The decrease and recover in the current density caused by the bubble accumulation confirmed that the $Ni_{0.7}Fe_{0.3}PS_3$-MXene || $Ni_{0.9}Fe_{0.1}PS_3$-MXene couple had high activity and stability for overall water splitting reaction.

A one-step CO_2 oxidation method was reported to synthesize niobium pentoxide/carbon/niobium carbide (MXene) hybrids (Nb_2O_5/C/Nb_2C), with high performances as photocatalysts for hydrogen evolution through water splitting [136]. In the Nb_2O_5/C/Nb_2C hybrids, the Nb_2O_5 nanoparticles were deposited uniformly on surface of the Nb_2C nanosheets, while amorphous carbon was also produced during the reaction. The optimal reaction time was 1 h, which led to Nb_2O_5/C/Nb_2C hybrid sample that had the highest hydrogen generation rate of 7.81 μmol/h g_{cat}. The performances were much superior to that of pure Nb_2O_5 by about 4 times. The enhanced performance of the Nb_2O_5/C/Nb_2C hybrids was ascribed to close connection between the Nb_2O_5 nanoparticles and the highly conductive Nb_2C nanosheets, as well as the effective separation of the photogenerated charge carriers at the Nb_2O_5/Nb_2C interfaces. Figure 4.48 shows photocatalytic mechanism of the Nb_2O_5/C/Nb_2C ternary hybrid.

4.2.4 Hydrogen Storage

$LiBH_4$ has been shown to have strong potential as materials for solid-state hydrogen storage applications. It was demonstrated that, if $LiBH_4$ was encapsulated in 2D Ti_3C_2 MXene, its hydrogen storage capability could be enhanced [137]. A simple impregnation process was used to realize the capsulation. As a result, the initial desorption temperature of $LiBH_4$ was greatly decreased, while the dehydrogenation/rehydrogenation kinetics were largely promoted. The initial desorption temperature of the $LiBH_4$-$2Ti_3C_2$ hybrid was reduced to 172.6 °C, whereas 9.6 wt% hydrogen was released at 380 °C in 1 h. Comparatively, the bare $LiBH_4$ released only 3.2 wt%

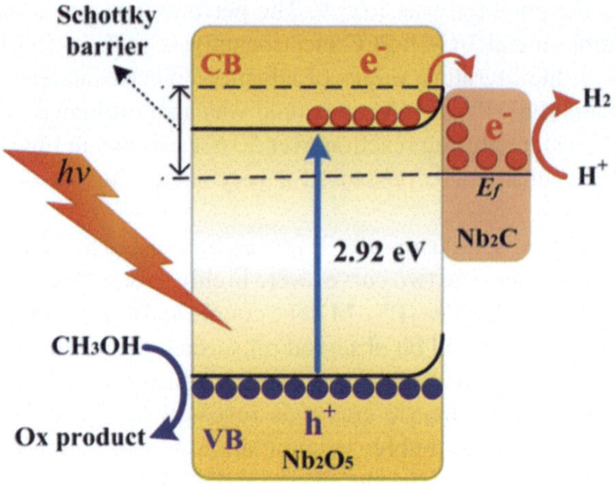

Fig. 4.48 Photocatalytic mechanism of the $Nb_2O_5/C/Nb_2C$ hybrids. Red and blue spheres represent the photo-induced electrons and holes, respectively, with E_f = Fermi level and Ox product = oxidation product. Reproduced with permission from [136]. Copyright © 2017, John Wiley & Sons

hydrogen. Moreover, after dehydrogenation, the samples could be partially rehydrogenated at 300 °C at the H_2 pressure of 95 bar. Due to the special layered structure of the Ti_3C_2 MXene, the grain growth and particle agglomeration of the $LiBH_4$ phase could be effectively prevented. At the same time, the $LiBH_4$ phase was stabilized by the Ti_3C_2 MXene, because of the confinement effect.

Figure 4.49a shows desorption kinetics of the LBH-2Ti_3C_2 hybrid sample, with isothermal desorption experiments to be recorded at different holding temperatures. The LBH-2Ti_3C_2 sample released 8.2, 11.3 and 12.6 wt% H_2 at 300, 350 and 380 °C, respectively. In comparison, the total amounts of hydrogen desorption of bare $LiBH_4$ were 0.9, 2.9 and 5 wt%, respectively. The enhanced desorption kinetics could be ascribed to the decrease in size of $LiBH_4$ particles, because they were encapsulated inside the layered nanosheets. FTIR spectra of the samples are shown in Fig. 4.49b. The intensities of the B–H bonds were decreased with increasing temperature. After heating at 380 °C, signals of the B–H bonds were nearly absent, owing to the dehydrogenation process of $LiBH_4$.

Reversibility of the LBH-2Ti_3C_2 hybrid was examined by checking the dehydrogenation/rehydrogenation process with three cycles, as illustrated in Fig. 4.50. The LBH-2Ti_3C_2 hybrid released 10.6 wt% H_2 at the first dehydrogenation cycle at 350 °C. Rehydrogenation of the hybrid was conducted at 300 °C at 95 bar H_2 for 24 h. After that, about 6.5 wt% H_2 was released at 350 °C in 6 h at the second dehydrogenation cycle, while only 5.5 wt% H_2 was obtained after the three cycles. Both the rehydrogenation pressure and temperature of the LBH-2Ti_3C_2 hybrid are much lower than those of the bare $LiBH_4$, which were 35 MPa and 600 °C).

Adsorption, dissociation and diffusion behaviours of H_2 molecules on 2D Ti_2C monolayers with and without carbon vacancies, have been studied by using first-principles calculations, based on density functional theory [138]. As the H_2 molecules were adsorbed on the 2D Ti_2C monolayers with an end-on configuration, the adsorption was of a weak molecular physisorption characteristic. In comparison, if a side-on

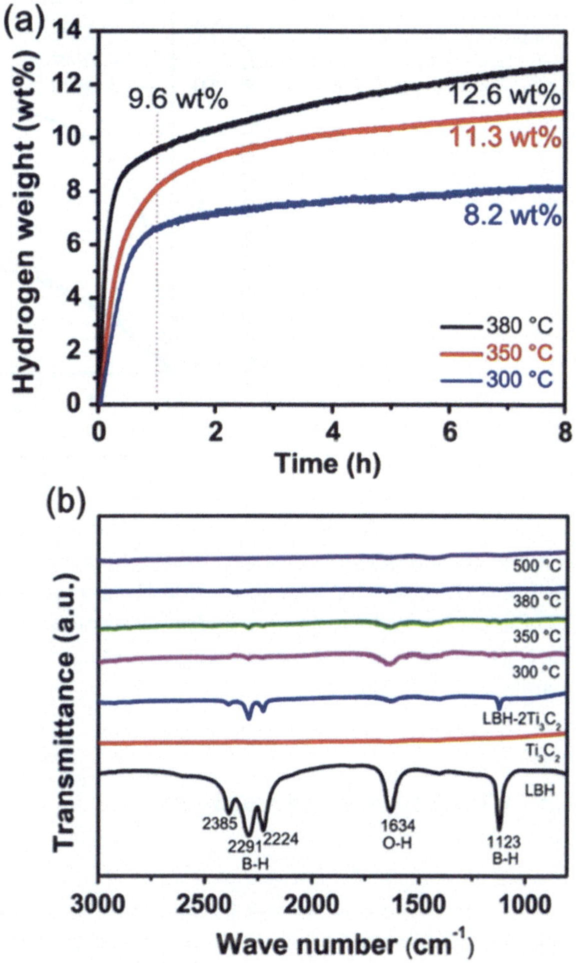

Fig. 4.49 **a** Dehydrogenation kinetic curves of the LBH-2Ti_3C_2 hybrid calculated according to the weight of $LiBH_4$. **b** FTIR spectra of pure $LiBH_4$, Ti_3C_2, LBH-2Ti_3C_2 hybrid and the dehydrogenated products. Reproduced with permission from [137]. Copyright © 2018, American Chemical Society

configuration was adopted, the H_2 molecule would decompose into two H atoms, which were preferably bonded to the Ti atoms at surface of the monolayer. The dissociation mechanism of the H_2 has been clarified, according to the chemical bonding analytic results. Furthermore, the diffusion behaviour of hydrogen on the surface of Ti_2C, with and without the vacancies of carbon, was deeply studied. The diffusion of hydrogen was enhanced with the presence of carbon vacancies.

Hydrogen storage capabilities of nitride-based MXenes, Ti_2N, were evaluated by using density functional calculations [139]. It was found that H_2 molecules could be adsorbed onto and desorbed from Ti_2N monolayers with certain adsorption strengths at the ambient conditions. Furthermore, the stability and H_2 adsorption behaviors were examined by calculating adsorption energies, charge populations and ab initio molecular dynamic simulations. The Kubas-type interaction between the

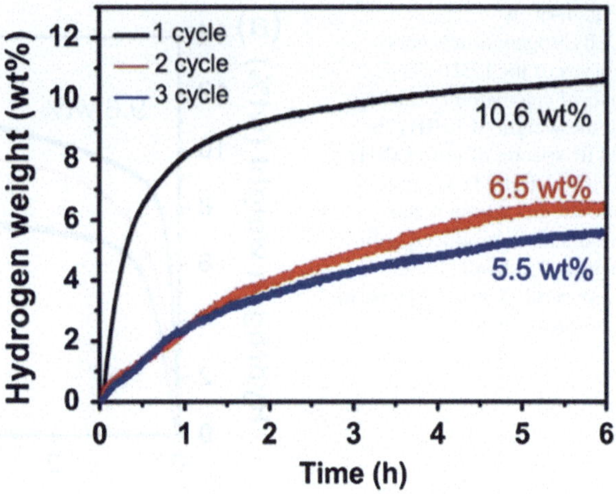

Fig. 4.50 Dehydrogenation cycle profiles of the LBH-2Ti_3C_2 hybrid. Reproduced with permission from [137]. Copyright © 2018, American Chemical Society

H_2-molecules with Ti_2N monolayers was studied with the charge transfers and electronic projected density of states. The optimal H_2 storage capacity of Ti_2N could reach 8.6 wt%, in which the reversible H_2 storage took about 3.4 wt%. Also, –OH decorated Ti_2N had a reversible hydrogen storage capacity of 2.7 wt%. In summary, 2D Ti_2N MXenes could be promising candidates for reversible H_2 storage materials at ambient conditions.

4.3 Conclusions and Outlook

Renewable energy is a hot research topic all around the world, thus triggering the requirement of energy storage technologies to the energies that are generated or collected but not necessarily used instantaneously. For energy storage devices, the electrode materials would play a critical role in determining energy storage capacities. Although significant progress has been made, the currently available electrode materials for energy storage devices still need further improvement in electrochemical performances. Analogue to graphene or reduced graphene oxide (rGO), 2D structured MXenes have quite extensively studied as electrode materials for various batteries as well as supercapacitors. This is simply because MXenes have a list of advantages. The most distinctive property is their high electrical conductivity. Also, the rich surface chemistry due to the wide availability of the functional groups enables them to be able host various cations, thus offering opportunity combine with other components for form hybrids with more advanced properties.

In electrolytes of all neutral, aqueous, and sulphuric acid, MXene-based materials exhibited outstanding electrochemical behaviors, indicating that they could promising candidates as the electrodes of supercapacitors. For lithium-ion batteries, MXene-derived electrodes had not only high specific capacities but also high cycling stabilities for the charge–discharge processes. Besides, they have also been explored as electrodes of other types of ion batteries due to the high tolerance of the interlayer spacing of MXenes. Furthermore, MXenes have been demonstrated to be suitable for the construction of battery–capacitor hybrid energy storage devices. Of course, there are still issues that need to be tackled for real applications. Among various problems, the development of stable synthetic approaches with which the type and property of the surface functional groups of the MXene nanosheets can be readily tailored to meet the specific requirements. Currently, the formation of the function groups on the surface of the MXene nanoflakes is just in an arbitrary manner. Comparatively, the utilization of MXenes as electrode materials high efficiency OER and HER processes is still underway, while the potentials for such applications are not very clear. In addition, although MXenes with various other compositions have been predicted or reported, the most intensive works are focusing on titanium carbides. Nevertheless, it is expected more and more attempts will be made to explore MXenes for energy storage applications.

Acknowledgements This work was supported by the National Natural Science Foundation of China (51762023 and 51962013), the Natural Science Foundation of Jiangxi, China (20192ACB20018), and Key R&D Program of Jiangxi Province (20171BBE50006, 20192ACB80007, and 20192ACB80004). Ling Bing Kong would like acknowledge Shenzhen Technology University (SZTU) for financial support through the Start-up Grant (2018) and grant from the Natural Science Foundation of Top Talent of SZTU (grant no. 2019010801002).

References

1. Anasori, B., Lukatskaya, M.R., Gogotsi, Y.: 2D metal carbides and nitrides (MXenes) for energy storage. Nat. Rev. Mater. **2**, 16098 (2017)
2. Chaudhari, N.K., Jin, H., Kim, B.Y., Baek, D.S., Joo, S.H., Lee, K.Y.: MXene: an emerging two-dimensional material for future energy conversion and storage applications. J. Mater. Chem. A **5**, 24564–24579 (2017)
3. Okubo, M., Sugahara, A., Kajiyama, S., Yamada, A.: MXene as a charge storage host. Acc. Chem. Res. **51**, 591–599 (2018)
4. Sun, S.J., Liao, C., Hafez, A.M., Zhu, H.L., Wu, S.P.: Two-dimensional MXenes for energy storage. Chem. Eng. J. **338**, 27–45 (2018)
5. Sun, Y.J., Chen, D.S., Liang, Z.Q.: Two-dimensional MXenes for energy storage and conversion applications. Mater. Today Energy **5**, 22–36 (2017)
6. Tang, H., Hu, Q., Zheng, M.B., Chi, Y., Qin, X.Y., Pang, H., et al.: MXene-2D layered electrode materials for energy storage. Progr. Nat. Sci. Mater. Int. **28**, 133–147 (2018)
7. Xiong, D.B., Li, X.F., Bai, Z.M., Lu, S.G.: Recent advances in layered $Ti_3C_2T_x$ MXene for electrochemical energy storage. Small **14**, 1703419 (2018)
8. Yoon, Y.H., Lee, K.S., Lee, H.Y.: Low-dimensional carbon and MXene-based electrochemical capacitor electrodes. Nanotechnology **27**, 172001 (2016)
9. Zhang, X., Zhang, Z.H., Zhou, Z.: MXene-based materials for electrochemical energy storage. J. Energy Chem. **27**, 73–85 (2018)
10. Naguib, M., Kurtoglu, M., Presser, V., Lu, J., Niu, J.J., Heon, M., et al.: Two-dimensional nanocrystals produced by exfoliation of Ti_3AlC_2. Adv. Mater. **23**, 4248–4253 (2011)

11. Tang, Q., Zhou, Z., Shen, P.W.: Are MXenes promising anode materials for Li ion batteries? Computational studies on electronic properties and Li storage capability of Ti_3C_2 and $Ti_3C_2X_2$ (X = F, OH) monolayer. J. Am. Chem. Soc. **134**, 16909–16916 (2012)
12. Er, D.Q., Li, J.W., Naguib, M., Gogotsi, Y., Shenoy, V.B.: Ti_3C_2 MXene as a high capacity electrode material for metal (Li, Na, K, Ca) ion batteries. ACS Appl. Mater. Interfaces **6**, 11173–11179 (2014)
13. Ren, C.E., Hatzell, K.B., Alhabeb, M., Ling, Z., Mahmoud, K.A., Gogotsi, Y.: Charge- and size-selective ion sieving through $Ti_3C_2T_x$ MXene membranes. J. Phys. Chem. Lett. **6**, 4026–4031 (2015)
14. Xie, Y., Naguib, M., Mochalin, V.N., Barsoum, M.W., Gogotsi, Y., Yu, X.Q., et al.: Role of surface structure on Li-ion energy storage capacity of two-dimensional transition-metal carbides. J. Am. Chem. Soc. **136**, 6385–6394 (2014)
15. Xu, K., Ji, X., Zhang, B., Chen, C., Ruan, Y.J., Miao, L., et al.: Charging/discharging dynamics in two-dimensional titanium carbide (MXene) slit nanopore: Insights from molecular dynamic study. Electrochim. Acta **196**, 75–83 (2016)
16. Ashton, M., Hennig, R.G., Sinnott, S.B.: Computational characterization of lightweight multilayer MXene Li-ion battery anodes. Appl. Phys. Lett. **108**, 023901 (2016)
17. Sun, D.D., Hu, Q.K., Chen, J.F., Zhang, X.Y., Wang, L.B., Wu, Q.H., et al.: Structural transformation of MXene (V_2C, Cr_2C, and Ta_2C) with O groups during lithiation: a first-principles investigation. ACS Appl. Mater. Interfaces **8**, 74–81 (2016)
18. Pan, H.: Electronic properties and lithium storage capacities of two-dimensional transition-metal nitride monolayers. J. Mater. Chem. A **3**, 21486–21493 (2015)
19. Chen, X.Z., Kong, Z.Z., Li, N., Zhao, X.J., Sun, C.H.: Proposing the prospects of Ti_3CN transition metal carbides (MXenes) as anodes of Li-ion batteries: a DFT study. Phys. Chem. Chem. Phys. **18**, 32937–32943 (2016)
20. Du, Y.T., Kan, X., Yang, F., Gan, L.Y., Schwingenschlogl, U.: MXene/graphene heterostructures as high-performance electrodes for Li-ion batteries. ACS Appl. Mater. Interfaces **10**, 32867–32873 (2018)
21. Yang, E.J., Ji, H.J., Kim, J.H., Kim, H.J., Jung, Y.S.: Exploring the possibilities of two-dimensional transition metal carbides as anode materials for sodium batteries. Phys. Chem. Chem. Phys. **17**, 5000–5005 (2015)
22. Mashtalir, O., Naguib, M., Mochalin, V.N., Dall'Agnese, Y., Heon, M., Barsoum, M.W., et al.: Intercalation and delamination of layered carbides and carbonitrides. Nat. Commun. 4:1716 (2013)
23. Sun, D.D., Wang, M.S., Li, Z.Y., Fan, G.X., Fan, L.Z., Zhou, A.G.: Two-dimensional Ti_3C_2 as anode material for Li-ion batteries. Electrochem. Commun. **47**, 80–83 (2014)
24. Kim, S.J., Naguib, M., Zhao, M.Q., Zhang, C.F., Jung, H.T., Barsoum, M.W., et al.: High mass loading, binder-free MXene anodes for high areal capacity Li-ion batteries. Electrochim. Acta **163**, 246–251 (2015)
25. Lin, Z.Y., Sun, D.F., Huang, Q., Yang, J., Barsoum, M.W., Yan, X.B.: Carbon nanofiber bridged two-dimensional titanium carbide as a superior anode for lithium-ion batteries. J. Mater. Chem. A **3**, 14096–14100 (2015)
26. Ren, C.E., Zhao, M.Q., Makaryan, T., Halim, J., Boota, M., Kota, S., et al.: Porous two-dimensional transition metal carbide (MXene) flakes for high-performance Li-ion storage. ChemElectroChem **3**, 689–693 (2016)
27. Byeon, A.Y., Zhao, M.Q., Ren, C.E., Halim, J., Kota, S., Urbankowski, P., et al.: Two-dimensional titanium carbide MXene as a cathode material for hybrid magnesium/lithium-ion batteries. ACS Appl. Mater. Interfaces **9**, 4296–4300 (2017)
28. Luo, J.M., Tao, X.Y., Zhang, J., Xia, Y., Huang, H., Zhang, L.Y., et al.: Sn^{4+} ion decorated highly conductive Ti_3C_2 MXene: promising lithium-ion anodes with enhanced volumetric capacity and cyclic performance. ACS Nano **10**, 2491–2499 (2016)
29. Simon, P.: Two-dimensional MXene with controlled interlayer spacing for electrochemical energy storage. ACS Nano **11**, 2393–2396 (2017)

30. Zou, G.D., Zhang, Z.W., Guo, J.X., Liu, B.Z., Zhang, Q.R., Fernandez, C., et al.: Synthesis of MXene/Ag composites for extraordinary long cycle lifetime lithium storage at high rates. ACS Appl. Mater. Interfaces **8**, 22280–22286 (2016)
31. Wang, F., Wang, Z.J., Zhu, J.F., Yang, H.B., Chen, X.J., Wang, L., et al.: Facile synthesis SnO_2 nanoparticle-modified Ti_3C_2 MXene nanocomposites for enhanced lithium storage application. J. Mater. Sci. **52**, 3556–3565 (2017)
32. Ahmed, B., Anjum, D.H., Gogotsi, Y., Alshareef, H.N.: Atomic layer deposition of SnO_2 on MXene for Li-ion battery anodes. Nano Energy **34**, 249–256 (2017)
33. Wu, X.H., Wang, Z.Y., Yu, M.Z., Xiu, L.Y., Qiu, J.S.: Stabilizing the MXenes by carbon nanoplating for eeveloping hierarchical nanohybrids with efficient lithium storage and hydrogen evolution capability. Adv. Mater. **29**, 1607017 (2017)
34. Naguib, M., Come, J., Dyatkin, B., Presser, V., Taberna, P.L., Simon, P., et al.: MXene: a promising transition metal carbide anode for lithium-ion batteries. Electrochem. Commun. **16**, 61–64 (2012)
35. Ahmed, B., Anjum, D.H., Hedhili, M.N., Gogotsi, Y., Alshareef, H.N.: H_2O_2 assisted room temperature oxidation of Ti_2C MXene for Li-ion battery anodes. Nanoscale **8**, 7580–7587 (2016)
36. Liu, F.F., Zhou, J., Wang, S.W., Wang, B.X., Shen, C., Wang, L.B., et al.: Preparation of high-purity V_2C MXene and electrochemical properties as Li-ion batteries. J. Electrochem. Soc. **164**, A709–A713 (2017)
37. Zhou, J., Gao, S.H., Guo, Z.L., Sun, Z.M.: Ti-enhanced exfoliation of V_2AlC into V_2C MXene for lithium-ion battery anodes. Ceram. Int. **43**, 11450–11454 (2017)
38. Mashtalir, O., Lukatskaya, M.R., Zhao, M.Q., Barsoum, M.W., Gogotsi, Y.: Amine-assisted delamination of Nb_2C MXene for Li-ion energy storage devices. Adv. Mater. **27**, 3501–3506 (2015)
39. Zhou, J., Zha, X.H., Zhou, X.B., Chen, F.Y., Gao, G.L., Wang, S.W., et al.: Synthesis and electrochemical properties of two-dimensional hafnium carbide. ACS Nano **11**, 3841–3850 (2017)
40. Zhao, S.S., Meng, X., Zhu, K., Du, F., Chen, G., Wei, Y.J., et al.: Li-ion uptake and increase in interlayer spacing of Nb_4C_3 MXene. Energy Storage Mater. **8**, 42–48 (2017)
41. Ali, A., Hantanasirisakul, K., Abdala, A., Urbankowski, P., Zhao, M.Q., Anasori, B., et al.: Effect of synthesis on performance of MXene/iron oxide anode material for lithium-ion batteries. Langmuir **34**, 11325–11334 (2018)
42. Kumar, R., Liu, J., Hwang, J.Y., Sun, Y.K.: Recent research trends in Li-S batteries. J. Mater. Chem. A **6**, 11582–11605 (2018)
43. Liu, J., Zhang, Q., Sun, Y.K.: Recent progress of advanced binders for Li-S batteries. J. Power Sources **396**, 19–32 (2018)
44. Liu, Y.J., He, P., Zhou, H.S.: Rechargeable solid-state Li-air and Li-S batteries: materials, construction, and challenges. Adv. Energy Mater. **8**, 1701602 (2018)
45. Zhang, J., Huang, H., Bae, J.W., Chung, S.H., Zhang, W.K., Manthiram, A., et al.: Nanostructured host materials for trapping sulfur in rechargeable Li-S batteries: structure design and interfacial chemistry. Small Methods 2 (2018)
46. Zheng, D., Wang, G.W., Liu, D., Si, J.Y., Ding, T.Y., Qu, D.Y., et al.: The progress of Li-S batteries-understanding of the sulfur redox mechanism: dissolved polysulfide ions in the electrolytes. Adv. Mater. Technol. **3**, 1700233 (2018)
47. Zhao, Y.M., Zhao, J.X.: Functional group-dependent anchoring effect of titanium carbide-based MXenes for lithium-sulfur batteries: a computational study. Appl. Surf. Sci. **412**, 591–598 (2017)
48. Rao, D.W., Zhang, L.Y., Wang, Y.H., Meng, Z.S., Qian, X.Y., Liu, J.H., et al.: Mechanism on the improved performance of lithium sulfur batteries with MXene-based additives. J. Phys. Chem. C **121**, 11047–11054 (2017)
49. Song, J.J., Su, D.W., Xie, X.Q., Guo, X., Bao, W.Z., Shao, G.J., et al.: Immobilizing polysulfides with MXene-functionalized separators for stable lithium-sulfur batteries. ACS Appl. Mater. Interfaces **8**, 29427–29433 (2016)

50. Liang, X., Garsuch, A., Nazar, L.F.: Sulfur cathodes based on conductive MXene nanosheets for high-performance lithium–sulfur batteries. Angew. Chem. Int. Ed. **54**, 3907–3911 (2015)
51. Liang, X., Rangom, Y., Kwok, C.Y., Pang, Q., Nazar, L.F.: Interwoven MXene nanosheet/carbon-nanotube composites as Li-S cathode hosts. Adv. Mater. **29**, 1603040 (2017)
52. Bao, W.Z., Xie, X.Q., Xu, J., Guo, X., Song, J.J., Wu, W.J., et al.: Confined sulfur in 3D MXene/reduced graphene oxide hybrid nanosheets for lithium-sulfur battery. Chem. A Eur. J. **23**, 12613–12619 (2017)
53. Dong, Y.F., Zheng, S.H., Qin, J.Q., Zhao, X.J., Shi, H.D., Wang, X.H., et al.: All-MXene-based integrated electrode constructed by Ti_3C_2 nanoribbon framework host and nanosheet interlayer for high-energy-density Li-S batteries. ACS Nano **12**, 2381–2388 (2018)
54. Kajiyama, S., Szabova, L., Sodeyama, K., Iinuma, H., Morita, R., Gotoh, K., et al.: Sodium-ion intercalation mechanism in MXene nanosheets. ACS Nano **10**, 3334–3341 (2016)
55. Bak, S.M., Qiao, R.M., Yang, W.L., Lee, S.S., Yu, X.Q., Anasori, B., et al.: Na-ion intercalation and charge storage mechanism in 2D vanadium carbide. Adv. Energy Mater. **7**, 1700959 (2017)
56. Zhao, M.Q., Xie, X.Q., Ren, C.E., Makaryan, T., Anasori, B., Wang, G.X., et al.: Hollow mXene spheres and 3D macroporous mXene frameworks for Na-ion storage. Adv. Mater. **29**, 1702410 (2017)
57. Chen, Z.P., Ren, W.C., Gao, L.B., Liu, B.L., Pei, S.F., Cheng, H.M.: Three-dimensional flexible and conductive interconnected graphene networks grown by chemical vapour deposition. Nat. Mater. **10**, 424–428 (2011)
58. Wu, Y.T., Nie, P., Jiang, J.M., Ding, B., Dou, H., Zhang, X.G.: MoS_2-nanosheet-decorated 2D titanium carbide (MXene) as high-performance anodes for sodium-ion batteries. ChemElectroChem **4**, 1560–1565 (2017)
59. Guo, X., Xie, X.Q., Choi, S.H., Zhao, Y.F., Liu, H., Wang, C.Y., et al.: Sb_2O_3/MXene ($Ti_3C_2T_x$) hybrid anode materials with enhanced performance for sodium-ion batteries. J. Mater. Chem. A **5**, 12445–12452 (2017)
60. Dong, Y.F., Wu, Z.S., Zheng, S.H., Wang, X.H., Qin, J.Q., Wang, S., et al.: Ti_3C_2 MXene-derived sodium/potassium titanate nanoribbons for high-performance sodium/potassium ion batteries with enhanced capacities. ACS Nano **11**, 4792–4800 (2017)
61. Tao, M.L., Zhang, Y.Q., Zhan, R.M., Guo, B.S., Xu, Q.J., Xu, M.W.: A chemically bonded $CoNiO_2$ nanoparticles/MXene composite as anode for sodium-ion batteries. Mater. Lett. **230**, 173–176 (2018)
62. Zhang, Y.Q., Zhan, R.M., Xu, Q.J., Liu, H., Tao, M.L., Luo, Y.S., et al.: Circuit board-like CoS/MXene composite with superior performance for sodium storage. Chem. Eng. J. **357**, 220–225 (2019)
63. Peng, S.J., Han, X.P., Li, L.L., Zhu, Z.Q., Cheng, F.Y., Srinivansan, M.U., et al.: Unique cobalt sulfide/reduced graphene oxide composite as an anode for sodium-ion batteries with superior rate capability and long cycling stability. Small **12**, 1359–1368 (2016)
64. Han, F., Zhang, C.Z., Sun, B., Tang, W., Yang, J.X., Li, X.K.: Dual-carbon phase-protective cobalt sulfide nanoparticles with cable-type and mesoporous nanostructure for enhanced cycling stability in sodium and lithium ion batteries. Carbon **118**, 731–742 (2017)
65. Naguib, M., Adams, R.A., Zhao, Y.P., Zemlyanov, D., Varma, A., Nanda, J., et al.: Electrochemical performance of MXenes as K-ion battery anodes. Chem. Commun. **53**, 6883–6886 (2017)
66. Mathis, T.S., Kurra, N., Wang, X.H., Pinto, D., Simon, P., Gogotsi, Y.: Energy storage data reporting in perspective-guidelines for interpreting the performance of electrochemical energy storage systems. Adv. Energy Mater. **9**, 1902007 (2019)
67. Lukatskaya, M.R., Mashtalir, O., Ren, C.E., Dall'Agnese, Y., Rozier, P., Taberna, P.L., et al.: Cation intercalation and high volumetric capacitance of two-dimensional titanium carbide. Science **341**:1502–1505 (2013)
68. Lukatskaya, M.R., Bak, S.M., Yu, X.Q., Yang, X.Q., Barsoum, M.W., Gogotsi, Y.: Probing the mechanism of high capacitance in 2D titanium carbide using in situ X-ray absorption spectroscopy. Adv. Energy Mater. **5**, 1500589 (2015)

69. Hu, M.M., Li, Z.J., Hu, T., Zhu, S.H., Zhang, C., Wang, X.H.: High-capacitance mechanism for $Ti_3C_2T_X$ MXene by in situ electrochemical Raman spectroscopy investigation. ACS Nano **10**, 11344–11350 (2016)
70. Ghidiu, M., Lukatskaya, M.R., Zhao, M.Q., Gogotsi, Y., Barsoum, M.W.: Conductive two-dimensional titanium carbide 'clay' with high volumetric capacitance. Nature **516**, 78–81 (2014)
71. Xu, S.K., Wei, G.D., Li, J.Z., Ji, Y., Klyui, N., Izotov, V., et al.: Binder-free $Ti_3C_2T_x$ MXene electrode film for supercapacitor produced by electrophoretic deposition method. Chem. Eng. J. **317**, 1026–1036 (2017)
72. Wang, X.Y., Fu, Q.S., Wen, J., Ma, X.Z., Zhu, C.C., Zhang, X.T., et al.: 3D $Ti_3C_2T_x$ aerogels with enhanced surface area for high performance supercapacitors. Nanoscale **10**, 20828–20835 (2018)
73. Hu, M.M., Hu, T., Cheng, R.F., Yang, J.X., Cui, C., Zhang, C., et al.: MXene-coated silk-derive d carbon cloth toward flexible electrode for supercapacitor application. J. Energy Chem. **27**, 161–166 (2018)
74. Zhang, C.F., Anasori, B., Seral-Ascaso, A., Park, S.H., McEvoy, N., Shmeliov, A., et al.: Transparent, flexible, and conductive 2D titanium carbide (MXene) films with high volumetric capacitance. Adv. Mater. **29**, 1702678 (2017)
75. Lukatskaya, M.R., Kota, S., Lin, Z.F., Zhao, M.Q., Shpigel, N., Levi, M.D., et al.: Ultra-high-rate pseudocapacitive energy storage in two-dimensional transition metal carbides. Nat. Energy **2**, 17105 (2017)
76. Dall'Agnese, Y., Lukatskaya, M.R., Cook, K.M., Taberna, P.L., Gogotsi, Y., Simon, P.: High capacitance of surface-modified 2D titanium carbide in acidic electrolyte. Electrochem. Commun. **48**, 118–122 (2014)
77. Li, J., Yuan, X.T., Lin, C., Yang, Y.Q., Xu, L., Du, X., et al.: Achieving high pseudocapacitance of 2D titanium carbide (MXene) by cation intercalation and surface modification. Adv. Energy Mater. **7**, 1602725 (2017)
78. Mashtalir, O., Lukatskaya, M.R., Kolesnikov, A.I., Raymundo-Pinero, E., Naguib, M., Barsoum, M.W., et al.: The effect of hydrazine intercalation on the structure and capacitance of 2D titanium carbide (MXene). Nanoscale **8**, 9128–9133 (2016)
79. Fu, Q.S., Wen, J., Zhang, N., Wu, L.L., Zhang, M.Y., Lin, S.Y., et al.: Free-standing $Ti_3C_2T_x$ electrode with ultrahigh volumetric capacitance. RSC Adv. **7**, 11998–12005 (2017)
80. Ghidiu, M., Kota, S., Halim, J., Sherwood, A.W., Nedfors, N., Rosen, J., et al.: Alkylammonium cation intercalation into Ti_3C_2 (MXene): effects on properties and ion-exchange capacity estimation. Chem. Mater. **29**, 1099–1106 (2017)
81. Wen, Y.Y., Rufford, T.E., Chen, X.Z., Li, N., Lyu, M.Q., Dai, L.M., et al.: Nitrogen-doped $Ti_3C_2T_x$ MXene electrodes for high-performance supercapacitors. Nano Energy. **38**, 368–376 (2017)
82. Yang, C.H., Que, W.X., Yin, X.T., Tian, Y.P., Yang, Y.W., Que, M.D.: Improved capacitance of nitrogen-doped delaminated two-dimensional titanium carbide by urea-assisted synthesis. Electrochim. Acta **225**, 416–424 (2017)
83. Tang, Y., Zhu, J.F., Yang, C.H., Wang, F.: Enhanced capacitive performance based on diverse layered structure of two-dimensional Ti_3C_2 MXene with long etching time. J. Electrochem. Soc. **163**, A1975–A1982 (2016)
84. Come, J., Xie, Y., Naguib, M., Jesse, S., Kalinin, S.V., Gogotsi, Y., et al.: Nanoscale elastic changes in 2D $Ti_3C_2T_x$ (MXene) pseudocapacitive electrodes. Adv. Energy Mater. **6**, 1502290 (2016)
85. Shpigel, N., Lukatskaya, M.R., Sigalov, S., Ren, C.E., Nayak, P., Levi, M.D., et al.: In situ monitoring of Gravimetric and viscoelastic changes in 2D intercalation electrodes. ACS Energy Lett. **2**, 1407–1415 (2017)
86. Ling, Z., Ren, C.E., Zhao, M.Q., Yang, J., Giammarco, J.M., Qiu, J.S., et al.: Flexible and conductive MXene films and nanocomposites with high capacitance. Proc. Natl. Acad. Sci. **111**, 16676–16681 (2014)

87. Boota, M., Anasori, B., Voigt, C., Zhao, M.Q., Barsoum, M.W., Gogotsi, Y.: Pseudocapacitive electrodes produced by oxidant-free polymerization of pyrrole between the layers of 2D titanium carbide (MXene). Adv. Mater. **28**, 1517–1522 (2016)
88. Zhu, M.S., Huang, Y., Deng, Q.H., Zhou, J., Pei, Z.X., Xue, Q., et al.: Highly flexible, freestanding supercapacitor electrode with enhanced performance obtained by hybridizing polypyrrole chains with MXene. Adv. Energy Mater. **6**, 1600969 (2016)
89. Boota, M., Pasini, M., Galeotti, F., Porzio, W., Zhao, M.Q., Halim, J., et al.: Interaction of polar and nonpolar polyfluorenes with layers of two-dimensional titanium carbide (MXene): Intercalation and pseudocapacitance. Chem. Mater. **29**, 2731–2738 (2017)
90. Ren, Y.Y., Zhu, J.F., Wang, L., Liu, H., Liu, Y., Wu, W.L., et al.: Synthesis of polyaniline nanoparticles deposited on two-dimensional titanium carbide for high-performance supercapacitors. Mater. Lett. **214**, 84–87 (2018)
91. Lu, X., Zhu, J.F., Wu, W.L., Zhang, B.: Hierarchical architecture of PANI@$TiO_2/Ti_3C_2T_x$ ternary composite electrode for enhanced electrochemical performance. Electrochim. Acta **228**, 282–289 (2017)
92. Solis, M.A.C., Wu, Z.P., Liu, C.G.: Microstrip-Strip Feed Transparent Ceramic Rectangular Dielectric Resonator Antenna. IEEE, New York (2013)
93. Tian, Y.P., Yang, C.H., Que, W.X., He, Y.C., Liu, X.B., Luo, Y.Y., et al.: Ni foam supported quasi-core-shell structure of ultrathin Ti_3C_2 nanosheets through electrostatic layer-by-layer self-assembly as high rate-performance electrodes of supercapacitors. J. Power Sources **369**, 78–86 (2017)
94. Zhao, M.Q., Ren, C.E., Ling, Z., Lukatskaya, M.R., Zhang, C.F., Van Aken, K.L., et al.: Flexible MXene/carbon nanotube composite paper with high volumetric capacitance. Adv. Mater. **27**, 339–345 (2015)
95. Zhou, Z.H., Panatdasirisuk, W., Mathis, T.S., Anasori, B., Lu, C.H., Zhang, X.X., et al.: Layer-by-layer assembly of MXene and carbon nanotubes on electrospun polymer films for flexible energy storage. Nanoscale **10**, 6005–6013 (2018)
96. Zang, X.N., Shen, C.W., Kao, E., Warren, R., Zhang, R.P., Teh, K.S., et al.: Titanium disulfide coated carbon nanotube hybrid electrodes enable high energy density symmetric pseudocapacitors. Adv. Mater. **30**, 1704754 (2018)
97. Yan, P.T., Zhang, R.J., Jia, J., Wu, C., Zhou, A.G., Xu, J., et al.: Enhanced supercapacitive performance of delaminated two-dimensional titanium carbide/carbon nanotube composites in alkaline electrolyte. J. Power Sources **284**, 38–43 (2015)
98. Dall'Agnese, Y., Rozier, P., Taberna, P.L., Gogotsi, Y., Simon, P.: Capacitance of two-dimensional titanium carbide (MXene) and MXene/carbon nanotube composites in organic electrolytes. J. Power Sources **306**:510–515 (2016)
99. Wang, Z.Y., Qin, S., Seyedin, S., Zhang, J.Z., Wang, J.T., Levitt, A., et al.: High-performance biscrolled MXene/carbon nanotube yarn supercapacitors. Small **14**, 1802225 (2018)
100. Yan, J., Ren, C.E., Maleski, K., Hatter, C.B., Anasori, B., Urbankowski, P., et al.: Flexible MXene/graphene films for ultrafast supercapacitors with outstanding volumetric capacitance. Adv. Func. Mater. **27**, 1701264 (2017)
101. Couly, C., Alhabeb, M., Van Aken, K.L., Kurra, N., Gomes, L., Navarro-Suarez, A.M., et al.: Asymmetric flexible MXene-reduced fraphene oxide micro-supercapacitor. Adv. Electron. Mater. **4**, 1700339 (2018)
102. Fan, Z.M., Wang, Y.S., Xie, Z.M., Wang, D.L., Yuan, Y., Kang, H.J., et al.: Modified MXene/holey graphene films for advanced supercapacitor electrodes with superior energy storage. Adv. Sci. **5**, 1800750 (2018)
103. Fu, J.J., Yun, J.M., Wu, S.X., Li, L., Yu, L.T., Kim, K.H.: Architecturally robust graphene-encapsulated MXene Ti_2CT_x@polyaniline composite for high-performance pouch-type asymmetric supercapacitor. ACS Appl. Mater. Interfaces **10**, 34212–34221 (2018)
104. Wang, Y., Dou, H., Wang, J., Ding, B., Xu, Y.L., Chang, Z., et al.: Three-dimensional porous MXene/layered double hydroxide composite for high performance supercapacitors. J. Power Sources **327**, 221–228 (2016)

105. Li, H., Musharavati, F., Zalenezhad, E., Chen, X., Hui, K.N., Hui, K.S.: Electrodeposited Ni-Co layered double hydroxides on titanium carbide as a binder-free electrode for supercapacitors. Electrochim. Acta **261**, 178–187 (2018)
106. Zhang, X.F., Liu, Y., Dong, S.L., Ye, Z.Y., Wei, Y.D.: Low-temperature synthesized nanocomposites with amorphous FeOOH on $Ti_3C_2T_x$ for supercapacitors. J. Alloy. Compd. **744**, 507–515 (2018)
107. Zhu, J.F., Tang, Y., Yang, C.H., Wang, F., Cao, M.J.: Composites of TiO_2 nanoparticles deposited on Ti_3C_2 MXene nanosheets with enhanced electrochemical performance. J. Electrochem. Soc. **163**, A785–A791 (2016)
108. Tian, Y.P., Yang, C.H., Que, W.X., Liu, X.B., Yin, X.T., Kong, L.B.: Flexible and free-standing 2D titanium carbide film decorated with manganese oxide nanoparticles as a high volumetric capacity electrode for supercapacitor. J. Power Sources **359**, 332–339 (2017)
109. Jiang, H.M., Wang, Z.G., Yang, Q., Hanif, M., Wang, Z.M., Dong, L.C., et al.: A novel $MnO_2/Ti_3C_2T_x$ MXene nanocomposite as high performance electrode materials for flexible supercapacitors. Electrochim. Acta **290**, 695–703 (2018)
110. Yuan, W.Y., Cheng, L.F., Zhang, B.X., Wu, H.: 2D-Ti_3C_2 as hard, conductive substrates to enhance the electrochemical performance of MnO_2 for supercapacitor applications. Ceram. Int. **44**, 17539–17543 (2018)
111. Zou, R., Quan, H.Y., Pan, M.H., Zhou, S., Chen, D.Z., Luo, X.B.: Self-assembled MXene ($Ti_3C_2T_x$)/α-Fe_2O_3 nanocomposite as negative electrode material for supercapacitors. Electrochim. Acta **292**, 31–38 (2018)
112. Xia, Q.X., Fu, J.J., Yun, J.M., Mane, R.S., Kim, K.H.: High volumetric energy density annealed-MXene-nickel oxide/MXene asymmetric supercapacitor. RSC Adv. **7**, 11000–11011 (2017)
113. Rakhi, R.B., Ahmed, B., Hedhili, M.N., Anjum, D.H., Alshareef, H.N.: Effect of postetch annealing gas composition on the structural and electrochemical properties of Ti_2CT_x MXene electrodes for supercapacitor applications. Chem. Mater. **27**, 5314–5323 (2015)
114. Krishnamoorthy, K., Pazhamalai, P., Sahoo, S., Kim, S.J.: Titanium carbide sheet based high performance wire type solid state supercapacitors. J. Mater. Chem. A **5**, 5726–5736 (2017)
115. Zhang, X.F., Liu, Y.F., Dong, S.L., Yang, J.Q., Liu, X.D.: Surface modified MXene film as flexible electrode with ultrahigh volumetric capacitance. Electrochim. Acta **294**, 233–239 (2019)
116. Li, L., Zhang, N., Zhang, M.Y., Wu, L., Zhang, X.T., Zhang, Z.G.: Ag-nanoparticle-decorated 2D titanium carbide (MXene) with superior electrochemical performance for supercapacitors. ACS Sustain. Chem. Eng. **6**, 7442–7450 (2018)
117. Lin, Z.F., Barbara, D., Taberna, P.L., Van Aken, K.L., Anasori, B., Gogotsi, Y., et al.: Capacitance of $Ti_3C_2T_x$ MXene in ionic liquid electrolyte. J. Power Sources **326**, 575–579 (2016)
118. Wang, X., Mathis, T.S., Li, K., Lin, Z., Vlcek, L., Torita, T., et al.: Influences from solvents on charge storage in titanium carbide MXenes. Nat. Energy **4**, 241–248 (2019)
119. Li, Y.B., Shao, H.F., Lin, Z., Lu, J., Liu, L.Y., Duployer, B., et al.: A general Lewis acidic etching route for preparing MXenes with enhanced electrochemical performance in non-aqueous electrolyte. Nat. Mater. (2020). https://doi.org/10.1038/s41563-020-0657-0
120. Li, P., Shi, W.H., Liu, W.X., Chen, Y.F., Xu, X.L., Ye, S.F., et al.: Fabrication of high-performance MXene-based all-solid-state flexible microsupercapacitor based on a facile scratch method. Nanotechnology **29**, 445401 (2018)
121. Yue, Y., Liu, N., Ma, Y.A., Wang, S.L., Liu, W.J., Luo, C., et al.: Highly self-healable 3D microsupercapacitor with MXene-graphene composite aerogel. ACS Nano **12**, 4224–4232 (2018)
122. Couly, C., Alhabeb, M., Van Aken, K.L., Kurra, N., Gomes, L., Navarro-Suarez, A.M., et al.: Asymmetric flexible MXene-reduced graphene oxide micro-supercapacitor. Adv. Electron. Mater. **4**, 1700339 (2018)
123. Chen, X., Wang, S.L., Shi, J.J., Du, X.Y., Cheng, Q.H., Xue, R., et al.: Direct laser etching free-standing MXene-MoS_2 film for highly flexible micro-supercapacitor. Adv. Mater. Interfaces **6**, 1901160 (2019)

124. Wu, C.W., Unnikrishnan, B., Chen, I.W.P., Harroun, S.G., Chang, H.T., Huang, C.C.: Excellent oxidation resistive MXene aqueous ink for micro-supercapacitor application. Energy Storage Mater. **25**, 563–571 (2020)
125. Guo, Z.L., Zhou, J., Zhu, L.G., Sun, Z.M.: MXene: a promising photocatalyst for water splitting. J. Mater. Chem. A **4**, 11446–11452 (2016)
126. Seh, Z.W., Fredrickson, K.D., Anasori, B., Kibsgaard, J., Strickler, A.L., Lukatskaya, M.R., et al.: Two-dimensional molybdenum carbide (MXene) as an efficient electrocatalyst for hydrogen evolution. ACS Energy Lett. **1**, 589–594 (2016)
127. Huang, B., Zhou, N.G., Chen, X.Z., Ong, W.J., Li, N.: Insights into the electrocatalytic hydrogen evolution reaction mechanism on two-dimensional transition-metal carbonitrides (MXene). Chem. A Eur. J. **24**, 18479–18486 (2018)
128. Xiu, L.Y., Wang, Z.Y., Yu, M.Z., Wu, X.H., Qiu, J.S.: Aggregation-resistant 3D MXene-based architecture as efficient bifunctional electrocatalyst for overall water splitting. ACS Nano **12**, 8017–8028 (2018)
129. Wang, H., Peng, R., Hood, Z.D., Naguib, M., Adhikari, S.P., Wu, Z.L.: Titania composites with 2D transition metal carbides as photocatalysts for hydrogen production under visible-light irradiation. ChemSusChem **9**, 1490–1497 (2016)
130. Li, Y.J., Deng, X.T., Tian, J., Liang, Z.Q., Cui, H.Z.: Ti3C2 MXene-derived Ti_3C_2/TiO_2 nanoflowers for noble-metal-free photocatalytic overall water splitting. Appl. Mater. Today **13**, 217–227 (2018)
131. Hao, N.X., Wei, Y., Wang, J.L., Wang, Z.W., Zhu, Z.H., Zhao, S.L., et al.: In situ hybridization of an MXene/TiO_2/NiFeCo-layered double hydroxide composite for electrochemical and photoelectrochemical oxygen evolution. RSC Adv. **8**, 20576–20584 (2018)
132. Peng, C., Wei, P., Li, X.Y., Liu, Y.P., Cao, Y.H., Wang, H.J., et al.: High efficiency photocatalytic hydrogen production over ternary Cu/TiO_2@$Ti_3C_2T_x$ enabled by low-work-function 2D titanium carbide. Nano Energy **53**, 97–107 (2018)
133. Ran, J.R., Gao, G.P., Li, F.T., Ma, T.Y., Du, A.J., Qiao, S.Z.: Ti_3C_2 MXene co-catalyst on metal sulfide photo-absorbers for enhanced visible-light photocatalytic hydrogen production. Nat. Commun. **8**, 13907 (2017)
134. Shao, M.M., Shao, Y.F., Chai, J.W., Qu, Y.J., Yang, M.Y., Wang, Z.L., et al.: Synergistic effect of 2D Ti_2C and g-C_3N_4 for efficient photocatalytic hydrogen production. J. Mater. Chem. A **5**, 16748–16756 (2017)
135. Du, C.F., Khang Ngoc, D., Liang, Q.H., Zheng, Y., Luo, Y.B., Zhang, J.L., et al.: Self-assemble and in situ formation of $Ni_{1-x}Fe_xPS_3$ nanomosaic-decorated MXene hybrids for overall water splitting. Adv. Energy Mater. **8**, 1801127 (2018)
136. Su, T.M., Peng, R., Hood, Z.D., Naguib, M., Ivanov, I.N., Keum, J.K., et al.: One-Step Synthesis of Nb_2O_5/C/Nb_2C (MXene) composites and their use as photocatalysts for hydrogen evolution. ChemSusChem **11**, 688–699 (2018)
137. Zang, L., Sun, W.Y., Liu, S., Huang, Y.K., Yuan, H.T., Tao, Z.L., et al.: Enhanced hydrogen storage properties and reversibility of $LiBH_4$ confined in two-dimensional Ti_3C_2. ACS Appl. Mater. Interfaces **10**, 19598–19604 (2018)
138. Wang, S., Du, Y.L., Liao, W.H., Sun, Z.M.: Hydrogen adsorption, dissociation and diffusion on two-dimensional Ti_2C monolayer. Int. J. Hydrogen Energy **42**, 27214–27219 (2017)
139. Li, Y.M., Guo, Y.L., Chen, W.G., Jiao, Z.Y., Ma, S.H.: Reversible hydrogen storage behaviors of Ti_2N MXenes predicted by first-principles calculations. J. Mater. Sci. **54**, 493–505 (2019)

Chapter 5
Other Applications

5.1 Introduction

Besides their applications in energy storage and conversion, MXenes and MXene-based hybrids and composites have also found a wide range of applications in other fields, such as environmental protection, chemical sensors, biosensors, microwave absorbers, EMI shielding and transparent thin films (electrodes or conductors), as well as other applications that are not covered in this book [1–9]. The applications of MXenes and their hybrids/composites in environmental protection will be first introduced, which include the removal of heavy metallic polluting ions (e.g., Hg^{2+}, Cd^{2+}, Cr^{3+} etc.), organic molecules (e.g., Rhodamine B and methyl orange) and radionuclide pollutants. For sensing applications, there will be chemical sensors, biochemical sensors, electrochemical sensors, piezoelectric sensors, gas sensors and so on. Electromagnetic inference (EMI) shielding and microwave absorption are the utilization of the MXene-based materials with different functionalities, where EMI is to use their high electrical conductivity and microwave absorbing is mainly achieved by combining with other components. Finally, MXene-based thin films have been employed as electrodes in flexible electronic devices.

5.2 Environmental Remediating Agents

Pollution has been a worldwide issue, which draws attentions from researchers of various fields. Heavy metallic ions, toxic organic compounds, bio-toxins and toxic gases are among the most serious pollutants. To address this problem, various physical, chemical and biological methods have been proposed and implemented, such as membrane filtrations, precipitations, adsorptions, solvent extractions, ion exchanges and so on [10–12]. Comparatively, adsorption is among the most cost-effective and simplest ones [13]. Also, adsorption has no problem of secondary pollution, because

Z. Xiao et al., *MXenes and MXenes-based Composites*, Engineering Materials,
https://doi.org/10.1007/978-3-030-59373-5_5

no harmful substances are produced in the process. To ensure the adsorption to effective and efficient, the adsorbents should have a sufficiently large specific surface area and appropriate functionalities, depending on characteristics of the adsorbates.

In this regard, porous materials, such as activated carbons, kaolinites, zeolites, chitosans, metal–organic frameworks and so on, are promising adsorbents to remove the environmental pollutants [14–17]. It is well known that materials with low dimensions usually have large specific surface area and thus can be used as adsorbents for removal of pollutants, such as ordered mesoporous silica, carbon nanotubes, graphenes and phosphorenes [18–21]. Therefore, it is reasonable to expect that 2D MXenes could be promising candidates as adsorbents for the applications in environmental pollutant purification [22].

5.2.1 Adsorption of Heavy Metallic Ions

Heavy metal ions, such as Pb(II), Cr(IV), Hg(II), Cd(II) and Cu(II), are highly toxic, so that they are very harmful to human beings and other living organisms. They are widely present in water and soils. Because these heavy metal ions cannot be degraded, the most effective way to remove them is through adsorption. MXenes can be used to adsorb the heavy metal ions through electrostatic and chemical interactions with the active sites on surfaces of the nanosheets. Significant progress has been made in the utilization of MXenes to eliminate heavy metal ions.

The level of Pb(II) in drinking water has to be strictly controlled. 2D layered MXene, $Ti_3C_2(OH)_xF_{2-x}$, could be intercalated through alkalization, in order for adsorption of heavy-metal ions. The kinetics of adsorption of heavy-metal ions and the effect of intercalated sites on the adsorption behaviours were theoretically studied by using first-principles with density functional theory [23]. As 1/9 of the 2D monolayer was covered by the heavy-metal ions, the alkalization-intercalated MXene, alk-MXene or $Ti_3C_2(OH)_2$), showed strong absorption capability of heavy-metal ions. The charge equilibrium was maintained, due to the formation of a hydrogen potential trap, where the hydrogen atoms surrounded the adsorbed heavy-metal atoms. It was also observed that the presence of –F groups on surface of the alk-MXene would reduce the ion adsorption capability, whereas the intercalation of Li, Na and K atoms would promote the adsorption effect. Meanwhile, the hydroxyl groups located vertically to the Ti atoms were especially strong in eliminating the metal ions, otherwise the effect was largely weakened.

Pb adsorption behaviours of various MXenes, $M_2X(OH)_2$, with M=Sc, Ti, V, Cr, Zr, Nb, Mo, Hf, Ta, and X=C, N, were comprehensively studied by using first principles calculations with generalized gradient approximation through density functional theory [24]. It was revealed that $M_2N(O_2H_{2-2x}Pb_x)$ had much lower formation energies than $M_2C(O_2H_{2-2x}Pb_x)$, implying that nitride-based MXenes possessed stronger Pb(II) adsorption effect than their carbide counterparts, which was attributed to the difference in the number of valence electron between C and N. Meanwhile, the mass fraction of $Ti_2C(OH)_2$ was the lowest for the removal of Pb(II) ions. However, neither

$Sc_2C(OH)_2$ nor $Zr_2C(OH)_2$ could be used to eliminate Pb(II) ions. The kinetics of Pb(II) adsorption was closely related to the potential wells, depending on the density of states and electron localization functions.

Density functional theory (DFT) calculation was used to study the adsorption properties of Pb(II) and Cu(II) on 2D Ti_3C_2, V_2C_1 and Ti_2C_1 [25]. The surface functional groups of –H, –OH and –F posed significant effects on the adsorption behaviours of the MXenes. All the three MXenes demonstrated strong adsorption of Pb(II) and Cu(II) ions. Specifically, Ti_2C_1 displayed a high adsorption capacity of 2560 mg/g in adsorbing Pb(II). The ab initio molecular dynamics simulation results indicated that Ti_2C_1 was still stable after adsorbing Pb(II) at room temperature. In addition, the MXenes possessed outstanding thermal stability, high capacity controllability and selective adsorption capability, which made them highly potential as candidate materials for the removal of heavy metal ions.

2D alk-MXene, $Ti_3C_2(OH/Ona)_xF_{2-x}$, obtained through chemical exfoliation combined with alkalization intercalation, had selective adsorption of Pb(II) ions, in the presence of Ca(II)/Mg(II) ions at a high concentration [26]. A very short time of 2 min was required to approach the sorption equilibrium. More importantly, the alk-MXene exhibited an extremely high efficiency in the removal of Pb(II) ions, with a sorption capacity of 4500 kg water per kilogram of the alk-MXene. The effluent content of Pb(II) was as low as 10 μg/L, which was much lower than the drinking water standard of 10 μg/L, according to the World Health Organization. It was confirmed that the sorption effect was associated with the –OH groups at the active sites of Ti, at which the Pb(II) ions were exchanged, due to the formation of a hexagonal potential trap.

A facial process was utilized to obtain $Ti_3C_2T_x$ nanosheet powder with a coupling agent silane (KH570), in order to develop adsorbent to remove Pb(II) ions [27]. In this case, the $Ti_3C_2T_x$ nanosheets were modified with KH570 to form $Ti_3C_2T_x$-KH570 hybrid, which exhibited sufficiently high specific surface area, high thermal stability and strong Pb(II) adsorption efficiency. Specifically, at a concentration of 3.2 g/L, the $Ti_3C_2T_x$-KH570 hybrid powder could nearly complete remove Pb(II) ions in aqueous solution. It was found that, at room temperature, the adsorption capacities of the $Ti_3C_2T_x$-KH570 hybrid and the pristine $Ti_3C_2T_x$ powders for Pb(II) ion were about 150 mg/g and <50 mg/g, respectively. The hydroxyl groups of the MXene nanosheets and carbonyl groups related to the methacryloxypropyl chains of the KH570 agent were responsible for the strong Pb(II) ion adsorption capability.

$Ti_3C_2T_x$ powder was dispersed in a mixture solvent of distilled water and ethanol, with pH value to be adjusted to be 4–4.5. After the addition of KH570, the mixture was heated in water bath at 70 °C with strong stirring, during which TiO_2 nanoparticles were formed at the surface of the $Ti_3C_2T_x$ nanosheets, due to the occurrence of oxidation. The samples were then collected, washed and dried properly. The $Ti_3C_2T_x$-KH570 hybrid with 10% KH570 was mainly adopted for experimental studies, which was confirmed by the SEM observation.

In the $Ti_3C_2T_x$-KH570 hybrid, the TiO_2 nanoparticles and the –OH groups were present on the surfaces of the $Ti_3C_2T_x$ nanosheets, which reacted with KH570. Owing to the acidic environment, the $–OCH_3$ groups were hydrated by the hydroxyl groups on the surfaces of either TiO_2 nanoparticles or the $Ti_3C_2T_x$ nanosheets in the solution of $YsiX_3$, because of the hydrophilic nature. As a result, the KH570 was linked onto the surface of the TiO_2 nanoparticles and the $Ti_3C_2T_x$ nanosheets. The coupling process of the silane had four stages: (i) formation of –SiOH due to the hydrolyzation of –SiX, (ii) presence of polysiloxane such Si–OH caused by the dehydration-related condensation, (iii) hydrogen bonding between the Si–OH group in the polysiloxane and the –OH groups on the surface of the $Ti_3C_2T_x$ nanosheets and (iv) covalent bonding owing to the dehydration reaction and linkage by the $Ti_3C_2T_x$ nanosheets during the consolidation process.

The dynamic process was of characteristic, in which just 1/3 of the silicon hydroxyls would react in between the silane agent and the $Ti_3C_2T_x$ nanosheets. In other words, 2/3 of the Si–OH groups were condensed among themselves or present freely in the system [28]. In this study, the self-condensation was facilitated. Additionally, the freely present –OH groups would be bonding with the Pb(II) ions. The $Ti_3C_2T_x$ nanosheet powder had a BET specific surface area of just 8.53 m^2/g, while the hybrid with 10% KH570 possessed a specific surface area of as high as 75.4 m^2/g, which was also optimal in terms of the content of KH570.

Figure 5.1 shows histogram of adsorption capacities of the samples with different addition quantities of KH570, with 10% of KH570 to be the optimized content. This is was because the adsorption capacity was reduced with increasing amount of KH570, due to the decrease in the unreacted functional groups on the surface of the $Ti_3C_2T_x$ nanosheets, although the content of the methacryloxy group was increased as the quantity of KH570 was increased. The decrease in the Pb(II) ion adsorption capability at high concentrations of KH570 was attributed to the fact that the –OH

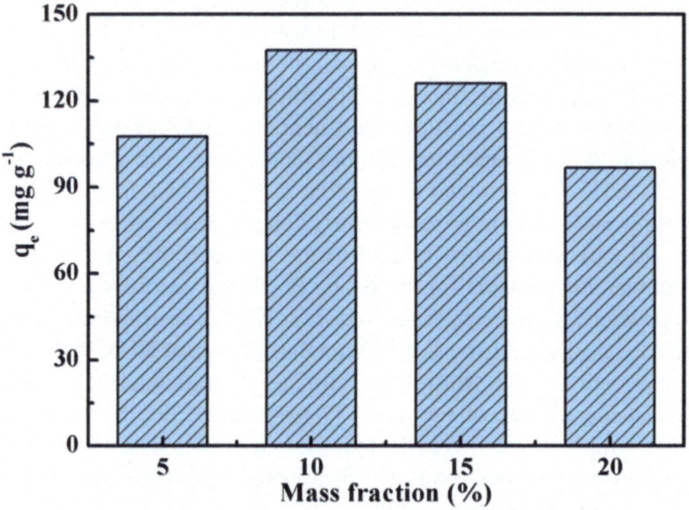

Fig. 5.1 Histogram of the Pb(II) ion removal efficiency of the $Ti_3C_2T_x$-KH570 hybrid powders with different contents of KH570. Reproduced with permission from [27]. Copyright © 2019, Springer

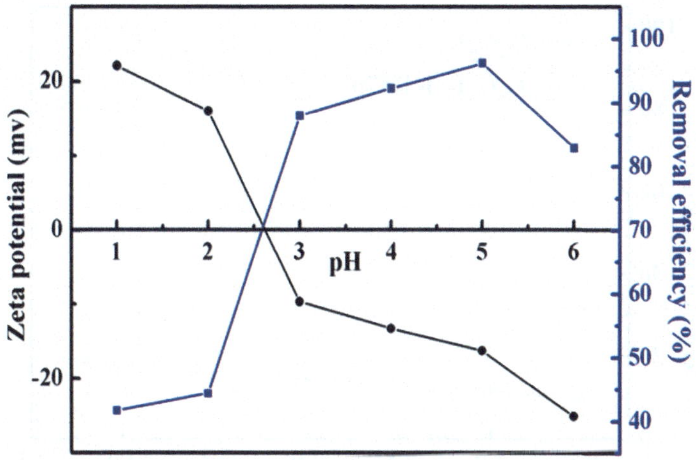

Fig. 5.2 Zeta potential and removal efficiency of the $Ti_3C_2T_x$-KH570 hybrid powder as a function of pH value of the solution. Reproduced with permission from [27]. Copyright © 2019, Springer

groups of KH570 had no contribution to the adsorption effect, since the number of the unreacted groups on the surface of the $Ti_3C_2T_x$ nanosheets was decreased.

The Pb(II) adsorption efficiency of the $Ti_3C_2T_x$-KH570 hybrid was also influenced by the pH value of the aqueous solutions, as revealed in Fig. 5.2, simply because the pH value altered the properties of the Pb(II) ions and the zeta potentials of sorbent. It was reported that, in the solutions of $Pb(NO_3)_2$, Pb(II) ions were ulphide to produce $Pb(OH)^+$ and $Pb(OH)_2$, as the pH value was close to 5.9, while $Pb(OH)_2$ was precipitated at pH value of >7.0 [29]. When the pH value was increased from 1.0 to 6.0, the surface charge density of the $Ti_3C_2T_x$-KH570 hybrid powder was monotonically decreased. At pH $\approx$ 2.6, zeta potential of the $Ti_3C_2T_x$-KH570 hybrid powder approached zero. In comparison, zeta potentials of the pristine $Ti_3C_2T_x$ nanosheet powder was always negative in the pH value range of 1.0–6.0. The Pb(II) eliminating capacity was slowly increased as the pH value was increased from 1.0 to 5.0, owing to the electron pair of oxygen to bond the Pb(II) ions through the coordinate covalent bonding [30]. Then, the presence of $Pb(OH)^+$ and $Pb(OH)_2$ resulted in decrease in Pb(II) removal capacity as the pH value reached to 6.0.

Figure 5.3 shows adsorption efficiency as a function of the dosage of the $Ti_3C_2T_x$-KH570 hybrid powder, in the range of 0.5–3.2 g/L, with the starting concentration of Pb(II) to 500 mg/L. The experiment was carried out at 30 °C, with the volume of the solution to be 20 ml. It was demonstrated that the Pb(II) ion removal efficiency of the $Ti_3C_2T_x$-KH570 hybrid powder was almost 100% at the dosage of 3 g/L. The increase in the adsorption capacity with increasing dosage of the $Ti_3C_2T_x$-KH570 hybrid powder was readily ascribed to the enlargement in the surface area and the increase in the number of the active sites. At the $Ti_3C_2T_x$-KH570 dosage of 3.2 g/L, the Pb(II) ion was completely removed from the aqueous solution. In contrast, the pristine $Ti_3C_2T_x$ power could only remove <20% Pb(II) over the dosage range.

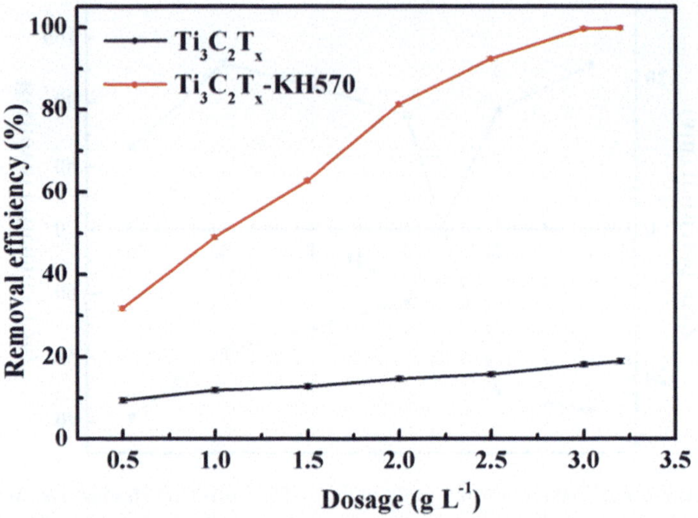

Fig. 5.3 Pb(II) ion adsorption capacities of the $Ti_3C_2T_x$-KH570 hybrid powder as a function of its dosage. Reproduced with permission from [27]. Copyright © 2019, Springer

MXene-alginate nanohybrid was prepared for the applications to remove Pb(II) and Cu(II) ions from wastewaters [31]. The MXene-alginate hybrid boosted the chelating efficiency of the Pb(II) and Cu(II) ions and enhanced their transport properties, thus leading to high adsorption capacities and rapid equilibrium capabilities. Specifically, the maximum adsorption capacities for Pb(II) and Cu(II) ions were 382.7 mg/g and 87.6 mg/g, respectively, while the adsorption equilibrium was reached in 1/4 h. $Ti_3C_2T_x$ powders with a quantity of 0.16 g were dispersed in aqueous solutions of sodium alginate with quantities of 68, 106, 160, 240, 370 and 640 mg dissolved in 20 mL deionized water, with strong stirring for 6 h. The mixtures were then freezed in an ultra-low temperature refrigerator for 12 h. MXene-alginate hybrid powders were freeze-dried in vacuum for one whole day.

Figure 5.4 shows adsorption capacities of the MXene-alginate hybrid powders for the Pb(II) and Cu(II) ions, in which the solidum alginate was used at different ratios. It was found that the adsorption capacities for both ions were increased with increasing ratio of the sodium alginate used in preparing the hybrid powders. The initial relatively fast increase in the adsorption capacity could be understood by the fact that the numbers of both the amino and carboxyl groups were increased as the content of sodium alginate was increased. In other words, the chelating efficiencies of the Pb(II) and Cu(II) ions were enhanced. The slightly slowed increase in the adsorption capacity with further increase in the ratio of sodium alginate was attributed to the variation in the combination profile of the two components. At high ratios, the sodium alginate was also attached onto surface of the $Ti_3C_2T_x$ nanosheets, besides staying in between their interlayer spaces. The turning point was observed at the ratio of sodium alginate to be 70%.

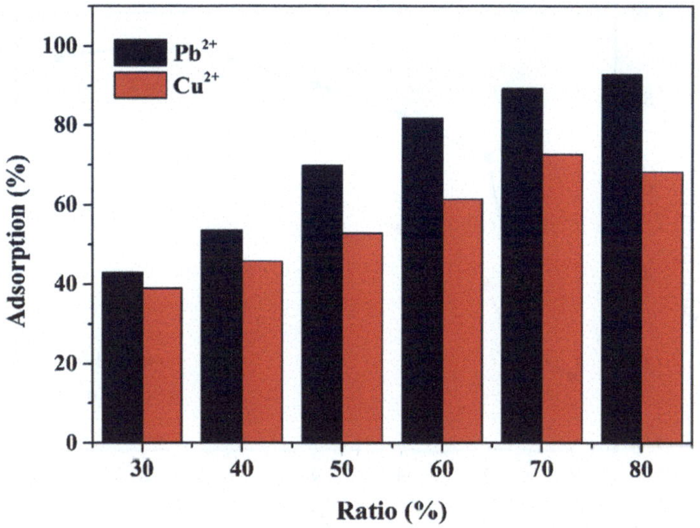

Fig. 5.4 Pb(II) and Cu(II) ion adsorption capacities of the MXene-alginate hybrids with different alginate ratios. Reproduced with permission from [31]. Copyright © 2019, Royal Society of Chemistry

Figure 5.5 depicts adsorption efficiencies of the MXene-alginate hybrid at different dosages. The adsorption effect was almost linearly increased with increasing

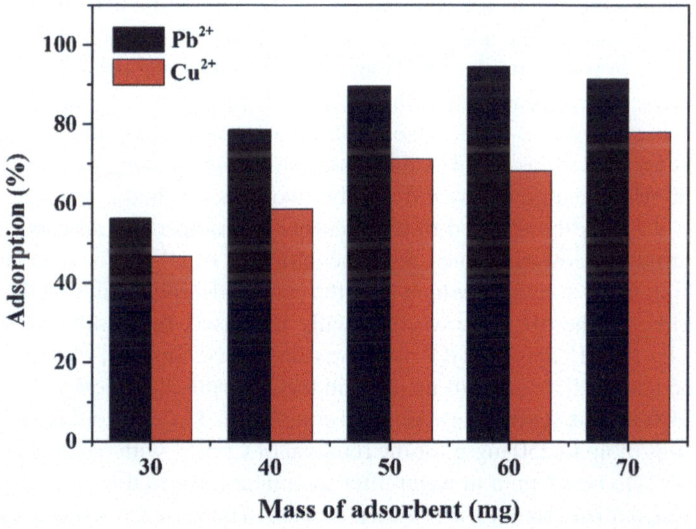

Fig. 5.5 Adsorption capacities of the MXene-alginate hybrid for Pb(II) and Cu(II) ions at different dosages. Reproduced with permission from [31]. Copyright © 2019, Royal Society of Chemistry

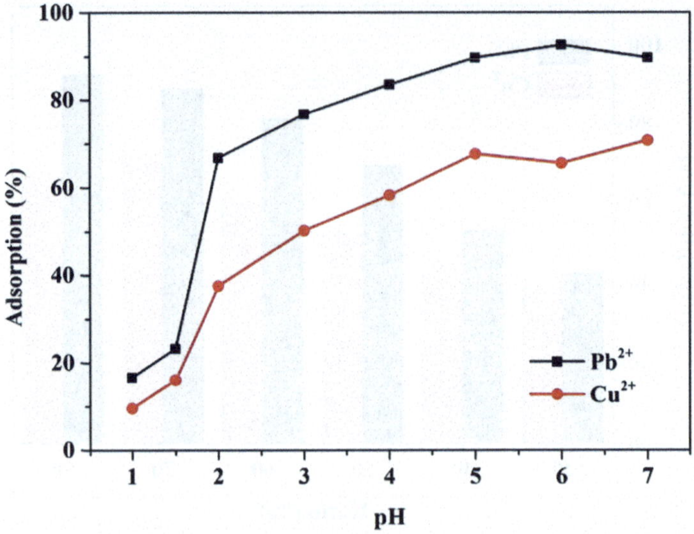

Fig. 5.6 Effect of pH on Adsorption efficiencies of the MXene-alginate hybrid for Pb(II) and Cu(II) ions as a function of pH value of the solutions. Reproduced with permission from [31]. Copyright © 2019, Royal Society of Chemistry

dosage up to 50 mg, after which it started to decline. 50 mg dosage was used for other experiments. The effect of pH value of the solutions on the adsorption capacity of the MXene-alginate hybrid is demonstrated in Fig. 5.6. Since the industrial wastewaters containing heavy metal ions are generally of acidic nature, the pH value was examined up to 7. Obviously, the adsorption capacity of the MXene-alginate nanohybrid was very weak in removing the Pb(II) and Cu(II) ions as the pH value was below 1. As pH value was increased to 2, the adsorption capacity was sharply increased. In the pH value range of 2–5, the adsorption capacity was nearly linearly increased, while it was levelled off as the pH value was above 5, suggesting that the adsorption effect of the MXene-alginate hybrid for the two ions reached a stable level. This observation was readily ascribed to the presence of competitive adsorptions of the hybrid materials. At low pH values, the concentrations of H^+ and/or H_3O^+ ions were relatively high in the solutions, which took the active adsorption sites on the surfaces of the hybrids. As the pH value was gradually increased, the number of the active sites taken by the H^+ and/or H_3O^+ ions was decreased. Instead, the metallic ions started to be adsorbed, leading to increase in the adsorption efficiency [32].

As the MXene was synthesize at optimal conditions, $Ti_3C_2T_x$ could reach adsorption capacities of up to 250 mg/g for the removal of Cr(VI), with residual concentrations of Cr(VI) to be <5 ppm in water after treatment [33]. In this case, the removal of Cr(VI) was realized through a two-step process. The toxic Cr(VI) was reduced to less toxic Cr(III) and then the Cr(III) ions were adsorbed by the MXene. The Cr(VI) ions were adsorbed onto the $Ti_3C_2T_x$ nanosheets through the electrostatic attraction, because the surface of the MXene was positively charged, while $Cr_2O_7^{2-}$ ions were negative. Meanwhile, as the pH value was sufficiently low, the –OH groups on the surface of $Ti_3C_2T_x$ would be protonated. After adsorption, electrons were transferred from $Ti_3C_2T_x$ to $Cr_2O_7^{2-}$ with the aid of H^+ ions, to form TiO_2 and Cr(III) ions. The Cr(III) ions could be attached to the Ti–O bonds on the MXene nanosheets, thus leading to Ti–O–Cr(III) items, with the mechanism to be schematically shown in Fig. 5.7.

Cr(IV) ion removal efficiencies of the $Ti_3C_2T_x$-10% suspension at pH = 5.0 are shown in Fig. 5.8a. Initially, the removal was conducted at a very high rate, since all the active sites on the surfaces of the $Ti_3C_2T_x$-10% nanosheets participated the removal reaction, while the number of Cr(VI) ions was also sufficiently high. As expected, the removal rate was gradually decreased, because both the number of the active sites and the concentration of the Cr(VI) ions were decreased. In addition, the removal rate was increased, as the content of the $Ti_3C_2T_x$-10% nanosheets was increased from 0.2 to 1.0 g/L.

Figure 5.8b shows the removal efficiencies at different pH values. The removal rate was increased with decreasing value of pH. At pH = 13, MXene had almost no capability to eliminate the Cr(VI) ions. According to the ICP-AES results, the total content of Cr in the treated solution was as low as 2.0 ppm, as the system was at pH = 5.0. The color of the solution was bleached gradually, confirming the reduction of Cr(IV) to Cr(III) and the adsorption of Cr(III), as evidenced by the XPS results of the samples in Fig. 5.8c, d.

A one-step process was employed to obtain urchin-like rutile TiO_2–C (u-RTC) nanohybrids, with a large fraction of (110) facets, by an *in-situ* oxidation of MXene ($Ti_3C_2(OH)_{0.8}F_{1.2}$) with the presence of $FeCl_3$ [34]. The u-RTC hybrid was derived from layered anatase TiO_2–C (l-ATC) intermediate phase, which had a high portion

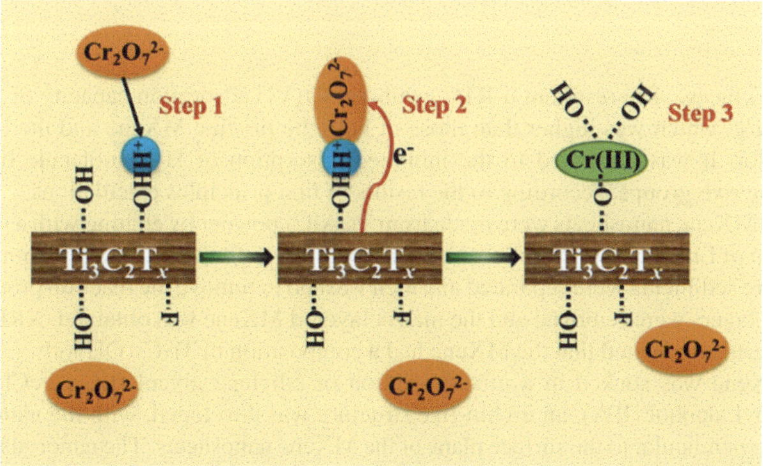

Fig. 5.7 Schematic diagram to illustrate removal mechanism of the Cr(VI) ions by the $Ti_3C_2T_x$ nanosheets from water. Reproduced with permission from [33]. Copyright © 2015, American Chemical Society

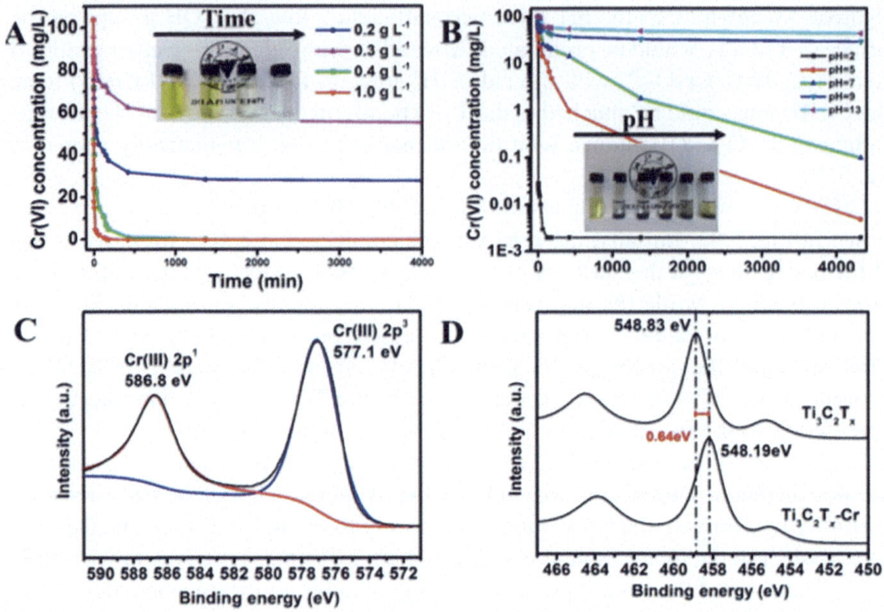

Fig. 5.8 **a** Efficiencies of the $Ti_3C_2T_x$-10% nanosheets to remove Cr(VI) in 400 mL solution of 104 mg/L at pH = 5.0. The inset shows the photographs of the samples after reaction for different times, indicating the change in color of the $Ti_3C_2T_x$-10% nanosheets with a concentration of 0.4 g/L. **b** Removal effects of Cr (VI) from 400 mL solution with a concentration of 104 mg/L at different pH values. The $Ti_3C_2T_x$-10% suspension has a concentration of 0.4 g/L. The inset shows the final solution color at different pH values. **c** Cr 2p XPS spectrum of the Cr-loading $Ti_3C_2T_x$-10% nanosheets. **d** Chemical shift of binding energy of Ti 2p before and after immersing in the Cr (VI) solution for 3 days. Reproduced with permission from [33]. Copyright © 2015, American Chemical Society

of (001) facets. The resultant u-RTC exhibited Cr(VI) adsorption capacity of about 225 mg/g, which was higher than those of both the pristine MXene and the l-ATC precursor. It was attributed to the inhibited adsorption of H_2O molecule by the bridging oxo groups, according to the results of first principles calculations.

The MXene nanosheets were made from Ti_3AlC_2 powder by etching with a mixed solution of LiF + 6 M HCl, followed by drying at 60 °C for 2 days. After etching, the resulting sediments were separated and then washed to remove the reaction products. The Al layers were removed and the matrix layered MXene was obtained. XRD and EDX results indicated that the MXene had a composition of $Ti_3C_2(OH)_{0.8}F_{1.2}$. After the MXene was socked in a mixed solution of ethylene glycol (EG), $FeCl_3$ and isopropyl alcohol (IPA), an urchin-like structure was developed, with the nanorods to be perpendicular to the surface plane of the MXene nanosheets. The nanorods were single crystals. Thereafter, layered rutile TiO_2–C/TiC nanocomposite was formed.

With the addition of EG at a level of about 0.5 wt%, anatase TiO_2 (A-TiO_2) started to be present. It implied that the MXene started to decompose, whereas the surface groups of Ti–OH and Ti–F were transferred to Ti–O ones, thus leading to the presence of TiO_2. As the incubation time was increased, both the content and the size of the A-TiO_2 phase were greatly increased. The surface of the MXene nanosheets was fully covered by cuboid nanosheets finally, i.e., a layer of anatase TiO_2/C nano-composite (or l-ATC) was formed.

In the presence of $FeCl_3$, nanosized rectangular sheets of A-TiO_2 and rod-like precipitates of R-TiO_2 coexisted, after reaction for 1 h. When the incubation time was further increased, the rod-like precipitates was changed to needle-like items. The aspect ratio was increases gradually, while the average diameter was decreased, a typical Ostwald-ripening process. In addition, as the reaction temperature was above 300 °C, the A-TiO_2 was entirely transferred into R-TiO_2, so that u-RTC was obtained.

To evaluate the adsorption performance of the u-RTC hybrid, synthetic Cr(VI) contaminated water samples with high concentrations of competing anions, including Cl^-, NO_3^- and SO_4^{2-}, were prepared as the feeding solution, with the effluent history is depicted in Fig. 5.9. The u-RTC hybrid displayed an outstanding Cr(VI) adsorption capability, with large water quantities of 370 and 1120 kg per kg u-RTC, according to the WHO standard of drinking water (0.01 mg/L) and the China wastewater discharge standard (0.1 mg/L). The feasibility of the u-RTC hybrid for water purification was

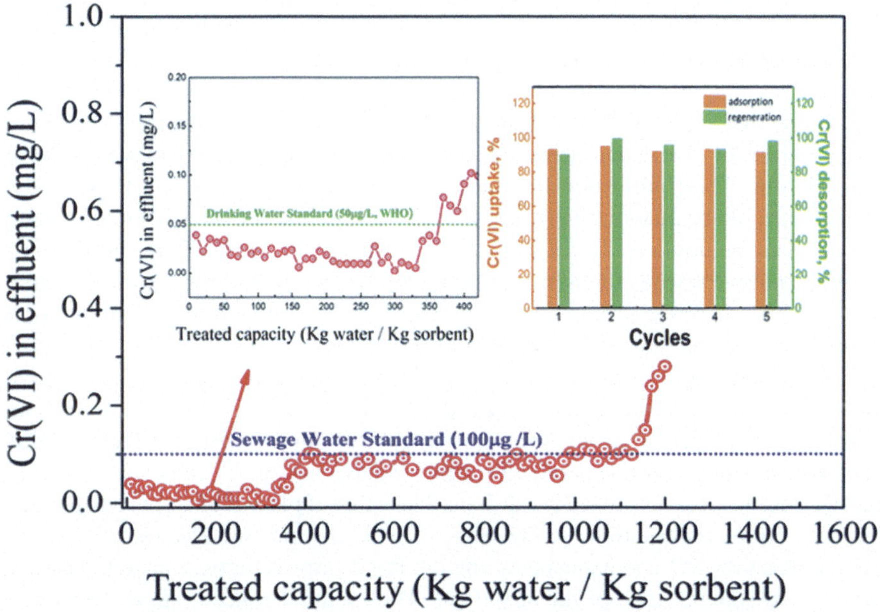

Fig. 5.9 The applied evaluation onto u-RTC with feeding synthetic Cr(VI) contaminated waters (dose: 2 g/L, initial Cr(VI) = 2 mg/L, SO_4^{2-}=Cl^-=NO_3^-=200 mg/L, pH = 5.5–6.8). The right inset shows sorption-regeneration cycles (regenerant: 5% NaOH solution at 298 K). Reproduced with permission from [34]. Copyright © 2016, Royal Society of Chemistry

also evidenced by real wastewater test. More significantly, the exhausted u-RTC hybrid could be recycled by regenerating with 5% NaOH solution.

An accordion-like layered Ti_3C_2 MXene was demonstrated to have outstanding adsorption capacity of Cr(VI) ions from aqueous solutions [35]. After etching with HF solution, Ti_3C_2 MXene was obtained, which had an enlarged specific surface area (SSA), as compared with the precursor Ti_3AlC_2. Due to the presence of numerous active sites, the Ti_3C_2 powder could be used eliminate potassium dichromate from solutions.

BiOBr-Ti_3C_2 nanohybrid was form by using self-assembly process through electrostatic interactions for the removal of Cr(VI) and degradation of rhodamine B (RhB) and 2,4-Dinitrophenol (2,4-DNP) [36]. The combination of Ti_3C_2 and BiOBr resulted in materials with enhanced degradation efficiency. For the removal of Cr(VI), the BiOBr-Ti_3C_2 hybrid had an efficiency that was higher than that of BiOBr by 6 times. Meanwhile, the BiOBr-Ti_3C_2 could be utilized to eliminate multiple component pollution items. For instance, RhB and Cr(VI) could be entirely got rid of in 80 min by using the BiOBr-Ti_3C_2 hybrid powder, in aqueous solutions. The improvement in the catalytic activity was attributed to the fact that Schottky junction was formed at the interface between BiOBr and Ti_3C_2, so the effectiveness of carrier separation was increased.

2D $Ti_3C_2T_x$ MXene has been used to adsorb and remove Cu(II) ions from aqueous solutions [37]. In this case, Cu^{2+} was reduced to Cu^+ (Cu_2O) by the oxygenated moieties in the layered structure of MXene. The delaminated $Ti_3C_2T_x$ (DL-$Ti_3C_2T_x$) displayed a stronger and faster elimination of Cu(II), as compared with the multi-layered one (ML-$Ti_3C_2T_x$). At optimal conditions, a maximum adsorption capacity ($Q_{exp,max}$) of 78.45 mg/g was achieved, while the Cu(II) ions were removed by 80% in just 1 min. The equilibrium time and the maximum uptake of Cu(II) onto the adsorbents could be quantitatively described by using a pseudo-second-order kinetic model and the Freundlich adsorption isotherm, respectively. According to the thermodynamic analytic results, the adsorption process had an endothermic characteristic. The DL-$Ti_3C_2T_x$ possessed an adsorption capacity (Q_e) to be higher than that of a commercially available activated carbon by 2.7 times.

$Ti_3C_2T_x$ nanosheets have been shown to have promising capability to remove Ba(II) from both synthetic and co-produced water samples [38]. The MXene nanosheets displayed a strong sorption capability, rapid kinetics, effective trace barium elimination and reversible adsorption behaviours, an efficiency of 9.3 mg/g and 100% removal at optimized conditions. The value of pH was an important parameter, with optimal levels to be in the range of 6–9, at which the electrostatic attractions between the MXene nanosheets and the Ba(II) ions were maximized. Another parameter was the dosage of the adsorbent. The eventual concentration Ba(II) was decreased with increasing dosage of the adsorbent. The maximum removal capacity was observed at 100 mg, which was closely related to the total number of adsorption sites available for the interactions with the Ba(II) ions.

Final concentration and the percentage removal of Ba(II) ions at different initial Ba concentrations, as a function of contact time, are shown in Fig. 5.10a, b, respectively. The removal efficiency was dependent on the initial concentration of Ba(II) ion. For instance, at the initial concentration of Ba(II) of 1 ppm, the Ba(II) ions were adsorbed by 95%. If the concentration was increased to 55 ppm, 20% Ba(II) ions were retained in the sample.

Kinetically, it took <1 h to completely eliminate the Ba(II) ions from solution. In all cases, the MXene demonstrated a very fast kinetics, with nearly 35% Ba (II) to be taken away in just a few seconds after the contact. The rapid kinetics was benefitted from the unique layered structure of the MXene nanosheets, which allowed for the diffusion and intercalation of Ba(II) ions. The high initial removal rate was similar to the profiles of chemical reactions. As the equilibrium approached, the removal reaction slowed down. The adsorption kinetics could be well fitted with the Lagergren pseudo-first order model and the pseudo-second-order model, as observed in Fig. 5.10c, d.

A magnetic $Ti_3C_2T_x$ MXene nanohybrid (MGMX nanocomposite) was prepared by using hydrothermal method, for aqueous adsorptive removal of Hg(II) ions [39]. The MGMX nanohybrid was highly stable, which was confirmed by the zeta-potential and dynamic light-scattering results. It displayed high Hg(II) removal capacities over

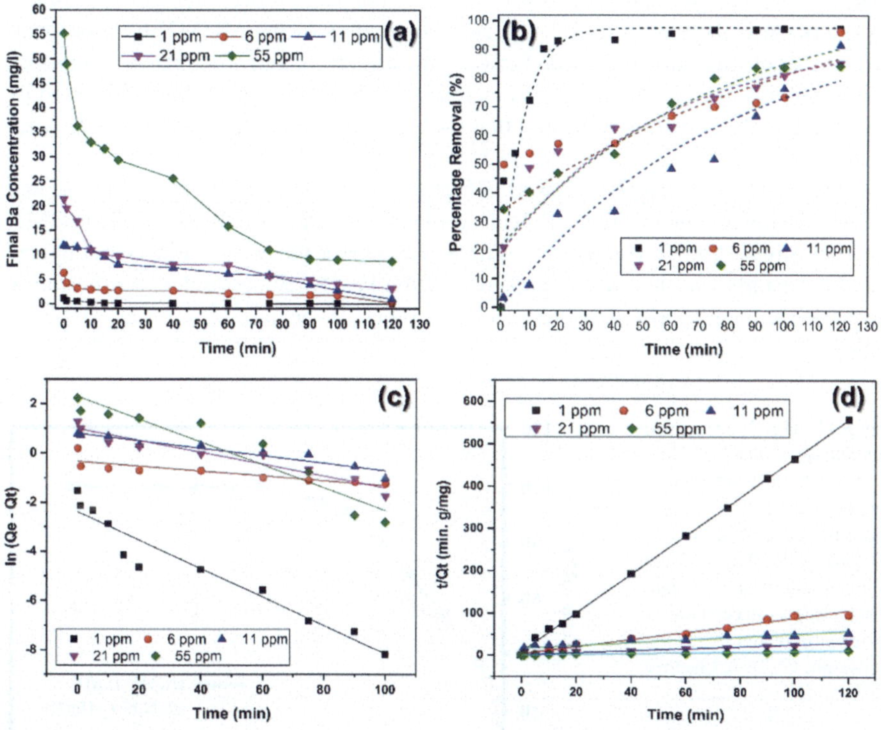

Fig. 5.10 Effect of reaction time on Ba(II) removal effect with the MXene samples at different pH levels: **a** final barium concentration, **b** percentage removal of barium, **c** MXene-barium adsorption data fitting to first order kinetic mode and **d** fitting to second order kinetic model. Reproduced with permission from [38]. Copyright © 2017, Elsevier

a wide range of pH value, with a maximum Hg(II) uptake capacity of 1128 mg/g. The adsorption characteristic was understood by using the Redlich-Peterson adsorption isotherm, pseudo second-order kinetics and thermodynamics models. More importantly, the MGMX nanohybrid could be recycled for several times.

To synthesize the MGMX nanohybrid samples, 200 mg $Ti_3C_2T_x$ powder was dispersed in iron acetate ($Fe(C_2H_3O_2)_2$) solution in deionized (DI) water with a concentration of 0.25 mg/mL, which was stirred under for 0.5 h in the protection of Ar. After that, the suspension was subject to hydrothermal reaction in the presence of ammonia as the catalyst at 180 °C for 4 h. The hybrid powder was collected, washed and freeze dried. The dried powder was heated at 400 °C for 4 h obtain MGMX, which contained Fe_2O_3 nanoparticles, with an average diameter of 64 nm and magnetization of 23.6 emu/g.

The functional groups on the $Ti_3C_2T_x$ nanosheets offered possible ion exchange reaction with Hg(II). The Fe_2O_3 nanoparticles had small dimensions and large specific surface area, which had a positive effect on the adsorption of the Hg(II) ions. There could be strong electrostatic attractions or repulsions between the negatively charged MGMX nanosheets and the positively charged Hg(II) ions. The surface charge of the MGMX nanohybrid was highly dependent on pH value of the systems. As the pH was decreased to 3.2, the surface charges of the MGMX became positive, so that electrostatic repulsion between the adsorbent and adsorbate was present. Therefore, adsorption efficiency was increased with increasing pH level. SEM–EDX elemental analysis results indicated that the amount of mercury was 38.5 wt% after adsorption, possibly in the form of HgO and $Hg(OH)_2$.

Figure 5.11 shows Hg(II) adsorption performances of the ml-$Ti_3C_2T_x$ MXene and MGMX samples. The ml-$Ti_3C_2T_x$ had a removal efficiency of 58.9%, while MGMX could eliminate 99.3% Hg(II) ions. The higher performance of the MGMX hybrid could be readily ascribed to the presence of the Fe_2O_3 nanoparticles. Furthermore, the MGMX nanohybrid also outperformed other 2D materials, such as graphene oxides.

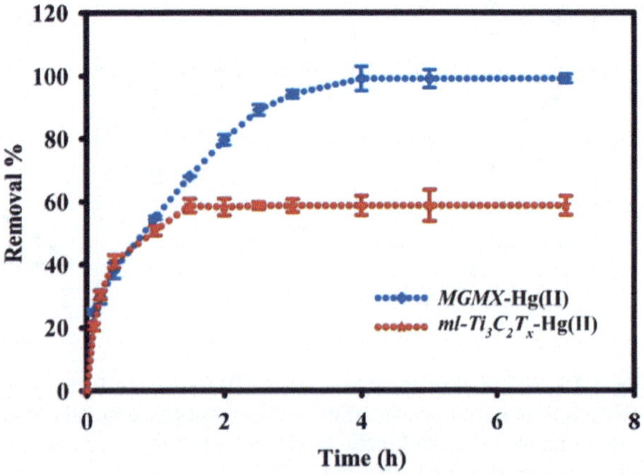

Fig. 5.11 Hg(II) adsorption efficiencies of the ml-$Ti_3C_2T_x$ and the MGMX nanohybrid, with experimental conditions: 10 mg/L of Hg(II), 0.025 g/L of adsorbent at pH = 6 and 298 K. Reproduced with permission from [39]. Copyright © 2017, Elsevier

$Ti_3C_2T_x$-alginate hybrid was also explored to remove Hg(II) ion [40]. The hybrid was made as core–shell aerogel spheres, with different ratios of MXene and sodium alginate (SA). The hybrid-based spheres exhibited strong adsorption efficiency for Hg(II) ion from aqueous solution, due to the special architecture, highly porous network structure and large specific surface area, as well as enriched functional groups and active sites on the surface of the MXene nanosheets. Specifically, the MX-$SA_{4:20}$ sample displayed the optimal Hg(II) ion adsorption efficiency of 932.84 mg/g. In addition, the highly efficient Hg(II) adsorption capability was well retained at high acidic levels, e.g., HNO_3 solutions with concentrations of 0.5–1.0 M, together with outstanding reproducibility. More importantly, the hybrid spheres were capable of removing multiple metallic ions.

MX-SA spheres, with different ratios of $Ti_3C_2T_x$ MXene and SA, were prepared. 50 mL 2.0 mg/mL $Ti_3C_2T_x$ suspension was mixed with 50 mL 20 mg/mL SA solution, with the aid of strong stirring. The MXene-alginate mixed suspension was then added drop-by-drop with a syringe needle into 200 mL $CaCl_2$ solution with a concentration of 1.0 M. The MX-$SA_{4:20}$ reaction mixture was further agitated for 1 h, followed by thorough washing with DI water till the pH values to be in the range of 6.5–7. The washed powders were dried with a freeze dryer. Hybrid powders with different contents of $Ti_3C_2T_x$ were denoted as MX-$SA_{1:20}$, MX-$SA_{2:20}$, MX-$SA_{3:20}$, MX-$SA_{6:20}$ and GO-$SA_{4:20}$.

For comparison, GO-$AS_{4:20}$, MX-$SA_{4:20}$ and SA_{20} samples were prepared in a similar way. Obviously, the MX-$SA_{4:20}$ exhibited an adsorption capacity of nearly 100%, while those of the GO-$AS_{4:20}$ and SA_{20} were just 34.6% and 11.5%, respectively, as seen in Fig. 5.12. Although GO has a similar multi-layered structure, it has much less surface functional groups for the heavy metallic ions to occupy. Furthermore, $Ti_3C_2T_x$ MXene nanosheets were intrinsically hydrophilic, chemically stable and multifunctional, which was advantageous over GO. More significantly, the MX-$SA_{4:20}$ hybrid spheres had a cage-like structure to host the Hg(II) ions, thus leading to highest ion adsorption capacity.

The reduction in the adsorption equilibrium time was dependent on the content of the MXene phase, while the concentration of the alginate was kept unchanged up to the content ratio of 4:20, i.e., MX-$SA_{4:20}$. The MX-$SA_{4:20}$ hybrid sample exhibited very rapid adsorption kinetics, with 90% metallic ions to be removed in just 1 h, while entire elimination of the Hg(II) ions was achieved in 2 h, as revealed in Fig. 5.13a. In comparison, the 2 h removal efficiencies of MX-$SA_{1:20}$ and MX-$SA_{2:20}$ possessed adsorption capacities of 44% and 72%, respectively.

The recovery of the adsorbed Hg(II) ions was examined by using HCl solution. For instance, MX-$SA_{4:20}$ hybrid sample with Hg(II) ions was dispersed in HCl solutions with concentrations of 2, 4, 6 and 8 M for 5 h, before analysis with ICP-OES. It was found the HCl solution with a concentration of 8 M was able to desorb the Hg(II) ions by about 99.7%. In other words, the total recovery of the adsorbed Hg(II) ions from the surface of the MX-$SA_{4:20}$ hybrid spheres could only be achieved using HCl solutions with sufficiently high acidic level. In addition, the MXene-SA hybrid showed strong removal efficiency for Hg(II) from tap water, as shown in Fig. 5.13b.

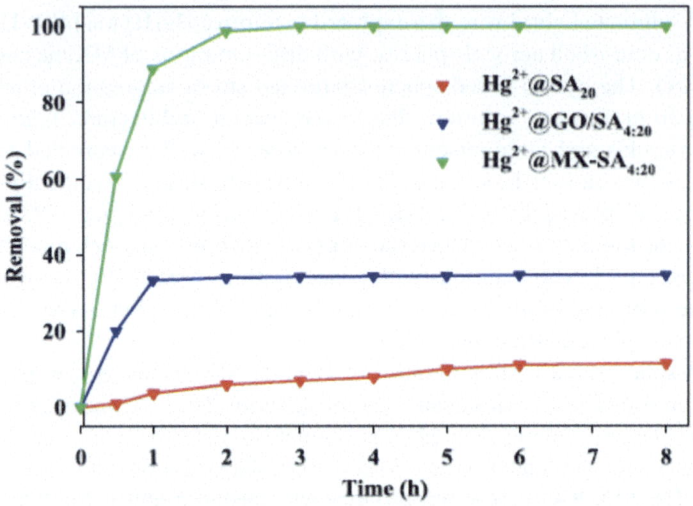

Fig. 5.12 Hg(II) adsorption capacities of the sodium alginate spheres (SA20), graphene oxide-sodium alginate spheres (GO-$SA_{4:20}$) and MX-$SA_{4:20}$ spheres, with the experimental parameters: 30 mL 15 ppm Hg(II), 42 mg adsorbing materials at pH = 5.5 and 298 K. Reproduced with permission from [40]. Copyright © 2019, Elsevier

A facial process was developed for the exfoliation of Ti_2AlC MAX, leading to either nanofibers (Alk-Ti_2C_{fibr}) or nanosheets (Alk-Ti_2C_{sheet}), the elimination of Cd(II) ion from aqueous solutions [41]. The MXene nanostructures displayed outstanding mechanical strength, high thermal stability and water resistance. They also contained oxygenated active sites, after the hydrothermal synthesis at alkaline conditions. Both the nanofiber and nanosheet-like structures were confirmed by using SEM and XRD. The nanostructures were further characterized by using TEM, Raman spectroscopy, Brunauer–Emmett–Teller (BET) surface area analysis, ζ-potential measurement and XPS, before and after metallic ion adsorbing process.

Specifically, the nanosized architectures showed an optimal Cd(II) ion adsorption efficiency of 326 mg/g [41]. The interactions between the hydroxyl groups of the MXene nanostructures and the Cd(II) ions were quantitatively characterized, including electrostatic interaction, adsorption-related oxidizing reaction and complex generation. The high efficient Cd(II) ion elimination was attributed to their Owing to their unique structure, high porosity, large specific surface area, and oxygenated functional groups, Alk-Ti_2C_{sheet} nanosheets were highly time-efficient for Cd^{2+} removal. Moreover, Alk-Ti_2C_{fibr} and Alk-Ti_2C_{sheet} nanostructures were tested for simulated groundwater, showing that synthesized nanostructures were capable for removing Cd^{2+} ions at the ppb level. The results obtained from this study suggested that nanostructures synthesized using this route could provide a new approach to prepare and exfoliate additional MAX phases for the removal of heavy metallic ions and other pollutants in the environment.

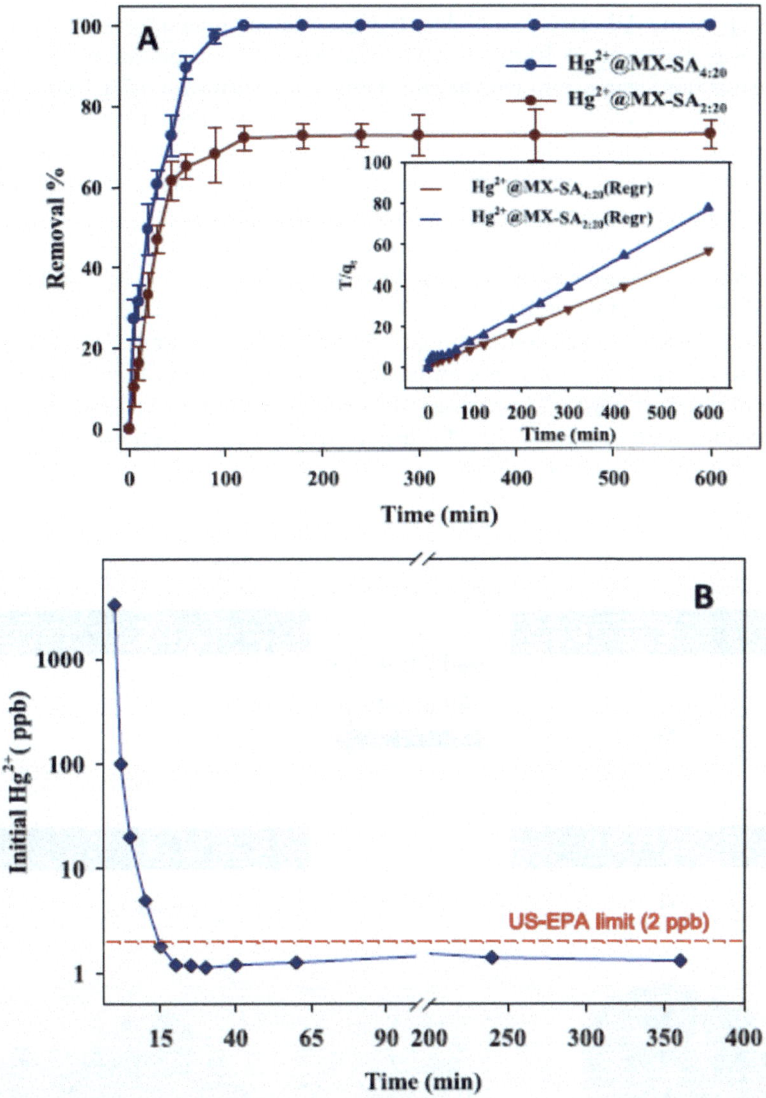

Fig. 5.13 **a** Kinetics profiles of the MX-SA$_{2:20}$ and MX-SA$_{4:20}$ samples to adsorb the Hg(II) ions, with both to follow the pseudo second-order kinetics regression, suggesting that the Hg(II) ions adsorbed onto the MX-SA$_{2:20}$ and MX-SA$_{4:20}$ to be governed by the chemisorption and rate-limiting characteristic of the adsorption reactions. **b** Kinetics to remove Hg(II) ions from tap-water with MX-SA$_{4:20}$. Reproduced with permission from [40]. Copyright © 2019, Elsevier

To prepare the Alk-Ti_2C_{fibr} and Alk-Ti_2C_{sheet} nanoarchitectures, 200 mg Ti_2AlC powder was dispersed in 40 mL NaOH solutions with concentrations of 1–10 M, forming reaction suspensions with the aid of magnetic stirring for 0.5 h. Hydrothermal reaction was conducted at temperatures of 120–200 °C, for time durations of 5–12 h. The reaction products were thoroughly washed with distilled water and alcohol, followed by vacuum drying process. Figure 5.14 shows a schematic diagram of the fabrication process to develop the MXene-based nanostructures.

Cd(II) ion solutions with concentrations of 19–561 ppm were used to examine the adsorption behaviour of the MXene-based nanostructures. Except for the MX-A obtained in 5 M NaOH at 100 °C for 12 h, all other samples demonstrated high adsorption capacities, especially at relatively low concentrations of the Cd(II) ion. It was found that 100 °C was not sufficiently high to etch out the Al layers from the parent phase of Ti_2AlC. In contrast, the nanosheet sample, Alk-Ti_2C_{sheet}, which was hydrothermally reacted at 200 °C, displayed distribution coefficients (K_d) in the range of $6.62 \times 10^2 - 4.79 \times 10^6$.

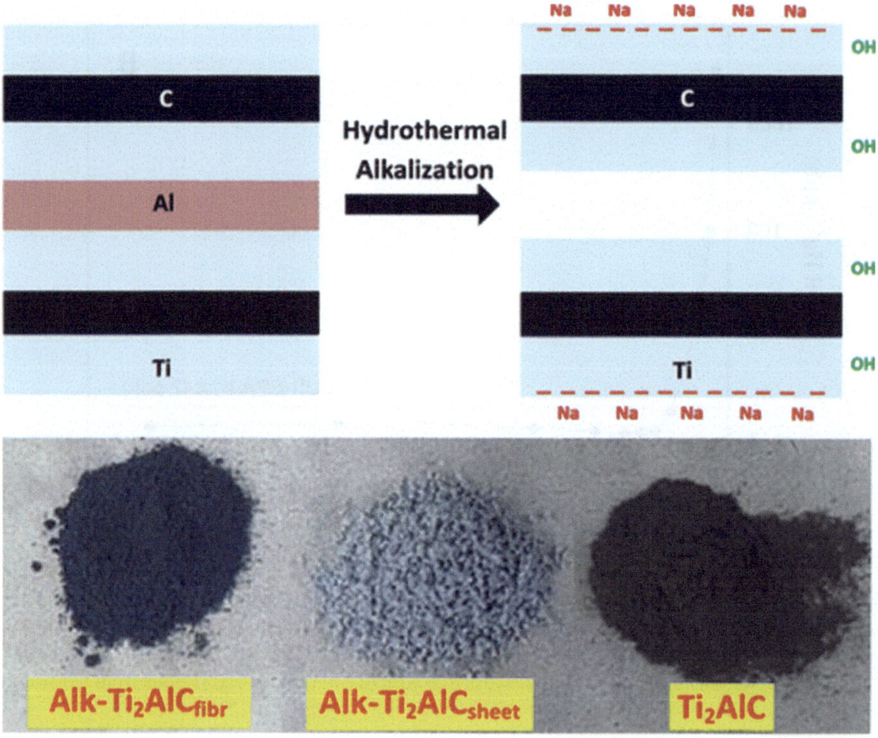

Fig. 5.14 Schematic d of diagram showing the etching process of the Al layer from the MAX phase of Ti_2AlC to obtain MXene nanofibers and nanosheets, together with photographs of the samples at different stages. Reproduced with permission from [41]. Copyright © 2019, American Chemical Society

According to Langmuir adsorption isotherm model, the theoretical adsorption capacity Alk-Ti_2C_{sheet} was much higher than those of graphene oxides and their various derivatives. Thus, the synthesized nanostructures are an appropriate adsorbent for the efficient capture of cadmium ions in the aqueous phase. The correlation coefficient (R^2) of the Langmuir model was larger than those of the Freundlich and Redlich-Peterson adsorption isotherm models, so that the former was more suitable to explain the experimental results, as illustrated in Fig. 5.15a. In describing the ulphide of liquid–solid adsorption with the Langmuir isotherm model, it is usually assumed that the adsorbent is covered by a monolayer, while the surface is homogeneous and all the active sites are nearly the same [26].

Figure 5.15b shows Cd(II) ion adsorption performance of the Alk-Ti_2C_{sheet} as a function of pH value. At pH = 2, the Cd(II) ion capture efficiency was as low as 44%, while it was sharply increased to about 99% at pH = 3. In the pH value range of 3–9, the adsorption efficiency was maintained essentially unchanged at a high level. The rapid increase in the adsorption capacity at higher pH values was attributed to the variation in properties of the Cd(II) ions and surface characteristics of the MXene nanostructure. At low pH values, the surface charge density of the Alk-Ti_2C_{sheet} was relatively low, so that hydrogen ions were competitively adsorbed instead of the Cd(II) ions.

The Cd(II) ion adsorption efficiencies of the Alk-Ti_2C_{sheet} with dosages of 2.5–10 mg/L at the Cd(II) ion concentration of 22 ppm are depicted in Fig. 5.15c. With increasing dosage of the MXene nanostructure, the Cd(II) ion adsorption capacity was monotonically increased. The influence of temperature on the adsorption efficiency of Cd(II) ion was not very strong in the range of 25–60 °C. However, the adsorption behaviour was slightly prompted at higher temperatures, as observed in Fig. 5.15d. The enhanced mass transport of the Cd(II) ion from the liquid to the surface of the Alk-Ti_2C_{sheet} nanosheets was responsible for the enhanced adsorption capacity.

It was found that Alk-Ti_2C_{sheet} had a stronger adsorption capability than Alk-Ti_2C_{fibr}. Specifically, at the Cd(II) concentration of 22 ppm, with MXene dosage of 10 mg, the times to achieve adsorption equilibrium were 1 h and 1.5 h when using Alk-Ti_2C_{sheet} and Alk-Ti_2C_{fibr}, respectively. Figure 5.15e shows adsorption efficiencies of the Alk-Ti_2C_{sheet} sample at different concentrations of Cd(II) ion, confirming the rapid adsorption kinetics. The Cd^{2+} elimination process could be expressed by the pseudo-first-order and second-order kinetics models. The Alk-Ti_2C_{sheet} sample exhibited an adsorption process with chemical characteristic and rate determination step.

It was observed that the presence of other ions, such as Cu(II), Pb(II), Zn(II) and Cr(III), had a negative effect on the Cd(II) adsorption with Alk-Ti_2C_{fibr} and Alk-Ti_2C_{sheet} nanoarchitectures. For instance, the Cd(II) elimination efficiencies of Alk-Ti_2C_{fibr} and Alk-Ti_2C_{sheet} were decreased to 35% and 66%, from 96.4% and 99.9%, respectively. Although the adsorption efficiency for Cd(II) ion was decreased due to the addition of other metallic ions, the results further confirmed both the Alk-Ti_2C_{fibr} and Alk-Ti_2C_{sheet} MXene nanostructures had enriched oxygenated biding sites, thus leading to the strong tendency to bind the divalent and trivalent metallic ions.

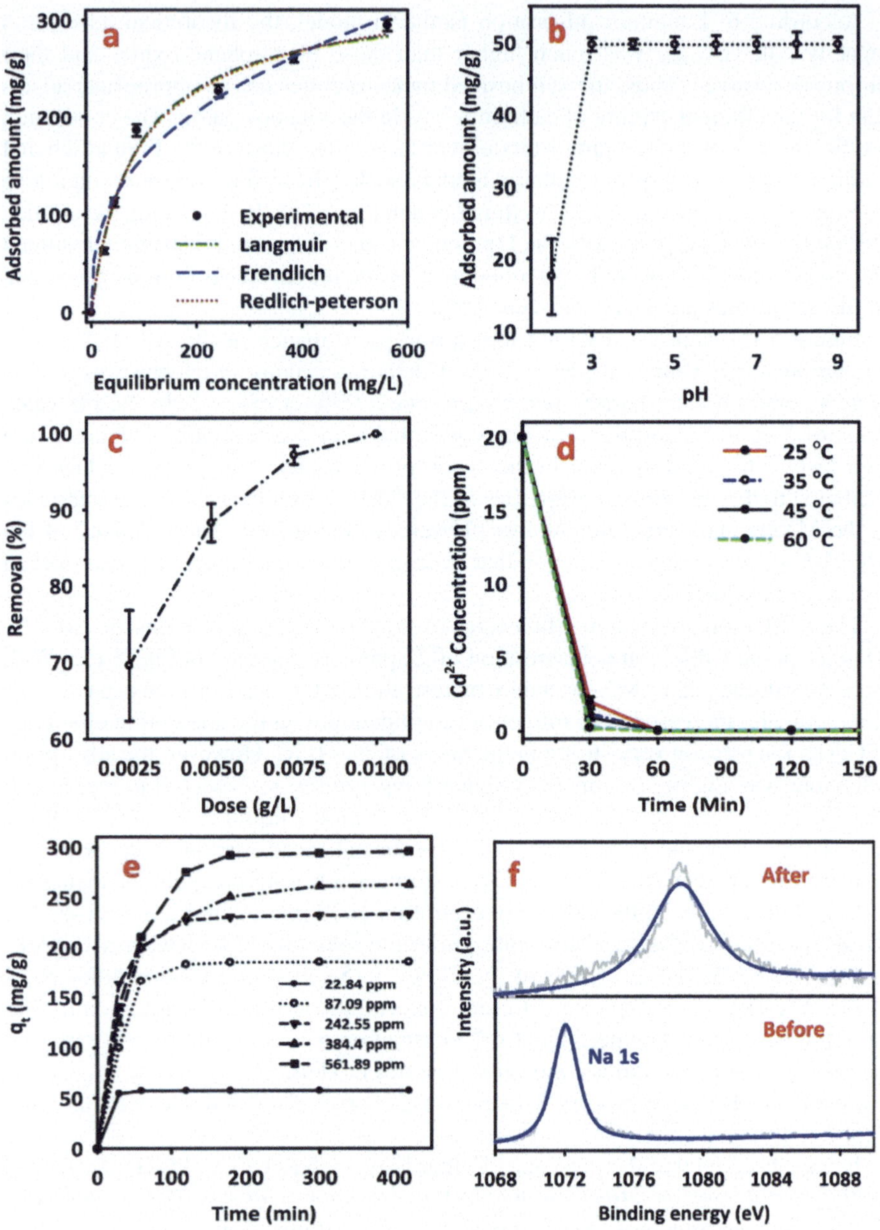

Fig. 5.15. Adsorption performances of the Alk-Ti2AlCsheet sample to remove Cd(II) ions in aqueous solutions: **a** adsorption isotherms, **b** adsorption capabilities versus pH value, **c** adsorption efficiencies versus the dosage of the MXene nanostructure, **d** adsorption effects at different temperatures, **e** adsorption kinetics at different initial metallic ion concentrations and **f** XPS component peak-fitting spectra of Na 1 s in the Alk-Ti2AlCsheet sample before and after the adsorption of Cd(II). Reproduced with permission from [41]. Copyright © 2019, American Chemical Society

The Alk-Ti_2C_{fibr} and Alk-Ti_2C_{sheet} MXene-based nanoarchitectures exhibited displayed a high recycling capability, which could be quickly recovered once they were treated with 0.1 M HCl solution for 5 h. After examining for 4 cycles, the adsorption capability of the Alk-Ti_2C_{sheet} sample retained to be 90.1%, while that of Alk-Ti_2C_{fibr} was lowered to 63.1%, as illustrated in Fig. 5.16a.

The mechanism of the adsorption of the Cd(II) ion by the MXene nanostructures has been proposed. The adsorption process of the Alk-Ti_2C_{sheet} for the Cd(II) ions in aqueous solution was through ion exchange and formation of complex on the MXene surface, as schematically depicted in Fig. 5.16b. pH value of the solution had a strong influence on the adsorption process. At the isoelectric point, $pH_{PZC} = 2.8$, below which the surface of Alk-Ti_2C_{sheet} was positively charged, so that the active sites and the Cd(II) ions were electrostatically repulsed. At pH > 2.8, the negative charge density on the surface of the MXene was increased, because of the increment in the number of negatively charged terminal groups, thus resulting in electrostatic attraction between the active sites and the Cd(II) ions. As a consequence, the Cd(II) adsorption capacities were 35% and 99% at pH values of 2 and 3, respectively.

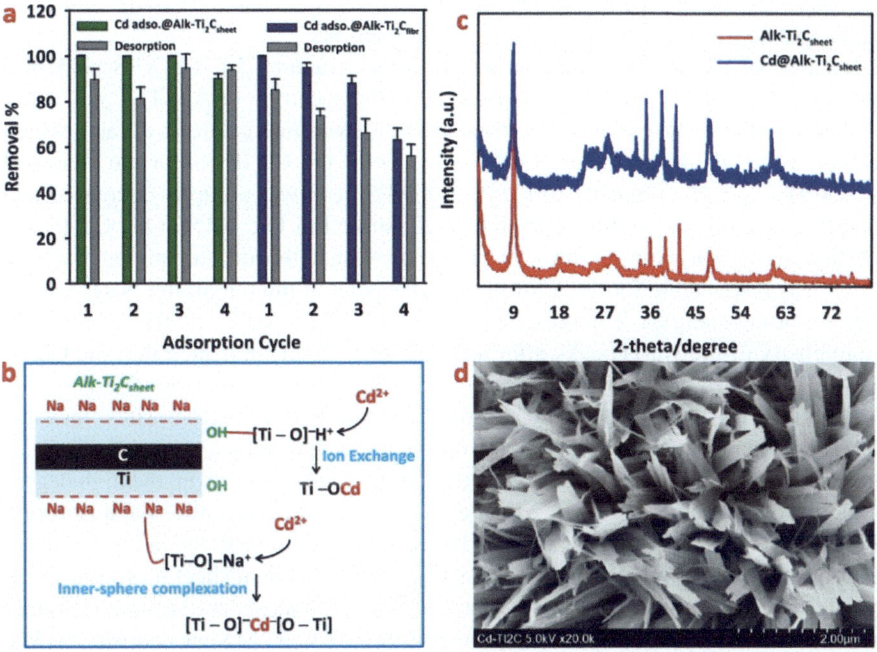

Fig. 5.16 **a** Cycling stability of Cd^{2+} removal onto Alk-Ti_2AlC_{fibr} and Alk-Ti_2AlC_{sheet} to remove Cd(II) ions. **b** Schematic diagram describing mechanism of Cd(II) ion adsorption with the Alk-Ti_2C_{sheet} sample. **c** XRD patterns of as-obtained Alk-Ti_2C_{sheet} and Cd(II)-adsorbed Alk-Ti_2AlC_{sheet}. **d** SEM image of the Alk-Ti_2AlC_{sheet} after Cd(II) adsorption. Reproduced with permission from [41]. Copyright © 2019, American Chemical Society

The MXene nanostructure contained $[Ti–O]^-Na^+$ and $[Ti–O]^-H^+$ groups on their surface, which was proved by using XPS, as seen in Fig. 5.15f. The strong adsorption of the metallic ions was achieved through the formation of complex items on the surface of the MXene nanoarchitectures. There are generally two surface complex structures, i.e., (i) ion-sphere complex with the metallic ions to be attached on the surface of the MXene phase through hydration and (ii) ion-self complex with water molecules without the link with the MXene nanostructures. The highly effective adsorption of the Cd(II) ions by the Alk-Ti_2C_{sheet} nanostructure was through the formation of ion-sphere complex on the MXene surface.

The Cd(II) ion adsorption of Alk-Ti_2C_{sheet} was additionally evaluated by using XPS analytic technique. After adsorption by Alk-Ti_2C_{sheet}, the Cd 3d had two peaks at binding energies of 411.48 eV and 418.26 eV, corresponding to $3d_{3/5}$ and $3d_{3/3}$, respectively. Without Cd(II) ion adsorption, the O 1 s peak of Alk-Ti_2C_{sheet} split into three peaks, at binding energies of 530.1, 530.69 and 532.23 eV, confirming the formation of $[Ti–O]^-Na^+$, Ti–O and Ti–OH, respectively. After Cd(II) ion adsorption, the O 1 s peaks upshifted to higher binding energies. The Ti 2p peak shift in the Alk-Ti_2C_{sheet} with the adsorption of Cd(II) ions suggested the presence of Ti–O–Cd bonds, i.e., there was strong interaction between the $[Ti–O]^-$ item and the Cd(II). Similarly, the Ti 2p peaks also shifted to higher binding energies.

Because of the sodium intercalation, into the structure, Na 1 s XPS peak at a binding energy of 1072.05 eV was observed in the spectrum of Alk-Ti_2C_{sheet}. After alkalization, the surface of the Ti_2C nanostructures was intercalated by Na(I), which promoted the adsorption of the Cd(II) ions. Once the Cd(II) ions were adsorbed by Alk-Ti_2C_{sheet}, the intensity of the Na 1 s XPS peak was abruptly decreased and upshifted to a higher binding energies, as illustrated in Fig. 5.15f. One Cd(II) ion substituted two Na(I) ions to form $Na^+(H_2O)_n$, owing to the ion exchanging reaction, which was responsible for the shift in the XPS peak in the remaining Na(I) ions to higher binding energies. The presence of the Cd 3d XPS peaks and the reduction in the intensity of the Na(I) peaks after adsorption of Cd(II) indicated that the Cd(II) ions took the sites of Na(I).

Because the electrostatic interaction between surface active sites and the $Na^+(H_2O)_n$ item was sufficiently strong, the charges at the surface of the Alk-Ti_2C_{sheet} nanostructures were mainly localized at the $[Ti–O]^-$ sites. Moreover, the high degree crystallinity of the Alk-Ti_2C_{sheet} MXene phase was not altered after the adsorption of the Cd(II) ions, as evidenced by the XRD results, as demonstrated in Fig. 5.16c. In addition, the Cd(II) ion adsorbed Alk-Ti_2C_{sheet} retain its nanosheet morphology, as revealed in Fig. 5.16d. In other words, the Alk-Ti_2C_{sheet} nanosheets had much stronger mechanical stability, as compared with the common $Ti_3C_2T_x$ MXene nanosheets [37].

5.2.2 *Removal of Organic Dyes*

The pollution of organic dyes in water has also been one of the most serious environmental and health concerns. Due to their stable molecular compositions, dyes are not biodegradable. Dyes could be present in water as either cations or anions, so that they could be removed by using appropriate adsorbents. It has been shown that organic dye pollutants can be eliminated by using MXenes. The adsorbed organic dyes could be further degraded through the photo-catalytic effect of UV irradiation [42, 43].

$Ti_3C_2T_x$ MXene has been studied for the applications in adsorption and photocatalytic decomposition of organic pollutants [44]. It was very efficient in adsorbing the cationic dye methylene blue (MB), with an adsorption capacity of about 39 mg/g. The adsorption of MB on the $Ti_3C_2T_x$ nanosheets followed the Langmuir and Freundlich isotherms and the Freundlich model, with fitting parameters ν and K_F to be 19.96 and 48.15, respectively. Figure 5.17 shows experimental data and the fitting results. According to the strength of binding (K_L), derived from the Langmuir formula, the dye was bonded onto the $Ti_3C_2T_x$ surfaces in aqueous solutions, in a strong and irreversible way. The interaction between $Ti_3C_2T_x$ and aqueous MB in dark included three stages, i.e., (i) active adsorption of MB molecules onto surface of the $Ti_3C_2T_x$ nanosheets in the first 2 h, (ii) decrease in the degree of ordering of the stacking disorder because of wedging of the layered structure and chemical MXene transformation in 18 h and (iii) oxidation of $Ti_3C_2T_x$ to form TiO_2 thereafter. Moreover, the degradation of MB and AB80 dye was accelerated as their solutions was irradiated with UV in the presence of $Ti_3C_2T_x$.

Dye elimination efficiencies of the $Ti_3C_2T_x$ sample are depicted in Fig. 5.17a. Obviously, after reaction for 20 h, the concentration of the AB80 solution was almost unchanged, while that of the MB solution was rapidly decreased in 8 h. Therefore, it was concluded that the $Ti_3C_2T_x$ MXene phase preferentially adsorb MB, a cationic dye, MB, simply because the electrostatic interaction between the molecules of MB and the surface of the MXene nanosheets was much stronger than that between the AB80 and the MXene phase in dark. Figure 5.17b shows adsorption isotherms of MB adsorption on the $Ti_3C_2T_x$ nanosheets.

Specifically, the value of K_L was relatively high, which is a measure of the adsorbate-adsorbent binding strength, implying that the MB molecules were firmly bonded with the $Ti_3C_2T_x$ nanolayers. The separation factor (R_L), derived from the Langmuir isotherm curve, as a function of the starting concentration of MB, is illustrated as the inset in Fig. 5.17b. The values of R_L were at the scale of 10^{-4} over the MB concentration range, suggesting that the adsorption was nearly irreversible [44]. Freundlich isotherm displayed higher Pearson's correlation coefficient © than the Langmuir one, which confirmed that the Freundlich model was more suitable to fit the experimental data, as mentioned above. This conclusion was also evidenced by the value of $\nu > 1$. In other words, the MXene nanosheets contained heterogeneous adsorption sites on their surfaces, due primarily to the presence of the functional groups of –OH, =O and –F.

As illustrated in Fig. 5.17a, as the 0.05 mg/mL MB solution was suspended with the $Ti_3C_2T_x$ nanosheet powder, the MB concentration started to decline rapidly in the first several hours and the concentration reduction lasted for a quite long time. To clarify the underlying mechanism of the gradual reduction in the concentration of MB, the variation in structure of the $Ti_3C_2T_x$ sample was carefully examined.

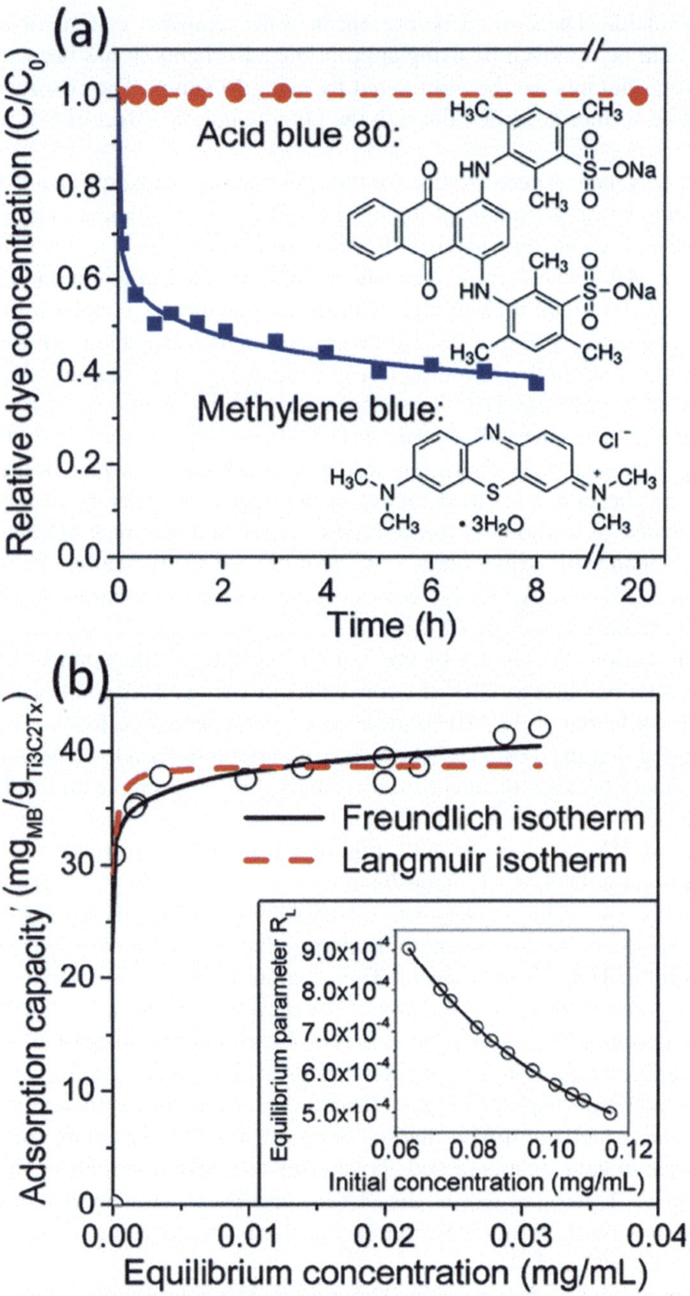

Fig. 5.17 **a** Time dependence of the concentrations of MB ($C_0 = 0.05$ mg/mL) and AB 80 ($C_0 = 0.06$ mg/mL) in aqueous solutions with the presence of $Ti_3C_2T_x$ nanosheets in dark, with the insets showing chemical structures of the dyes. **b** Adsorption isotherms of MB on $Ti_3C_2T_x$, with the inset showing dependence of the equilibrium parameter R_L with the initial concentration of MB. Reproduced with permission from [44]. Copyright © 2014, Royal Society of Chemistry

According to XRD patterns of the recycled MXene nanosheet samples, as demonstrated in Fig. 5.18. After adsorption reaction for 2 h in dark, no shift in the position of the (0002) peak was observed, while 20 h adsorption resulted in a slight shift to lower angle. Meanwhile, the peak was largely widened and weakened, implying that the ordered structure in the [0001] direction was slightly damaged. It was highly possible that the $Ti_3C_2T_x$ laminated structure was intercalated by the MB molecules.

With estimation from the XRD patterns, the *c*-lattice parameter (*c*-LP) was enlarged by about 0.2 nm. However, the MB molecule has a dimension of $1.7 \times 0.76 \times 0.325$ nm^3. In this case, if the MB molecules were intercalated into the interlayer spaces of the $Ti_3C_2T_x$ nanosheets in parallel to the surface, the *c*-LP would be enlarged by 0.325 nm, which was much larger than the experimental result. This implied that the MB molecules were not entirely intercalated into the MXene interlayer space. Therefore, further studies should be conducted to clarify the difference in the variation in *c*-LP between the expected result and the experimental observation.

A hot alkaline solution treatment process was used to enlarge the interlayer spacing and tailor the surface functional groups of $Ti_3C_2T_x$ MXene, in order to enhance its adsorption performance to organic dye pollutants [45]. After being treated with LiOH solution, the interlayer spacing of $Ti_3C_2T_x$ MXene was enlarged by 29%, while the surface functional group –F was replaced by –OH. Comparatively, the LiOH and

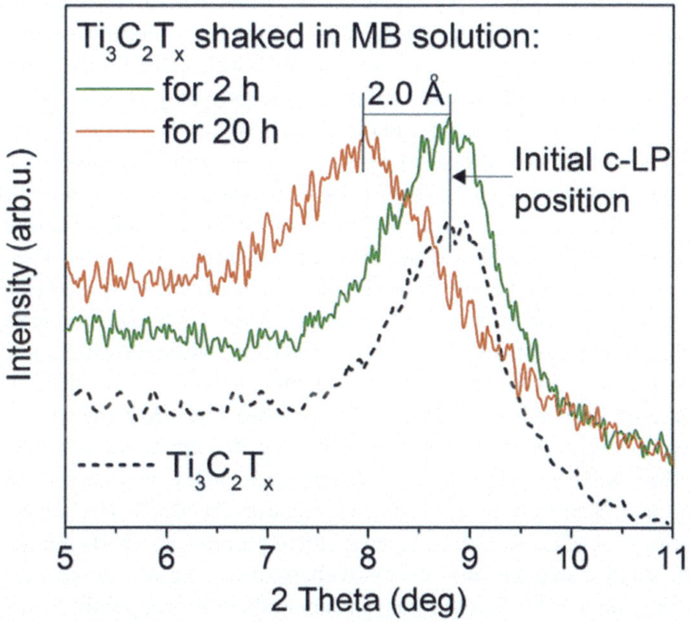

Fig. 5.18 XRD patterns of the (0002) peak of $Ti_3C_2T_x$ before and after merging in 0.05 mg/mL MB aqueous solution for 2 and 20 h. Reproduced with permission from [44]. Copyright © 2014, Royal Society of Chemistry

NaOH treated $Ti_3C_2T_x$ MXenes showed much higher absorption capacities. Specifically, the NaOH–$Ti_3C_2T_x$ MXene exhibited an adsorption capacity of 189 mg/g to MB, which was the highest. This outstanding performance was attributed to the synergistic effect of surface adsorption and intercalation adsorption. The experimental data could be well fitted by the Langmuir isotherm.

TiO_2/Ti_3C_2 hybrid was synthesized by using a hydrothermal reaction route as photocatalyst to remove methyl orange (MO) [46]. The photocatalytic activities of pure TiO_2 and the TiO_2/Ti_3C_2 nanohybrid for the photo-degradation of MO, under the irradiation of UV, were compared. It was demonstrated that the TiO_2/Ti_3C_2 hybrid showed much stronger activity than pure TiO_2, because the TiO_2/Ti_3C_2 had higher efficiency in the electron–hole separation with the UV irradiation. Specifically, for the dosages of 0.001, 0.002 and 0.003 mol, MO was degraded by nearly 98% in 0.5 h for the/Ti_3C_2 nanohybrid. The photocatalysis efficiency of 0.004 mol TiO_2/Ti_3C_2 nanocomposites is relatively worse. In comparison, the degradation rate with pure TiO_2 was just 77%, while that value with pure Ti_3C_2 was 42%.

TiO_2/Ti_3C_2 nanohybrid, with the TiO_2 nanosheets to expose (001) facets on the surface of layered Ti_3C_2, was synthesized by using hydrothermal method in which the TiO_2 nanophase was obtained due to the partial oxidation of Ti_3C_2 [47]. The *in-situ* formation of the TiO_2 nanosheets had more perfect interface with the MXene nanolayers. Since the (001) facets of TiO_2 were highly active, TiO_2/Ti_3C_2 nanohybrid exhibited high efficiency in the photogeneration of electron–hole pairs. At the same time, the carrier separation was facilitated through the hole trapping effect due to the interfacial Schottky junction, serving as a reservoir of holes. Specifically, the effective charge separation effect and the exposed active facets synergistically offered a high efficiency for the photocatalytic degradation of MO dye.

CeO_2/Ti_3C_2 hybrid was prepared by using a one-step hydrothermal reaction process, with CeO_2 nanorods to be uniformly distributed on the surface of the Ti_3C_2 nanosheets [48]. The nanohybrid showed outstanding photocatalytic activity for the photodegradation of Rhodamine B under the irradiation of UV-light. The enhancement in the photocatalytic activity was readily ascribe to the increased efficiency in the utilization of the solar energy. After irradiation for 1.5 h, the dye was degraded by 75% in the presence of the CeO_2/Ti_3C_2 nanohybrid, while the efficiencies of pure CeO_2 and Ti_3C_2 were 24% and 63%, respectively.

A hydrothermal process was employed to prepare Ti_3C_2 MXene (*h*-Ti_3C_2), with less-toxic etching agents of $NaBF_4$ and HCl, for the applications in removal of methylene blue and methyl orange dyes from water [49]. Conventional HF etching process was also used to obtain *t*-Ti_3C_2. Comparatively, the *h*-Ti_3C_2 was advantageous in various aspects, including lager *c*-lattice parameter, interlayer spacing and BET specific surface area, because the hydrothermal etching was proceeded in a mild way. Therefore, the *h*-MXenes exhibited stronger adsorption capabilities to eliminate methylene blue and methyl orange dyes. In addition, the method was also applied to prepare Nb_2C.

Figure 5.19a shows dye adsorption characteristics of two MXenes. Neither *h*-Ti_3C_2 nor *t*-Ti_3C_2 had adsorption capability to remove MO, because the two Ti_3C_2 MXenes were negatively charged, so that they were repulsed with the anionic MO

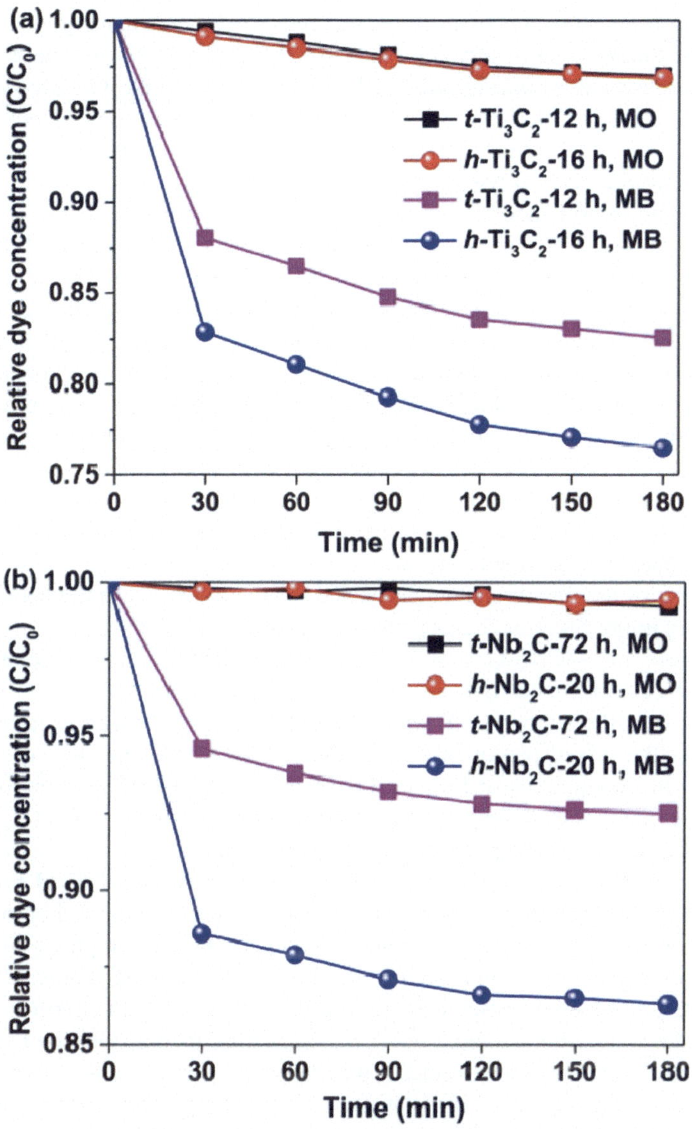

Fig. 5.19 Relative dye concentrations versus time of the two MXene samples: **a** Ti_3C_2 and **b** Nb_2C. Reproduced with permission from [49]. Copyright © 2018, Elsevier

dye, as discussed earlier. In comparison, the Ti_3C_2 MXenes showed strong adsorption to the cationic dye of MB. After adsorption reaction for 3 h, the relative concentrations of MB were 76.4% and 82.7% for the *h*-Ti_3C_2–16 h and *t*-Ti_3C_2–12 h, respectively. The adsorption capacity was estimated to be 24 $mg_{MB}/g_{adsorbent}$ for a testing period of 2 h.

Dye adsorption behaviors of the Nb_2C MXenes are illustrated in Fig. 5.19b. Nearly no adsorption of MO was observed for the Nb_2C MXene, which was similar to that of Ti_3C_2. The relative dye concentration of MB after 3 h over *h*-Nb_2C-20 h was 86.2%, while that over *t*-Nb_2C-72 h was 92.4%. Comparatively, the adsorption performance of Nb_2C was lower than that of Ti_3C_2, which was mainly attributed to the fact that the relative atomic mass of Nb ($m_{Nb} = 92.9$) was larger than that of Ti ($m_{Ti} = 47.8$).

GO-MXene hybrid membranes with promising substrate rejection and permeability were prepared for wastewater treatment [4]. The performance of the hybrid membranes could be readily tailored by varying the ratios of GO and MXene. The sample with a GO/MXene mass ratio of 1/4 displayed water flux of about 72 L/m^2·h·bar, while the value of the GO membrane was only 6.5 L/m^2·h·bar, because of the unique 2D interlayer channel structure and the hydrophilic property of the MXene phase. In addition, the GO/MXene hybrid membranes possessed high long-term stability in water. The efficiency to reject the commonly encountered organic dyes, such as NR, MB, CV, BB, was over 99.5%. The membranes were also very effective in eliminating natural organic items in raw water resources, such as HA and BSA. The superior water flux of the GO/MXene hybrid membranes over that of the GO ones was ascribed two factors, i.e., (i) the proper increase in the interlayer spacing of the membrane materials and (ii) the reduction in the density of the oxygen-containing functional groups in the materials.

To prepare the GO/MXene hybrid membrane samples, commercially available mixed cellulose ester (MCE) membranes with a pore diameter of 0.45 μm were utilized as the supports, on which GO/MXene hybrid layers were deposited by filtrating the mixed precursor suspensions. The GO-MXene suspensions were prepared with total quantity of 1 mg in 100 mL DI water. GO/MXene hybrid samples GO/MXene membrane samples, with quantity ratios of 1/0, 4/1, 2/1, 1/1, 1/2, 1/4, 0/1, were fabricated for the purpose of systematic study.

Figure 5.20 shows representative surface and cross-sectional SEM images the GO/MXene hybrid and the component membranes. The GO sample showed slightly wrinkled surface, with a thickness of about 0.55 μm. All the samples were compositionally homogeneous as evidenced by the Ti element mapping profile, confirming that the GO and MXene phases were uniformly assembled in the hybrids. The GO sample had a water contact angle of 44.7°, suggesting its high surface hydrophilic level, because the GO nanolayer surface was densely decorated with various oxygen-related functional groups, such as –COOH, –OH and epoxy groups. The water contact angle was gradually increased with increasing content of MXene in the hybrids, where the increase in hydrophobic level was closely related to the incorporation of the MXene phase.

XRD patterns of the dried and soaked GO/MXene membranes were employed to clarify the interactions between the 2D nanolayers and the molecules of water. The GO sample had *d*-spacing of 0.73 nm, while the MXene sample exhibited a *d*-spacing of 1.45 nm. As expected, as the composition of MXene was increased, the *d*-spacing of the hybrid samples was enlarged. Because MXene nanosheets had larger thickness than the GO nanosheet, the spacing between the every two adjacent nanolayers from center to center in the hybrid membranes was gradually increased.

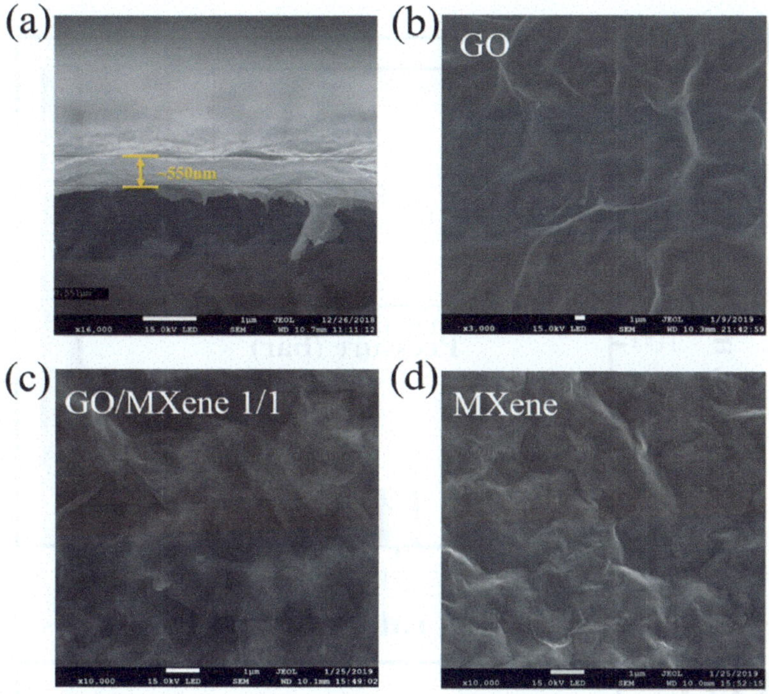

Fig. 5.20 SEM images of the samples: **a** cross-section view of GO/MXene hybrid membrane with mass ratio of 1/1, **b** surface view of GO, **c** surface view of GO/MXene and **d** MXene membranes (scale bar = 1 μm). Reproduced with permission from [4]. Copyright © 2020, Elsevier

Similarly, the *d*-spacing of the hybrid membranes would be increased after they were wetted.

In the wet membrane, the π–π interaction between the nanolayers was greatly reduced, due to the interaction between the molecules of water and the GO or MXene nanosheets, closely related to the hydrogen bonding and electrostatic charge. Meanwhile, the intensity of the XRD diffraction peaks of GO was decreased, as the content of MXene was increased in the hybrid membranes, since the long-range stacking of the GO nanolayers was interrupted owing to the incorporation of the MXene phase.

Figure 5.21 depicts filtration efficiencies of the hybrid membranes with different compositions, by measuring water flux and rejection rate of organic pollutants in water. With increasing content of the MXene phase, the water permeability through the membrane was increased, due to several reasons. This can be explained as follows. On one hand, the π–π bonding strength between the GO nanolayers was decreased due to the introduction of the MXene phase. Accordingly, the interlayer spacing of the MO nanolayers was enlarged, thus leading to enhanced flux. On the other hand, the strength of the hydrogen bonding between the molecules of water and the oxygen-related functional groups at oxidized sites of the GO nanolayers was reduced,

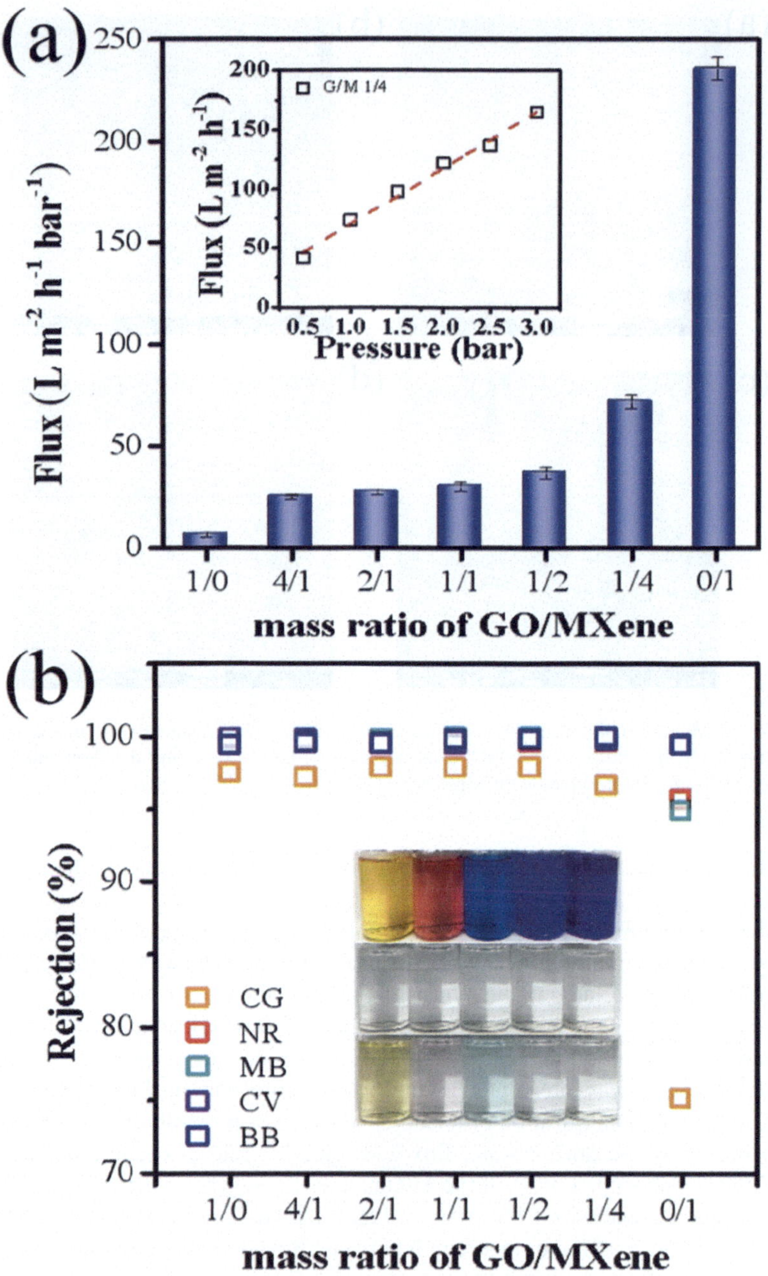

Fig. 5.21 Filtration performances of the GO/MXene hybrid membranes: **a** variation of flux of the GO/MXene hybrid samples with different ratios with the inset showing the flux as a function of the pressure gradient (ΔP) and **b** variation of substrate rejection of the GO/MXene samples for several dyes (CG, NR, MB, CV, BB) with the inset showing the photographs of the dyes dissolved in water before (top) and after filtration through the hybrid membrane with mass ratio of 1/4 (middle) and the MXene membrane (bottom). Reproduced with permission from [4]. Copyright © 2020, Elsevier

resulting in a decrease in the resistance to the water flow and hence an increase in flux.

As water molecules travelled through the nanosized capillary channels inside GO hybrid membranes, it was possible to form hydrogen bonding between the water molecules and the oxygen-related functional groups, so that the membranes were more resistive to the water flow, especially at the oxidized sites on the surface of the nanosheets. As revealed by the pressure-filtration experimental data, the water flux had a linear relationship with the gradient of differential pressure (ΔP), which passed through the GO/MXene hybrid membranes, as shown as the inset in Fig. 5.21a. In other words, it was a laminar flow, over the pressure range of 0.5–3 bar. At the same time, the capillary nanochannels of the hybrid membranes were sufficiently strong.

Dye molecular sieving performance or substrate selectivity of the GO/MXene hybrid membranes were characterized, in terms of vacuum filtration of dye solutions. Figure 5.21b shows rejection rates of the GO/MXene hybrid membranes for five dye molecules, which were selected according to their charging characteristics. The hybrid membranes exhibited rejection efficiency to CG was about 97%, which was the lowest value among the five dye examples. Similarly, the rejection rate of the pure MXene membrane to CG was very low, because of the relatively large interlayer spacing. Therefore, the high rejection rate of the hybrid membranes was primarily ascribed to the size excluding effect. For the rest four dyes, rejection efficiencies were all close to 100%, owing to the concurrent effect of size exclusions and electrostatic interactions, between the molecules of the dyes and the oxygen-related functional groups. The MXene membrane had also rejection efficiencies of 99.5% to CV and BB, because both of them possessed relatively heavy molecular weights and large ion radius.

An *in-situ* fabrication process was devised to prepare MXene-Fe_3O_4 hybrids, which exhibited strong adsorption efficiency to remove MB [50]. The characteristic laminated 2D structure was well retained in the nanohybrid, while it was superparamagnetic in nature with a magnetization of 20.3 emu/g. According to Zeta potential results, the surface of the MXene-Fe_3O_4 hybrid nanosheets was negatively charged in neutral solutions, thus leading to high elimination efficiency for cationic dyes. In addition, a removal efficiency of 91.9% was observed at a high temperature of 55 °C, which was accompanied by decolorization of the mixture. The adsorption process of MB followed the Freundlich isotherm model at high temperatures of 40 and 55 °C, while Langmuir isotherm model was valid at low temperature of 25 °C. Figure 5.22 shows elimination efficiencies of the MXene-Fe_3O_4 hybrid powder to remove MB from the aqueous solution at different temperatures.

Thermodynamic data revealed that the adsorption process of MB by the MXene-Fe_3O_4 nanohybrid was exothermic and chemisorptive characteristics [50]. As a result, the removal efficiency of the MXene-Fe_3O_4 for MB was boosted as the temperature was increased. It was found that the Ti–OH groups on surface of the MXene-Fe_3O_4 nanohybrid were crucial in the decolorization process of MB, due to the formation of hydrogen bonding, Ti–OH$\cdots$N. Therefore, electrostatic interaction was main driving force at high temperatures, whereas surface adsorption through electrostatic attraction was responsible for the elimination of MB at low temperature of 25 °C.

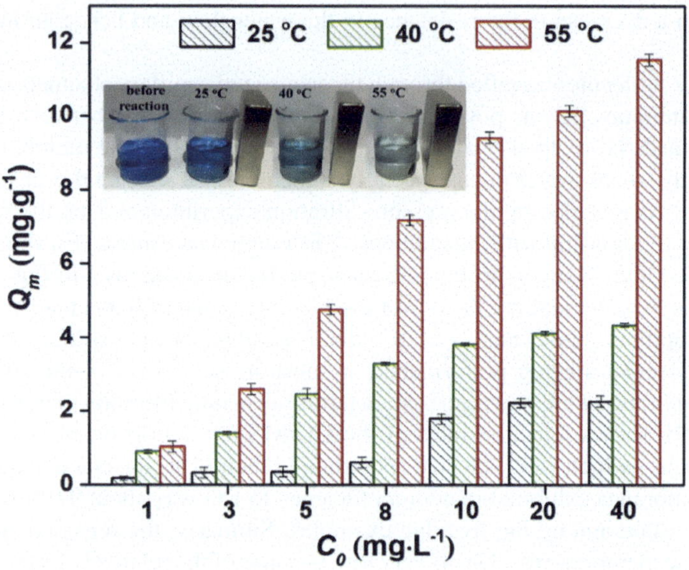

Fig. 5.22 MB adsorption efficiencies of the MXene-Fe_3O_4 hybrid powder at different temperatures (25, 40 and 55 °C), with the inset showing the photographs of the MB aqueous solutions before and after the adsorption process at different temperatures (experimental conditions: MXene-Fe_3O_4 dosage = 1 g/L, experiment time duration = 24 h). Reproduced with permission from [50]. Copyright © 2019, Elsevier

Ti_3C_2–OH/In_2S_3/CdS ternary hybrid was synthesized by using a simple hydrothermal process, for the application as visible-light photocatalysis to degrade Rhodamine B (RhB) and methyl orange (MO) [51]. The hybrid sample had a specific surface area of about 68 m^2/g, with a strong photocatalytic effect and removal efficiency to get rid of RhB when irradiating with visible light. The photocatalytic performance was much higher than those of the pristine CdS and In_2S_3. Moreover, the Ti_3C_2–OH/In_2S_3/CdS hybrid powder was also of strong photocatalytic activity to degrade the mixture of RhB and MO. Meanwhile, the photocatalytic efficiency was higher than that of its GO/In_2S_3/CdS counterpart. The active species, hole (h^+) and $^{\bullet}O_2^-$, were crucial components in promoting the reaction between the 2D hybrid and the organic dyes. In this case, the rapid separation of the charge carriers and charge transfer of the Ti_3C_2–OH nanosheets and the concurrent effect of CdS/In_2S_3 were responsible for the superior photocatalytic effect of the ternary hybrid.

5.2.3 Elimination of Radionuclide Pollutants

MXenes have been shown to have strong capability to withstand radiation and chemical compatibility to molten salts, so that they could be used as materials for the applications in nuclear waste treatment. The adsorption capability of uranyl items, with a general formula of $UO_2(L_1)_x(L_2)_y(L_3)_z$, where L_1, L_2 and L_3 represented the ligands of H_2O, OH and CO_3, respectively, on hydroxylated $V_2C(OH)_2$ MXene nanosheets, was studied by using density functional theory simulation methods [52]. It was found that the uranyl items were strongly bonded to the hydroxylated MXene nanosheets, with binding energies in the range from − 3.3 to − 4.6 eV, which thus indicated that the $V_2C(OH)_2$ MXenes were potential adsorbent to remove uranyl ions. This was simply because two U–O bonds could be formed with the hydroxylated MXene nanosheets. Furthermore, the axial oxygen atoms in the uranyl ions formed hydrogen bonds with the hydroxylated V_2C, which enhanced the adsorption capability of the MXene. In comparison, U–F bonds were much weaker than the U–O ones, implying that the –F functional group of the MXene was not effective in removing uranyl ions.

Hydroxylated $Ti_3C_2(OH)_2$ was studied as an example of MXenes, for the adsorption of uranyl ions ($UO_2{}^{2+}$) in aqueous environments, by using density functional theory simulations [53]. It was predicted that the $UO_2{}^{2+}$ ions were bonded with $Ti_3C_2(OH)_2$ nanosheets in aqueous solution, no matter whether there were anionic ligands of OH^-, Cl^- and $NO_3{}^-$. The bidentate coordination of $UO_2{}^{2+}$ tended to sit on surface of the MXene nanosheets, as compared with other adsorption configurations. Meanwhile, the $UO_2{}^{2+}$ ions were preferentially connected with the deprotonated –O adsorption sites, instead of the hydroxylated surface. Both chemical and hydrogen bonds were present in the adsorption process. Theoretical prediction indicated that, if the stable adsorption configuration was adopted, the adsorption capacity of $[UO_2(H_2O)_5]^{2+}$ could reach 595.3 mg/g.

It is understood that the interlayer spacing of the $Ti_3C_2T_x$ MXenes was still small when they were used to adsorb the relatively large hydrated radionuclide ions, thus leading to low adsorption capacity. However, the ulphidered $Ti_3C_2T_x$ MXenes could be readily delaminated into nanoflakes, so that the adsorption capacity could be significantly increased. Meanwhile, the adsorption behaviours of the MXene for the removal of U(VI) ions had been further enhanced through hydration [54]. It has been theoretically confirmed that the interlayer spacing of MXenes was enlarged, due to the formation of hydrophilic groups and the presence of van der Waals weak binding, which resulted in enhanced adsorption capacity of U(VI) ions.

V_2CT_x MXene was demonstrated to be uranium (U(VI)) sorbent, with a high uptake capacity of 174 mg/g [55]. Meanwhile, the absorbent had fast sorption kinetics and high selectivity. The experimental data abided a heterogeneous adsorption model, because the MXene functioned with its heterogeneous adsorption sites. According the density functional theory calculation results, combined with the outcome of X-ray absorption fine structure characterizations, the U(VI) ions tended to be coordinated with the –OH groups at the V-sites on surface of the MXene nanosheets, through the formation of the bidentate inner-sphere complexes.

Figure 5.23a shows pH-dependent U(VI) sorption behavior of the multilayered V_2CT_x. The uptake capacity increased rapidly with increasing pH value of the solutions and approached the maximum level of 174 mg/g at pH = 5.0. In fact, U(VI) hydrolyzed as the pH value was high than 5.0 and concentration reached 100 mg/L. In this case, the V_2CT_x nanosheets were negatively charged with the pH values in the range of 2.5–7.0. As the pH was increased, the zeta potential of V_2CT_x suspensions were more and more negative, suggesting the electrostatic interaction between the sorbent and U(VI) ions were gradually enhanced. As a result, the removal efficiency of U(VI) was increased. XRD results indicated that both the position of (0002) of V_2CT_x was unchanged after the sorption of U(VI) was conducted for 4.5 h, while the degree of ordering of the layered structures was slightly damaged.

Figure 5.23b shows sorption kinetic properties as the initial U(VI) concentration was 100 mg/L, at pH = 4.5, with sorption reactions for 5–360 min. The sorption kinetics consisted of two steps, i.e., (i) rapid sorption in the early 20 min and (ii) slow sorption approaching an equilibrium in 4.5 h. Obviously, the experimental sorption kinetics followed the pseudo-first-order kinetic model and pseudo-second-order kinetic model. The sorption capacity given by the model fitting was 143 mg/g, which

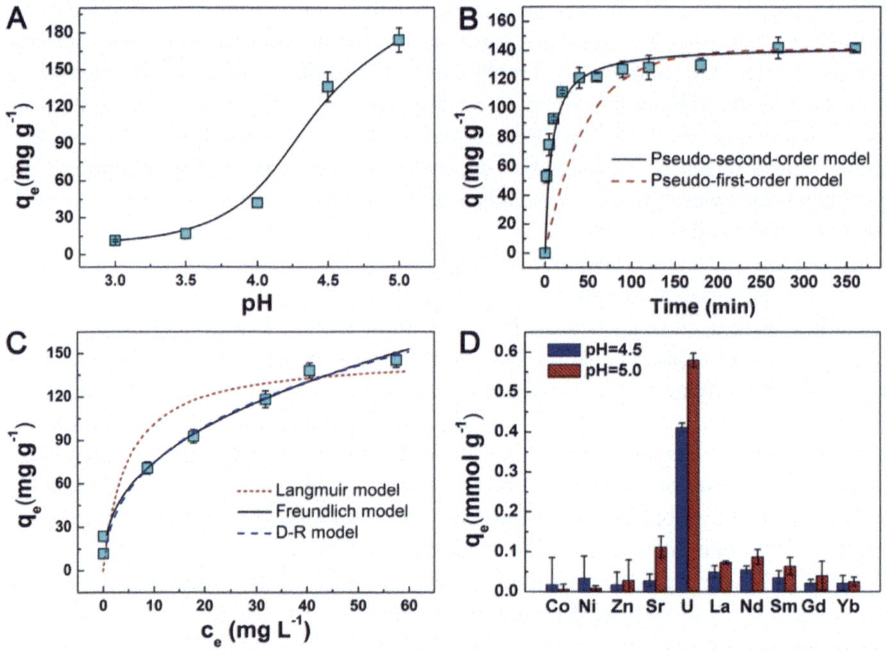

Fig. 5.23 U(VI) sorption performances from aqueous solution onto multilayered V_2CT_x: **a** pH, **b** contact time, **c** initial U(VI) concentration and **d** competing metal cations. Detailed experimental parameters are: **a** $m_{sorbent}/V_{solution}$ = 0.4 mg/mL, $[U]_{initial}$ = 100 mg/L, **b** pH = 4.5 ± 0.1, $m_{sorbent}/V_{solution}$ = 0.4 mg/mL, $[U]_{initial}$ = 100 mg/L, **c** $[U]_{initial}$ = 5–120 mg/L, $m_{sorbent}/V_{solution}$ = 0.4 mg/mL, pH = 4.5 ± 0.1 and **d** $[U]_{initial}$ = $[M]_{initial}$ = 0.42 mmol/L, $m_{sorbent}/V_{solution}$ = 0.4 mg/mL. Reproduced with permission from [55]. Copyright © 2016, American Chemical Society

was almost the same as the experimental equilibrium uptake capacity of 142 mg/g. Therefore, the sorption of U(VI) by the V_2CT_x MXene nanosheets was chemical adsorption-controlled process.

The sorption isotherm is shown in Fig. 5.23c. The U(VI) sorption displayed rapid increment at the concentration of U(VI) to be <10 mg/L, which corresponded to a sorption capacity of 70 mg/g. After that, the sorption rate was decreased. Three models, including Langmuir, Freundlich and Dubinin-Radusckevich (D-R), were employed to reveal the sorption mechanisms. Among them, the Freundlich and D-R models well described the sorption process, as seen in Fig. 5.23c, because the V_2CT_x MXene nanosheets contained numerous heterogeneous adsorption sites.

The sorption of U(VI) by the V_2CT_x MXene was examined in terms of selectivity, by using various competing cations, such as Co(II), Ni(II), Zn(II), Sr(II), La(III), Nd(III), Sm(III), Gd(III) and Yb(III), at pH levels of 4.5 and 5.0. Figure 5.23d shows the experimental results. The V_2CT_x nanosheets exhibited U(VI) sorption capacities of 0.41 and 0.58 mmol/g, or 98 and 138 mg/g, at pH levels of 4.5 and 5.0, respectively. In comparison, the sorption capacities of the competing ions were all <0.1 mmol/L.

5.2.4 Elimination of Gaseous Contaminants

Gaseous contaminants are serious pollutants that are harmful to both humans and other living organisms [56–58]. There are two groups of gaseous pollutants, i.e., (i) toxic inorganic gases, such as NO_x, SO_x, H_2S, NH_3, CO, CO_2, etc., and (ii) various volatile organic compounds (VOCs). The application of MXenes as sorbents to remove gaseous pollutants has been explored, especially theoretically.

Adsorption behaviors of NH_3, H_2, CH_4, CO, CO_2, N_2, NO_2 and O_2 on monolayer Ti_2CO_2 MXene nanosheet was studied by using first-principles simulations, as candidates for the gas adsorption [59]. It was found that NH_3 was the only one to be chemically adsorbed on Ti_2CO_2, with an effective charge transfer of 0.17 electron. Interestingly, the adsorption of NH_3 molecules on the Ti_2CO_2 nanosheet could be enhanced by applying strains to the MXene, while the adsorption of others was nearly unaffected. Meanwhile, the adsorbed NH_3 molecules could be released by removing the applied strains. In other words, the adsorption process was highly reversible.

M_2C MXenes, with M=Ti, Zr, Hf, V, Nb, Ta, Cr, Mo and W, were explored for the adsorption of CO_2 [60]. By using density functional theory studies with dispersion, the authors we show the M_2C MXenes could strongly adsorb CO_2 molecules, with CO_2 adsorption capacity to be sufficiently high at low partial pressures of CO_2 and high temperatures. Specifically, for Hf_2C, there was a charge transfer from MXene to CO_2, with about two electrons. The loading capacity of CO_2 was in range of 2.34–8.25 mol/kg.

The same authors also studied the performances of nitride MXenes for the capture and storage CO_2 at a wide range of pressure and temperature by using. An integrated theoretical framework [61]. The nitride MXenes were M_2N, with M=Ti, Zr, Hf, V, Nb, Ta, Cr, Mo and W. The adsorption and desorption rates of CO_2 molecules on surface of the MXenes were obtained by using transition state theory and density functional theory calculations. It was found that strong adsorption and activation of CO_2 could occur at temperatures of >737 °C, for some MXenes. According to the theoretical predictions, Ti_2N, Zr_2N, Hf_2N, V_2N, Nb_2N and Ta_2N were all capable of activating CO_2. The adsorption capacity of CO_2 was in the range of 2.32–7.96 mol/kg.

5.2.5 *Removal of Other Pollutants*

MXenes have also been evaluated as adsorbents to remove other pollutants. For example, a sandwich-like MXene-iron oxide (MXI) hybrid was developed for the sequestration of phosphates in water environment [62]. The MXI hybrid was obtained by etching out the Al layer and intercalating with magnetic ferric oxide nanoparticles. A short separation time of 2 min was observed for the MXI hybrid to adsorb phosphate, with adsorption capacities of 2100 and 2400 kg/kg, in theory and experiment, respectively. The adsorption efficiency was closely related to the presence of the magnetic iron oxide nanoparticles. The nanoparticles were intercalated into the interlayer spacing of the MXene nanosheets, so as to enlarge the interlayer spacing. As a result, high phosphate adsorption efficiency was maintained, because of the complexation with the magnetic nanoparticles and the Ti–O terminal groups in the MXenes.

The MXI hybrid was characterized with XRD, SEM/TEM and XPS. The MXene had a composition of $Ti_3C_2OH_{0.8}F_{1.2}$, according to the XPS result. TEM images revealed that thickness of the 2D MXene was about 20 nm, corresponding to about 15 layers. The iron oxide nanoparticles, with an average particle size of 18 nm, were derived from mixed solutions of $FeSO_4$ and $FeCl_3$, together with NaOH as the precipitant. The *c*-parameter of the MXene was 1.99 nm, which was increased to 2.994 nm, as MXene/Fe_3O_4 ratio was 2:1. The mole ratio of Fe_3O_4/Fe_2O_3 was in the range of 4.45–3.95.

HRTEM images revealed that the average size of the Fe_3O_4 nanoparticles was 18.2 nm. Meanwhile, anatase TiO_2 nanoparticles with a cuboid-like and a size of 12.5 nm were also present. In other words, the MXI nanohybrid consisted of MXene, Fe_3O_4 and a small fraction of Fe_2O_3. Furthermore, the MXI possessed a relatively weak magnetic behavior, with saturation magnetization of about 0.3 emu/g and coercive force of 432 G. The adsorption of phosphate onto the MXI nanohybrid was only effective in acidic environment, at pH values in the range of 2.5–6.0. The phosphates were present as negatively charged $HPO_4^{2-}/H_2PO_4^-$, over a wide range of pH level. The zero-potential charge (pH_{zpc}) onto MXI was observed at pH value of about 6.82. Therefore, the positively charged MXI adsorbed the negatively charged $HPO_4^{2-}/H_2PO_4^-$ in acidic solutions.

The MXI nanohybrid was highly selective in the sorption of the phosphates in the presence of competing ions, as shown in Fig. 5.24a–c. In this case, the special sandwich structure of the MXI nanohybrids, together with the surface active sites of Ti–OH and the intercalated iron oxides, was responsible for the highly selective

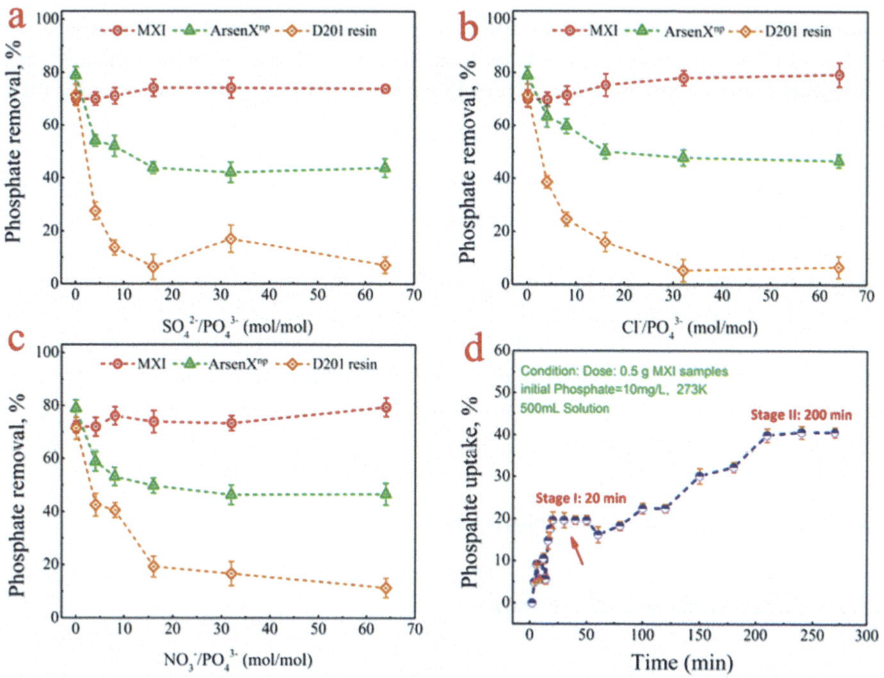

Fig. 5.24 Phosphate adsorption behaviours of the MXI hybrid: **a**–**c** common anions (SO_4^{2-}, Cl^- and NO_3^-) and **d** sorption kinetics for phosphate sequestration at various time intervals. Reproduced with permission from [62]. Copyright © 2016, Royal Society of Chemistry

adsorption of the phosphates. SEM observation results indicated that the layered structure was well retained, whereas the surface of the MXI nanosheets was decorated with the absorbed phosphate ions.

Figure 5.24d shows kinetics of the adsorption process. Obviously, the phosphate adsorption onto the MXI nanosheets experienced two stages, i.e., fast initial adsorption and prolonged sorption equilibrium. The earlier rapid sorption corresponded to the surface adsorption of the phosphate ions onto the surface of the magnetic MXI hybrid nanosheets, whereas the slowing-down process was readily related to the inner-layer diffusion of the ions into the laminar structures. The two sorption stages confirmed the bimetallic cooperative adsorption profile of the iron oxide intercalated and Ti terminated hydroxide hybridized MXI.

5.2.6 *Removal of Urea*

It is interestingly found that $Ti_3C_2T_x$ MXene could be used to adsorb urea, making it possible to make wearable artificial kidney with the 2D nanomaterials [63]. Such wearable artificial kidney could be utilized to conduct continuous ambulatory dialysis for more than 3 million patients with end-stage renal disease, where the efficient removal of urea is a key problem. The MXene based device exhibited removal efficiency of 99% from aqueous solution, while it was also as high as 94% from the dialysate at an initial urea concentration of 30 mg/dL. At room temperature, with the maximum urea adsorption capacity reached 10.4 mg/g. At a testing temperature of 37 °C, the urea removal efficiency from dialysate was increased by about one time, with the maximum urea adsorption capacity approaching 21.7 mg/g. Meanwhile, $Ti_3C_2T_x$ demonstrated high hemocompatibility, without inducing cell apoptosis or reducing the metabolizing cell fraction. Also, no impact on cell viability was observed at concentrations of up to 200 μg/mL. Due to the biocompatibility of $Ti_3C_2T_x$ and the high selectivity to adsorb urea from dialysate, this achievement offered a promising potential to develop wearable artificial kidney for dialysate regeneration.

5.3 Chemical/Biological Sensors

5.3.1 Electrochemical Sensors

2D nanomaterials have been extensively explored for the applications in sensors and biosensors [64–69]. As expected, MXenes should also be promising candidates for such applications, due to their unique physical, chemical and mechanical properties, as discussed previously [70]. When $Ti_3C_2T_x$ MXenes were into electrodes, they acted as transducers to immobilize biological receptors onto the surface of the nanosheets, thus having strong sensing performances. Generally, it is difficult to facilitate direct electron transfer (DET) of enzymes and proteins, because of the deeply rooted location of the redox active centers.

However, enzymes and proteins could be easily immobilized onto the surfaces of MXenes [71]. As a result, they could be used as the supports to realize DET, with boosted electrode kinetics, reduced detection limits, increased sensitivity and enhanced selectivity. Furthermore, the overall working electrode surface area could be enlarged, while and overpotential during electroanalysis was decreased, by decorating the surfaces of the MXene nanosheets having immobilized enzymes with metallic or metallic oxide nanoparticles. By doing this, the overall catalytic activity of the MXenes based sensors would be significantly improved.

The transducers based on the $Ti_3C_2T_x$/nanoparticle hybrids exhibited stronger tendency in the incorporation with the biological enzyme protein receptors onto the surface the MXene nanosheets. The biosensors were fabricated by using drop-casting process. In addition, the presence of the $Ti_3C_2T_x$ nanosheets enhanced the catalytic activity of the matrix materials, thus well retaining the bioactivity of the enzyme proteins, because their original conformations were not affected. The $Ti_3C_2T_x$ MXene biosensors could be used to electrochemically detect various biomolecular items. For instance, a glucose biosensor was based on $Ti_3C_2T_x$/Nafion/AuNP nanohybrid, which was coated on glassy carbon electrode (GCE). It was able to immobilize glucose oxidase (Gox).

Because of the complicate scaffolds of the Gox, DET between the redox centers of the Gox and working electrode very difficult, as mentioned earlier. The redox active flavin adenine dinucleotide (FAD) of Gox was located inside the prosthetic protein shells. As a result, the DET of Gox was prevented as the conventional electrodes were employed. In this study, $Ti_3C_2T_x$ MXene was incorporated in the biosensing matrix, the device had a sufficiently high electrical conductivity, which promoted the electron transfer between the active redox centers of Gox and the interfaces of the electrodes. Moreover, the introduction of Au nanoparticles further enhanced the sensitivity of the sensing devices, since the protection shell of the proteins was weakened by the Au nanoparticles. In addition, the overpotential for the detection of H_2O_2 was reduced, as a by-product produced during the oxidation of the glucose.

Because the interfering molecules possessed negative charges, the permeability of the negatively charged Nafion was decreased, so that the MXene biosensors demonstrated high selectivities. Specifically, the Gox-Au-MXene-naf/GCE biosensing devices exhibited a high sensitivity of 4.2 μA m/M cm^2, when used to detect glucose with amperometries. In terms of stability, as the response current of the Gox-Au-MXene-naf/GCE biosensors at a glucose concentration of 3.0 mM, it was retained by 93% after storage for two months.

Similar Ti_3C_2 MXene sensors were developed to promote the electron transfer of ulphider (Hb), with biosensing effect to H_2O_2 [72] and $NaNO_2$ [73], without the need of mediators. In the first case, the device could be used to detect H_2O_2 with a high linearity over a wide concentration range of 0.1–260 μM, while the detection limit was as low as 20 nM [72]. In the second one, the biosensors demonstrated a high efficiency in detecting nitrite, with a wide linear range of concentration, i.e., 0.5–11,800 μM, together with an low detection limit of 0.12 μM [73]. In these biosensors, the Ti_3C_2 MXenes offered a compatible micro-environment to the bioreceptor molecules, such as Hb. At the same time, they helped to enlarge the effective area of the immobilized surfaces, so that the rate of collision between the substrate and Hb was increased.

TiO_2 nanoparticles were employed to decorate organ-like Ti_3C_2 MXene, forming Ti_3C_2/TiO_2 nanohybrid, for the application to immobilize ulphider (Hb), as a mediator-free biosensor [74]. According to the spectroscopic and electrochemical dada, it was found that Ti_3C_2/TiO_2 nanohybrid was an excellent immobilizing matrix and highly biocompatible with redox proteins, thus leading to high bioactivities and stabilities to proteins. The unique organ-like structure of the Ti_3C_2/TiO_2 nanohybrid facilitated the direct electron transfer of Hb.

As a result, the nanohybrid based biosensors exhibited outstanding sensing behaviors to H_2O_2, with a wide linear range of 0.1–380 μM. The corresponding sensitivity was as high as 447.3 μA/mM·cm^2, while the detection limit was as low as 14 nM. In this case, since the TiO_2 nanoparticles were had high biocompatibilities, the surface of the Ti_3C_2/TiO_2 nanohybrid was a safe microenvironment for Hb. Accordingly, biosensors displayed an excellent long-term stability. For instance, the sensing response could be retained by about 95% after storage for 60 days. These Ti_3C_2/TiO_2 nanohybrid based biosensors could be used for applications in environmental analysis and biomedical detections.

Figure 5.25 shows amperometric response curves of the glassy carbon electrodes at − 0.35 V with the addition of H_2O_2 in constantly stirred PBS at pH = 7.0. The Nafion/Hb/TiO_2–Ti_3C_2/GCE senor had a very short response time of <3 s. Meanwhile, the Nafion/Hb/TiO_2–Ti_3C_2/GCE device exhibited a linear current versus the concentration of H_2O_2. Without the presence of TiO_2, the Nafion/Hb/Ti_3C_2/GCE device had a poorer sensing performance, with a linear range of 0.1–260 μM and a sensitivity of 321.7 μA/mM·cm^2.

The superior performances of the Nafion/Hb/TiO_2–Ti_3C_2/GC electrode could be explained from several aspects. First of all, the organ-like structure of the Ti_3C_2/TiO_2 hybrid offered a proper microenvironment, in which both the activity and stability of the proteins were well retained. Also, the Ti_3C_2/TiO_2 hybrid had a large specific surface area and a unique organ-like structure, which ensured the substrate to be confined. Accordingly, they could readily access to the enzymes that were immobilized on surface of the Ti_3C_2/TiO_2 nanosheets. As a result, there would be more effective collisions between the substrate and the redox proteins, thus resulting in enhanced sensing performances of these special biosensors.

Ti_3C_2 MXene and graphene oxide (GO) were incorporated to form Ti_3C_2/GO nanohybrid, which was applied to prepare sensors of H_2O_2 by using inkjet-printing process [75]. The Ti_3C_2/GO hybrid-based sensor could stabilize the biological activity of hemoglobin (Hb). As the sensors were utilized to detect H_2O_2, a dynamic range of 2–100 μM and a low detection limit of 1.95 μM were observed. At the same time, the device showed a high sensitivity and outstanding selectivity.

$Ti_3C_2T_x$ MXene nanosheets were decorated with Pt nanoparticles (PtNPs) in order to develop materials with enhanced electrochemical performances [76]. The $Ti_3C_2T_x$/PtNPs hybrid was coated on the surface of GCE, which resulted in much higher and more stable redox effects, over an anodic potential window, when compared with the pristine $Ti_3C_2T_x$ MXene modified GCE. The H_2O_2 sensor based on the GCE modified with $Ti_3C_2T_x$/PtNPs possessed a limit of detection (LOD) of 448 nM, with a potential at which the reduction started at +250 mV (vs. Ag/AgC1). More importantly, the $Ti_3C_2T_x$/PtNPs based sensors were sensitive to various small redox molecules, including ascorbic acid (AA), dopamine (DA), uric acid (UA) and acetaminophen (APAP), along with high selectivity and LOD of nM scale.

$Ti_3C_2F_x$ was treated with a high anodic potential of +200 mV to oxidize the outer surface, which was used to oxidize NADH [77]. The oxidation process led to the formation of TiO_2 nanolayers, which could be dissolved by F^- ions. As a result, the oxidized MXene was less electroactive than the pristine one. It was found that oxidation current of NADH was decreased for the oxidized $Ti_3C_2F_x$ than the pristine $Ti_3C_2F_x$. Moreover, the device was highly sensitive to H_2O_2 when exposing a large cathodic potential of − 500 mV to $Ti_3C_2F_x$ with chronoamperometry.

Ti_3C_2 MXene was also employed to immobilize tyrosinase (Tyr) enzyme, for biosensors to detect phenol [78]. The MXene Ti_3C_2 enzymatic device offered a highly efficient interface, at which the phenol was oxidized to o-quinone, as monitored with

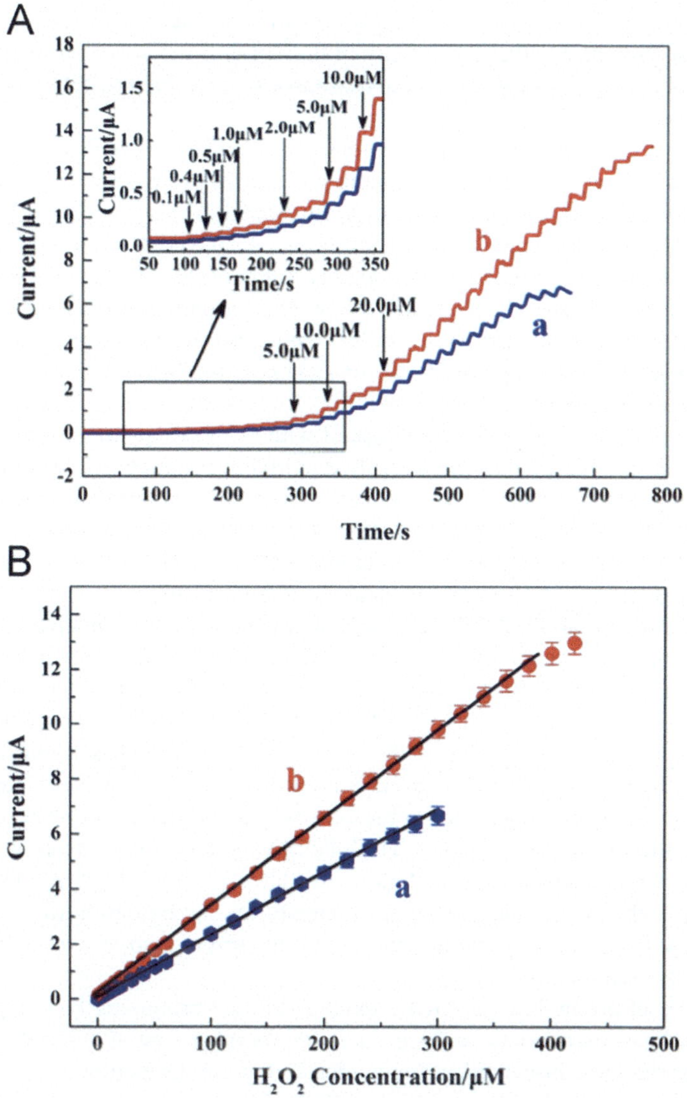

Fig. 5.25 **a** Current–time responses of the Nafion/Hb/Ti_3C_2/GCE (curve a) and Nafion/Hb/TiO_2–Ti_3C_2/GCE (curve b) at − 0.35 V to successive addition of H_2O_2 in stirred 0.1 M with pH = 7.0 PBS. **b** Steady state current vs. H_2O_2 concentration of Nafion/Hb/Ti_3C_2/GCE (curve a) and Nafion/Hb/TiO_2–Ti_3C_2/GCE (curve b). Reproduced with permission from [74]. Copyright © 2018, Elsevier

amperometry. The biosensors were highly sensitive to phenol, because the surface of MXene so compatible that the activity of the Tyr enzyme was well retained after it was immobilized. The response of the biosensor was retained by 89% after testing for six weeks to 1.5 μM phenol.

Ti_3C_2 MXene alkaline intercalated with KOH could be used to detect toxic heavy metallic ions by using square wave stripping voltammetry (SWASV) [79]. Electroactivity of the alk-Ti_3C_2/GCE allowed for detecting metallic ions, such as Cd(II), Pb(II), Cu(II) and Hg(II). The target heavy metallic ions displayed distinct redox peaks at optimized conditions, corresponding to detect limits of 98, 41, 32 and 130 nM, for Cd(II), Pb(II), Cu(II) and Hg(II), respectively. The formation of the [Ti–O]–H^+ and [Ti–O]–K^+ moieties in the alk-Ti_3C_2 MXene enhanced the exchange of the cations, thus promoting their adsorption. As a consequence, the heavy metallic ions were easily reduced to metals.

If the –F functional groups were replaced with –OH groups in the alk-Ti_3C_2, its hydrophilicity was increased, which resulted in higher conductivity, thus promoting the electron transfer processes of the heavy metals. The alk-Ti_3C_2/GCE sensor showed high stability after storage at 4 °C for three weeks. The response currents for Cd(II), Pb(II), Cu(II) and Hg(II) were retained by 90.5%, 94.3%, 96.3% and 93.8%, respectively. Figure 5.26 shows performances of the sensors.

Furthermore, the alk-$Ti_3C_2T_x$/GCE sensor was also highly sensitive to bromate ions (BrO_3^-), a serious contaminant of drinking water [80]. The sensor worded through the reduction of BrO_3^-, due to the special electrocatalytic characteristics, with a detection limit of as low as 41 nM. The reduction of BrO_3^- was evidenced by the presence of TiO_2 nanocrystals at the surface of the MXene based sensor.

Acetocholinesterase (AchE) enzyme immobilized $Ti_3C_2T_x$ MXene nanosheets were used to develop amperometric biosensors, for the detection of organophosphate pesticide malathion [81]. The $Ti_3C_2T_x$ MXene based biosensors were characterized by various advantages, including low detection limit, high reproducibility, outstanding stability and anti-interference capability, the oxidation of malathions. After storage for more than one month, the initial current response to malathion of the sensor was retained by 85%.

The AchE-chitosan-$Ti_3C_2T_x$/GCE device exhibited promising biocompatibility and high surface conductivity, making it suitable for other type of pesticides, such as including herbicides, insecticides, fungicides and so on. Detection of malathion in water samples displayed high recoveries of 95.0–105.0%, demonstrating high analytical applicability of MXene materials for the applications in agricultural analysis and environment safety. Figure 5.27 shows sensing performances of the device.

A field effect transistor (FET) sensor was fabricated with MXene, which was used to monitor dopamine (DA) in spiked hippocampal neurons [82]. The $Ti_3C_2T_x$-MXene FET sensor was examined in conductance in the form of micropatterns through gating controls. The MXene based biosensor worked as n-type or p-type FET, as the gate voltage was higher or lower than + 0.3 V. The apparent field effect represented to the front gate modulation, while the materials were highly semiconductive, so that the MXene based biosensor possessed sufficiently high sensitivity for real applications.

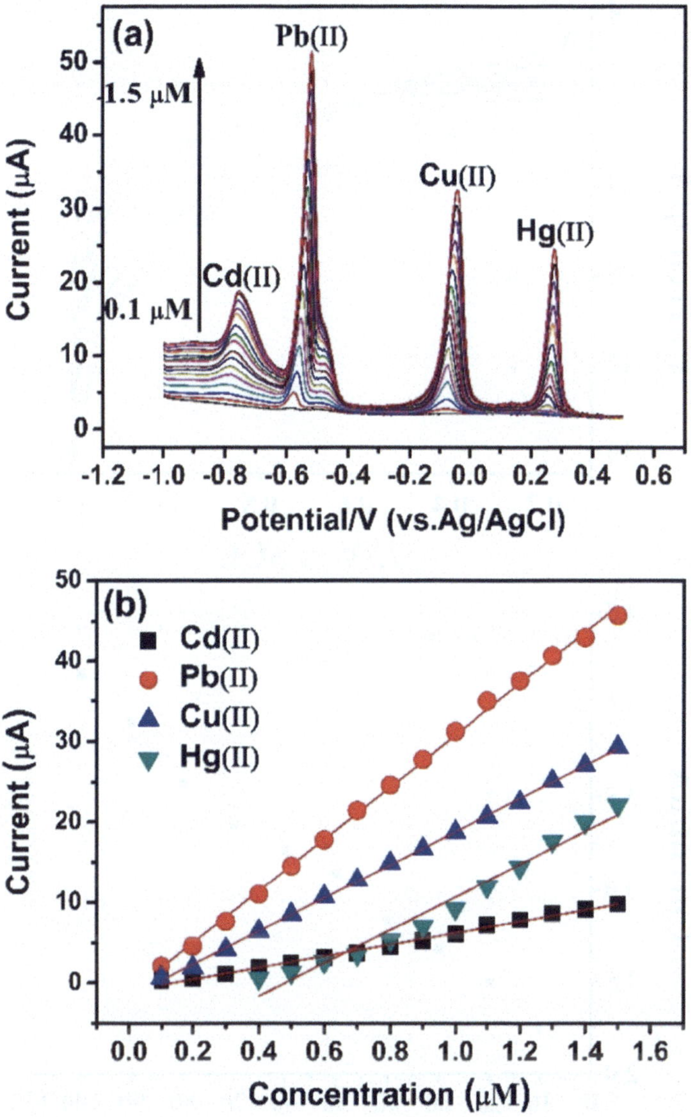

Fig. 5.26 **a** SWASV response curves of the alk-Ti_3C_2 modified GCE for simultaneous detection of Cd(II), Pb(II), Cu(II) and Hg(II), over the concentration range of 0.1–1.5 μM. **b** Respective calibration curves of Cd(II), Pb(II), Cu(II) and Hg(II) from the data in panel **a**. Reproduced with permission from [79]. Copyright © 2017, Elsevier

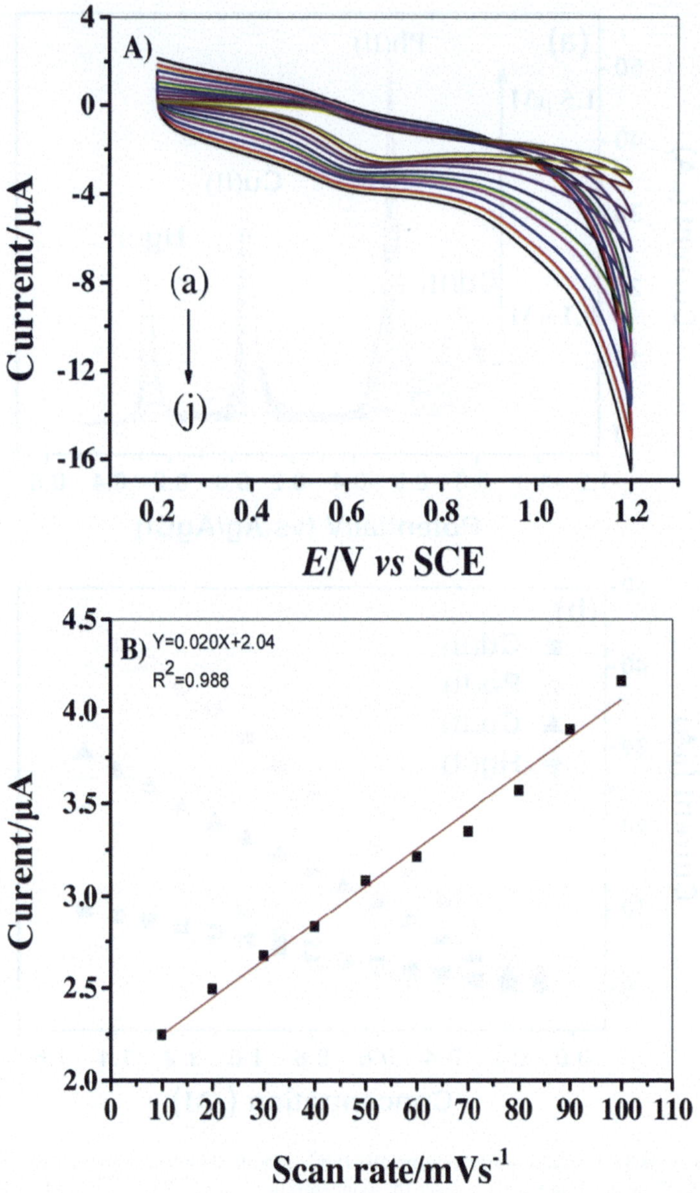

Fig. 5.27 **a** CV curves of the AchE/CS-$Ti_3C_2T_x$/GCE sensor in 0.1 M PBS (pH = 8.0) containing 1.0 mM ATCl at various scan rates (10 (a)-100 (j) mVs^{-1}). **b** oxidation peak csurrent *vs* scan rate of the device. Reproduced with permission from [79]. Copyright © 2017, Elsevier

There were pretty strong π–π interactions between the DA molecules and functionalized surface of the MXene nanosheet to ensure the sensing performance of the device to DA. In this case, the DA molecules interacted with the functional groups on surface of the MXene nanosheets, which resulted in the concentration of holes, so that the conductivity of the sensor was increased accordingly. Furthermore, the MXene micropatterns could be used for calcium imaging. As a result, the neural activity could be evaluated.

A mechanical milling process was reported to prepare d-Ti_3C_2-Bi nanohybrid, which was used as electrochemical sensors in the form of microgrid [83]. In the formation of the d-Ti_3C_2-Bi nanohybrid, Bi(III) was first deposited on the surface of delaminated Ti_3C_2 nanolayers through electrostatic interaction, followed by *in-situ* conversion to Bi nanorods (NRs). Sensing performances of the sensors are shown in Fig. 5.28. The sensor based samples processed with optimal parameters displayed linear sensitivities to Pb(II), Cd(II) and Zn(II) in the concentration range of 1–20 μg/L. The peak potentials of Pb(II), Cd(II) and Zn(II) were at − 0.54 V, − 0.76 V

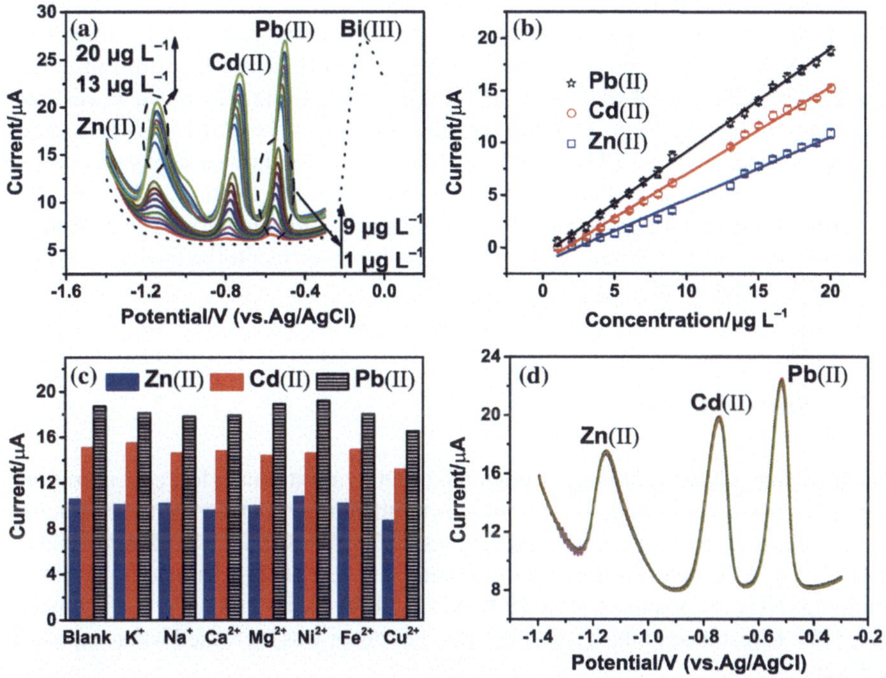

Fig. 5.28 Sensing characteristics of the d-Ti_3C_2-Bi/MSA samples: **a** SWASV response for simultaneous analysis of Pb(II), Cd(II) and Zn(II) from 1 to 20 μg/L, **b** calibration curves for Pb(II), Cd(II) and Zn(II), **c** effect of interference ions on current for 20 μg/L Pb(II), Cd(II) and Zn(II), **d** peak current collected from each SWASV response for 20 μg/L Pb(II), Cd(II) and Zn(II). Experimental details include deposition potential of − 1.4 V, deposition time of 3 min and supporting electrolyte of 0.1 M acetate buffer with pH = 4.5. Reproduced with permission from [83]. Copyright © 2019, Springer

and − 1.15 V versus Ag/AgCl, respectively, which were sufficiently separated one another. Also, the sensitivity values for Pb(II), Cd(II) and Zn(II) were 0.98, 0.84 and 0.60 μA·L/μg, with corresponding detection limits of 0.2, 0.4 and 0.5 μg/L, respectively. It was believed that homogeneous distribution of the Bi NRs, high electrical conductivity of the MXene nanostructure and the superior diffusion effect related to the microgrid structure were concurrently responsible for the outstanding sensing performances of the nanohybrid materials to the cations.

More recently, a multiple component hybrid-based electrochemical sensor was developed, which consisted of MnO_2/Mn_3O_4 from MOFs and MXene-AuNPs, with outstanding sensitivity to organophosphorus pesticides (Ops) [84]. The 3D MnO_2/Mn_3O_4 nanohierarchical structure was well retained after that of the Mn-MOFs, which displayed vertical alignment with nanosheets of high degree of orientation. The nanostructures were well incorporated with MXene-AuNPs nanolayers, leading to large specific surface areas. As a result, the materials exhibited highly electrochemically sensitive and strongly biocompatibile to the environment. The optimized platform, i.e., AchE-Chit–MXene-AuNPs-MnO_2/Mn_3O_4-GCE, possessed high sensitivity to methamidophos, over the concentration range from 10^{-12} to 10^{-6} M, covering as large as six orders of magnitude, demonstrating a h linearity with $R = 0.995$. Meanwhile, the biosensors displayed a sufficiently low limit of detection (LOD), with a level of 1.34×10^{-13} M. Such a LOD was much lower than the maximum residue limits (MRLs) for methamidophos, i.e., 0.01 mg/kg, as regulated by European Union (EU). In addition, the biosensors showed strong recovering behaviour in detecting methamidophos, which was in the range of 95.2–101.3%. Therefore, these nanohybrids could be used for applications in various electrochemical sensors to detect or monitor pesticide residuals as well as similar environmentally harmful chemical items.

5.3.2 Gas Sensors

The detection of volatile organic compounds (VOCs) at extremely low concentrations (e.g., ppb) has been a challenge for various applications, such as removal of pollutions, identification of toxic gases and diagnosis of diseases. Gas sensing performance of $Ti_3C_2OH_2$ MXene to gases, including acetone (CH_3COCH_3), ethanol (C_2H_5OH), ammonia (NH_3), propanal (C_2H_5CHO), NO_2 and SO_2, was studied at room temperature [85]. Resistance of the $Ti_3C_2OH_2$ MXene sensor was increased when exposure to the gases.

Gas sensing performance of the $Ti_3C_2T_x$ based sensors to 100 ppm CH_3COCH_3, C_2H_5OH, NH_3, C_2H_5CHO, NO_2, SO_2 and 10,000 ppm CO_2, in N_2, is shown in Fig. 5.29a. The sensitivity was expressed as the relative change in the electrical resistance of the device upon the injection of the gases, against the background resistance, i.e., $\Delta R/R_b$ (%). In this case, the $Ti_3C_2T_x$ sensors experienced an increase in resistance, irrelevant to the properties of the gases, implying that the charge carrier transport of the channel was blocked, once gas molecules were adsorbed, no matter

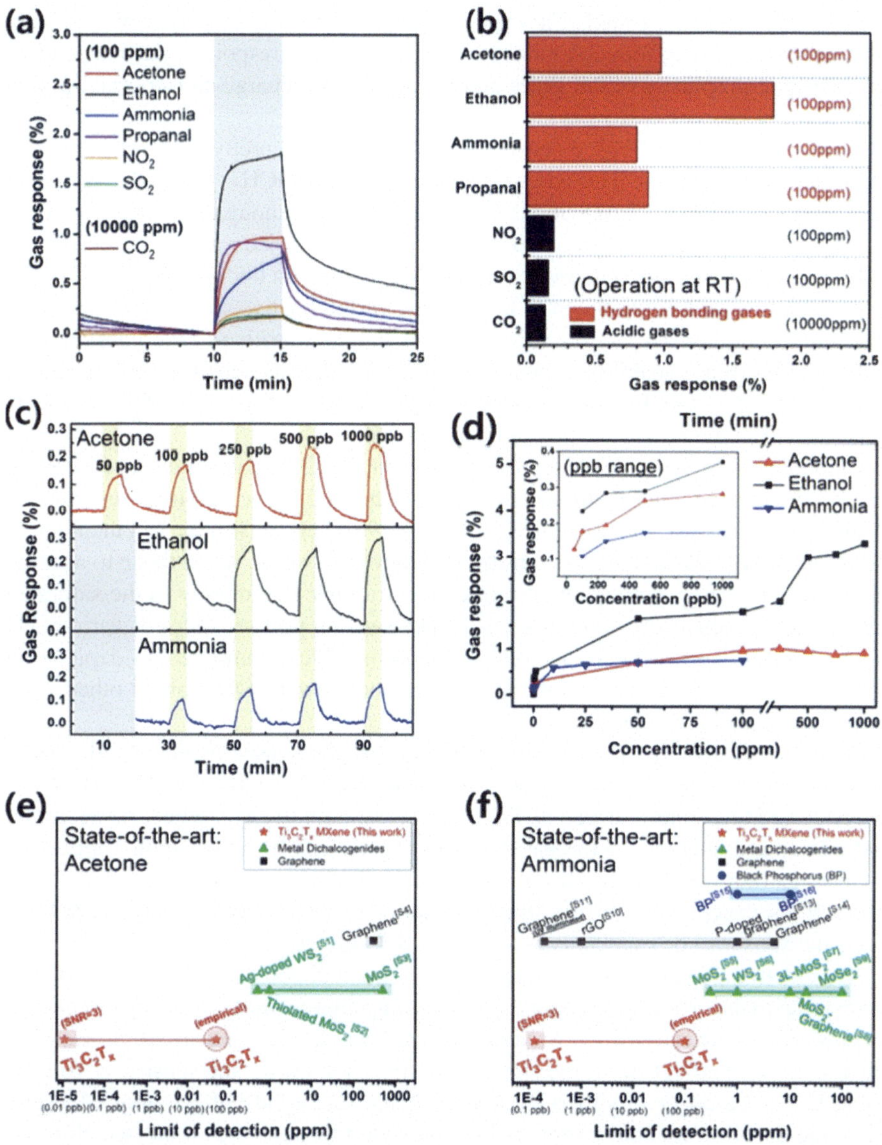

Fig. 5.29 Gas sensing characteristics of the $Ti_3C_2T_x$ MXene based sensors at room temperature. **a** Resistance variations and **b** maximal resistance changes upon exposure to 100 ppm acetone, ethanol, ammonia, propanal, NO_2, SO_2, and 10,000 ppm CO_2 at room temperature. **c** Resistance variations *versus* time upon exposure to highly diluted acetone (top), ethanol (middle) and ammonia (bottom) in ppb concentration range (50–1000 ppb). **d** Maximal resistance changes in a wide range of diluted target gases (0.05–1000 ppm). Limit of detection (LOD) for room temperature sensors based on 2D materials to detect **e** acetone and **f** ammonia, showing the $Ti_3C_2T_x$ MXene based sensors to have the lowest LOD. The *Y*-axes in **e** and **f** have no quantitative meaning, while the lines were drawn at different heights of the *Y*-axis only to spatially differentiate the 2D materials. Reproduced with permission from [85]. Copyright © 2018, American Chemical Society

what kind of gas was tested. This special gas sensing response was distinctively different from those of semiconductive gas sensors, whose responses were dependent on electronic properties of the gases and the type of the charge carriers (i.e., p-type or n-type).

The $Ti_3C_2T_x$ MXene based sensors also demonstrated pretty high selectivity, as illustrated in Fig. 5.29b. The gas sensitivities to CH_3COCH_3, C_2H_5OH, NH_3 and C_2H_5CHO were 0.97, 1.7, 0.8 and 0.88%, respectively. Among them, the sensitivity to CH_3COCH_3 was highest. In comparison, the sensing device was very insensitive to NO_2, SO_2 and CO_2. The sensitivity was also dependent on thickness of the $Ti_3C_2T_x$ films, which was closely related to the number of total active sites at the surface. It was found that the $Ti_3C_2T_x$ based sensor was highly sensitive to the gases having hydrogen bonding capability to acidic gases, reflecting its sensing mechanism. In other words, the –OH functional groups on the surface of the $Ti_3C_2T_x$ nanosheets were critical in the sensing behavior of the device.

Figure 5.29c shows variation in resistance of the $Ti_3C_2T_x$ based sensor responding to low concentration gases, in order to identify its LOD. The maximal sensitivities to different gases are demonstrated in Fig. 5.29d. The sensitivity to ethanol was monotonically linear to the concentration in the whole range, while those to acetone and ammonia were saturated at certain concentrations, due mainly to the saturated number of the active sites on the surface. The sensing performances of various 2D material-based gas sensors to acetone and ammonia were compared, as depicted in Fig. 5.29e, f. The $Ti_3C_2T_x$ sensor displayed a much lower LOD than all other types of 2D materials-based sensors.

Monolayer Ti_2CO_2 MXene was employed as gas sensor to detect NH_3 [86]. Adsorption behaviors of various gases, including NH_3, H_2, CH_4, CH_3COCH_3, CO_2, N_2, NO_2 and O_2, on the surface of Ti_2CO_2, was systematically studied. It was found that adsorption energies of the gases were only negligibly varied upon application of biaxial strain with magnitudes of 0–4% on the Ti_2CO_2 layer, because of the weak interactions, except NH_3. The significant variation in the adsorption energy of MXene made the sensor to be especially selective to NH_3 molecules.

The electronic interactions between the NH_3 molecules and the MXene nanosheets led overlap in the orbital, corresponding to a strong charge transfer of 0.174 electrons. Moreover, conductivity of the MXene interface was greatly increased, as the NH_3 molecules were adsorbed, in the strain state. In other words, the sensitivity to NH_3 could be increased or modified by adjusting the magnitude of the applied strain. Also, the desorption could be reversibly controlled through the strain, which made the Ti_2CO_2 MXene to be a media to realize the sensing, capturing and storage NH_3.

The interaction between NH_3 molecules and O-terminated MXenes, with a formula of M_2CO_2, M=Sc, Ti, Zr and Hf, with different charge states, was theoretically evaluated by using first -principles simulations [87]. It was demonstrated that NH_3 molecules could be adsorbed on the surface of M_2CO_2 with strong charge transfer, thus making them highly potential for the applications to sense and store NH_3. Furthermore, the adsorbed NH_3 could be readily released by controlling the concentration of the electrons that were injected into the MXene. More specifically, Zr_2CO_2 was highly sensitive and selective to NH_3. The adsorption energy was largely decreased from 0.81 to 0.20 eV, if additional two electrons were injected into a 3 × 3 sheet of Zr_2CO_2.

Ti_3AlC_2 was coated on flexible polyimide substrate by using a simple solution casting method to fabricate gas sensors at room temperature [88]. The $Ti_3C_2T_x$ based sensors were characterized in terms of sensitivity to ethanol, methanol, acetone and ammonia gases. It was found that the device exhibited p-type sensing characteristics. The sensitivities were highest and lowest response to ammonia and acetone gases, respectively. The LOD to acetone gas was theoretically predicted, with a level of about 9.27 ppm. The sensing mechanism was established by considering the interactions between the majority charge carriers of $Ti_3C_2T_x$ and the gas molecules.

Figure 5.30 shows dynamic gas sensing characteristics of the $Ti_3C_2T_x$ sensor to 100 ppm ethanol, methanol, acetone and ammonia gases in ambient atmosphere. The fresh $Ti_3C_2T_x$ had a thickness of 30 nm and a fresh resistance of t about 10 kΩ at room temperature. It has been established that conduction properties of MXene could be tailored through the surface functional groups, with conductivity that could be varied from metallic to narrow band gap semiconductor [89]. The functional groups resulted in dipole polarization of the surface and reduced the electrical conductivity, so that the MXene could be developed into materials with gas sensing behaviors. The $Ti_3C_2T_x$ base sensor was periodically exposed to gases with a concentration of 100 ppm for

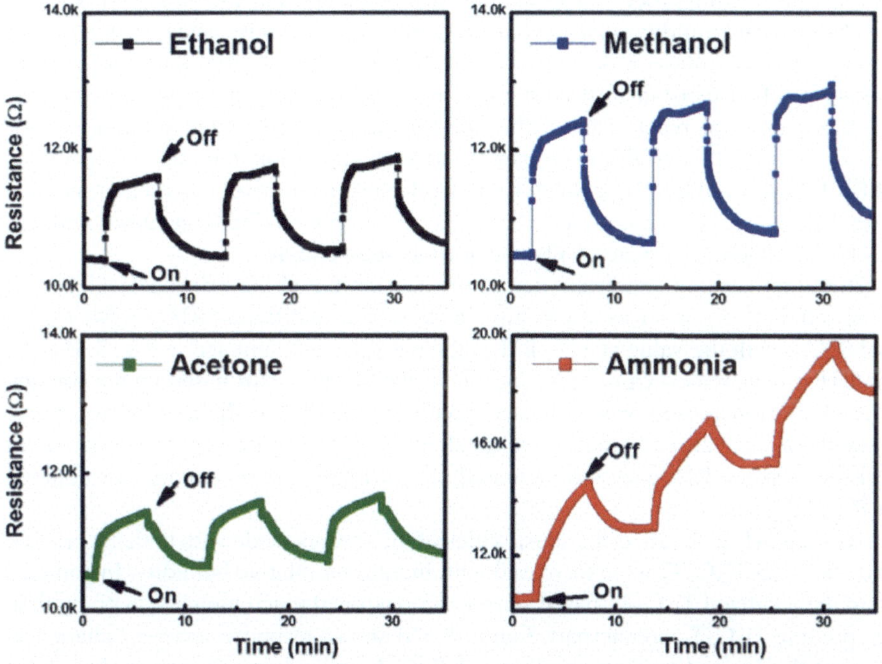

Fig. 5.30 Gas-sensing performances of the sensor device based on $Ti_3C_2T_x$ to 100 ppm ethanol, methanol, acetone and ammonia gases bubbled at room temperature. Reproduced with permission from [88]. Copyright © 2017, American Chemical Society

5 min, which then became fresh with clean air with nitrogen to oxygen ration of 80:20. As observed, the resistance of the $Ti_3C_2T_x$ sensor was increased, acting as a p-type characteristic. The sensing profile was highly reproducible.

Both pristine and alkalized $Ti_3C_2T_x$ MXenes were employed as gas and humidity sensing materials, in the form of room temperature planar-structured sensors prepared by using a simple dip coating process [90]. It was found that both the gaseous and humidity sensitivities of the MXene-based sensors were enhanced due to the intercalation of the alkali metallic ions (i.e., Na^+) and the increased ratio of [–O]/[–F] on the surface of the $Ti_3C_2T_x$ nanosheets. Specifically, the sensor based on the alkalized $Ti_3C_2T_x$ displayed a variation in response signal by about 60 times in nearly the whole relative humidity (RH) range of 11–95% and a sensitivity of 29% to 100 ppm NH_3 at 25 °C.

To fabricate gas/humidity sensors, interdigitated electrodes of Au/Ni with a dimension of 8 mm × 8 mm were deposited on Al_2O_3 ceramic substrate with a size of 8.2 mm × 8.2 mm. The Au/Ni electrodes were prepared by using sputtering, with the thicknesses of the Au and Ni layers to be 0.1 μm and 1 μm, respectively. The distances in between the adjacent Au negs were about 80 μm, while the width was about 70 μm. The as-obtained substrates were thoroughly cleaned with the aid of ultrasonication with deionized water and ethanol. The cleaned substrates were coated with uniform layers of pristine and alkalized $Ti_3C_2T_x$ nanosheets by using a simple drip-coating process. The samples were then dried in vacuum at 60 °C.

The gas and humidity sensing characteristics were tested by using a static testing system. A bias voltage was applied to the sensors with a digital multimeter when measuring their resistance. Both the response and recovery times were defined as the times taken to reach 90% of the full responses. Steady relative humidities of 11, 33, 54, 75, 85 and 95% were generated with supersaturated aqueous solutions of LiCl, $MgCl_2$, $Mg(NO_3)_2$, NaCl, KCl, and KNO_3, respectively. Gas environments with different gas concentrations were create by simply mixing target gases with air. All the measurements were conducted at room temperature.

Humidity sensing characteristics of the sensors based the pristine $Ti_3C_2T_x$ and alkalized $Ti_3C_2T_x$, in terms of variation in resistance, at different RH (33, 54, 75, 85, and 95%), with the value at 11% RH as the baseline, are depicted in Fig. 5.31a. As the RH was increased from 11%, the resistance of the sensor based on the pristine $Ti_3C_2T_x$ was increased, where the sensitivity was defined as the ratio of $R_{RH}/R_{11\%}$. In contrast, the sensor based on the alkalized $Ti_3C_2T_x$ had a negative response in resistance as the RH level was increased. Accordingly, the sensitivity was denoted as $R_{11\%}/R_{RH}$.

As seen in Fig. 5.31b, c, the sensitivities of the sensors made with both the pristine and alkalized $Ti_3C_2T_x$ were increased with increasing relative humidity. In comparison, the alkalized $Ti_3C_2T_x$-based sensor demonstrated much higher sensitivity than the pristine $Ti_3C_2T_x$ counterpart. Figure 5.31d shows response-recovery characteristics of the device based on the alkalized $Ti_3C_2T_x$ in the humidity range of 11–98% at room temperature. The stability of the alkalized $Ti_3C_2T_x$-based sensor at 98% RH is illustrated in Fig. 5.31e.

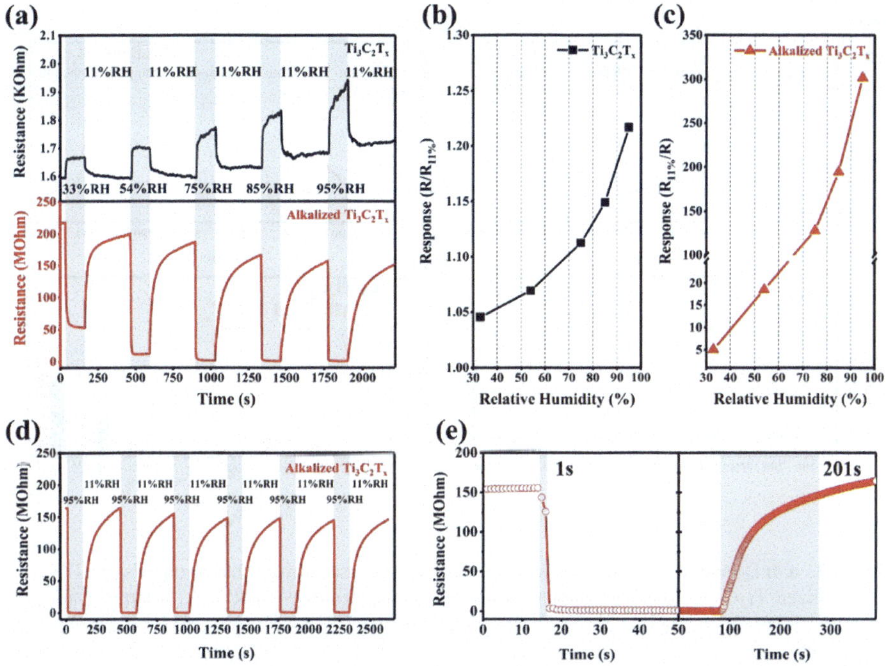

Fig. 5.31 **a** Dynamic response-recovery curves of the sensor devices based on pristine $Ti_3C_2T_x$ and alkalized $Ti_3C_2T_x$ to the variations in relative humidity. **b, c** Sensitivities of the sensors based on pristine $Ti_3C_2T_x$ and alkalized $Ti_3C_2T_x$ to as a function of relative humidity. **d** Detailed response-recovery curves of the sensor based on alkalized $Ti_3C_2T_x$ to the variation in RH between 11% RH and 98% at room temperature. **e** Response and recovery times of the sensor based on the alkalized $Ti_3C_2T_x$ between 11% RH and 98% RH at room temperature. Reproduced with permission from [90]. Copyright © 2019, American Chemical Society

Gas sensing performances to NH_3 were evaluated in the concentration range of 10–500 ppm, with the sensitivity to be defined as $\Delta R/R_{air} \times 100\%$. Response-recovery curves of the devices based on the pristine and alkalized $Ti_3C_2T_x$ MXene nanosheets are shown in Fig. 5.32a, b, respectively. Similarly, the resistance of the pristine $Ti_3C_2T_x$-based sensor was increased upon exposure to NH_3, while that of the alkalized $Ti_3C_2T_x$ one was reduced. Comparatively, the sensitivity of the alkalized $Ti_3C_2T_x$-based device was much higher. For instance, to 100 ppm of NH_3, the sensitivity of the alkalized $Ti_3C_2T_x$ sensor was twice that of the one based on the pristine $Ti_3C_2T_x$.

Figure 5.32c shows the stability of sensitivity to 100 ppm NH_3 with five cycles, confirming that the alkalized $Ti_3C_2T_x$-basd gas sensor was highly reproducible. Additional experiment results revealed that the MXene materials possessed negligible device-to-device variation. At the same time, selectivity of the sensors was also evaluated. Figure 5.32d shows sensitivities of the two groups of sensors to different gases

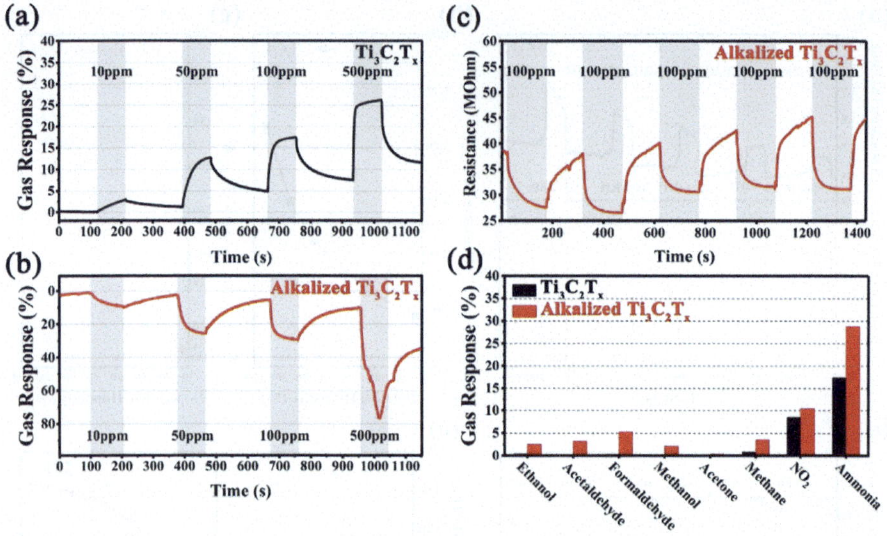

Fig. 5.32 **a**, **b** Dynamic response-recovery curves of the sensor devices based on pristine $Ti_3C_2T_x$ and alkalized $Ti_3C_2T_x$ to NH_3 vapor at different concentrations 38% RH. **c** Dynamic response-recovery curves of the sensor based on the alkalized $Ti_3C_2T_x$ to 100 ppm NH_3. **d** Sensitivities of the sensor devices based on pristine $Ti_3C_2T_x$ and alkalized $Ti_3C_2T_x$ to different gases with a concentration of 100 ppm. Reproduced with permission from [90]. Copyrights 2019, American Chemical Society

measured at the same concentration of 100 ppm. Obviously, the selectivity to NH_3 with respect to NO_2 was enhanced for the alkalized $Ti_3C_2T_x$-based sensing device.

It has been well accepted that the oxygen functionalization of $Ti_3C_2T_x$ will trigger the transformation of MXene phase from metallic conduction to a semiconduction, which could be utilized for potential applications as gas and humidity sensing devices. The resistance of the sensor based on the alkalized $Ti_3C_2T_x$ was reduced when the device was exposed to water vapor and NH_3 gas, which meant that the conduction was increased. The sensing response of the alkalized MXene was different from that of the pristine one, because of the the difference in the type of the carriers in the two groups of MXenes with different functional group properties.

The sensing mechanisms of the alkalized $Ti_3C_2T_x$-based sensor to water and NH_3 were proposed by considering the adsorption of the two molecules and the variation in conduction cause by electron transferring, as schematically demonstrated in Fig. 5.33. The intercalation of the Na^+ ions in the MXene nanosheets was critical in facilitating the sensing effect to humidity. As illustrated in route ①, H_2O molecules could interact with one Na^+ ion for form clusters of $[Na(H_2O)_m]^+$. The formation of the clusters would increase the conduction of the materials, thus leading reduction in resistance of the sensing device.

In route ②, the N atom in the s were adsorbed onto the surface of the $Ti_3C_2T_x$ nanosheets through interactions between the N atom of theNH_3 molecule and the

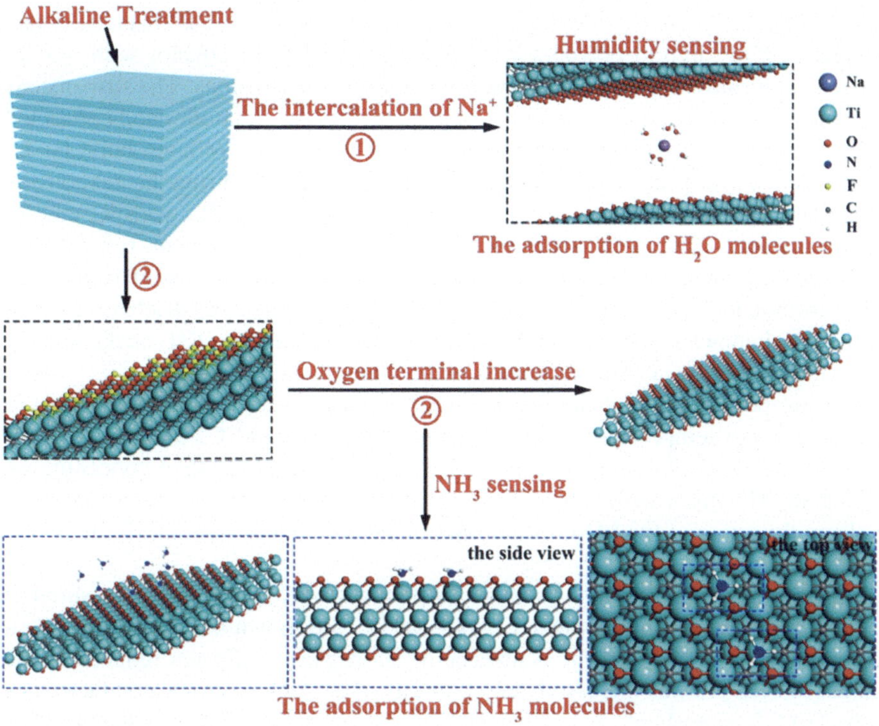

Fig. 5.33 Schematic diagram of the adsorption ulphide of the H_2O and NH_3 molecules on the surface of the alkalized $Ti_3C_2T_x$ nanosheets. Reproduced with permission from [90]. Copyright © 2019, American Chemical Society

Ti atom of the MXene. After alkaline intercalation, more –O terminal groups were produced, so that the density of the N–Ti bonding site was increased, resulting in an enhancement in the adsorption capacity to NH_3 molecules.

An *in-situ* hydrothermal growth method was used to construct $W_{18}O_{49}$-$Ti_3C_2T_x$ hybrids, consisting of 1D $W_{18}O_{49}$ nanorods (NRs) and 2D MXene nanosheets, as sensing materials to detect acetone [91]. The $W_{18}O_{49}$-$Ti_3C_2T_x$ hybrid-base sensors were highly sensitive to acetone at very low concentrations, with high selectivity and long-life stability. Specifically, the devices showed a limit of detection (LOD) of as low as 0.17 ppm, with response and recover times of 5.6 s and 6 s, respectively. As compared with the pristine $W_{18}O_{49}$ NRs and $Ti_3C_2T_x$ nanosheets, the hybrid exhibited much higher sensitivity to acetone vapor, which was ascribed to the fact that the active $W_{18}O_{49}$ NRs were homogeneously distributed on the surface of the $Ti_3C_2T_x$ nanosheets. Meanwhile, the absence of the F-containing functional groups in the $Ti_3C_2T_x$ phase owing to the solvothermal reaction and the interface effect between the $W_{18}O_{49}$ NRs and the $Ti_3C_2T_x$ nanosheets were also of contribution to the enhanced sensitivity.

To fabricate the sensor devices, 20 mg $W_{18}O_{49}$ NR, $Ti_3C_2T_x$ nanosheet and $W_{18}O_{49}$-$Ti_3C_2T_x$ hybrid powders were dispersed in 10 mL ethanol with the aid of strong sonication for 0.5 h to obtain homogeneous suspensions. The suspensions were then casted on ceramic substrates deposited with Au electrodes and connected with Pt wires. The coatings were instantly dried with a 275 W lamp bulk. Ni–Cr coils were used as heater to control the working temperature during the gas sensing measurement. The sensitivity of the sensors was given by R_a/R_g, with R_a and R_g to the resistance values of the sensors in the air and gas environment, respectively. Response (τ_{res}) and recovery times (τ_{rec}) were recorded as the times used to approach 90% variation in the resistance as a gas sample was injected in and discharged out of the testing chamber, respectively. The effect of relative humidity (RH) on the sensing properties of the devices was examined at the same time.

Sensing performance of the pristine $Ti_3C_2T_x$-based sensor had no response to acetone at room temperature and high temperatures of $\geq$300°C, while it was only weakly sensitive at the temperatures of 200–250 °C, at the acetone concentration of 20 ppm. The insensitive behaviour of the pristine $Ti_3C_2T_x$ could be attributed to the large thickness of the MXene nanosheets and the high content of –F functional groups.

Sensing behaviors of the sensors to 20 ppm acetone, which were measured at working temperatures in the range of 200–400 °C, are shown in Fig. 5.34a–e. At a given temperature, the intrinsic resistance of the devices in air was reduced as the content of $Ti_3C_2T_x$ was increased, due primarily to the high electrical conductivities of the $Ti_3C_2T_x$ nanosheets. For instance, the resistance value of the pristine $W_{18}O_{49}$-based sensor was 18.4 MΩ at 200 °C, which the corresponding values of the $W_{18}O_{49}$-$Ti_3C_2T_x$-1%, $W_{18}O_{49}$-$Ti_3C_2T_x$-1.5% $W_{18}O_{49}$-$Ti_3C_2T_x$-2% $W_{18}O_{49}$-$Ti_3C_2T_x$-2.5% devices were 15.6, 9.5, 8.2 and 4.7 MΩ, respectively. Also, the resistances of all the devices in air were decreased as the working temperature was raised, because of the thermally excited conduction of the materials.

Due to the n-type semiconducting characteristic of the materials, the resistances of the gas sensors were all decreased, when they were exposed to acetone. Figure 5.34f shows sensitivities of the devices based on different materials as a function of working temperature to 20 ppm acetone. The $W_{18}O_{49}$-$Ti_3C_2T_x$-2% hybrid-based sensor had the highest sensitivity at all working temperatures. The optimal working temperature was 350 °C, at which the sensitivities were maximized. However, the sensitivities at 300 °C were also promising. Both the response and recovery times were decreased with increasing working temperature, which were stabilized as the working temperature was above 300 °C, as illustrated in Fig. 5.34g. In comparison, the pristine $W_{18}O_{49}$-based device had longer response and recovery times at 300 °C, as observed in Fig. 5.34h.

Figure 5.35a shows dynamic response-recovery characteristics of the sensor devices based on different materials measured at the working temperature of 300 °C. For all the devices, the sensitivity was gradually increased as the concentration of acetone was increased over the whole concentration range studied, where the sensor based on $W_{18}O_{49}$-$Ti_3C_2T_x$-2% displayed the highest sensing performance. In addition, all the sensors based on the hybrids with MXene contents of 1.0–2.0% were

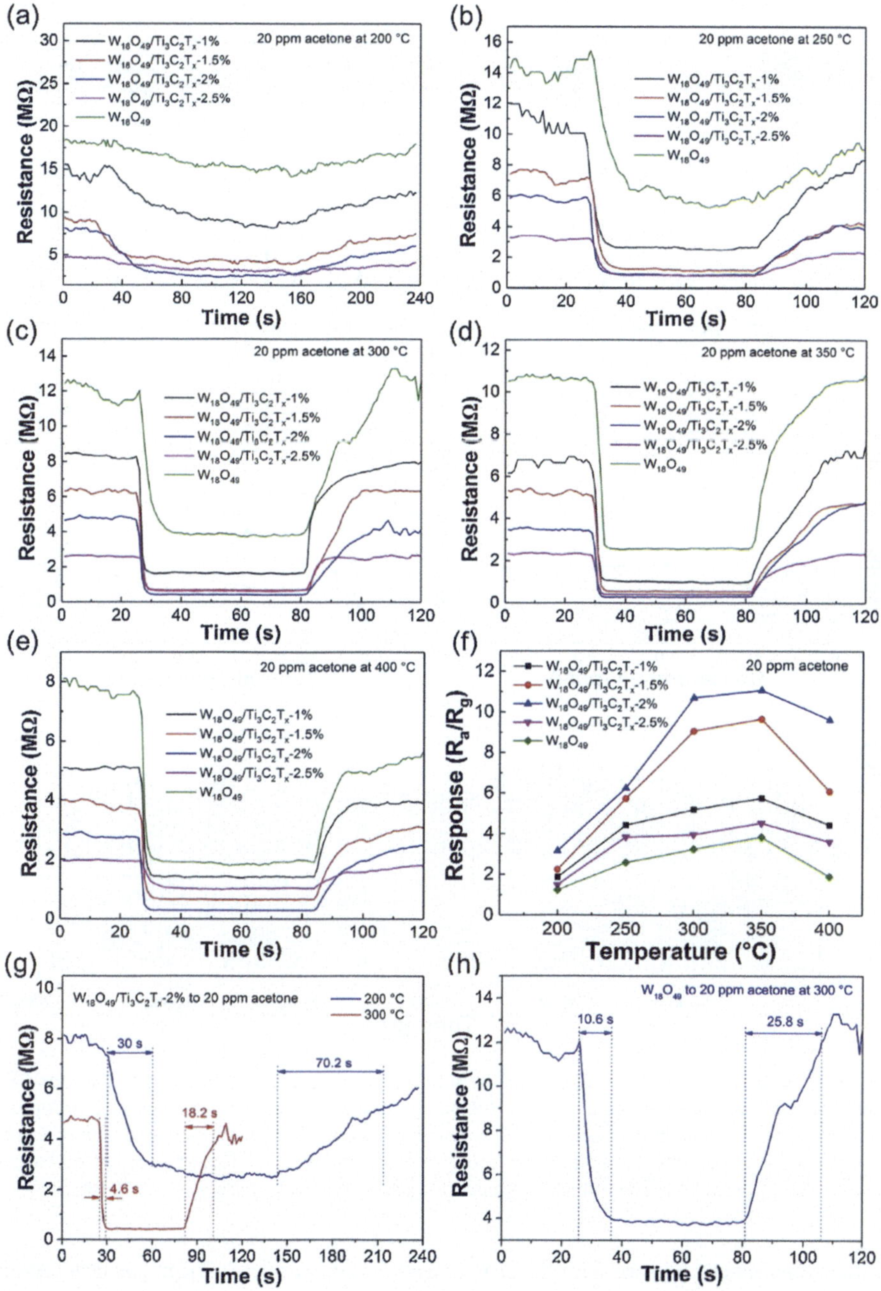

Fig. 5.34 Response-recovery curves of the sensor devices based on different materials to 20 ppm acetone at various working temperatures: **a** 200 °C, **b** 250 °C, **c** 300 °C, **d** 350 °C and **e**, 400 °C. **f** Sensitivity curves of the sensors to 20 ppm acetone at different working temperatures. **g** Response and recovery curves of W18O49-Ti3C2Tx-2% hybrid-based sensor to 20 ppm acetone at 200 and 300 °C. **h** Response and recovery curves of the pristine W18O49-based sensor to 20 ppm acetone at 300 °C. Reproduced with permission from [91]. Copyright © 2019, ElsevierReproduced with permission from [91]. Copyright © 2019, Elsevier

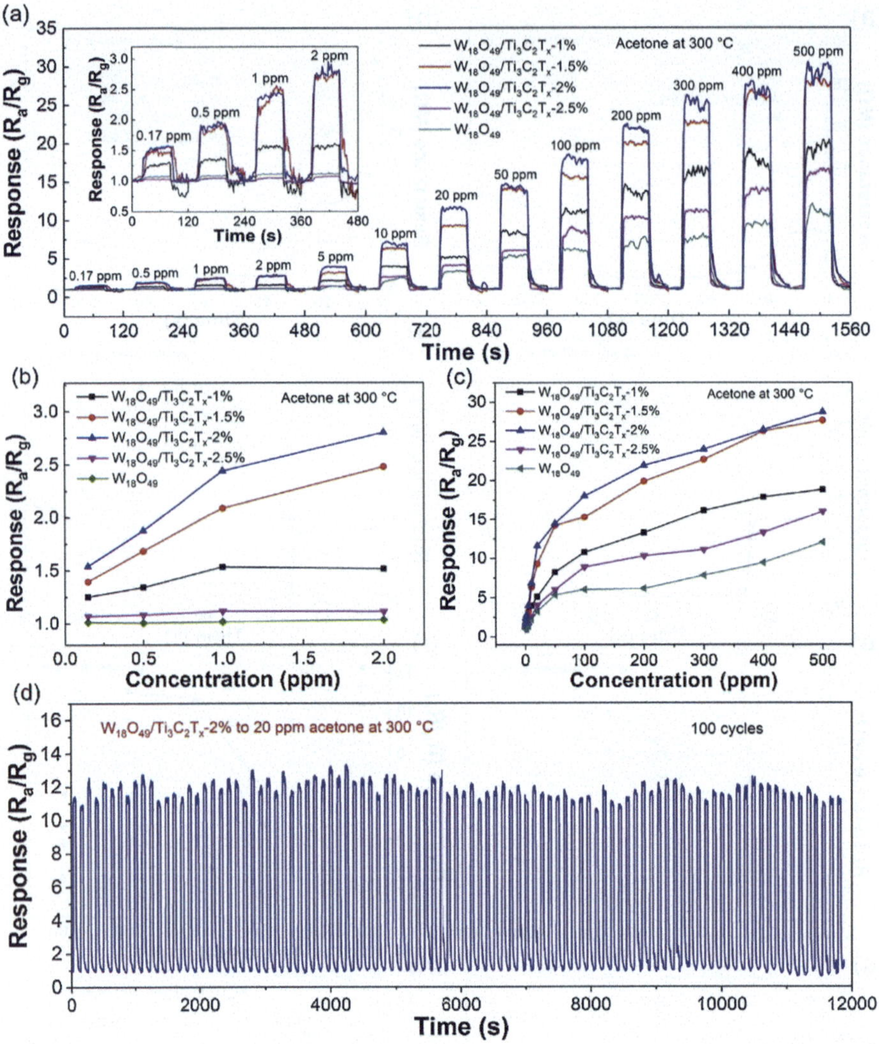

Fig. 5.35 **a** Response-recovery curves of the sensors based on different materials to acetone at 300 °C in the concentration range of 0.17–500 ppm, with the inset showing the enlarged response-recovery curves in the acetone concentration range of 0.17–2 ppm. **b, c** Sensitivities of the sensors as a function of the concentration of acetone. **d** Cycling curves of the $W_{18}O_{49}$-$Ti_3C_2T_x$-2% hybrid-based sensor to 20 ppm acetone. Reproduced with permission from [91]. Copyright © 2019, Elsevier

stabily sensitive to acetone over the concentration range of 0.17–2 ppm, as illustrated as the inset in Fig. 5.35a. In comparison, the ones made with the pristine $W_{18}O_{49}$ and $W_{18}O_{49}/Ti_3C_2T_x$-2.5% were almost not sensitive to acetone if the gas concentration was below 2 ppm.

As observed in Fig. 5.35b, the sensitivities of the devices based on $W_{18}O_{49}$-$Ti_3C_2T_x$-1%, $W_{18}O_{49}$-$Ti_3C_2T_x$-1.5% and $W_{18}O_{49}$-$Ti_3C_2T_x$-2% to 0.17 ppm acetone were 1.2, 1.4 and 1.5, respectively. Figure 5.35c shows sensitivities the gas sensors versus the concentration of acetone in the range of 5–500 ppm. Specifically, the $W_{18}O_{49}$-$Ti_3C_2T_x$-2% hybrid-based device has sensitivity level of 4.2–5 ppm acetone, which was obviously higher than those of other sensors. In addition, in the acetone concentration range of 100–500 ppm, the sensitivity was almost linearly increased.

Sensing stability of the $W_{18}O_{49}$-$Ti_3C_2T_x$-2% hybrid-based device was examined. Figure 5.35d shows the sensing responses of the device to 20 ppm acetone for 100 cycles, where response pattern was nearly kept unchanged, confirming that the sensor was high stable when working repeatedly. More importantly, the laminated microstructure and the flat surfaces of the MXene nanosheets were not altered after cycling for 100 times. The highly stable performance of the $W_{18}O_{49}/Ti_3C_2T_x$-2% hybrid-base sensor device was resulted in two factors. On one hand, as the $W_{18}O_{49}$ and $Ti_3C_2T_x$ were combined to form hybrid, the functional groups were got rid of from surface of the $Ti_3C_2T_x$ nanosheets, as evidenced by XPS results. On the other hand, the surface of the $Ti_3C_2T_x$ nanosheets was entirely coated by the $W_{18}O_{49}$ NRs.

Figure 5.36a depicts response behaviors of different sensor devices to 20 ppm ethanol at 300 °C. Although the device based on the $W_{18}O_{49}$-$Ti_3C_2T_x$-2% hybrid

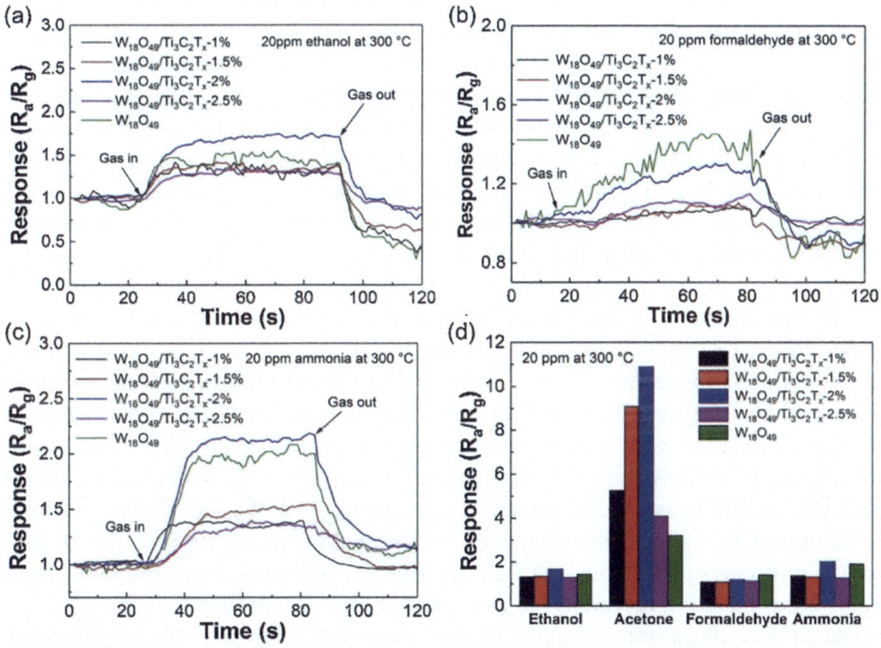

Fig. 5.36 Response-recovery curves of the sensors to different gases with a concentration of 20 ppm at 300 °C: **a** ethanol, **b** formaldehyde and **c** ammonia. **d** Sensitivity histogram of the sensors to different gases. Reproduced with permission from [91]. Copyright © 2019, Elsevier

sample had the highest sensitivity, the sensing response was relatively low. For formaldehyde, the two sensors made of $W_{18}O_{49}$ and $W_{18}O_{49}$-$Ti_3C_2T_x$-2% were slightly sensitive, while all other ones were almost insensitive, as revealed in Fig. 5.36b. Comparatively, all the devices displayed noticeable sensitivity to 20 ppm ammonia, whereas the $W_{18}O_{49}$-$Ti_3C_2T_x$-2% hybrid-based sensor had the highest sensitivity of 2.2, as illustrated in Fig. 5.36c. In addition, the response and recovery times were all longer than those to acetone. Sensitivities of the devices to gases of acetone, ethanol, formaldehyde and ammonia are plots as histogram in Fig. 5.36d, confirming high sensitivity to acetone.

It was found that RH had a pretty strong effect on sensing ulphide of the $W_{18}O_{49}$-$Ti_3C_2T_x$-2% hybrid-based sensor, with the sensing performance to 20 ppm acetone at 300 °C as an example. Generally, the sensitivity was decreased with increasing RH level. Specifically, at the values of RH to be ≤60%, the reduction in sensitivity was quite pronounced, while the negative effect was slowed down in the RH range of 60–98%. However, even at 98% RH, the sensitivity of the device made of $W_{18}O_{49}$-$Ti_3C_2T_x$-2% was sufficiently high to 20 ppm acetone, with a value of 8.7, although it was declined by about 25%, as compared with that at 20% RH.

This observation was understood by considering the competitive effect water molecules over the O_2 molecules in trapping electrons, so that the oxygen-related ionic species (e.g., O_2^-, O^- and O^{2-}) adsorbed on the surface of the $W_{18}O_{49}$-$Ti_3C_2T_x$-2% hybrid nanosheets would have reduced possibility to react with the gases to be tested. As a result, the change in resistance of the sensor was decrease accordingly. Meanwhile, the adsorption of the H_2O molecules on the surfaces of items in the hybrid was saturated as the RH value reached a level, i.e., 60%, above which the effect of RH on the sensing behaviour of the sensors was weakened. Accordingly, the variation in sensitivity of the sensor in the RH range 60–90% was slowed down.

Because the $Ti_3C_2T_x$-based device was insensitive to acetone, the enhanced sensing behaviour of the hybrid-based sensors was mainly related to the interfacial characteristics between the active $W_{18}O_{49}$ NRs and the $Ti_3C_2T_x$ nanosheets. The $W_{18}O_{49}$-$Ti_3C_2T_x$ nanosheets adsorbed O_2 molecules in air, which took electrons from the semiconducting materials, thus generating O_2^- items, at the working temperatures of <150 °C, while they converted to O^- and O^{2-} as the working temperature was raised above 150 °C. As a result, the resistance of the sensor devices was increased, since the concentration of electron was decreased. Once acetone was introduced, it would react with the oxygen-related items, so that electrons were released, thus leading to increase in conduction of the $W_{18}O_{49}$-$Ti_3C_2T_x$ hybrid material and hence decrease in resistance of the devices. In addition, the $W_{18}O_{49}$ phase contained high concentration of oxygen vacancies, which enabled to adsorb both the O_2 and acetone molecules. Therefore, the $W_{18}O_{49}$-$Ti_3C_2T_x$ hybrids were expected to have larger-scale change in resistivity, so as to exhibit high sensitivities.

The MXene phase offered a platform to enhance the sensing effect by forming hybrids with the active component $W_{18}O_{49}$. Because the surface of the $Ti_3C_2T_x$ nanosheets was enriched with functional groups of –O, –OH and –F, which served as active sites for the $W_{18}O_{49}$ to nucleate and grow into NRs. Since the $W_{18}O_{49}$ NRs were uniformly distributed on the surface of the MXene nanosheets, they were exposed more efficiently to O_2 and acetone molecules, as compared with the pristine $W_{18}O_{49}$. Also, the elimination of the –F groups from the surface of the MXene nanosheets during the solvothermal reaction step promoted the adsorption of the acetone molecules, due to the formation of hydrogen bonds with the –O/ –OH functional groups.

Moreover, owing to the high electrical conductivity of the $Ti_3C_2T_x$ nanosheets, the electrons at the conduction band of $W_{18}O_{49}$ might be transferred to $Ti_3C_2T_x$, resulting in the formation of Schottky barrier at the $W_{18}O_{49}$-$Ti_3C_2T_x$ interface. As the devices were exposed in air, the hight of Schottky barrier was increased, so that and a new electron depletion layer was generated at the surface of $W_{18}O_{49}$. Upon exposure to acetone, electrons were released and the Schottky barrier was back to its initial hight and the electron depletion layer was absent, so that the resistance of the hybrid-based devices was decreased.

In addition, there was an optimal content of MXene in the hybrids, in terms of achieving high sensing performances to acetone. Initially, the sensitivity was gradually enhanced with increasing content of the $Ti_3C_2T_x$ phase, because of the increase in the degree of dispersion of the $W_{18}O_{49}$ NRs. However, as the concentration of $Ti_3C_2T_x$ was >2 wt%, -F groups were present. As a consequence, the sensitivity of the $W_{18}O_{49}$-$Ti_3C_2T_x$-based devices was reduced. Furthermore, a content of $Ti_3C_2T_x$ would inevitably lead to stacking/agglomeration of the nanosheets, which in turn sacrificed more active sites for the adsorption–desorption reaction of the acetone molecules.

In a recent study, vanadium carbide MXene (i.e., V_2CT_x) gas sensor was developed, which was highly sensitive to nonpolar gases at room temperature [92]. The V_2CT_x-based sensors were made of single-/few-layered V_2CT_x nanosheets that were coated on polyimide films, by using a drop casting process. To measure sensitivities of the devices, after the baseline was stabilized, the V_2CT_x sensor was exposed to the different gases for 10 min, as demonstrated in Fig. 5.37c–f. Then, synthetic air was flowed in the testing chamber to allow the sensors to recover. The resistances of the V_2CT_x films with a thickness of 0.2 μm were in the range of 7–10 kΩ at room temperature. The as-coated V_2CT_x films were semiconductor.

Figure 5.37a shows response curves of the gas sensor different gases at a concentration of 100 ppm. In all cases, the resistance of the V_2CT_x-based device was increased, because all the gases were of reducing characteristics. The resistances were recovered after the gases were not injected. The V_2CT_x phase was a p-type semiconductor, while the sensor had resistances at the scale of kΩ. The oxygen-related functional groups, e.g., –OH, on the surface of the V_2CT_x nanosheets promoted the adsorption of the gases. Figure 5.37b shows histogram of sensitivities of the sensor to different gases with a concentration of 100 ppm. The sensor exhibited highest sensitivity to hydrogen, indicating its high selectivity. The response and recovery times of the V_2CT_x-based sensor to hydrogen were 2 min and 7 min, respectively.

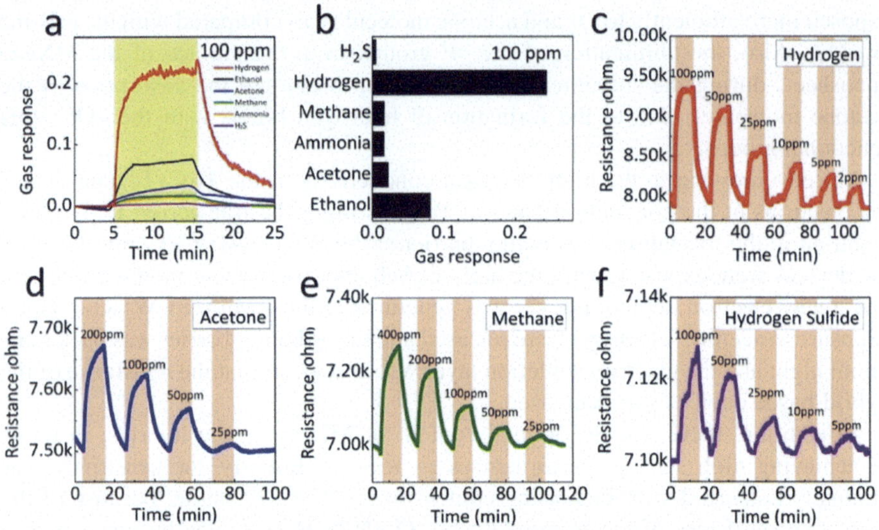

Fig. 5.37 Gas sensing performances of the V_2CT_x sensor at room temperature: **a** variation in compiled resistance and **b** sensitivities to 100 ppm gases (hydrogen, ethanol, acetone, methane, ammonia and hydrogen sulphide) at room temperature. Real-time sensing responsive characteristics of the V_2CT_x gas sensors at different concentrations of various gases: **c** hydrogen, **d** acetone, **e** methane and **f** hydrogen ulphide. Reproduced with permission from [92]. Copyright © 2019, American Chemical Society

5.3.3 Piezoelectric Sensors

With the rapid development of flexible wearable electronics, the devices should have large stretchability, high gauge factors (GF) and broad range of strains. It has been observed that MXene based nanocomposites or nanohybrids exhibited high sensitivities of up to GF of 773 and tuneable sensing range of 30–130% strain. As a result, MXenes could find applications in utilization of phonations and body movement. For instance, $Ti_3C_2T_x$/SWCNTs nanohybrids were developed by using layer by layer (LbL) assembling method through spray coating, in order to fabricate strain sensors with a low LOD of 0.1% [93]. The $Ti_3C_2T_x$/SWCNTs nanohybrid based strain sensor was used to assemble skin attachable wearable devices, for potential applications in capturing physiological signals and monitoring human motions.

The sensing mechanism was closely related to the tunnelling distances between the MXene nanolayers and the interconnected SWCNTs. As an external pressure was applied to the device, the overlapping areas and distances of the two components were decreased and increased accordingly. Once the change in resistivity resulted from the varied interlayer distance could be detected, the MXene nanohybrid based sensor started to work. In this case, the piezoresistive sensor was a pressure sensor, which converted the external pressure into resistance signal.

Piezoresistive effect of the $Ti_3C_2T_x$/SWCNT thin film could be tailored by controlling the number of LbL spray coating and the concentration of the $Ti_3C_2T_x$ and CNTs suspensions. The hybrid thin films were denoted as $MXene_1$–CNT_1, $MXene_1$–$CNT_{0.2}$, $MXene_1$–$CNT_{0.1}$, CNT_1–$MXene_1$, CNT_1–$MXene_{0.2}$ and CNT_1–$MXene_{0.1}$. The subscripts stood for the concentrations of the MXene or SWCNTs suspensions. For instance, "1" was to mean that the concentration was 1 mg/mL. Figure 5.38a shows representative piezoresistive behaviors of the stretchable film based on $MXene_1$–CNT_1 spray coated for 10 cycles, at applied strains of 1–80%.

The relative resistance $(R–R_0)/R_0$ was slightly decreased in the strain range of 0–30%, with a GF of 64.6, due to the decrease in electrical conductivity. This was because the mechanical microcrack junctions were disconnected owing to the tensile stress applied to the $Ti_3C_2T_x$ layers. Moreover, both the gap and density of the cracks were further increased as the strain was further increased. As the applied strain was increased to the range of 40–70%, an abrupt increase in resistance took place, corresponding to a high GF of 772.6, suggesting a rapid reduction in electrical conductivity. In this case, disconnection in between the conductive SWCNTs, which thus enlarged the effect of the microcracks. Therefore, the $Ti_3C_2T_x$/SWCNT nanohybrid film strain sensor was superior to the conventional brittle metallic foils and semiconductor based strain sensors, with GF of about 2 and sensing strain range of $\varepsilon < 5\%$ [94].

As the applied strain approached 70%, an abrupt increment in resistance was observed, because of the accelerated disconnection of the mechanical junctions in the $Ti_3C_2T_x$ layers and the reduced conductive pathways in the SWCNTs. The $I–V$ curves were linear as the strain was below 30%, implying that the relative resistance was constant. In the region of large repeated strains, the variations in relative resistance for the maximum strains of 5%, 10%, 20% and 50% were 2.8, 3.5, 10.3 and 67.1, respectively, as seen in Fig. 5.38b. As observed in Fig. 5.38c, a very small strain of 0.1% could trigger the response in relative resistance of response, corresponding to a minimum displacement of 10 μm. In other words, the $Ti_3C_2T_x$/SWCNTs hybrid strain sensor was able to identify small strains.

Figure 5.38d shows frequency responses and output signals of the $Ti_3C_2T_x$/SWCNT hybrid film strain sensors. In the frequency range of 0.1–2 Hz, the electrical responses of the device were very stable. The variations of the electrical signals were nearly independent on the frequency, which implied that the adhesion between the $Ti_3C_2T_x$/SWCNT hybrid film and the latex rubber was sufficiently strong. Also, the output current signals were very close to the dynamic strain inputs at the frequency of 0.5 Hz, as illustrated in Fig. 5.38e. The pattern of the current response was nearly identical to that of the input strain, suggesting the quick response of the sensors.

Figure 5.38f shows the results for a cyclic strain, from 0 to 20% and then back to 0%, at the frequency of 1 Hz. The $Ti_3C_2T_x$/SWCNTs hybrid film strain sensor exhibited a stable response to the loading–unloading cycles, together with a high reproducibility and high signal-to-noise ratios for more than 5000 cycles. Meanwhile, the sensing current of the sensor was almost unchanged after the cycling testing, indicating the high durability and stability of the device. It was found that

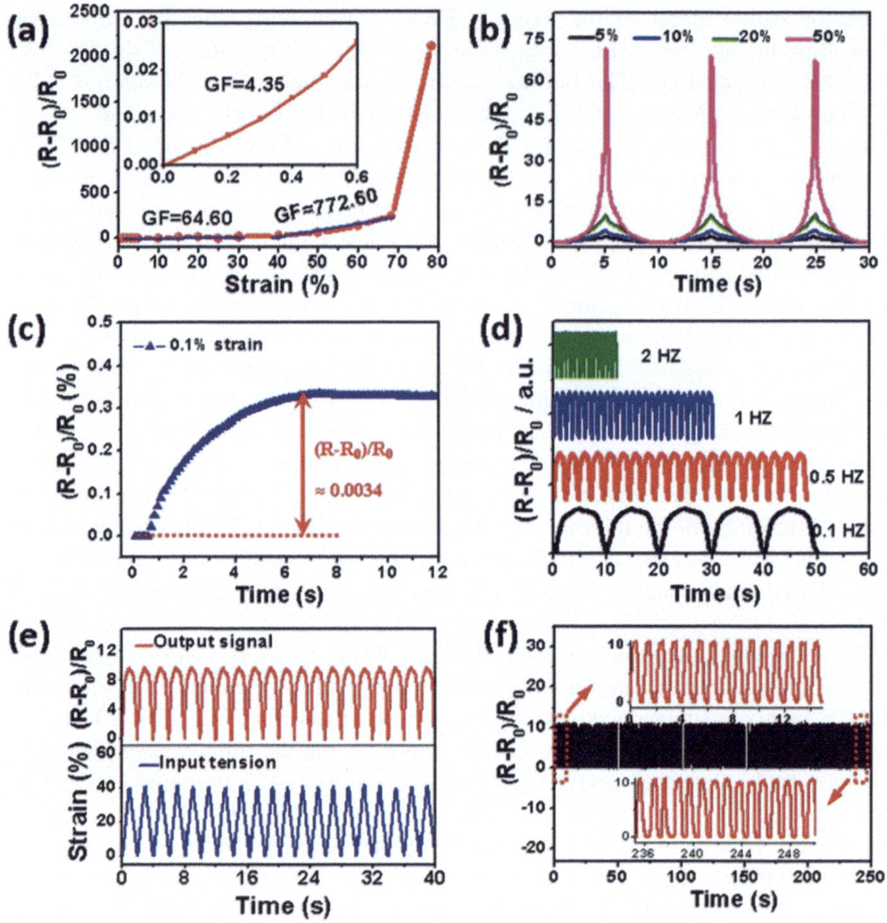

Fig. 5.38 **a** Relative resistance–strain curve of the strain sensor based on $Ti_3C_2T_x$ MXene/SWCNT/latex at a stretching rate of 5%/min, with the inset showing the curve below the strain of 0.6%. **b** Variations in relative resistance as a function of time at the strain frequency of 0.1 Hz for four applied strains. **c** Changes in relative resistance as a function of time at the low strain of 0.1% at a stage moving speed of 0.01 mm/s. **d** Relative resistance responses of the strain sensor at different frequencies at 20% strain. **e** Time retention curves of the change in resistance and strain with time. **f** Durability tests at the strain of 20% at 1 Hz. The resistance change curves were recorded after 1000 cycles and 50 cycles in each recording. Reproduced with permission from [93]. Copyright © 2018, American Chemical Society

this $Ti_3C_2T_x$/SWCNTs hybrid film strain sensor had much higher performances than RGO_1–CNT_1 devices.

The $Ti_3C_2T_x$/CNT/latex strain sensors had been used to develop skin attachable flexible devices to detect real time slight physiological signal and monitor large-scale motion of human beings, which was demonstrated experimentally. For instance, the complex skin stretching action over the human throat when speaking the words "carbon", "sensor" and "MXene" repeatably could be captured as the current response. This capability to recognize the phonation of humans could find applications in phonation rehabilitation and human/machine interactions. They could also be applied to response human body motions, such as that the knee muscle, during walking, running and jumping. All these actions could be detected by the strain sensor in the form of variation in the relative sensor resistance.

Ti_3C_2 MXene was used to fabricate piezoresistive sensors, where the distance between adjacent the MXene nanolayers was reduced, when the devices were subject to external pressures, thus leading to reduction in resistance of the sensors [95]. The sensors could be used in human body, including eye corners cheeks, throats and so on, so as to monitor the stimuli, such as eye blinkings, cheek bulgings and throat swallowings. Moreover, knee bending action was sensed by monitoring the variation in current with time. For high integration, the sensors could be linked to sulphides devices, so that the current signals were converted into wireless electromagnetic radiations with high conductivities. Furthermore, the MXene sensors demonstrated high sensitivities and high stabilities in responding to the subtle activities of human bodies, making them highly potential in flexible portable wearable sensing devices in healthcares and sulphides analysis.

By using the density-functional theory, it was predicted that M_2CO_2 MXenes with M=Sc, Y and La could display large Poisson's ratio and in-plane piezoelectricity, which close to those of 2H-MoS_2 [96]. The high piezoelectric effect of the M_2CO_2 MXenes would be especially useful, when piezoelectric devices with ultra-small thickness, such as cantilevers and diaphragms, are required. When the MXene is used for diaphragm, the piezoelectric effect related charges were concentrated in six locations, depending on the lattice symmetries. Specifically, the in-plane and out-of-plane piezoelectric charges were located near the circular clamped boundaries and inside the pressure-generated bubbles, respectively. The results are very useful to identify where the piezoelectric charges are concentrated, thus enabling mechanical energy harvesting and piezoelectric sensing applications.

5.3.4 Photoluminescence (PL) Sensors

Photoluminescence has been widely used for applications in the fields of optical, biomedical and cellular imaging. For example, highly sensitive optical MXene based quantum dots (MQDs) have been shown to be promising for the applications in bio-imaging devices [97]. The narrow band gaps of $Ti_3C_2T_x$, with 0.05 and 0.1 eV for –F and –OH functional groups respectively, could be broadened through quantum effects, thus resulting in strong photoluminescence emissions. By using hydrothermal processing, Ti_3C_2-MQDs were obtained at different temperatures, with 100 °C, 120 °C and 150 °C to be denoted as MQDs-100, MQDs-120 and MQDs-150. Respectively.

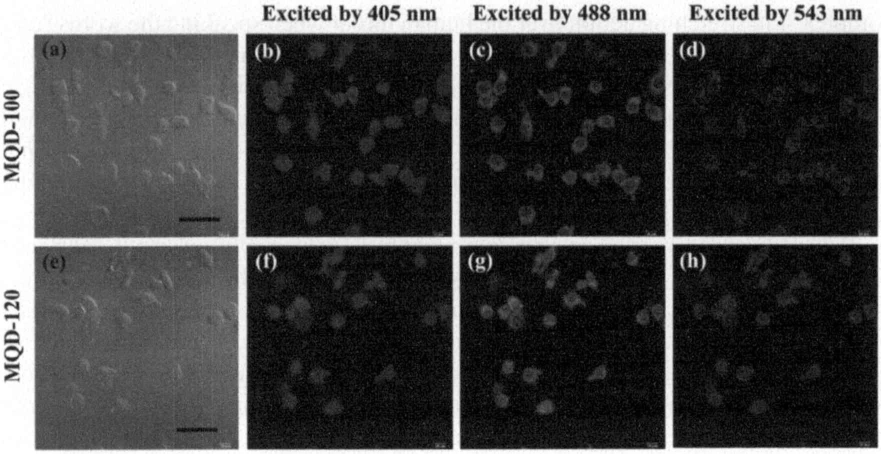

Fig. 5.39 **a, e** Bright-field images of RAW264.7 cells. Confocal imaging (405, 488 and 543 nm) of RAW264.7 cells incubated with **b–d** MQD-100 and **f–h** MQD-120. Reproduced with permission from [93]. Copyright © 2018, American Chemical Society

Both the MQDs-100 and MQDs-120 were employed to develop biocompatible sulphide imaging probes, which could realize the photoluminescent detection of RAW264.7 cell lines, with results to be shown in Fig. 5.39a, e. In-vitro bioimaging of the RAW264.7 cells was conducted with a confocal microscopy. After the cells were incubated for 4 h, the endocytosis process made the MQDs to be readily incorporated. As illustrated in Fig. 5.39b–d, the colors of the confocal images were bright blue, green and red, for the excitations at 405, 488, and 543 nm, respectively, confirming high potentials of the MQDs for the applications in biolabeling through the endocytosis process. Confocal luminescence images of RAW264.7 cell with the presence of MQD-120 are depicted in Fig. 5.39f–h. It was found that the MQDs were mainly located in the membrane and cytoplasm of the cells, while the luminescent spots found in the nucleus region were extremely low. This strongly suggested that the MQDs had penetrated the cells, while not approaching the nucleus. As a result, no genetic disruption occurred to the cells.

In addition, it was found that the MQDs exhibited high biocompatibility with cells. The MQDs-150 could be used to monitor Zn^{2+} ions through the luminescence quenching mechanism, which was attributed to the amorphous characteristic of the MQDs. The presence of Zn^{2+} suddenly quenched the PL intensity of MQD-150. In contrast, it was nearly unaltered when other metallic ions, including Fe^{3+}, Co^{2+}, Ni^{2+}, Cr^{2+}, Pb^{2+}, Al^{3+}, Cu^{2+}, Mn^{2+} and Sn^{2+}, were added. This observation implied that MQD-150 was highly selective to Zn^{2+} ions. Also, both the MQD-100 and MQD-120 samples were not responsive to the metallic ions.

The quenching effect was only available when there were specific interactions between the surface functional groups of the MQDs and the metallic ions [98–100]. The strong quenching behavior of MQD-150 probably because its structure was carbon-rich. The quenching mechanism was explored by examining the luminescent quenching levels of MQD-150 with different concentrations of Zn^{2+} ions. It was demonstrated that the intensity of the PL was gradually reduced as the concentration of Zn^{2+} was increased. Therefore, it was induced that the selectivity to Zn^{2+} could be attributed to the fact that there were unique coordination interactions between Zn^{2+} ions and the –OH groups of the MQD-150. Nevertheless, MQD-150 was a potential candidate to develop Zn^{2+} sensors.

Ti_3C_2 MQDs were also employed to develop intracellular photoluminescent ratiometric pH sensors for the applications to detect cell metabolisms [101]. As an example, MCF-7 cells were used as the target. Because both the inter-band transition of carriers and surface defect sites of the Ti_3C_2-MQDs were dependent on the QD size, excitation wavelengths were observed. Specifically, the system had a blue photoluminescent response that was dependent on pH value, corresponding to a quantum yield of 7.13%. Therefore, the Ti_3C_2 MQDs based ratiometric photoluminescence probe could be used to detect intracellular pH level of the targets.

Exosomes have been utilized as non-invasive biomarkers to detect and diagnose diseases. For instance, Ti_3C_2 MXenes were combined with Cy3 labelled CD63 aptamer (Cy3-CD63 aptamer) to form a nanohybrid, as a self-standard ratiometric fluorescence resonance energy transfer (FRET) nanoprobe, which was used to quantitatively detect exosomes [102]. The Ti_3C_2 nanosheets adsorbed the Cy3-CD63 aptamer through hydrogen bond, as well as the interaction of metal chelate, so as to have a high selectivity. In this case, the fluorescence signal from the Cy3-CD63 aptamer could be quenched rapidly, due to the effect of FRET between the Cy3 and Ti_3C_2 nanosheets.

Since the exosomes could be combined with the aptamer and then released from the surface of Ti_3C_2, due to the high affinity of the aptamer to CD63 protein on surface of the exosomes, the fluorescence of Cy3 could be recovered. At the same time, self-fluorescence signal of the Ti_3C_2 was almost unchanged during the process. As a result, it could serve as the standard reference. With this self-standard turn-on FRET biosensing platform, the limit of detection (LOD) of exosome was about 1.4×10^3 particles/mL. This value was which was lower than that of the conventional ELISA technique by >1000 times. The unique fluorescence sensor could also be applied to detect multiple biomarkers on surfaces of the exosomes. Furthermore, the exosomes could be different, when combined with the fluorescent confocal scanning microscope images.

$Ti_3C_2T_x$ MXene has been combined with Nafion to obtain hybrid film used to fabricate solid-state ECL sensor [103]. The presence of $Ti_3C_2T_x$ ensured high conductivity of the hybrid film and also increased the quantity of $Ru(bpy)_3{}^{2+}$ adsorbed on surface of the electrode. Tripropylamine was examined as a representative ECL co-reactant. The limit of detection (LOD) was as low as 5 nM. The sensor exhibited high reproducibility and life stability. Furthermore, the high performance of the sensor was confirmed by the discrimination of the single-nucleotide mismatch in human urine.

Ti_3C_2 MXene-based photoluminescent quantum dots (N,P-MQDs) decorated with phosphorus and nitrogen were synthesized by using a hydrothermal reaction process [104]. This hybrid quantum dots exhibited green fluorescence at the wavelength of 560 nm, in which the photoluminescence quantum yield (PLQY) was about 20%. Both photostabilities and pH resistance capacity were the highest among the similar materials. Density functional theory (DFT) was employed to simulate the mechanisms of the shift and enhancement in the fluorescent effect. Moreover, the N,P-MQDs could be utilized as fluorescent probes to label macrophages. Additonally, the N,P-MQDs-based device was highly sensitive to Cu^{2+} ions, offering a label-free fluorescence platform for the precise detection of Cu^{2+} ion. The approach could be extended to other sensing devices.

5.3.5 Miscellaneous

The special metallic conduction behavors have made MXenes to be useful in developing terahertz (THz) detectors. THz waves are able to penetrate sulphide materials. The application of MXenes as THz detectors has been examined theoretically by using density functional theory (DFT) calculations on Ti_3C_2 [105]. It was predicted that Ti_3C_2 exhibited outstanding light extinction and optical absorption characteristics at THz frequencies. Furthermore, stacked Ti_3C_2 nanosheets could have higher sensitivity than monolayer ones, due to their stronger optical absorption behaviors. In addition, the Ti_3C_2 nanostacks had higher electronic density of states near the Fermi energy level than stacked graphene nanoflakes. They also exhibited a sufficiently high thermoelectric figure of merit (ZT) value for the applications as THz detectors.

MXenes have been evaluated for the applications in sensing system namely surface enhanced Raman spectroscopy (SERS). Ti_3C_2 based hybrids with report Ag, Au and Pd nanoparticles were prepared can be used as substrates for SERS [106]. The surface plasmon resonance (SPR) frequency of nanoparticles was calculated. The calculation and experimental results were in a good agreement at the SPR wavelengths of 440 and 558 nm, for the Ag and Au nanoparticles, respectively. For Pd nanoparticles with a planar morphology, the predicted SPR wavelength was 160 nm. This was because the non-spherical Pd nanoparticles were coupled with the 2D MXene nanosheets, thus leading to a largely red-shifted SPR band peaked at 230 nm. The noble metallic nanosized particles attached onto the MXene nanosheets were SERS sensitive to methylene blue (MB), with an estimated enhancement factor to be as high as 10^5.

In a separate study, $Ti_3C_2T_x$ was used to enhance the Raman signals of organic dyes both on substrates and in solutions [107]. The MXene SERS substrates were prepared by using a spray-coating method. The substrates were employed to monitor the conventional dyes, achieving a theoretical enhancement factor of 10^6, when used to detect Rhodamine 6G at an extremely low concentration of 10^{-7} M. More significantly, the SERS substrates contained no noble metallic nanoparticles. The dual sulphide of electromagnetic and chemical enhancements of the $Ti_3C_2T_x$ SERS substrates facilitated their potential for biochemical molecular and SERS sensors.

Multilayered Ti_2NT_x were obtained through sonication in DMSO, with SERS activities in the forms of paper-, silicon- and glass-based SERS substrates [108]. The maximum Raman enhancement factor was as high as 10^{12}, when using Rhodamine 6G as the model dye, with the excitation at the wavelength of 532 nm. The simple paper-based SERS substrate could be used to detect explosives at trace levels.

5.4 Electrocatalytic Applications

Electrolytic effect or solar water splitting has been among the most extensively studied, because of its significance in the development of new energies. Basically, it consists of two reactions, i.e., (i) oxygen evolution reaction (OER) ($2H_2O \rightarrow 4\,e^- + 4H^+ + O_2$) and (ii) hydrogen evolution reaction (HER) ($2H^+ + 2e^- \rightarrow H_2$). Similar to the oxygen reduction reaction (ORR) at the cathode of fuel cells, it is necessary to use expensive noble metallic catalysts. Therefore, the commercial applications of such effects have been hindered. In this regard, it is urgent to explore cost-effective yet highly efficient electrocatalysts to replace the noble metallic ones.

A free-standing hybrid film based on delaminated HF-etched $Ti_3C_2T_x$ and exfoliated graphitic carbon nitride (g-C_3N_4) nanosheets was prepared by using the vacuum filtration method [109]. When using as OER, the $Ti_3C_2T_x$/g-C_3N_4 hybrid film had a sharp onset potential at 1.44 V, versus reversible hydrogen electrode (RHE), while a low operating potential of 1.65 V delivered a current density of 10.0 mA/cm^2, in 0.1 M KOH solution. The outstanding performance was ascribed to the presence of the Ti-N_x motifs which served as electroactive sites. Also, the high surface area related to the hierarchically porous structure leading to a high electrochemical double layer capacitive effect, the hydrophilicity and the high porosity. All these together ensured a good wettability of the active surfaces, so to facilitate a smooth transportation of reactants and a rapid elimination of products. The frees-standing hybrid film was also demonstrated to be a promising cathode in rechargeable zinc (Zn)-air batteries. At the current density of 20 mA/cm^2, it had a discharging (ORR) at 0.87 V and a charging at 2.55 V.

$TiO_2/Ti_3C_2T_x$ hybrids were synthesized by using a hydrothermal process at 95 °C for 4 h, which demonstrated an enhancement in the photocatalytic HER at about 17.8 μmol/h·$g_{catalyst}$ by nearly 400%, as compared with the performance of the pure rutile TiO_2 [110]. The optimized content of $Ti_3C_2T_x$ was 5 wt%, as illustrated in Fig. 5.40. As the $Ti_3C_2T_x$ nanosheets were decorate with rutile TiO_2 nanoparticles, the two components were closely contacted, thus ensuring the sufficient light absorption of the TiO_2, rapid charge transport from TiO_2 to $Ti_3C_2T_x$, due to the formation of a depletion layer at the interface between $Ti_3C_2T_x$ and TiO_2. The depletion layer was known as Schottky barrier, which ensured the electron–hole separation, thus enhancing the photocatalytic activity of the nanohybrids. At optimal conditions, the TiO_2/Ti_2CT_x and TiO_2/Nb_2CT_x nanohybrids with 5 wt% content of MXenes were synthesized. The TiO_2/Nb_2CT_x nanohybrid exhibited the highest effect as co-catalysts of the photocatalytic HER. Nb_2CT_x exhibited the weakest photoluminescence emission, since it had the highest work function amongst the MXenes.

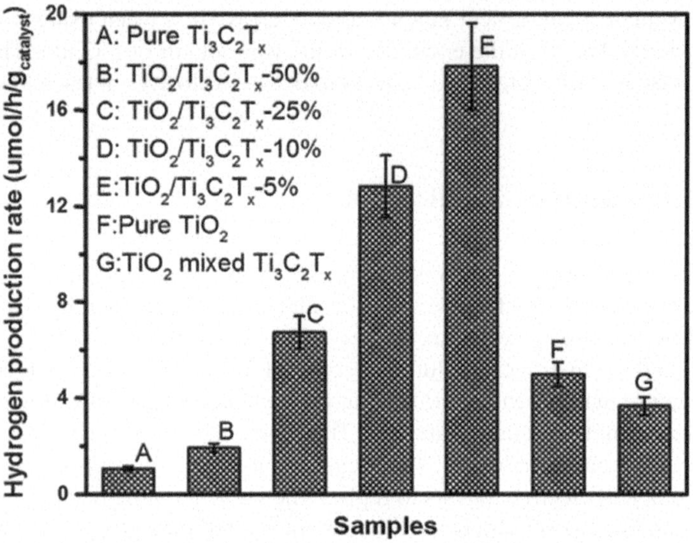

Fig. 5.40 Photocatalytic hydrogen production rates of the pure TiO_2, $TiO_2/Ti_3C_2T_x$ (5 wt %), $TiO_2/Ti_3C_2T_x$ (10 wt %), $TiO_2/Ti_3C_2T_x$ (25 wt %), $TiO_2/Ti_3C_2T_x$ (50 wt %), pure $Ti_3C_2T_x$ and a manually mixed $TiO_2/Ti_3C_2T_x$ sample with 5 wt % $Ti_3C_2T_x$. Reproduced with permission from [110]. Copyright © 2016, John Wiley and Sons

HER effect of V_2CO_2 with surface to be fully oxygen terminated was theoretically studied by using the first-principles calculations method [111]. It was observed that pure V_2CO_2 was not suitable as HER catalyst because of the poor catalytic performance, which was closely related to the strong hydrogen adsorption. Interestingly, once proper transition metal atoms (e.g., Fe, Co or Ni) were incorporate at the surface, the adsorption free energy of –H group could be nearly completely reduced, similar to that observed of the surface of Pt (111). Due to the boosted charge transfer between the transitional metal atoms and the oxygen atoms on the surface, the HER catalytic effect was significantly enhanced. In addition, the enhanced HER sulphide was well maintained for a relatively long time. Even more, it was possible to further improve the long-term catalytic stability through strain engineering technologies.

Zheng et al. prepared a group of MXene-Ag hybrid based composites for oxygen reduction reaction (ORR) catalytic applications [112]. The composites were derived from $AgNO_3$ solution and alkalization-intercalated MXene (alk-MXene, $Ti_3C_2(OH/Ona)_2$) solution with different concentrations of polyvinylpyrrolidone (PVP), through a simple mixing process at room temperature. The MXene-$Ag_{0.9}Ti_{0.1}$ bimetallic nanowire hybrids had a urchin-like morphology. Shaped composite exhibits unexpected electrocatalytic activity for the oxygen reduction reaction, with an onset potential of 0.921 V and a half-wave potential of 0.782 V. The presence of PVP triggered the formation of Ag seeds with fivefold nano-twins, thus leading to the development of the Ag/Ti ($Ag_{0.9}Ti_{0.1}$) bimetallic nanowires. Owing to the special bimetallic nanowire structures, a four-electron transfer process was readily facilitated. The nanocomposite catalysts possessed high current densities and high stabilities, because of the rich oxygen adsorption sites and the shortened diffusion paths for adsorbed oxygen items.

HF-etched $Ti_3C_2T_x$ has been employed as the support of platinum (Pt) nanoparticles, serving as the cathode catalyst in proton exchange membrane fuel cells [113]. Typically, Pt/$Ti_3C_2T_x$ hybrid with 20 wt% Pt nanoparticles was obtained 320 mg $Ti_3C_2T_x$, 240 mg trisodium citrate dehydrate and 160 mg chloroplatinic acid, which were dissolved in ethylene glycol. The reaction was conducted during the refluxing of mixture, containing at 160 °C in N_2. In this case, a large number of –OH and –F groups were formed at the surface of the HF-etched $Ti_3C_2T_x$, which acted as anchor sites to ensure the dispersion and binding of the Pt nanoparticles. Meanwhile, due to the strong interactions of Pt–$Ti_3C_2T_x$, electronic structure of the Pt nanoparticles could be modified. As a result, the oxidation and dissolution of the Pt catalytic nanoparticles were effectively avoided.

Ti_3C_2 MXene nanoparticles have been used to develop metal oxide/Ti_3C_2 heterostructures, in order to serve as photoanodes for photoelectrochemical water splitting [114]. It was found that both the photoactivity and stability could be significantly improved. For example, as compared with the pristine TiO_2, the TiO_2/Ti_3C_2 heterostructures exhibited a bias photon-to-current efficiency (ABPE) that was increased by about six times. According to the electron spin resonance (ESR), electrical impedance spectroscopic (EIS) and Mott-Schottky measurement results, the increased photoelectrochemical activity was attributed to two factors, i.e., (i) the presence of Ti_3C_2 which acted as the oxygen evolution cocatalyst and (ii) the strong interfacial interactions between TiO_2 nanoparticles and Ti_3C_2 nanosheets.

TiO_2 nanorod arrays were grown on FTO glass substrates by using a hydrothermal method. To prepare the reaction solution, tetrabutyltitanate was mixed with a HCl solution with a volume ratio of V_{HCl}:$V_{H2}O$ to be 1:1. The reaction was conducted at 150 °C for 20 h. To prepare TiO_2/Ti_3C_2 photoanodes, Ti_3C_2 MXene suspension was dropped onto the TiO_2 coated FTO substrate, followed by drying at room temperature overnight.

The potential application of MXene for photocatalytic reduction of carbon dioxide (CO_2) towards hydrocarbons has been theoretically explored, in order to reduce the emission of CO_2 and mitigate the issue of greenhouse effect [115]. The theoretical study was conducted by using first-principles computations, while the reduction of CO_2 was at the oxygen vacancies on MXene monolayers. Three MXenes, including Ti_2CO_2, V_2CO_2 and $Ti_3C_2O_2$, were studied and compared. It was found that Ti_2CO_2 had the strongest catalytic effect of CO_2 reduction. The reaction pathway was $CO_2 \rightarrow HCOO \rightarrow HCOOH$, which exhibited an energy barrier of 0.53 eV. As compared with various other single-carbon organic compounds, HCOOH was the most favorite, demonstrating the high selectivity of the hybrid catalyst. In addition, the authors also demonstrated that more oxygen vacancies could be created on the O-terminated surface of the MXene nanosheets the treatment with CO and H_2. The results could be used as a reference in the design of catalysts for the reduction of CO_2.

There have also been experimental studies on MXenes as the catalysts of CO_2 reduction. For instance, Low et al. coated TiO_2 nanoparticles onto highly conductive

MXene Ti_3C_2 nanolayers, by using an *in-situ* growth method combined with simple calcination step [116]. The nanostructures exhibited a rice crust-like morphology, in which the TiO_2 nanoparticles were uniformly distributed on the surfaces of Ti_3C_2. At optimized conditions, the TiO_2/Ti_3C_2 hybrid catalyst possessed a $CO_2 \rightarrow CH_4$ production rate of 0.22 μmol/h, which was higher than that achieved by using the commercial TiO_2 (P25) by a factor of nearly four times.

According to the $^{13}CO_2$ isotopic testing and *in-situ* diffuse reflectance Fourier transform infrared (FTIR) spectroscopic results, the photocatalytic CO_2 reduction mechanism of the TiO_2/Ti_3C_2 hybrid catalyst was revealed. Both the experimental results and the theoretical predictions supported the proposed photocatalytic enhancement mechanism. This outstanding photocatalytic effect of the hybrid catalyst was ascribed to the special nanostructure and the effective components. In this case, due to the fluffy rice crust-like structure of the hybrid, there were tremendous surface-active sites to be available for the reduction reaction of CO_2. Meanwhile, Ti_3C_2 had high electrical conductivity to effectively transfer the photogenerated electrons, while the recombination with photogenerated holes could be avoided. This TiO_2/Ti_3C_2 hybrid catalyst could be extended to those for photoconversion applications.

Similarly, a 2D/2D heterojunction consisting of ultrathin Ti_3C_2 and Bi_2WO_6 nanosheets was reported for CO_2 reduction [117]. The heterojunction was obtained through the *in-situ* growth of Bi_2WO_6 ultrathin nanosheets on the surface of Ti_3C_2 ultrathin nanosheets. Owing to the shortened charge transport paths and the enlarged interface contact area, theTi_3C_2/Bi_2WO_6 heterojunctions possessed strong bulk-to-surface and interfacial charge transport capabilities. At the same time, the enlarged specific surface area and enhanced porous structure of the Ti_3C_2/Bi_2WO_6 heterojunction significantly promoted the adsorption CO_2 molecules, thus resulting in improved photocatalytic performance for the CO_2 reduction. The optimal production yield of CH_4 and CH_3OH with the Ti_3C_2/Bi_2WO_6 hybrid was about five times higher than that achieved with pristine Bi_2WO_6 ultrathin nanosheets. Energy level structure of the Bi_2WO_6 and Ti_3C_2 and the process of photoinduced electron transfer at the interface of the heterojunction are schematically shown in Fig. 5.41. This kind of 2D/2D heterojunctions opened a new way to the development of cheap photocatalyst.

5.5 EMI Shielding or Microwave Absorption

MXenes have been employed as fillers of composites for electromagnetic interference (EMI) shielding and microwave absorption applications [118]. EMI is to utilize the high electrical conductivity of MXenes, while microwave absorption is to make use dielectric properties of the MXene based composites. Although there are no direct evidences, it has been assumed that the surface functional groups, intrinsic defects and dangling bonds all could have contributions to the absorption of MXenes. Also, abundant interfaces between the MXene nanosheets and the polymeric matrix

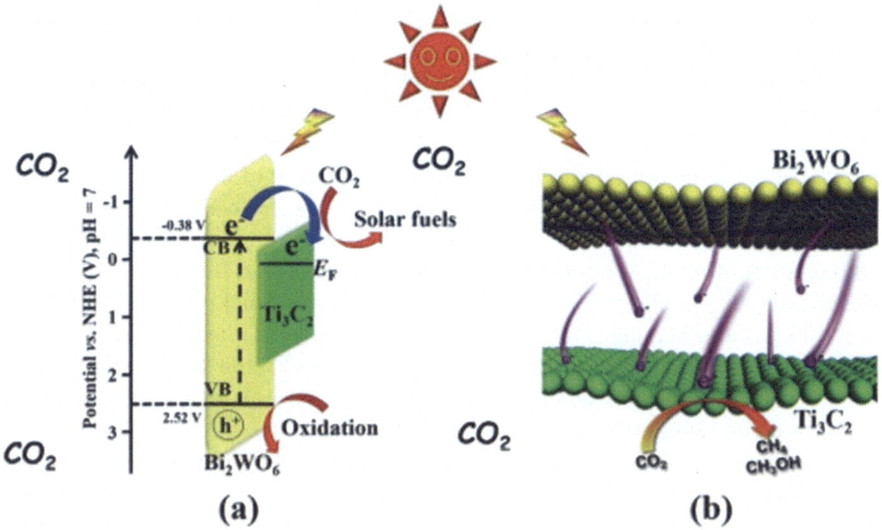

Fig. 5.41 **a** Energy level structure of the Bi_2WO_6 and Ti_3C_2. **b** The process of photoinduced electron transfer at the interface of the heterojunction. Reproduced with permission from [117]. Copyright © 2018, John Wiley and Sons

would be formed, which is responsible for the presence of electron relaxation polarization and interfacial polarization to cause dielectric loss. Although a quite large number of new publications have been available just before the submission of the manuscript, indicating that EMI application of MXene has been a very hot research topic [119–128].

Shahzad et al. reported EMI effect of $Ti_3C_2T_x$ films with relatively small thicknesses [129]. They used a high LiF to Ti_3AlC_2 molar ratio of 7.5:1 to obtain delaminated $Ti_3C_2T_x$ nanosheets, without the application of sonication. Films of the as-prepared HCl/LiF-derived delaminated $Ti_3C_2T_x$ nanosheets exhibited an EMI SE performance of 57 dB at a thickness of as thin as 8 μm. Furthermore, composite with 10 wt% sodium alginate (SA) had even high EMI SE value of 92 dB at a thickness of 45 μm. The presence of polarization losses from the induced dipoles related to the surface terminational groups and the surface free electrons of $Ti_3C_2T_x$ could reflect the EM waves. The nanolamellar structure might have enabled internal multiple reflection effect, which led to scattering of the EM waves.

Liu et al. reported an simple method to synthesize freestanding flexible hydrophobic MXene foams that showed promising EMI shielding effect [130]. The MXene foams were mechanically strong, due to the assembling of the MXene nanosheets into films, which were foamed with hydrazine. Unlike the MXenes with hydrophilic characteristics, the MXene foams were hydrophobic, with strong water resistance and high durability. Specifically, the lightweight MXene foams demonstrated an EMI shielding effectiveness of about 70 dB, whereas the MXene films possessed an effectiveness of 53 dB. The significantly enhanced EMI shielding effect could be readily attributed to the increased connectivity and thus increased electrical conductivity of the highly porous foams.

In a similar study, Han et al. further examined the effect of post-etching heat treatment on surface structures of $Ti_3C_2T_x$, which thus influenced dielectric properties and EM wave absorption behaviors of the final composites [131]. Once the Al layers were removed after the etching with HF, both the concentration of the surface functional groups and the electrical conductivity of the MXene were increased. As a consequence, the composites possessed much higher complex permittivity. It was found that oxidation of the MXenes could take place at certain heat treatment conditions. For instance, as the HF-etched ulphidered $Ti_3C_2T_x$ was quickly heated at 800 °C for 2 h in Ar, at a heating rate of 10 °C/min, the surface Ti layers were oxidized. The oxidation led to the presence of to form anatase and rutile TiO_2 nanoparticles, resulting in localized sandwich structures. Meanwhile, amorphous carbon nanosheets were formed and supported by the layered $Ti_3C_2T_x$ nanoflakes.

A freeze-casting process was reported to fabricate ultralight MXene aerogels, with densities of $\rho < 10$ mg/cm^3, without the presence of external supporters, which showed promising performances as EMI shielding materials [132]. This simple method could be easily scaled up to obtain MXene aerogels with sufficiently large dimensions for real applications. The aerogels possessed micrometer-sized pores. The EMI shielding effectiveness could be up to 75 dB, corresponding to a highest specific shielding effectiveness of 9904 dB·cm^3/g, due to their ultralow densities.

Ultrathin $Ti_3C_2T_x$ (U-$Ti_3C_2T_x$) MXene nanosheets (NS), with the layer thickness to be controlled through atomic-layer tailoring, were used to construct nanocomposites with SiO_2 for EMI shielding applications [133]. The U-$Ti_3C_2T_x$ NS composite with 80 wt% concentration displayed an EMI shielding effectiveness of 58.1 dB at the thickness of 1 mm. It was confirmed experimentally that electron transport played a significant role in determining the EMI shielding effect of the nanocomposites. Other factors, such as defect-induced polarizations and terminated functional groups, also had contributions to the EMI shielding behaviors. More importantly, the EMI shielding efficiency could be tailored by controlling the localized conductivity of the networks.

A simple chemical etching approach was reported to prepare Ti_2CT_X MXene nanosheets with sandwich structures and enriched surface functional groups for EMI applications [134]. An outstanding EMI shielding efficiency of 70 dB at X-band, with a thickness of just 0.8 mm, which was much beyond most of the EMI shielding materials based on graphene nanocomposites. The highly conductive MXene network ensured strong reflection of the microwaves at the surface of the materials, while the laminated structures facilitated additional attenuations of the microwaves for absorption.

The same authors designed 3D highly porous aerogels of 2D MXene with a long-range ordered aligned lamellar nanostructure, which exhibited EMI shielding effect with tailorable reflection and absorption [135]. The MXene-based aerogels were mechanically strong, compressible and lightweight, which were obtained by using a bidirectional freeze-casting, from MXene colloidal suspensions. The contributions of reflection and absorption to the EMI shielding effect could be readily adjusted by controlling the orientational profiles. Three groups of aerogels based on MXenes of $Ti_3C_2T_x$, Ti_2CT_x and Ti_3CNT_x were studied, which exhibited EMI shielding efficiencies (SE) of 70.5 dB, 69.2 dB and 54.1 dB, respectively. In this case, they had a thickness of 1 mm and a low density of 11.0 mg/cm^3, corresponding to a highest specific shielding efficiency of about 8800 dB cm^3/g. As the MXene aerogels were compressed to different degrees, the EMI effect was shifted from reflection to absorption gradually, thus offering an simple way to adjust the function of the materials for different applications.

Sun et al. developed an electrostatic assembly process to prepare highly conductive MXene@polystyrene nanocomposites, with negatively charged MXene nanosheets on positively charged polystyrene microspheres, which were then compressed with a mould [136]. Because MXene nanosheets were highly conductive, while the network in the polystyrene matrix offered a highly efficient conduction path, the nanocomposites had an electrical conductivity of as high as 1081 S/m, at a very low percolation threshold of 0.26 vol%. As a result, the MXene@polystyrene nanocomposites exhibited an outstanding EMI shielding effectiveness of 54 dB, over the X-band. The EMI shielding level could be further optimized to be 62 dB, as the content of MXene was slightly increased to 1.90 vol%. Meanwhile, the nanocomposite showed sufficiently high mechanical strength.

Freestanding ultrathin flexible $Ti_3C_2T_x$/poly(3,4-ethylenedioxythiophene)–poly(styrenesulfonate) (PEDOT:PSS) composite films, with a brick-and-mortar structure, were developed by using a biomimetic method through vacuum-assisted filtration, as EMI shielding materials [137]. The composite films had a $Ti_3C_2T_x$ to PEDOT:PSS weight ratio of 7:1 and a high conductivity of 340.5 S/cm. With a thickness of as thin as 11.1 μm, the sample could achieve a high EMI shielding effectiveness of 42.10 dB, corresponding to a specific EMI shielding efficiency of 19 497.8 dB·cm^2/g. At the same time, the composite films demonstrated outstanding mechanical properties, with tensile strength increasing from 5.62 to 13.71 Mpa. Accordingly, the ruptured strain was increased from 0.18 to 0.29%, with reference to the pure $Ti_3C_2T_x$ MXene films.

HF-etched multi-layered $Ti_3C_2T_x$ nanosheets were incorporated with epoxy to form composites with a filler concentration of 50 wt% [138]. The composites were prepared by using the simple mixing procedure. The sample with a thickness of 1.4 mm exhibited a reflection loss (RL) of −11 dB, corresponding to microwave absorption efficiency of about 92%, over the microwave frequency range of 12.4–18 GHz. However, the intrinsic dielectric characteristics of the MXene based composites restrict the bandwidth of the microwave absorption.

Wang et al. fabricated a new type of textiles based on MXene nanosheets that were coated with *in-situ* polymerized polypyrrole (Ppy), which showed dual functions of EMI shielding and Joule heating [139]. The textiles had a high electrical conductivity and hydrophobic ulphide, due to the presence of silicone coating. The optimal electrical conductivity was 1000 S/m, resulting in an EMI shielding effectiveness of 90 dB, at the thickness of 1.3 mm. The hydrophobic ulphide made the textiles to be water-resistant. Such multifunctional nanocomposites could find applications in the field of flexible electronic devices.

Porous Ti_2CT_x MXene/poly(vinyl alcohol) (PVA) composite foams were fabricated, with few-layered Ti_2CT_x (f-Ti_2CT_x) MXene and PVA, by using a simple freeze-drying process, as EMI shielding materials [140]. At the optimal conditions, the specific shielding effectiveness was 5136 dB·cm^2/g, at a very filler concentration of 0.15 vol%. In this case, reflection loss (SE_R) was <2 dB, implying that the nanocomposite foams were of strong microwave absorption capabilities. The absorption ulphide was attributed to the good impedance matching, related to the porous structures, internal reflection and polarization effect (dipole and interfacial polarization).

MXene/graphene composites with high flexibility and low density were fabricated by using simple spray coating and solvent casting methods [141]. The composites had a conductivity of 13.68 S/cm, corresponding to a low sheet resistance of 3.1 Ω/sq. As a result, outstanding EMI shielding was achieved, with an effectiveness of 53.8 dB (99.999%), contributed by reflection of 13.10 dB and absorption of 43.38 dB, over 8–12.4 GHz. The single coated carbon fabric had an absolute shielding effectiveness of about 35,370 dB·cm^2/g. In addition, the composites were highly hydrophobic, with a contact angle of 126°, wetting energy of − 43 mN/m, spreading coefficient of − 116 mN/m and a low work of adhesion of 30 mN/m. In addition, the samples exhibited an excellent thermal stability up to 400 °C.

Highly conductive 3D $Ti_3C_2T_x$ porous architectures were constructed, with the assistance of graphene oxide (GO), by using a hydrothermal treatment, combined with directional freezing and freeze-drying, for EMI shielding applications [142]. The hybrid aerogel foams demonstrated oriented cellular microstructures. In this case, the GO nanosheets acted as a skeleton, whereas the $Ti_3C_2T_x$ nanosheets were closely adhered on the cell walls. The hybrid-based epoxy nanocomposites had an electrical conductivity of 695.9 S/m, leading to an EMI shielding effectiveness of 50 dB at the X-band. The content of $Ti_3C_2T_x$ was as low as 0.74 vol%.

$Ti_3C_2T_x$ MXene-carbon nanotubes (CNTs) composite films were fabricated by using a spin spray layer-by-layer (SS-LbL) assemble method for potential applications in EMI shielding [143]. The LbL MXene-CNTs composite films were semi-transparent and highly conductive, with an electrical conductivity of 130 S/cm. Owing to the high electrical conductivity, the composite films exhibited a high specific shielding effectiveness of 58 187 dB·cm^2/g. Such flexible and semi-transparent composite films could be used as next-generation EMI shielding materials.

Ultrathin highly flexible $Ti_3C_2T_x$ (d-$Ti_3C_2T_x$, MXene)/cellulose nanofibers (CNFs) composite papers with a nacre-like lamellar structure were prepared through the self-assembly during vacuum filtration [144]. The d-$Ti_3C_2T_x$/CNF composite papers had a highest electrical conductivity of 739.4 S/m, resulting in a specific EMI shielding efficiency of 2647 dB·cm^2/g at a thickness of as thin as 47 μm. Furthermore, due to the interaction between the 1D CNFs and the 2D d-$Ti_3C_2T_x$ MXene, the nacre-like d-$Ti_3C_2T_x$/CNF composite papers exhibited both high strength and toughness at the same time, with a tensile strength of 135.4 Mpa and fracture strain of 16.7%, together with folding-resistant stabilities of up to 14 000 times.

$Ti_3C_2T_x$ nanosheets were incorporated into electrospun cellulose nanofibers to develop highly flexible d-$Ti_3C_2T_x$-r-CNFs composite films with strong EMI shielding performances [145]. The newly developed d-$Ti_3C_2T_x$-r-CNFs composite films were obtained by using a simple vacuum filtration process. At optimal conditions, the composite films exhibited an EMI shielding efficiency of 42.7 dB, at a thickness of only 15 μm, over the frequency range of 2–18 GHz. More importantly, the d-$Ti_3C_2T_x$-r-CNFs composite films displayed enhanced mechanical strength as compared with the pristine r-CNFs ones. The composite films were also of a high cycling stability. As a result, they could find applications as EMI shielding materials where high flexibility and ultralightweight are required.

A 3D $Ti_3C_2T_x$ MXene-C hybrid foam (MCF) was reported with strong EMI effect, which was made by using sol–gel process combined with post-thermal reduction [146]. EMI shielding performance of the MCF samples was measured by using composites with epoxy prepared through vacuum-assisted impregnation and curing step. With a small content of 4.25 wt% (MCF-5), the MCF-5/epoxy composite sample achieved an EMI shielding efficiency of 46 dB. The sample had an electrical conductivity of 184 S/m. The EMI shielding performance was much higher than the sample without the presence of MXene. Moreover, the materials had Young's modulus of 3.96 GPa and hardness of 0.31 GPa, which were increased by 13% and 11%, respectively.

MXenes have also been incorporated with various magnetic components to develop composite materials with higher EMI shielding and microwave absorption performances [147–151]. For example, Fe_3O_4 nanoparticles have been attached on MXenes to form 2D hierarchical composites of ($TiO_2/Ti_3C_2T_x/Fe_3O_4$), which displayed advanced microwave absorption capabilities [148]. The composites with varied [$TiO_2/Ti_3C_2T_x$]/[Fe_3O_4] ratios were prepared by using a moderate hydrothermal process. According to microstructural characterization results, both the Fe_3O_4 nanoparticles of and TiO_2 particles were attached on the surfaces of MXenes. By optimizing the ratio of [$TiO_2/Ti_3C_2T_x$]/[Fe_3O_4], microwave absorption performances readily modified, so that maximum reflection loss (R_L) could be achieved., the sample $TiO_2/Ti_3C_2T_x/Fe_3O_4$-5 showed optimal microwave adsorption behaviors.

Figure 5.42a shows calculated R_L curves of the $TiO_2/Ti_3C_2T_x/Fe_3O_4$-5 composite at different thicknesses. The R_L peaks shifted to lower frequency end, as the thickness of the sample was increased. Figure 5.42b shows the curve of thickness © versus peak frequency (f_m), according to the quarter-wavelength cancellation principle, with the yellow dots to represent the matching thicknesses, which were directly obtained from the peak frequencies and corresponding absorber thicknesses. Obviously, the simulated matching thickness versus absorption peak abided by the quarter-wavelength cancellation law. The normalized characteristic impedance of the $TiO_2/Ti_3C_2T_x/Fe_3O_4$-5 composite is demonstrated in Fig. 5.42c. At the matching thickness of 1.9 mm, the R_L peak was − 57.3 dB at 10.1 GHz, whereas the related Z was nearly equal to 1.0, suggesting that the normalized characteristic impedance played an important role in determining the microwave absorption properties of the composites.

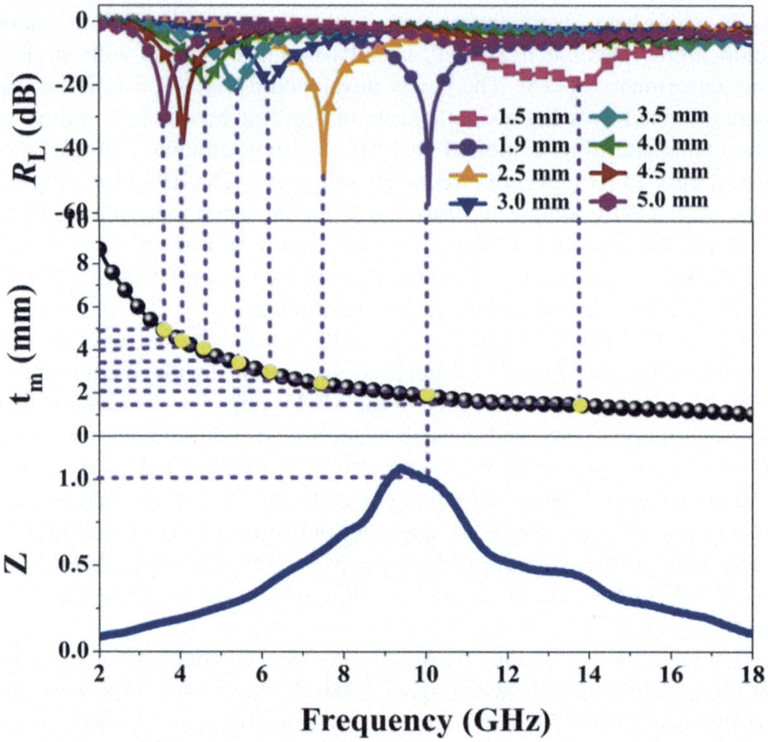

Fig. 5.42 **a** Reflection loss curves of the $TiO_2/Ti_3C_2T_x/Fe_3O_4$-5 composites at different thicknesses. **b** Simulated curves of thickness © versus peak frequency (f_m) according to the 1/4 wavelength cancellation principle. **c** Normalized input impedance (Z) of the $TiO_2/Ti_3C_2T_x/Fe_3O_4$-5 composite. Reproduced with permission from [148]. Copyright © 2018, Elsevier

MXenes were incorporated with Ag nanowires (AgNWs) to form MXene-AgNWs hybrids that were highly electrically conductive, thus making them useful for EMI protection [152]. The MXene-AgNWs hybrids could be coated on textiles by using layer-by layer (LbL) assembly process. Besides high EMI efficiency, the hybrid coatings were of superhydrophobic ulphide and highly sensitive to humidity. The MXene-AgNWs hybrids exhibited special nanostructure that was similar to leaves. The high electrical conductivity was ascribed to the fact that both of the components were of highly conductive to electricity. The AgNWs were closely protected by the MXene nanosheets, so that their oxidation resistance was significantly increased. At the same time, the nanosheet structure was beneficial to the attachment of the AgNWs onto the textile fibers. The MXene-AgNWs hybrid functionalized flexible textiles showed a sheet-resistance of 0.8 Ω/sq, enabling an EMI shielding efficiency of 54 dB at the X-band with a very small thickness of 0.12 mm.

Similarly, low-density MXene-AgNWs hybrid-based sponges were developed by using a dip-coating method combined with unidirectional freeze-drying process for EMI shielding applications [153]. A commercially available melamine formaldehyde sponge (MF) was used as the framework to deposit AgNWs layer, thus forming BMF-AgNWs nanostructure first. After that, MXene nanosheets were deposited onto the inside walls of the BMF-AgNWs nanostructure, by using unidirectional freeze-drying process, from MXene aqueous suspension. Specifically, the BMF-AgNWs displayed an EMI shielding efficiency of 40 dB, which was further increased to 52.6 dB for the AgNWs-MXene hybrid-based sponge. The sponge had a density of as low as about 0.5 mg/cm^3 owing to the highly porous structure. The high performance EMI effect was directly attributed to the high conductivity of the AgNWs and MXene nanosheets.

EMI shielding performances of four types of BMF/AgNW sponges, compressed at different volume ratios, i.e., 100% V_0, 51.2% V_0, 42.2% V_0 and 29.6% V_0, with to be $_0$: the initial volume of the MF sponge, compared. Due to the triaxial compression, the porosity of the MF sponge was slightly decreased from 99.5% to 98.4%, for the 29.6% V_0 volume ratio, corresponding to an increase in density from 7.5 mg/cm^3 to 25.3 mg/cm^3. As observed in Fig. 5.43a, as the electrical conductivities were very close, the sponges with buckled structures had higher EMI shielding efficiency than the pristine one at the X band of 8.2–12.4 GHz. The EMI shielding efficiency was proportional to the ratio of the applied volume and was maximized at 29.6% V_0. Because of the enlarged surface/interface areas in the buckled sponges, the EMI shielding efficiency was enhanced. With the incorporation of MXene, the EMI shielding performance was further increased to 52.6 dB, as revealed in Fig. 5.43b.

The total EMI shielding efficiency (SE_T) consisted of microwave absorption (SE_A), reflection at the surface/interface (SE_R) and the multiple reflections inside the materials (SE_{MR}). Figure 5.43c shows average values of SE_T, SE_A and SE_R, which were obtained at 8.2 GHz, for the samples of BMF/AgNWs with different contents of AgNWs, BMF-MXene and BMF-AgNWs-MXene. The value of SE_T was increased gradually, as electrical conductivity of the BMF/AgNWs hybrid was increased. For example, the BMF/AgNW-1 (1 to mean the number of the AgNWs coating) sample had an electrical conductivity of 1.4 S/cm, with a level of SE_T to 17.0 dB at 8.2 GHz. In comparison, the conductivity of the BMF/AgNW-4 sample was 15.3 S/cm, having a value of SE_T to 40 dB at 8.2 GHz. Even though the BMF-MXene sponge had a low conductivity of 0.1 S/cm, the value of SE_A was also about 32 dB, suggesting that the MXene phase played an important role in determining the EM shielding efficiency of the hybrid materials. Without surprising, the BMF-AgNWs-MXene hybrid sponge exhibited the maximal SE_A value of about 37 dB.

The power coefficients of *R* (reflection) and *A* (absorption) were extracted from the experimental *S* data, with which the power balance of the microwaves to interact with the sponges could be derived, as illustrated in Fig. 5.43d. It was found that the contribution of *R* was > 80% in most of the sample, whereas 96% of the shielding efficiency of the BMF-AgNWs-MXene sample was contributed by *R*. This is mainly because the hybrid-based sponges were of high electrical conductivity, so that the incident microwaves were mainly reflected at the surface of the materials.

Highly flexible conductors were derived from $Ti_3C_2T_x$ MXene nanosheets and single-walled carbon nanotubes (SWCNTs), which were deposited on latex, in the

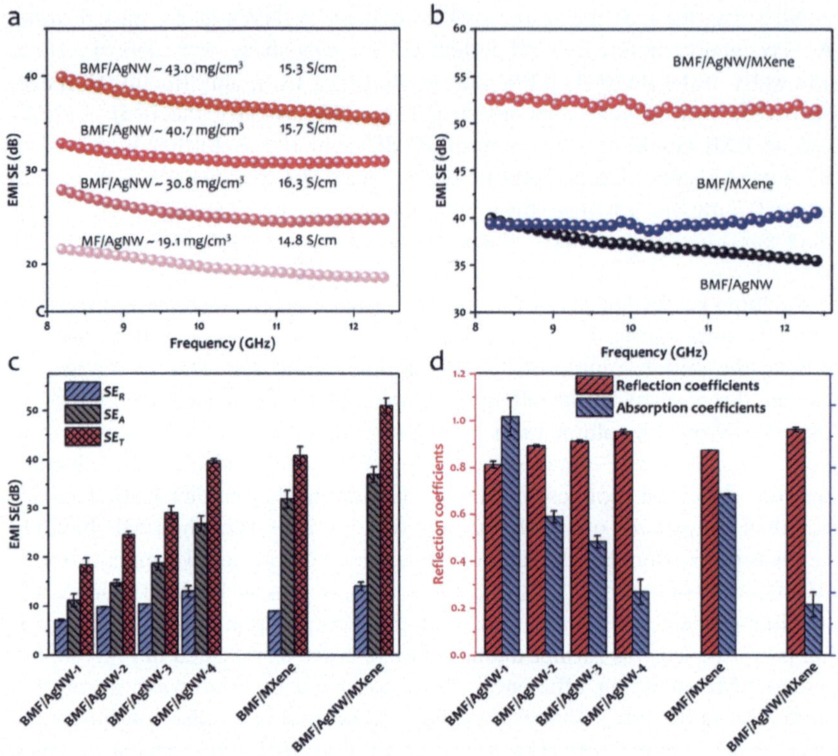

Fig. 5.43 **a** EMI shielding efficiency (at X band) of the MF-AgNWs and BMF-AgNWs sponges. **b** EMI efficiencies of the BMF-MXene, BMF-AgNWs and BMF-AgNWs-MXene sponges. **c** Average values of SE_T, SE_R and SE_A of the BMF-AgNWs sponges with different conductivities, BMF-MXene and BMF-AgNWs-MXene at 8.2 GHz. **d** Average values of *R* and *A*. Reproduced with permission from [153]. Copyright © 2019, Royal Society of Chemistry

form of crumpled layers, for applications as wearable electronics with high stretchability, including EMI shielding and antennas [123]. The SWNTs-MXene-latex (S-MXene-latex) hybrids were durable to withstand a strain of 800%, while they were highly resistive to fatigue with cycling for 500 times. The EMI effect was almost unchanged at the maximum strain, with shielding efficiencies of 30, 47 and 52 dB, for the sheets of one-, five- and ten-layers, respectively. A dipole antenna made with the highly stretchable S-MXene-latex film was durable against uniaxial stretched of up to 150%. The combination of the EMI shielding and stretchable antenna function, the S-MXene hybrid system acted as a wireless device that was able to transmit and protect the human body at the same time.

The $Ti_3C_2T_x$ MXene suspension had a concentration of 5 mg/mL, while SWCNTs dispersed in SDS aqueous solution with a SWNTs/SDS weight ratio of 1:20, leading to a concentration of the SWNTs to 0.1 mg/mL. To prepare S-MXene films, GO layer with a thickness of 100 nm was deposited, by filtrating 500 μL 0.5 mg/mL

GO dispersion with PVDF membranes that had a pore diameter of 0.22 μm. MXene and SWNTs mixed suspension with a solid leading of 1.0 mg at a MXene/SWNTs mass ratio of 3:2 was then filtrated on top of the GO layer. The as-obtained films were thoroughly washed with DI water to eliminate SDS, resulting in GO- SWNTs-MXene membranes. The S-MXene membranes were peeled off from the PVDF porous supports, which were subsequently transferred onto latex balloons that were pretreated with plasma. Because the substrates were expanded during the film deposition, the membranes would be crumpled after being dried.

Due to the laminated structure, the S-MXene nanosheets could be crumpled by applying prestrains to the latex substrates before the film deposition. After the release of the pre-strains after deposition, the substrates recovered their original state, while the S-MXene layers were crumpled. The degree of the crumple was readily controlled by the level of the strains experienced by the latex substrates. The higher the degree of the crumple, the high the stretchability of the S-MXene films. As the areal strain was increased from 0 to 800%, the degree of the crumples for the S-MXene films was gradually enhanced, with the texture length scale increasing from about 7 to 20 μm, corresponding the average height of the surface texture reducing from 7.0 to 3.2 μm. The crumpled S-MXene films could be inflated and deflated alternatively for up to 500 times, without the occurrence of fracture. The multilayer characteristics of the nanostructured S-MXene films were fatigue-free against fracture/being testing.

It was found that the presence of the SWCNTs was crucial to achieve S-MXene films with high streachability. The pristine MXene films were prone to fracture and crack when they were subject to the durance testing experiments. For too thick MXene films, the top layer was detached from the latex substrates, due to mainly two reasons. On one hand, as the thickness was increased, the S-MXene layers more and more rigid, i.e., both the flexibility and mechanical strength were declined, where the contracting forces applied to the films due to the latex to return the initial state were sufficiently strong to cause deformation of films. On the other hand, the degree of the crumple was increased with increasing film thickness, so that the deformation of the latex substrates would require more and more energies. The incorporation of the SWCNTs could be used to solve such problems. With the presence of the SWCNTs, they served as threads to string up the MXene nanosheets. As a result, the MXene nanosheets would not be detached one another in the films. The optimal content of SWCNTs was 40 wt%.

Figure 5.44a shows a schematic diagram for measuring electrical conduction/resistance of the S-MXene crumpled hybrid films, which could not be measured by using the standard method. As illustrated in Fig. 5.44b, it was observed that the I–V curves were always linear, confirming that the highly flexible S-MXene conductive films still maintained with the Ohmic contact. The resistance values of the films at the strains of 0% and 800% were 16.6 Ω and 23.5 Ω, corresponding to a resistance increment rate of about 41.6%.

Figure 5.44c shows resistance-strain characteristics of various samples with the conductor layer thickness of about 1 μm. The planar S-MXene sample could withstand relatively small areal strains of <10%, above which it was damaged, so that the resistance was rapidly increased. When there were no SWCNTs, the resistance of the

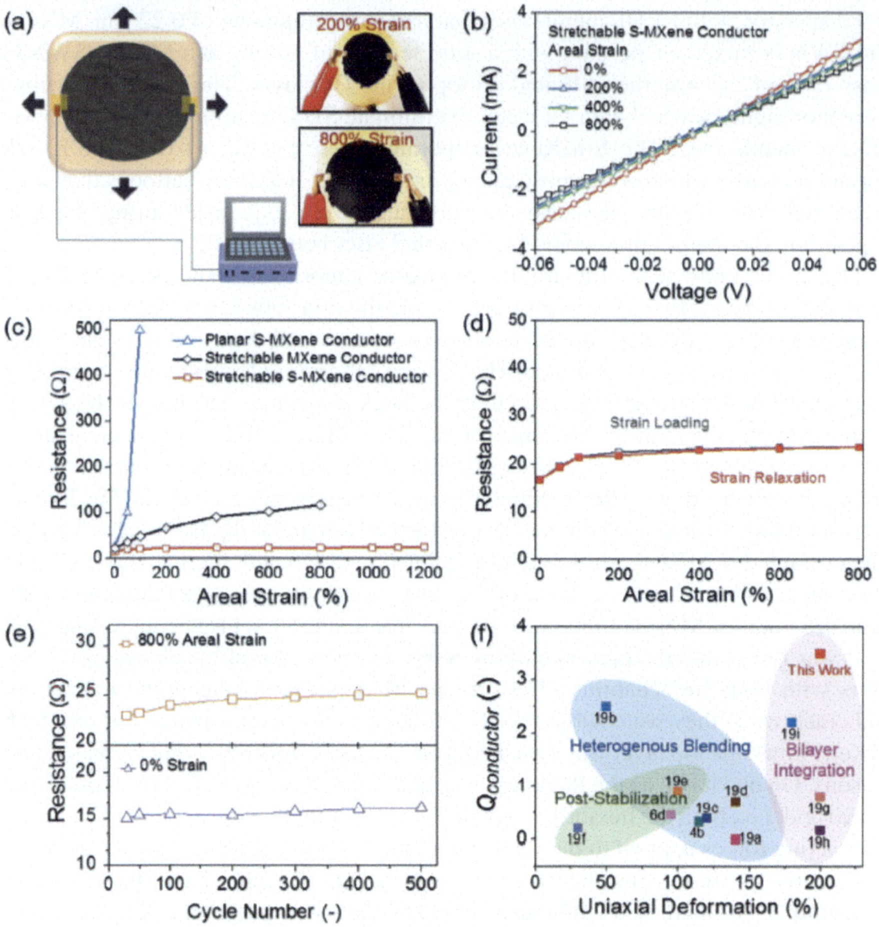

Fig. 5.44 **a** Schematic diagram and photographs of the setup to measure electrical resistances of the stretchable S-MXene conductors at different areal strains with an electrochemical workstation. **b** I–V curves of the stretchable S-MXene conductor at different areal strains. **c** Resistance curves of the planar S-MXene, stretchable MXene and stretchable S-MXene conductors at different areal strains. **d** Resistance values of the stretchable S-MXene conductor as a function of areal strain. **e** Resistances of the stretchable S-MXene conductor at the relaxation and stretching states as a function of fatigue testing cycle at areal strains of 0% and 800%. **f** Performances of the S-MXene conductor and the similar ones reported in the literature. Reproduced with permission from [123]. Copyright © 2019, John Wiley and Sons

sample was increased from 21 Ω to 117 Ω, as the areal strain was increased from 0 to 800%. In contrast, for the S-MXene films, the resistance was only increased from 17 to 25 Ω, the areal strain of up to1200%. The invariable resistance of the S-MXene film was attributed to two factors, i.e., (i) the crumpled S-MXene layers were highly flexible with well retained conduction network during the stretching process and (ii) the introduction of SWCNTs that were mechanically strong and electrically conductive. Furthermore, resistance profiles of strain loading and releasing processes were nearly the same, as observed in Fig. 5.44d.

Figure 5.44e shows electrical resistance values of the stretchable S-MXene conducting films, for the relaxation and stretching states during the fatigue testing, with areal strains of up to 800%. After cycling test for 500, the resistances of the sample were increased by less than 7% and 9%, at the stretching and relaxation states, respectively. Therefore, it was confirmed that S-MXene conductive sheets were of sufficiently high mechanical strength and electrical conduction characteristics. The variation in resistance value of the samples was probably ascribe to low degree of disconnection of the S-MXene coatings with latex substrates at the early stage of fatigue testing experiment.

Figure 5.44f depicts the performances of the stretchable S-MXene conducting films, together with the literature data. Comparatively, the S-MXene conductors had relatively high quality factor, defined as $Q_{\text{conductor}} = [(L_\varepsilon - L_0)/L_0]/[(|S_{11,\varepsilon}| - |S_{11,0}|)/|S_{11,0}|]$ [123]. For instance, the $Q_{\text{conductor}}$ was 3.5 at the uniaxial strain of 200%, while the electrical conductivity was as high as 2900 S/cm. Most of the conductors were prepared by using physical embedding conducting inclusions with polymers as matrix. In comparison, the bilayered S-MXene coatings were incorporated with elastomeric (latex) substrates, in which electrical conduction network was well retained. As a result, the samples were highly flexible, mechanically strong and electrically conductive, enabling EMI shielding layers and wearable antennas with high strechability.

Figure 5.45a shows a schematic diagram of the stretchable S-MXene EMI shielding sheet. When the microwaves approached the surface of the S-MXene layer, the waves would experience partial reflection at the surface, due to the high conductivity of the hybrid materials. The microwaves that entered the interior of the materials could be absorbed, due to the effect of the local dipoles related to the surface function groups, –F, –O and –OH in the MXene phase. Further wave attenuation was encountered at the interfaces and surfaces of MXene nanosheets and the SWCNTs.

EMI shielding performances of the single-layer sample at different strains are shown in Fig. 5.45b. The EMI shield efficiency of 30 dB (99.9%) was irrespective with the areal strain. The strain-independent EMI shielding efficiency was comparable with the performance of the as-prepare S-MXene sample, while their thickness was same, i.e., 1.0 μm, which could be used for real applications. Because of the high flexibility of the stretchable S-MXene sheets, the EMI shielding efficiency was kept almost unchanged after cycling test for 500 times, with areal strains of up to 800%, as revealed in Fig. 5.45c. In addition, the S-MXene conductors could not be thicker than 1 μm. For example, the sample with a thickness of 2 μm, the strain independent EMI shielding efficiency was absent.

Figure 5.45d shows EMI shielding performances of the filtrated S-MXene samples with different thicknesses and the stacked S-MXene conductor sheets with different layers. The EMI shielding efficiency was also examined at the areal strain of 800%. The samples with 2-, 5-, and 10-layers exhibited EMI shielding efficiencies of 38 dB, 47 dB and 52 dB, respectively. These performances were comparable with those of

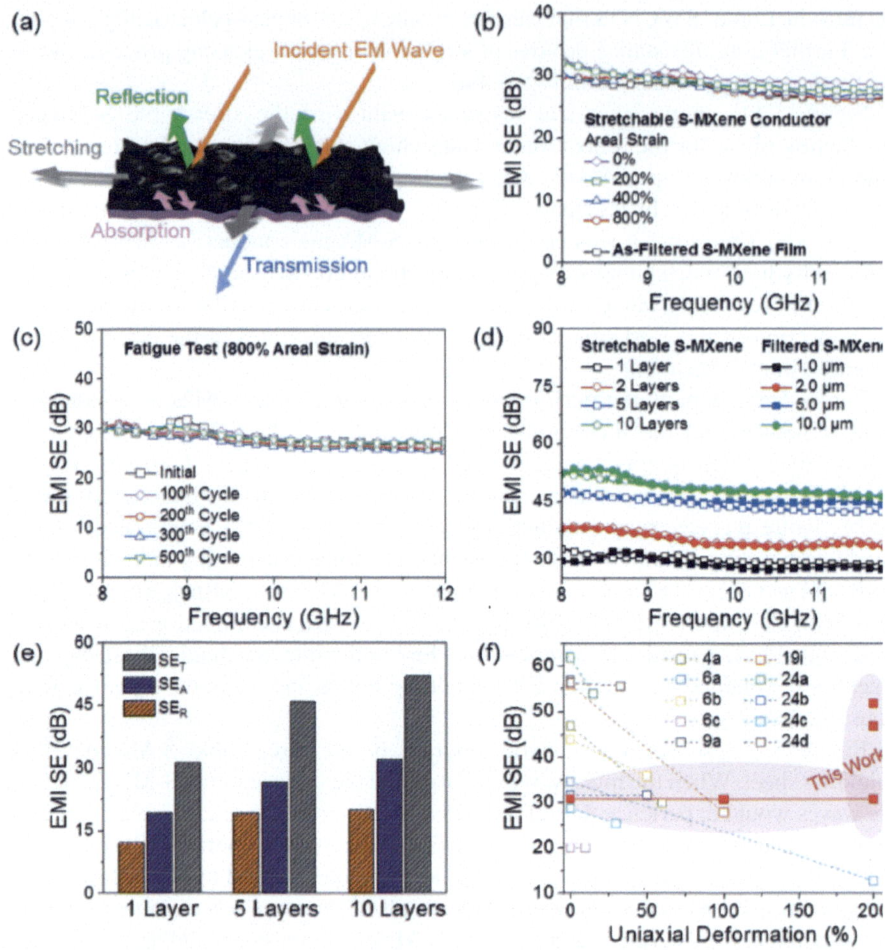

Fig. 5.45 **a** Schematic diagram of the EMI shielding mechanism of the stretchable S-MXene conductor. **b** EMI shielding efficiencies of the single layer of S-MXene conductor, with a thickness of 1 μm, at different areal strains and the as-prepared S-MXene film (1 μm). **c** EMI shielding efficiencies of the stretchable S-MXene conductor with fatigue testing for up to 500 cycles at the areal strain of 800%. **d** EMI shielding efficiencies of the samples with 1, 2, 5 and 10 stack(s) of S-MXene conductor(s) at the areal strain of 800% strain and the as-filtered S-MXene membranes with the ulp thicknesses (1-, 2-, 5-, 10 μm). **e** Fractions of EM absorption (SEA) and EM reflection (SER) to total EMI shielding efficiencies (SET) of the S-MXene conductors, with 1, 5 and 10 stacks, at the areal strain of 800% at the frequency of 8.9 GHz. **f** Performances of the S-MXene conductors and the similar ones reported in the open literature. Reproduced with permission from [123]. Copyright © 2019, John Wiley and Sons

as-prepared S-MXene membranes, with similar thicknesses, i.e., 2 μm, 5 μm and 10 μm, respectively. The fractions of microwave absorption (SEA) and reflection (SER) in the EMI efficiency of the S-MXene conducting films measured at 8.9 GHz are illustrated in Fig. 5.45c. It was thus suggested that the EMI shielding effect was mainly contributed by the absorption effect of the multilayer nanostructured S-MXene samples. Figure 5.45f indicated that the EMI shield materials reported in the literature could only deformation levels of <100%, or the EMI shielding efficiency was abruptly declined at relatively high strain, implying the advantageous aspect of the flexible S-MXene hybrid materials.

Figure 5.46a shows a schematic diagram of the stretchable dipole antenna made with the S-MXene membranes, denoted as S-MXene antennas. Typically, the S-MXene antenna was constructed with two pieces of stretchable S-MXene conducting sheets, which were symmetrically arranged by separating with an insulation gap. The antenna was fed with a 3.5-mm Sub Miniature version A (SMA) connector by using a coaxial cable. Before the S-MXene antenna was tested, the 3D radiation patterns were simulated with CST (Computer Simulation Technology) software, as shown in Fig. 5.46b, to demonstrate the potential applications of the S-MXene conductors in wireless communications. It was found that the S-MXene dipole antennas exhibited omnidirectional radiation patterns, where strongest electric fields were in the XY-plane at the uniaxial strain of 100%, implying that the S-MXene antennas possessed promising directivity. The strechability of the S-MXene antennas was evaluated by applying uniaxial stretching to the two S-MXene antenna branches at different magnitudes, as illustrated in Fig. 5.46a. As the uniaxial strain was increased, periodic wrinkles were gradually developed from the crumpled textures due to the strain-induced deformation, as observed in Fig. 5.46c.

Figure 5.46d shows experimental and simulated reflections $|S_{11}|$ and resonant frequencies of the S-MXene antenna at different uniaxial strains. The magnitudes of $|S_{11}|$ at the uniaxial strains of 0, 50, 100 and 150% were −36 dB, −33 dB, −33 dB and −30 dB, corresponding variations of $|S_{11}|$, i.e., $(|s_{11,0}| - |s_{11,\varepsilon}|)/|s_{11,\varepsilon}|$, of 9%, 9%, and 20%, respectively [123]. The $|S_{11}|$ parameter of the S-MXene antenna suggested that the reflected power was less than 0.1% at all the stretching levels. Meanwhile, with the uniaxial strains increasing from 0 to 150%, the resonant frequencies of the S-MXene antenna was reduced from 1.575 to 1.375 GHz in a linear manner. The experimental results were well consistent with the theoretical prediction with the CST software. After fatigue testing for 500 times, both the reflection and the resonant frequency of the S-MXene antenna were almost unchanged at the uniaxial strain of 100%, as demonstrated in Fig. 5.46e. As illustrated in Fig. 5.46f, the S-MXene antenna exhibited the Q_{antenna} levels of 11.1 and 7.5, at the uniaxial strains of 100% and 150%, respectively, which surpassed those of the similar devices reported in the open literature.

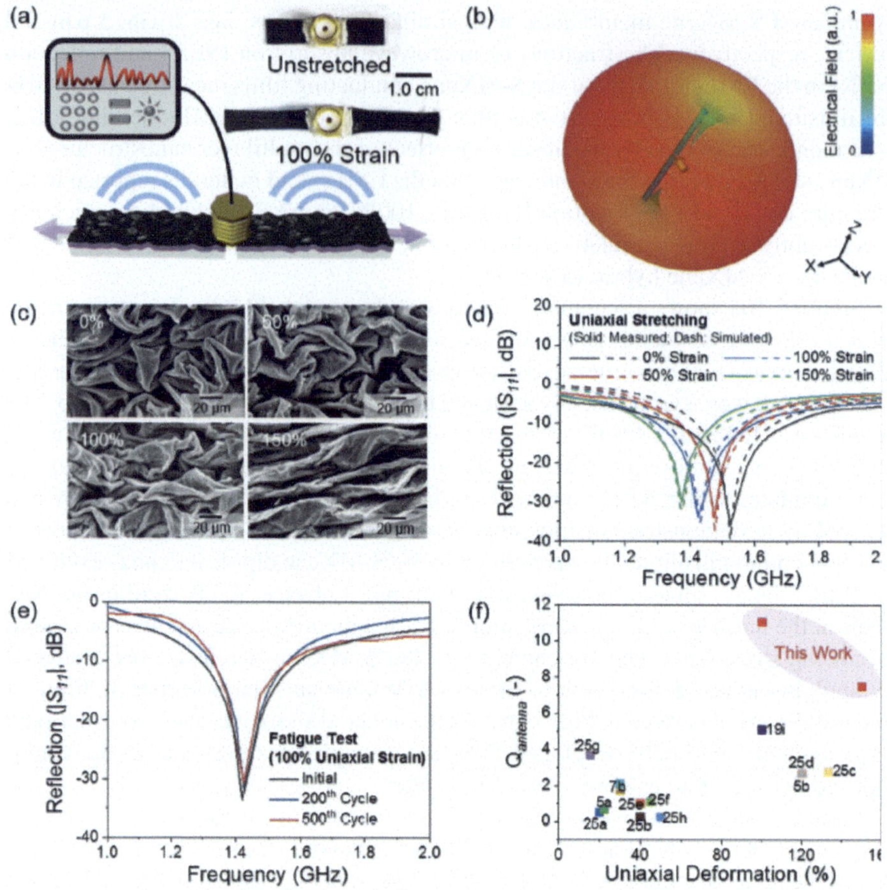

Fig. 5.46 **a** Schematic diagram and photographs (right) of the S-MXene dipole antenna at different stretching states. **b** Simulated 3D radiation patterns of the S-MXene antenna at the uniaxial strain of 100%. **c** SEM images of the S-MXene coating for the antenna wing at different uniaxial strains. **d** Experimental and theoretical reflections $|S_{11}|$ and resonant frequencies of the S-MXene antenna at different uniaxial strains. ©Performances of the stretchable S-MXene antenna after fatigue testing for 200 and 500 cycles at uniaxial strains of 100%. **f** Performances of the S-MXene antennas and similar antenna reported in the open literature. Reproduced with permission from [123]. Copyright © 2019, John Wiley & Sons

5.6 Transparent Conductive Films

As mentioned earlier, MXenes have high electrical conductivity, so that they have been acknowledged to be useful as transparent thin films, similar to graphene, for certain applications such as flexible electronic devices. It has been theoretically predicted that optical properties of MXenes-based films could be modified or adjusted by tailoring the surface functional groups [154].

Optical and dielectric properties of $Ti_3C_2T_x$ (T = –F, –O and –OH) have been theoretically studied [155]. It was found that the optical properties of MXene were highly dependent on properties of the surface functional groups. In the visible range, the oxidized samples had stronger absorptions, while the presence of –F group reduced the absorption. Over the ultraviolet band, both the absorption and reflectivity of the MXene were enhanced by all the functional groups. The functional groups also had influences on dielectric properties of MXenes. Therefore, the optical properties of MXenes could be modulated by tailoring the type and concentration of the surface functional groups.

One of the applications of MXenes based transparent thin films could be electrodes. A highly conductive intercalated $Ti_3C_2T_x$ thin film, with a thickness of 19 nm and high optical transparency of 85% has been demonstrated [156]. The MXene thin film had a dimension of 1×1 cm^2, where the Ti_3C_2 was derived from Ti_3AlC_2 films deposited by using a sputtering process, with etching solutions based on HF or NH_4HF_2. Optical properties of the thin films are shown in Fig. 5.47. The film had a metallic conductivity to about100 K, below which the resistivity increased as the temperature was decreased Meanwhile, negative magnetoresistance was observed, which was in an agreement with the characteristics of a weak localization phenomenon of other defective 2D materials. This transparent MXene thin film electrode could find applications in electronic, photonic and sensing devices.

Dillon et al. prepared relatively large scale optically transparent $Ti_3C_2T_x$ thin films with high in-plane conductivities, corresponding to bulk conductivity of 6500 S/cm, by a spin-casting method [157]. In these films, the basal planes of the $Ti_3C_2T_x$ nanosheets were highly oriented, in a parallel way to the substrates, as evidenced by the high intensity of (002) peak in the XRD pattern. The high electrical conductivity was ascribed to the metal-like free-electron density and the high degree of coplanar alignment of the MXene nanosheets. According to optical characterization results, an onset of free-electron plasma oscillations above 1130 nm was observed. Therefore, the MXene $Ti_3C_2T_x$ thin films could find applications in plasmonic field at near-infrared wavelengths However, conductivity of the films was almost immediately decreased upon exposure to O_2 or moisture. This issue should be tackled before they are used real applications.

Similarly, a facile spray coating process was developed to deposit transparent conductive thin films with delaminated Ti_3C_2 MXene nanoflakes [158]. The Ti_3C_2 MXene films were highly uniform, with thickness of 5–70 nm, which were deposited at ambient conditions. Sheet resistances (R_s) were in the range of 0.5–8 kΩ/sq, corresponding to optical transmittances of 90–40%, with figures of merit to be 0.5–0.7. The films exhibited strong mechanical flexibility against repeated bending testing. In addition, optoelectronic properties of the Ti_3C_2 MXene thin films could be tailored by introducing various cations through chemical or electrochemical intercalations. Reversible variation in transmittance over the UV–visible band was observed, as the films were electrochemically intercalated/deintercalated with tetramethylammonium hydroxide, making them potential as transparent electrodes in electronic, electrochromic and sensor device applications.

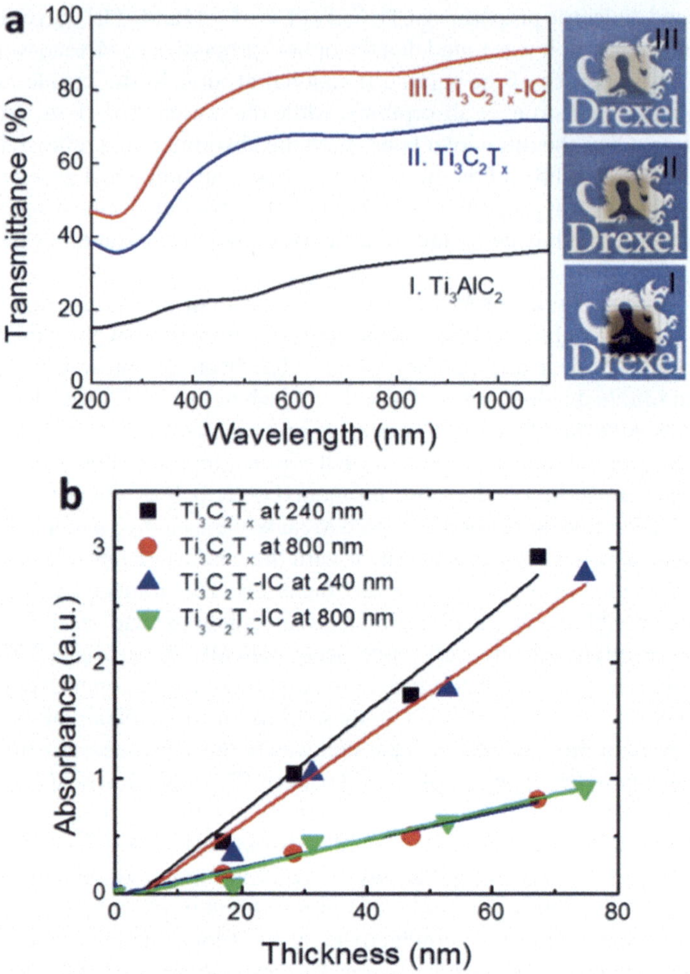

Fig. 5.47 **a** Transmittance spectra and photographs (on right) of (I) Ti_3AlC_2, (II) $Ti_3C_2T_x$ and (III) $Ti_3C_2T_x$-IC films with a thickness of 15 nm. **b** Light absorbances at wavelengths of 240 and 800 nm as a function of thickness of the $Ti_3C_2T_x$ and $Ti_3C_2T_x$-IC films. Reproduced with permission from [156]. Copyright © 2014, American Chemical Society

$Ti_3C_2T_x$ delaminated monolayers were fabricated as ultrathin transparent conduction films by using a solution deposition method [159]. The $Ti_3C_2T_x$ films displayed a lowest sheet resistance of 437 Ω/sq, with an optical transmittance of 77% at the wavelength of 550 nm. According to the results of field effect transistor experiment, the films exhibited metallic conducting ulphide, so that they could be used as electrode materials. More interestingly, the number of layers of the $Ti_3C_2T_x$ nanoflakes was simply dependent on the rate of spin-casting process, with 3, 4 and 5 layers for 800, 600 and 400 rpm, respectively. In addition, after thermal annealing at 100–200 °C, sheet resistance of the thin films could be further reduced, due to the removal of remnant adsorbed water and intercalated items in the $Ti_3C_2T_x$ layers.

A simple vacuum-free solution method was used to form $Ti_3C_2T_x$ MXene thin films by using electrohydrodynamic atomization technique [160]. The delaminated $Ti_3C_2T_x$ nanosheets were dispersed in ethanol for obtain suspensions, which were used to coat $Ti_3C_2T_x$ thin films. Three samples with thicknesses 135, 248 and 337 nm were prepared on glass substrates by controlling the deposition time, i.e., 20 min (S1), 40 min (S2) and 60 min (S3) were fabricated, respectively. All the films were annealed at 400 °C for 2 h in an inert environment. Resistivities of S1, S2 and S3 were 3.4×10^{-4} Ω·cm, 130×10^{-4} Ω·cm and 210×10^{-4} Ω·cm, respectively, while the transparency of S1 was 86.7% at the wavelength of 550 nm. In addition, S1 exhibited the highest diode property, with a current of 120 mA at 3 V. These $Ti_3C_2T_x$ thin films have potential applications in photo-electronics, energy storage devices and sensors.

Similarly, Aïssa et al. obtained an electrosprayed thin film of $Ti_3C_2T_x$ on glass substrates, with a thickness of 250 nm a low resistivity of 3.4×10^{-4} Ω·cm, which was comparable with that of indium tin oxide (ITO) thin film [161]. Furthermore, the authors attempted to increase the electrical conductivity and other properties by introducing a small quantity of graphene. For example, with 0.25 wt% graphene, the conductivity and Hall mobility were enhanced by two and three times, respectively. If the content of graphene was increased to 2.5 wt%, the resistivity could be reduced by about 30 times. In this case, graphene nanosheets acted as linkers to enhance the connectivity of the MXene nanosheets.

$Ti_3C_2T_x$ monolayer nanoflakes with sufficiently large lateral sizes were employed to fabricate field-effect transistors (FETs) as conductive channels, by using a simple drop-casting process on Si/SiO_2 substrates [162]. Delaminated HCl/LiF etched $Ti_3C_2T_x$ was dispersed to form suspensions to deposit thin films, followed e-beam lithography to develop FET devices, where Au was as use as metal contacts. The average values of resistivity and field-effect electron mobility of the devices were 2310 ± 570 Ω/sq and 2.6 ± 0.7 cm^2/V·s. Accordingly, the resistivity of a single layer of $Ti_3C_2T_x$ with thickness of about 1 nm was 2.31 ± 0.57 μΩ·m, corresponding to a conductivity of 4600 ± 1100 S/cm.

A aqueous colloidal spin-cast process was reported to fabricate Ti_2CT_x thin films, form MXene suspensions, with high conductivity and high optical transparency [163]. After the normalization by the number of transition metal atoms, the MXene film demonstrated the highest figure of merit (FOM). At about 2.7×10^5 cm^{-1}, absorbance coefficient of the Ti_2CT_x thin film was very close to that of that of $Ti_3C_2T_x$. The quantitative relationships between conductance and transparency of the films provided a reference for future studies. Also, the concentration of colloidal solution and the rate of spin-casting were important factors that influenced optical and electrical performances of the MXene based thin films.

V_2CT_x based MXene flexible transparent thin films were prepared to study their optoelectronic properties [164]. The MXene films were derived from aqueous colloidal suspensions obtained by dispersing V_2CT_x nanosheets in tetrabutylammonium hydroxide (TBAOH). It was found that the thin spin cast films had a close value of DC conductivity 3250 ± 100 S/cm, as compared with the thick films. The films with a thickness of 10 nm exhibited an optical absorption spectrum without distinct transitions over 500–2700 nm. The absorption coefficient was 1.22 ± 0.05 × 10^5 cm^{-1}, at the wavelength of 550 nm, which was more transparent than the Ti-based MXene films by nearly two times.

A simple process was employed to prepare ultrathin $Ti_3C_2T_x$ MXene films with thickness of 10 nm via interfacial assembly [165]. Due to their nanoflake structure, MXene nanosheets tend to self-assembly to form films through plane-to-plane stacking. The 10 nm thick MXene film had a relatively low sheet resistance of 310 Ω/□. The films could be easily transferred to different either flat or non-flat substrates because of the surface tensions. In addition, the MXene films could be deposited through multi-steps, thus obtaining films with different thicknesses. Furthermore, the films exhibited gas sensing effects, while the sensitivity was increased with decreasing size of the MXene nanoflakes, which suggested that the sensing behaviour was closely related to the interfaces of the MXene nanosheets.

Figure 5.48 shows a schematic diagram to demonstrate the fabrication process fo the MXene films. Ti_3C_2 aqueous suspensions with solid contents of 0.005–0.25 mg/mL, which were put in Petri dishes with diameters in the range of 5–20 cm. After that, ethyl acetate was introduced with such a quantity that its concentration was 10 vol%. Because ethyl acetate has solubilities of 6–8 vol% in water, ethyl acetate was floating on the surface of water as a thin layer. Then, the ethyl acetate would evaporate owing to the high vapor pressure, thus generating temperature gradient. The continuous evaporation of ethyl acetate resulted in high surface tension on the surface of water.

In this case, the MXene nanoflakes were thus self-assembled into films locations that were exposed owing to the Marangoni effect. Once the evaporation of the ethyl acetate was finished, MXene films with relatively low packing density were formed on the water surface. Then, a small quantity of 0.1 M HCl solution was dropped in to the suspensions, so that the electrostatic repulsion force in between the MXene nanoflakes was weakened. As a result, the floating MXene films became much denser. The MXene films were sufficiently strong to be transferred onto various substrates.

Multilayered films of $Ti_3C_2T_x$ MXene-poly(vinylidene fluoride) (PVDF), which had high dielectric constants and very low dielectric loss tangent, were prepared

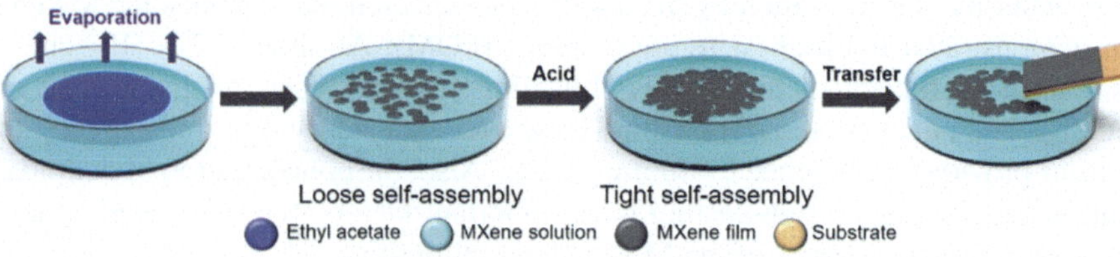

Fig. 5.48 Schematic diagram illustrating the interfacial self-assembly process to fabricate nanometer-scale MXene films. Reproduced with permission from [165]. Copyright © 2019, American Chemical Society

by using spin-coating and spray-coating processes combined with hot-press method [166]. Specifically, the 4MXene-5PVDF, with MXene and PVDF to be four and five layers, respectively, displayed a dielectric constant of 41 and a dielectric loss tangent of 0.028, which were recorded at 1 kHz. The dielectric constant of the MXene/PVDF films was relatively stable against frequency, with a value of 32.2 at the high frequency of 1 MHz, with a retention of 78.4%. Due to the multi-layered structure, the films had enriched interfacial polarizations, which is responsible for the high dielectric constant. Meanwhile, the electrical conduction across the films was suppressed due to the special ulphidered structure, resulting in very low dielectric loss tangent.

To make the films, the powder of PVDF was dissolved in a solvent of DMF with the aid of rigorous stirring for 6 h at about room temperature, forming PVDF solution at a mass concentration of 15%. The solution was deposited on cleaned glass substrates by using spin-coating at a speed of 300 rpm for 0.5 min, followed by drying at 80 °C for 8 h. Meanwhile, 10 mL MXene suspension with a concentration of 0.5 mg/mL coated onto PVDF film substrates by using spraying with a nozzle size of 0.2 mm at a working pressure of 50 psi. Both the MXene-PVDF hybrid and pure PVDF films were peeled off from the glass substrates, which were cut into small pieces with a dimension of 1.5 × 1.5 cm^2. The as-deposited films were then put together layer-by-layer, followed by hot-pressing treatment at 180 °C for 15 min, while the applied pressure was 10 Mpa.

AgNWs-MXene hybrid films were fabricated by using facial solution deposition process, which had high optical transparency, high conductivity, strong hydrophilic ulphide [167]. The method was cost-effective and could be readily scaled up for real industrial applications. Specifically, the AgNWs-MXene films displayed a sheet resistance of 15.9 Ω/□, while the optical transparency was as high as 92.5%, together with outstanding mechanical integrity and chemical stabilities. It was confirmed that the AgNWs-MXene hybrid films could be used as electrodes of flexible alternating current (AC) electroluminescent devices, with sufficiently high mechanical strength against rigorous flexibility testing. For instance, the luminescent performance of the device was unchanged when it was bended by 180°.

$Ti_3C_2T_x$ MXene nanosheets were dispersed in water with a concentration of 0.05 mg/mL. PET substrates were thoroughly cleaned by soaking in ethanol, isopropanol and DI water, with the aid of strong ultrasonication 10 min. The cleaned substrates were treated with O_2 plasma for 10 min, so that their surface became strongly hydrophilic in nature. Then, AgNWs suspensions with a solid content of 0.8 mg/mL were coated onto the substrates by spraying, by using an OPHIR airbrush (AC070) with a nozzle diameter of 0.2 mm. The applied pressure was set to be 50 psi, while the nozzle-substrate distance was kept to be 15 cm. The AgNWs layers were covered by the MXene layer also by using spraying under similar conditions. The AgNWs-MXene hybrid films were heated in vacuum at 70 °C to complete remove the residual solvents.

Optical properties of the hybrid films were characterized by using UV–vis–NIR spectrophotometer, while their electrical properties were measured by using four-probe resistance measurement. Figure 5.49a shows optoelectronic performances of the AgNWs-MXene hybrid films versus the loading level of the AgNWs. As the

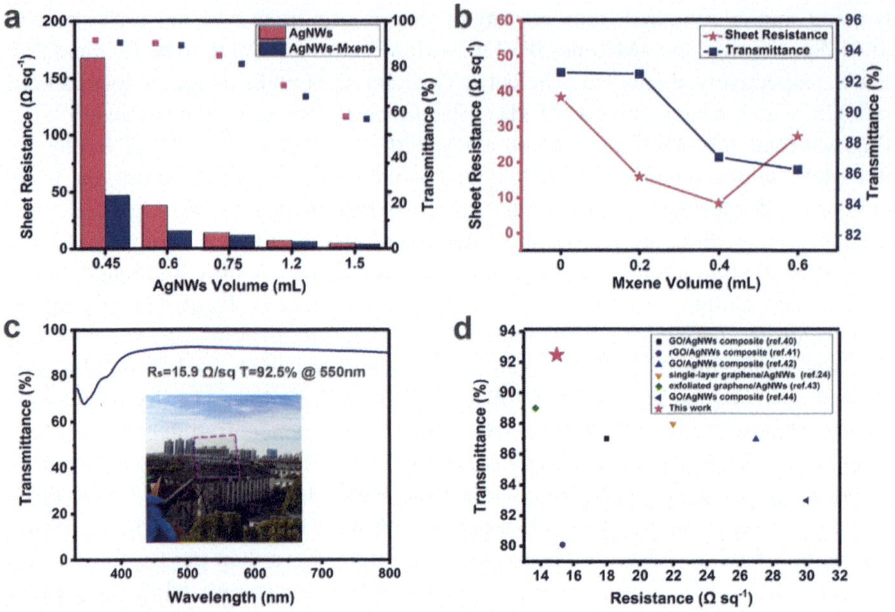

Fig. 5.49 **a** Electrical conductivities and optical transparencies of the pure AgNWs films and the AgNWs-MXene hybrid-based films as a function of the content of AgNWs. **b** Sheet resistances and transmittances of the hybrid-based electrodes with different contents of MXene nanosheets. **c** Optical transmittance curve of the AgNWs-MXene TCE with a photograph of the hybrid electrode as inset. **d** Optoelectronic performances of the AgNWs-MXene hybrid film-based electrode together with literature data. Reproduced with permission from [167]. Copyright © 2019, American Chemical Society

concentration of AgNWs was raised, sheet resistance of the films was largely reduced from 167.9 to 4.4 Ω/□. However, the optical transparency was also decreased, although the magnitude was not very significant. Furthermore, with the presence of MXene, electrical conductivity of the hybrid films was tremendously enhanced, with the improvement that could be up to four orders of magnitude. In addition, the content of the AgNWs could not be high, otherwise, the effect of MXene on the conducting behaviour of the hybrid films was less pronounced. Very importantly, if the content of MXene was sufficiently low, it had very weak effect on optical transparency of the hybrid films.

It was found that optoelectronic properties of the hybrid films were closely related to coverage of the MXene nanosheets. Figure 5.49b depicts resistance values of the AgNWs-MXene hybrid films as a function of the content of MXene (concentration of the MXene suspension). The sheet resistance was decreased and then started to rise. The sheet resistance was minimized at the coverage with 0.4 mL. Specifically, the lowest sheet resistance was 8.4 Ω/□, while the optical transparency was 87%. The return of the sheet resistance was simply attributed to the fact that the electrical conduction was mainly contributed by the network of AgNWs. At low content of MXene nanosheets, their function was to cure the defect points of the AgNWs network, thus leading reduction in the resistance. Further increase in the content of MXene, the curing effect was saturated, while the effective effect on conduction was increased, resulting in the rise of resistance of the hybrid films.

As mentioned above, the MXene phase mitigated the defects of the AgNWs, while the AgNWs could enhance the electrical conductivity of the MXene films. The complementation of the two components resulted in hybrid with outstanding electrical conductions, with a sheet resistance of about 16 Ω/□, along with an optical transparency of > 90%. Figure 5.49 shows optical transmittance curve and photograph of the hybrid-based electrodes. The incorporation of the AgNWs and MXene nanosheets led to hybrid materials with superior performances over those of their graphene counterparts, as observed in Fig. 5.49d.

Mechanical flexibility of transparent conductive film-based electrodes is a critical factor in real applications when encountering large deformations. The flexibility of the hybrid films was examined by monitor the resistance against repeated bending texting. The bending was expressed as the ratio of $(L_0 - L)/L_0$, with L_0 and L to be the initial length and length after bending of the electrode, respectively, as shown as the inset in Fig. 5.50a. Figure 5.50a shows sheet resistance of the sample as a function of the bending, where the 100% compression ratio resulted in only a variation of just 9%. Furthermore, the resistance value of the hybrid-based electrode was kept almost unchanged after bending for up to 1500 cycles, demonstrating outstanding high flexibility. In comparison, the resistance of the electrode based on pure AgNWs was varied relatively more severely against bending text. For instance, the resistance was largely increased by more eight times after bending for just 350 cycles at the angle of 120°, whereas it became nearly insulator, as observed in Fig. 5.50b.

The composition stability of the electrodes based on the AgNWs-MXene hybrid and pure AgNWs was examined by exposing them in air for 30 days. The hybrid-based electrode exhibited almost unchanged in resistance during the testing period, while the resistance value of the one made of pure AgNWs increased by about 80%, as revealed in Fig. 5.50c. According to SEM characterization results, the surface profile the of the hybrid sample was nearly unaltered after exposure for one-month time. In contrast, the pure AgNWs were heavily corroded.

The hybrid was also highly stable against low-temperature heat treatment. For example, at being heat at 75 °C for two days, the hybrid-based electrode was highly stable in terms of resistance, whereas the pure AgNWs were severely degraded, with the resistance value showing a rapid rise with increasing time, as illustrated in Fig. 5.50d. Additionally, the two groups of electrodes were also evaluated in terms of their anticorrosion resistance to hydrogen sulphide. In that experiment, the two samples were put in a chamber, in which 5 mL 0.5 M thioacetamide (TAA) used to generate hydrogen ulphide gas. Similarly, the hybrid-based electrode displayed much higher stability than the one made of pure AgNWs against the hydrogen sulphide gas.

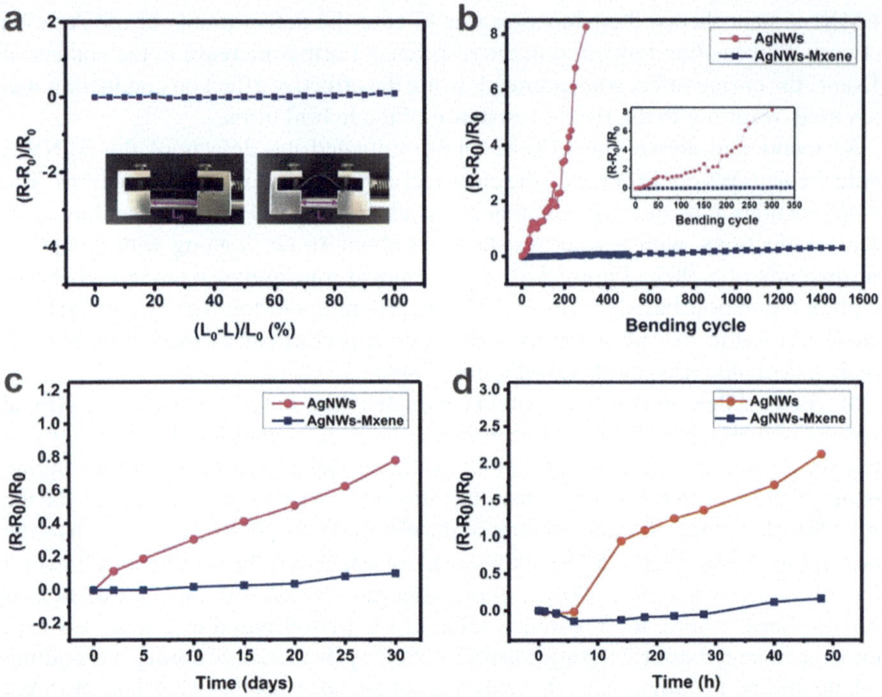

Fig. 5.50 Variations in resistance of the pure AgNWs film-based electrode and the AgNWx-MXene hybrid film-based electrode: **a** different compression ratios, **b** different bending cycles by above 120°, **c** exposure to air for 30 days and **d** at the high temperature of 75 °C for two days. Reproduced with permission from [167]. Copyright © 2019, American Chemical Society

5.7 Conclusions and Outlook

Similar to those in energy storage and conversion, the applications of MXenes and MXene-based hybrids/composites discussed in this chapter are also to use their multi-layered structure, high specific surface area, high electrical conductivity, superior mechanical strength and flexibility in combination with other components, as well as their enriched surface functional groups. Although the achievements are evidently remarkable, systematic studies and quantitative understandings of the mechanisms in most cases should be paid more attentions. For example, the lamellar architectures with sufficiently large interlayer space and appropriate functional groups enabled the MXenes to have strong tendency to adsorb various cations, thus leading to their capability of heavy ion removal. However, the capability to design MXene-based structures or functionalities for specific cations is still not available. The reported works in the open literature are mainly at the trial-and-error level. Gas sensors are based their semiconducting behaviours, thus leading opposite responses to reducing and oxidation gases, when variation in resistance is used to characterize the sensitivity of the sensing devices. In this respect, the strategies that have been used gas sensors based on the conventional semiconducting oxides are also applicable the MXene-based gas sensors. EMI shielding effect of MXenes is mainly attributed to their relatively high electrical conductivity, making the working mechanism similar to that of metallic and other conducive materials, while how to tailor the conductivity for given applications deserves further explorations. In terms of microwave absorption, some explanations of the mechanisms should be elaborated according from the electromagnetic principle point of view. When using as transparent and flexible electrodes, the optical transparency and electrical conductivity are inevitably contractive each other, which is seemly unsolvable at the moment. In this respect, it is urgent to develop effective routes to obtain MXene nanosheets with as large as lateral size and aspect ratio. Meanwhile, more applications will be explored in the near future, noting that there have been already examples that are not covered in this book. Nevertheless, this is a prosperous research topic among various 2D nanomaterials.

Acknowledgements This work was supported by the National Natural Science Foundation of China (51762023 and 51962013), the Natural Science Foundation of Jiangxi, China (20192ACB20018), and Key R&D Program of Jiangxi Province (20171BBE50006, 20192ACB80007, and 20192ACB80004). Ling Bing Kong would like acknowledge Shenzhen Technology University (SZTU) for financial support through the Start-up Grant (2018) and grant from the Natural Science Foundation of Top Talent of SZTU (grant no. 2019010801002).

References

1. Akkus, U.O., Balci, E., Berber, S.: $Mo_2TiC_2O_2$ MXene-based nanoscale pressure sensor. Phys. E-Low-Dimen. Syst. Nanostruct. **116**, 113762 (2020)
2. Blanco, E., Rosenkranz, A., Espinoza-Gonzalez, R., Fuenzalida, V.M., Zhang, Z.Y., Suarez, S., et al.: Catalytic performance of 2D-Mxene nano-sheets for the hydrodeoxygenation (HDO) of lignin-derived model compounds. Catal. Commun. **133**, 105833 (2020)
3. Gao, Y.J., Cao, Y.Y., Zhuo, H., Sun, X., Gu, Y.B., Zhuang, G.L., et al.: Mo_2TiC_2 MXene: a promising catalyst for electrocatalytic ammonia synthesis. Catal. Today **339**, 120–126 (2020)
4. Liu, T., Liu, X.Y., Graham, N., Yu, W.Z., Sun, K.N.: Two-dimensional MXene incorporated graphene oxide composite membrane with enhanced water purification performance. J. Membr. Sci. **593**, 117431 (2020)
5. Rafiq, S., Awan, S., Zheng, R.K., Wen, Z.C., Rani, M., Akinwande, D., et al.: Novel room-temperature ferromagnetism in Gd-doped 2-dimensional $Ti_3C_2T_x$ MXene semiconductor for spintronics. J. Magn. Magn. Mater. **497**, 165954 (2020)
6. Yao, Y., Lan, L.Y., Liu, X.X., Ying, Y.B., Ping, J.F.: Spontaneous growth and regulation of noble metal nanoparticles on flexible biomimetic MXene paper for bioelectronics. Biosens. Bioelectron. **148**, 111799 (2020)
7. Li, S., He, J.J., Nachtigall, P., Grajciar, L., Brivio, F.: Control of spintronic and electronic properties of bimetallic and vacancy-ordered vanadium carbide MXenes via surface functionalization. Phy. Chem. Chem. Phy. **21**, 25802–25808 (2019)
8. Shi, Y.Q., Liu, C., Liu, L., Fu, L.B., Yu, B., Lv, Y.C., et al.: Strengthening, toughing and thermally stable ultra-thin MXene nanosheets/polypropylene nanocomposites via nanoconfinement. Chem. Eng. J. **378**, 122267 (2019)
9. Tao, N., Zhang, D.P., Li, X.L., Lou, D.Y., Sun, X.Y., Wei, C.W., et al.: Near-infrared light-responsive hydrogels via peroxide-decorated MXene-initiated polymerization. Chem. Sci. **10**, 10765–10771 (2019)

10. Gu, B.H., Ku, Y.K., Jardine, P.M.: Sorption and binary exchange of nitrate, sulfate, and uranium on an anion-exchange resin. Environ. Sci. Technol. **38**, 3184–3188 (2004)
11. Fu, F.L., Xie, L.P., Tang, B., Wang, Q., Jiang, S.X.: Application of a novel strategy-Advanced Fenton-chemical precipitation to the treatment of strong stability chelated heavy metal containing wastewater. Chem. Eng. J. **189**, 283–287 (2012)
12. Zhang, Q., Wang, N., Zhao, L.B., Xu, T.W., Cheng, Y.Y.: Polyamidoamine dendronized hollow fiber membranes in the recovery of heavy metal ions. ACS Appl. Mater. Interfaces. **5**, 1907–1912 (2013)
13. Kerisit, S., Liu, C.X.: Molecular dynamics simulations of uranyl and uranyl carbonate adsorption at aluminosilicate surfaces. Environ. Sci. Technol. **48**, 3899–3907 (2014)
14. Bhattacharyya, K.G., Sen, G.S.: Adsorption of a few heavy metals on natural and modified kaolinite and montmorillonite: a review. Adv. Coll. Interface. Sci. **140**, 114–131 (2008)
15. Sen Gupta, S., Bhattacharyya, K.G.: Adsorption of heavy metals on kaolinite and montmorillonite: a review. Phy. Chem. Chem. Phy. **14**, 6698–6723 (2012)
16. Adeyemo, A.A., Adeoye, I.O., Bello, O.S.: Metal organic frameworks as adsorbents for dye adsorption: overview, prospects and future challenges. Toxicol. Environ. Chem. **94**, 1846–1863 (2012)
17. Yuan, L.Y., Tian, M., Lan, J.H., Cao, X.Z., Wang, X.L., Chai, Z.F., et al.: Defect engineering in metal-organic frameworks: a new strategy to develop applicable actinide sorbents. Chem. Commun. **54**, 370–373 (2018)
18. Wang, S.B., Sun, H.Q., Ang, H.M., Tade, M.O.: Adsorptive remediation of environmental pollutants using novel graphene-based nanomaterials. Chem. Eng. J. **226**, 336–347 (2013)
19. Perreault, F., de Faria, A.F., Elimelech, M.: Environmental applications of graphene-based nanomaterials. Chem. Soc. Rev. **44**, 5861–5896 (2015)
20. Zhao, G.X., Wen, T., Chen, C.L., Wang, X.K.: Synthesis of graphene-based nanomaterials and their application in energy-related and environmental-related areas. RSC Adv. **2**, 9286–9303 (2012)
21. Yuan, L.Y., Bai, Z.Q., Zhao, R., Liu, Y.L., Li, Z.J., Chu, S.Q., et al.: Introduction of bifunctional groups into mesoporous silica for enhancing uptake of thorium(IV) from aqueous solution. ACS Appl. Mater. Interfaces. **6**, 4786–4796 (2014)
22. Zhang, Y.J., Wang, L., Zhang, N.N., Zhou, Z.J.: Adsorptive environmental applications of MXene nanomaterials: A review. RSC Adv. **8**, 19895–19905 (2018)
23. Guo, J.X., Peng, Q.M., Fu, H., Zou, G.D., Zhang, Q.R.: Heavy-metal adsorption behavior of two-dimensional alkalization-intercalated MXene by first-principles calculations. J. Phys. Chem. C **119**, 20923–20930 (2015)
24. Guo, J.X., Fu, H., Zou, G.D., Zhang, Q.R., Zhang, Z.W., Peng, Q.M.: Theoretical interpretation on lead adsorption behavior of new two-dimensional transition metal carbides and nitrides. J. Alloy. Compd. **684**, 504–509 (2016)
25. Guo, X., Zhang, X.T., Zhao, S.J., Huang, Q., Xue, J.M.: High adsorption capacity of heavy metals on two-dimensional MXenes: an ab initio study with molecular dynamics simulation. Phy. Chem. Chem. Phy. **18**, 228–233 (2016)
26. Peng, Q.M., Guo, J.X., Zhang, Q.R., Xiang, J.Y., Liu, B.Z., Zhou, A.G., et al.: Unique lead adsorption behavior of activated hydroxyl group in two-dimensional titanium carbide. J. Am. Chem. Soc. **136**, 4113–4116 (2014)
27. Du, Y.C., Yu, B., Wei, L.Q., Wang, Y.L., Zhang, X.M., Ye, S.F.: Efficient removal of Pb(II) by $Ti_3C_2T_x$ powder modified with a silane coupling agent. J. Mater. Sci. **54**, 13283–13297 (2019)
28. Wang, G.Q., Xing, W., Zhuo, S.P.: Nitrogen-doped graphene as low-cost counter electrode for high-efficiency dye-sensitized solar cells. Electrochim. Acta **92**, 269–275 (2013)
29. Guo, D.M., An, Q.D., Xiao, Z.Y., Zhai, S.R., Yang, D.J.: Efficient removal of Pb(II), Cr(VI) and organic dyes by polydopamine modified chitosan aerogels. Carbohyd. Polym. **202**, 306–314 (2018)
30. Rangel-Mendeza, J.R., Monroy-Zepeda, R., Leyva-Ramos, E., Diaz-Flores, P.E., Shirai, K.: Chitosan selectivity for removing cadmium (II), copper (II), and lead (II) from aqueous phase: pH and organic matter effect. J. Hazard. Mater. **162**, 503–511 (2009)

31. Dong, Y.J., Sang, D.S., He, C.D., Sheng, X.F., Lei, L.W.: Mxene/alginate composites for lead and copper ion removal from aqueous solutions. RSC Advances. **9**, 29015–29022 (2019)
32. Huang, Y.G., Wang, Z.Q.: Preparation of composite aerogels based on sodium alginate, and its application in removal of Pb^{2+} and Cu^{2+} from water. Int. J. Biol. Macromol. **107**, 741–747 (2018)
33. Ying, Y.L., Liu, Y., Wang, X.Y., Mao, Y.Y., Cao, W., Hu, P., et al.: Two-dimensional titanium carbide for efficiently reductive removal of highly toxic chromium(VI) from water. ACS Appl. Mater. Interfaces. **7**, 1795–1803 (2015)
34. Zou, G.D., Guo, J.X., Peng, Q.M., Zhou, A.G., Zhang, Q.R., Liu, B.Z.: Synthesis of urchin-like rutile titania carbon nanocomposites by iron-facilitated phase transformation of MXene for environmental remediation. J. Mater. Chem. A. **4**, 489–499 (2016)
35. Tang, Y., Yang, C.H., Que, W.X.: A novel two-dimensional accordion-like titanium carbide (MXene) for adsorption of Cr(VI) from aqueous solution. J. Adv. Dielectr. **8**, 1850035 (2018)
36. Huang, Q.S., Liu, Y.T., Cai, T., Xia, X.N.: Simultaneous removal of heavy metal ions and organic pollutant by $BiOBr/Ti_3C_2$ nanocomposite. J. Photochem. Photobiol. A-Chem. **375**, 201–208 (2019)
37. Shahzad, A., Rasool, K., Miran, W., Nawaz, M., Jang, J.S., Mahmoud, K.A., et al.: Two-dimensional $Ti_3C_2T_X$ MXene nanosheets for efficient copper removal from water. ACS Sustain. Chem. Eng. **5**, 11481–11488 (2017)
38. Fard, A.K., McKay, G., Chamoun, R., Rhadfi, T., Preud'Homme, H., Atieh, M.A.: Barium removal from synthetic natural and produced water using MXene as two dimensional (2-D) nanosheet adsorbent. Chem. Eng. J. **317**, 331–342 (2017)
39. Shahzad, A., Rasool, K., Miran, W., Nawaz, M., Jang, J.S., Mahmoud, K.A., et al.: Mercuric ion capturing by recoverable titanium carbide magnetic nanocomposite. J. Hazard. Mater. **344**, 811–818 (2018)
40. Shahzad, A., Nawaz, M., Mortahida, M., Jang, J., Tahir, K., Kim, J., et al.: $Ti_3C_2T_x$ MXene core-shell spheres for ultrahigh removal of mercuric ions. Chem. Eng. J. **368**, 400–408 (2019)
41. Shahzad, A., Nawaz, M., Mortahida, M., Tahir, K., Kim, J., Lim, Y., et al.: Exfoliation of titanium aluminum carbide (211 MAX phase) to form nanofibers and two-dimensional nanosheets and their application in aqueous-phase cadmium sequestration. ACS Appl. Mater. Interfaces. **11**, 19156–19166 (2019)
42. Rasool, K., Pandey, R.P., Rasheed, P.A., Buczek, S., Gogotsi, Y., Mahmoud, K.A.: Water treatment and environmental remediation applications of two-dimensional metal carbides (MXenes). Mater. Today **30**, 80–102 (2019)
43. Sinopoli, A., Othman, Z., Rasool, K., Mahmoud, K.A.: Electrocatalytic/photocatalytic properties and aqueous media applications of 2D transition metal carbides (MXenes). Curr. Opin. Solid State Mater. Sci. **23**, 100760 (2019)
44. Mashtalir, O., Cook, K.M., Mochalin, V.N., Crowe, M., Barsoum, M.W., Gogotsi, Y.: Dye adsorption and decomposition on two-dimensional titanium carbide in aqueous media. J. Mater. Chem. A. **2**, 14334–14338 (2014)
45. Zheng, W., Zhang, P.G., Tian, W.B., Qin, X., Zhang, Y.M., Sun, Z.M.: Alkali treated $Ti_3C_2T_x$ MXenes and their dye adsorption performance. Mater. Chem. Phys. **206**, 270–276 (2018)
46. Gao, Y.P., Wang, L.B., Zhou, A.G., Li, Z.Y., Chen, J.K., Bala, H., et al.: Hydrothermal synthesis of TiO_2/Ti_3C_2 nanocomposites with enhanced photocatalytic activity. Mater. Lett. **150**, 62–64 (2015)
47. Peng, C., Yang, X.F., Li, Y.H., Yu, H., Wang, H.J., Peng, F.: Hybrids of two-dimensional Ti_3C_2 and TiO_2 exposing (001) facets toward enhanced photocatalytic activity. ACS Appl. Mater. Interfaces. **8**, 6051–6060 (2016)
48. Zhou, W.J., Zhu, J.F., Wang, F., Cao, M.J., Zhao, T.: One-step synthesis of $Ceria/Ti_3C_2$ nanocomposites with enhanced photocatalytic activity. Mater. Lett. **206**, 237–240 (2017)
49. Peng, C., Wei, P., Chen, X., Zhang, Y.L., Zhu, F., Cao, Y.H., et al.: A hydrothermal etching route to synthesis of 2D MXene (Ti_3C_2, Nb_2C): enhanced exfoliation and improved adsorption performance. Ceram. Int. **44**, 18886–18893 (2018)

50. Zhu, Z.B., Xiang, M.X., Shan, L.L., He, T., Zhang, P.: Effect of temperature on methylene blue removal with novel 2D-Magnetism titanium carbide. J. Solid State Chem. **280**, 86–93 (2019)
51. Fang, H.J., Pan, Y.S., Yin, M.Y., Xu, L.F., Zhu, Y., Pan, C.L.: Facile synthesis of ternary Ti_3C_2-OH/In_2S_3/CdS composite with efficient adsorption and photocatalytic performance towards organic dyes. J. Solid State Chem. **280**, 71–81 (2019)
52. Zhang, Y.J., Zhou, Z.J., Lan, J.H., Ge, C.C., Chai, Z.F., Zhang, P.H., et al.: Theoretical insights into the uranyl adsorption behavior on vanadium carbide MXene. Appl. Surf. Sci. **426**, 572–578 (2017)
53. Zhang, Y.J., Lan, J.H., Wang, L., Wu, Q.Y., Wang, C.Z., Bo, T., et al.: Adsorption of uranyl species on hydroxylated titanium carbide nanosheet: a first-principles study. J. Hazard. Mater. **308**, 402–410 (2016)
54. Wang, L., Tao, W.Q., Yuan, L.Y., Liu, Z.R., Huang, Q., Chai, Z.F., et al.: Rational control of the interlayer space inside two-dimensional titanium carbides for highly efficient uranium removal and imprisonment. Chem. Commun. **53**, 12084–12087 (2017)
55. Wang, L., Yuan, L.Y., Chen, K., Zhang, Y.J., Deng, Q.H., Du, S.Y., et al.: Loading actinides in multilayered structures for nuclear waste treatment: The first case study of uranium capture with vanadium carbide MXene. ACS Appl. Mater. Interfaces. **8**, 16396–16403 (2016)
56. Krupa, S.V., Legge, A.H.: Passive sampling of ambient, gaseous air pollutants: an assessment from an ecological perspective. Environ. Pollut. **107**, 31–45 (2000)
57. Samaddar, P., Son, Y.S., Tsang, D.C.W., Kim, K.H., Kumar, S.: Progress in graphene-based materials as superior media for sensing, sorption, and separation of gaseous pollutants. Coord. Chem. Rev. **368**, 93–114 (2018)
58. Soreanu, G., Dixon, M., Darlington, A.: Botanical biofiltration of indoor gaseous pollutants - A mini-review. Chem. Eng. J. **229**, 585–594 (2013)
59. Yu, X.F., Li, Y.C., Cheng, J.B., Liu, Z.B., Li, Q.Z., Li, W.Z., et al.: Monolayer Ti_2CO_2: a promising candidate for NH3 sensor or capturer with high sensitivity and selectivity. ACS Appl. Mater. Interfaces. **7**, 13707–13713 (2015a)
60. Morales-Garcia, A., Fernandez-Fernandez, A., Vines, F., Illas, F.: CO2 abatement using two-dimensional MXene carbides. J. Mater. Chem. A. **6**, 3381–3385 (2018)
61. Morales-Salvador, R., Morales-Garcia, A., Vines, F., Illas, F.: Two-dimensional nitrides as highly efficient potential candidates for CO_2 capture and activation. Phys. Chem. Chem. Phys. **20**, 17117–17124 (2018)
62. Zhang, Q.R., Teng, J., Zou, G.D., Peng, Q.M., Du, Q., Jiao, T.F., et al.: Efficient phosphate sequestration for water purification by unique sandwich-like MXene/magnetic iron oxide nanocomposites. Nanoscale. **8**, 7085–7093 (2016)
63. Meng, F.Y., Seredych, M., Chen, C., Gura, V., Mikhalovsky, S., Sandeman, S., et al.: MXene sorbents for removal of urea from dialysate: A step toward the wearable artificial kidney. ACS Nano **12**, 10518–10528 (2018)
64. Sinha, A., Lu, X., Wu, L., Tan, D., Li, Y., et al.: Voltammetric sensing of biomolecules at carbon based electrode interfaces: a review. TrAC-Trends Anal. Chem. **98**, 174–89 (2018)
65. Sinha, A., Tan, B., Huang, Y., Zhao, H., Dang, X., et al.: MoS2 nanostructures for electrochemical sensing of multidisciplinary targets: a review. TrAC-Trends Anal. Chem. **102**, 75–90 (2018)
66. Kim, H.U., Kim, H.Y., Kulkarni, A., Ahn, C., Jin, Y., Kim, Y., et al.: A sensitive electrochemical sensor for in vitro detection of parathyroid hormone based on a MoS_2-graphene composite. Sci. Rep. **6** (2016)
67. Wu, D.H., Wu, M.Y., Yang, J.H., Zhang, H.W., Xie, K.F., Lin, C.T., et al.: Delaminated $Ti_3C_2T_x$ (MXene) for electrochemical carbendazim sensing. Mater. Lett. **236**, 412–415 (2019)
68. Mohammadniaei, M., Nguyen, H.V., Tieu, M.V., Lee, M.H.: 2D materials in development of electrochemical point-of-care cancer dcreening devices. Micromachines. **10**, 662 (2019)
69. Zhou, S.J., Gu, C.X., Li, Z.Z., Yang, L.Y., He, L.H., Wang, M.H., et al.: $Ti_3C_2T_x$ MXene and polyoxometalate nanohybrid embedded with polypyrrole: ultra-sensitive platform for the detection of osteopontin. Appl. Surf. Sci. **498**, 143889 (2019)

70. Sinha, A., Zhao, H., Huang, Y., Lu, X., Chen, J., et al. MXene: an emerging material for sensing and biosensing. TrAC-Trends Anal. Chem.**105**, 424–35 (2018)
71. Rakhi, R.B., Nayuk, P., Xia, C., Alshareef, H.N.: Novel amperometric glucose biosensor based on MXene nanocomposite. Sci. Rep. **6**, 36422 (2016)
72. Wang, F., Yang, C.H., Duan, C.Y., Xiao, D., Tang, Y., Zhu, J.F.: An organ-like titanium carbide material (MXene) with multilayer structure encapsulating hemoglobin for a mediator-free biosensor. J. Electrochem. Soc. **162**, B16–B21 (2015)
73. Liu, H., Duan, C.Y., Yang, C.H., Shen, W.Q., Wang, F., Zhu, Z.F.: A novel nitrite biosensor based on the direct electrochemistry of hemoglobin immobilized on MXene-Ti_3C_2. Sens. Actuat. B-Chem. **218**, 60–66 (2015)
74. Wang, F., Yang, C.H., Duan, M., Tang, Y., Zhu, J.F.: TiO2 nanoparticle modified organ-like Ti_3C_2 MXene nanocomposite encapsulating hemoglobin for a mediator-free biosensor with excellent performances. Biosens. Bioelectron. **74**, 1022–1028 (2015)
75. Zheng, J.S., Diao, J.L., Jin, Y.Z., Ding, A.L., Wang, B., Wu, L.Z., et al.: An inkjet printed Ti_3C_2-GO eectrode for the electrochemical sensing of hydrogen eroxide. J. Electrochem. Soc. **165**, B227–B231 (2018)
76. Lorencova, L., Bertok, T., Filip, J., Jerigova, M., Velic, D., Kasak, P., et al.: Highly stable $Ti_3C_2T_x$ (MXene)/Pt nanoparticles-modified glassy carbon electrode for H_2O_2 and small molecules sensing applications. Sens. Actuat. B-Chem. **263**, 360–368 (2018)
77. Lorencova, L., Bertok, T., Dosekova, E., Holazova, A., Paprckova, D., Vikartovska, A., et al.: Electrochemical performance of $Ti_3C_2T_x$ MXene in aqueous media: towards ultrasensitive H_2O_2 sensing. Electrochim. Acta **235**, 471–479 (2017)
78. Wu, L., Lu, X., Wu, Z.S., Dong, Y., Wang, X., et al.: 2D transition metal carbide MXene as a robust biosensing platform for enzyme immobilization and ultrasensitive detection of phenol. Biosens. Bioelectr. **107**, 69–75 (2018)
79. Zhu, X.L., Liu, B.C., Hou, H.J., Huang, Z.Y., Zeinu, K.M., Huang, L., et al.: Alkaline intercalation of Ti_3C_2 MXene for simultaneous electrochemical detection of Cd(II), Pb(II), Cu(II) and Hg(II). Electrochim. Acta **248**, 46–57 (2017)
80. Rasheed, P.A., Pandey, R.P., Rasool, K., Mahmoud, K.A.: Ultra-sensitive electrocatalytic detection of bromate in drinking water based on Nafion/Ti_3C_2Tx (MXene) modified glassy carbon electrode. Sens. Actuat. B-Chem. **265**, 652–659 (2018)
81. Zhou, L.Y., Zhang, X.M., Ma, L., Gao, J., Jiang, Y.J.: Acetylcholinesterase/chitosan-transition metal carbides nanocomposites-based biosensor for the organophosphate pesticides detection. Biochem. Eng. J. **128**, 243–249 (2017)
82. Xu, B.Z., Zhu, M.S., Zhang, W.C., Zhen, X., Pei, Z.X., Xue, Q., et al.: Ultrathin MXene-micropattern-based field-effect transistor for probing neural activity. Adv. Mater. **28**, 3333–3339 (2016)
83. Zhu, X., Liu, B., Li, L., Wu, L., Chen, S., Huang, L., et al.: A micromilled microgrid sensor with delaminated MXene-bismuth nanocomposite assembly for simultaneous electrochemical detection of lead(II), cadmium(II) and zinc(II). Microchim. Acta. **186** (2019)
84. Song, D.D., Jiang, X.Y., Li, Y.S., Lu, X., Luan, S.R., Wang, Y.Z., et al.: Metal—organic frameworks-derived MnO_2/Mn_3O_4 microcuboids with hierarchically ordered nanosheets and Ti_3C_2 MXene/Au NPs composites for electrochemical pesticide detection. J. Hazard. Mater. **373**, 367–376 (2019)
85. Kim, S.J., Koh, H.J., Ren, C.E., Kwon, O., Maleski, K., Cho, S.Y., et al.: Metallic $Ti_3C_2T_X$ MXene gas sensors with ultrahigh signal-to-noise ratio. ACS Nano **12**, 986–993 (2018)
86. Yu, X.F., Li, Y.C., Cheng, J.B., Liu, Z.B., Li, Q.Z., Li, W.Z., et al.: Monolayer Ti_2CO_2: a promising dandidate for NH_3 sensor or capturer with high sensitivity and selectivity. ACS Appl. Mater. Interfaces. **7**, 13707–13713 (2015b)
87. Xiao, B., Li, Y.C., Yu, X.F., Cheng, J.B.: MXenes: Reusable materials for NH_3 sensor or capturer by controlling the charge injection. Sens. Actuat. B-Chem. **235**, 103–109 (2016)
88. Lee, E., Mohammadi, A.V., Prorok, B.C., Yoon, Y.S., Beidaghi, M., Kim, D.J.: Room temperature gas sensing of two-dimensional titanium carbide (MXene). ACS Appl. Mater. Interfaces. **9**, 37184–37190 (2017)

89. Anasori, B., Lukatskaya, M.R., Gogotsi, Y.: 2D metal carbides and nitrides (MXenes) for energy storage. Nat. Rev. Mater. **2**, 16098 (2017)
90. Yang, Z.J., Liu, A., Wang, C.L., Liu, F.M., He, J.M., Li, S.Q., et al.: Improvement of gas and humidity sensing properties of organ-like MXene by alkaline treatment. ACS Sens. **4**, 1261–1269 (2019)
91. Sun, S.B., Wang, M.W., Chang, X.T., Jiang, Y.C., Zhang, D.Z., Wang, D.S., et al.: $W_{18}O_{49}/Ti_3C_2T_x$ Mxene nanocomposites for highly sensitive acetone gas sensor with low detection limit. Sens. Actuat. B-Chem. **304**, 127274 (2020)
92. Lee, E., VahidMohammadi, A., Yoon, Y.S., Beidaghi, M., Kim, D.J.: Two-dimensional vanadium carbide MXene for gas sensors with ultrahigh sensitivity toward nonpolar gases. ACS Sens. **4**, 1603–1611 (2019)
93. Cai, Y.C., Shen, J., Ge, G., Zhang, Y.Z., Jin, W.Q., Huang, W., et al.: Stretchable $Ti_3C_2T_x$ MXene/carbon nanotube composite based strain sensor with ultrahigh sensitivity and tunable sensing range. ACS Nano **12**, 56–62 (2018)
94. Barlian, A.A., Park, W.T., Mallon, J.R., Jr., Rastegar, A.J., Pruitt, B.L.: Review: Semiconductor piezoresistance for microsystems. Proc. IEEE **97**, 513–552 (2009)
95. Ma, Y.N., Liu, N.S., Li, L.Y., Hu, X.K., Zou, Z.G., Wang, J.B., et al.: A highly flexible and sensitive piezoresistive sensor based on MXene with greatly changed interlayer distances. Nat. Commun. **8**, 1207 (2017)
96. Tan, J., Wang, Y.H., Wang, Z.T., He, X.J., Liu, Y.L., Wang, B., et al.: Large out-of-plane piezoelectricity of oxygen functionalized MXenes for ultrathin piezoelectric cantilevers and diaphragms. Nano Energy. **65**, 104058 (2019)
97. Xue, Q., Zhang, H.J., Zhu, M.S., Pei, Z.X., Li, H.F., Wang, Z.F., et al.: Photoluminescent Ti_3C_2 MXene quantum dots for multicolor cellular imaging. Adv. Mater. **29**, 1604847 (2017)
98. Huang, C.C., Yang, Z.S., Lee, K.H., Chang, H.T.: Synthesis of highly fluorescent gold nanoparticles for sensing mercury(II). Angew. Chem-Int. Ed. **46**, 6824–6828 (2007)
99. Guo, Y.M., Zhang, L.F., Zhang, S.S., Yang, Y., Chen, X.H., Zhang, M.C.: Fluorescent carbon nanoparticles for the fluorescent detection of metal ions. Biosens. Bioelectron. **63**, 61–71 (2015)
100. Kumar, D., Talreja, N.: Nickel nanoparticles-doped rhodamine grafted carbon nanofibers as colorimetric probe: naked eye detection and highly sensitive measurement of aqueous Cr^{3+} and Pb^{2+}. Korean J. Chem. Eng. **36**, 126–135 (2019)
101. Chen, X., Sun, X.K., Xu, W., Pan, G.C., Zhou, D.L., Zhu, J.Y., et al.: Ratiometric photoluminescence sensing based on Ti_3C_2 MXene quantum dots as an intracellular pH sensor. Nanoscale. **10**, 1111–1118 (2018)
102. Zhang, Q.X., Wang, F., Zhang, H.X., Zhang, Y.Y., Liu, M.L., Liu, Y.: Universal Ti_3C_2 MXenes based self-standard ratiometric fluorescence resonance energy transfer platform for highly sensitive detection of exosomes. Anal. Chem. **90**, 12737–12744 (2018)
103. Fang, Y.F., Yang, X.C., Chen, T., Xu, G.F., Liu, M.L., Liu, J.Q., et al.: Two-dimensional titanium carbide (MXene)-based solid-state electrochemiluminescent sensor for label-free single-nucleotide mismatch discrimination in human urine. Sens. Actuators B-Chem. **263**, 400–407 (2018)
104. Guan, Q.W., Ma, J.F., Yang, W.J., Zhang, R., Zhang, X.J., Dong, X.X., et al.: Highly fluorescent Ti_3C_2 MXene quantum dots for macrophage labeling and Cu^{2+} ion sensing. Nanoscale. **11**, 14123–14133 (2019)
105. Jhon, Y.I., Seo, M., Jhon, Y.M.: First-principles study of a MXene terahertz detector. Nanoscale. **10**, 69–75 (2018)
106. Satheeshkumar, E., Makaryan, T., Melikyan, A., Minassian, H., Gogotsi, Y., Yoshimura, M.: One-step solution processing of Ag, Au and Pd@MXene hybrids for SERS. Sci. Rep. **6**, 32049 (2016)
107. Sarycheva, A., Makaryan, T., Maleski, K., Satheeshkumar, E., Melikyan, A., Minassian, H., et al.: Two-dimensional titanium carbide (MXene) as surface-enhanced Raman scattering substrate. J. Phys. Chem. C **121**, 19983–19988 (2017)

108. Soundiraraju, B., George, B.K.: Two-dimensional titanium nitride (Ti_2N) MXene: Synthesis, characterization, and potential application as surface-enhanced Raman scattering substrate. ACS Nano **11**, 8892–8900 (2017)
109. Ma, T.Y., Cao, J.L., Jaroniec, M., Qiao, S.Z.: Interacting carbon nitride and titanium carbide nanosheets for high-performance oxygen evolution. Angew. Chem. Int. Ed. **55**, 1138–1142 (2016)
110. Wang, H., Peng, R., Hood, Z.D., Naguib, M., Adhikari, S.P., Wu, Z.L.: Titania composites with 2 D transition metal carbides as photocatalysts for hydrogen production under visible-light irradiation. Chemsuschem **9**, 1490–1497 (2016)
111. Ling, C.Y., Shi, L., Ouyang, Y.X., Chen, Q., Wang, J.L.: Transition metal-promoted V_2CO_2 (MXenes): a new and highly active catalyst for hydrogen evolution reaction. Adv. Sci. **3**, 1600180 (2016)
112. Zhang, Z.W., Li, H.N., Zou, G.D., Fernandez, C., Liu, B.Z., Zhang, Q.R., et al.: Self-reduction synthesis of new MXene/Ag composites with unexpected electrocatalytic activity. ACS Sustain. Chem. Eng. **4**, 6763–6771 (2016)
113. Xie, X.H., Chen, S.G., Ding, W., Nie, Y., Wei, Z.D.: An extraordinarily stable catalyst: Pt NPs supported on two-dimensional $Ti_3C_2X_2$ (X = OH, F) nanosheets for oxygen reduction reaction. Chem. Commun. **49**, 10112–10114 (2013)
114. Yu, X.L., Wang, T., Yin, W.C., Zhang, Y.H.: Ti_3C_2 MXene nanoparticles modified metal oxide composites for enhanced photoelectrochemical water splitting. Int. J. Hydrogen Energy **44**, 2704–2710 (2019)
115. Zhang, X., Zhang, Z.H., Li, J.L., Zhao, X.D., Wu, D.H., Zhou, Z.: Ti_2CO_2 MXene: a highly active and selective photocatalyst for CO_2 reduction. J. Mater. Chem. A. **5**, 12899–12903 (2017)
116. Low, J.X., Zhang, L.Y., Tong, T., Shen, B.J., Yu, J.G.: TiO_2/MXene Ti_3C_2 composite with excellent photocatalytic CO_2 reduction activity. J. Catal. **361**, 255–266 (2018)
117. Cao, S.W., Shen, B.J., Tong, T., Fu, J.W., Yu, J.G.: 2D/2D heterojunction of ultrathin MXene/Bi_2WO_6 nanosheets for improved photocatalytic CO_2 reduction. Adv. Func. Mater. **28**, 1800136 (2018)
118. Cao, M.S., Cai, Y.Z., He, P., Shu, J.C., Cao, W.Q., Yuan, J.: 2D MXenes: electromagnetic property for microwave absorption and electromagnetic interference shielding. Chem. Eng. J. **359**, 1265–1302 (2019)
119. Fan, Z., Wang, D., Yuan, Y., Wang, Y., Cheng, Z., Liu, Y., et al.: A lightweight and conductive MXene/graphene hybrid foam for superior electromagnetic interference shielding. Chem. Eng. J. 381 (2020)
120. Hu, S.J., Li, S.B., Xu, W.M., Zhang, J., Zhou, Y., Cheng, Z.X.: Rapid preparation, thermal stability and electromagnetic interference shielding properties of two-dimensional Ti3C2 MXene. Ceram. Int. **45**, 19902–19909 (2019)
121. Jin, X., Wang, J., Dai, L., Liu, X., Li, L., Yang, Y., et al.: Flame-retardant poly(vinyl alcohol)/MXene multilayered films with outstanding electromagnetic interference shielding and thermal conductive performances. Chem. Eng. J. 380 (2020)
122. Kumar, S., Kumar, P., Singh, N., Verma, V.: Steady microwave absorption behavior of two-dimensional metal carbide MXene and Polyaniline composite in X-band. J. Magn. Magn. Mater. **488** (2019)
123. Li, Y., Tian, X., Gao, S.P., Jing, L., Li, K.R., Yang, H.T., et al.: Reversible crumpling of 2D titanium carbide (MXene) nanocoatings for stretchable electromagnetic shielding and wearable wireless communication. Adv. Funct. Mater. 1907451 (2019)
124. Lipton, J., Weng, G.M., Alhabeb, M., Maleski, K., Antonio, F., Kong, J., et al.: Mechanically strong and electrically conductive multilayer MXene nanocomposites. Nanoscale. **11**, 20295–20300 (2019)
125. Liu, J., Liu, Z., Zhang, H.B., Chen, W., Zhao, Z., Wang, Q.W., et al.: Ultrastrong and highly conductive MXene-based films for high-performance electromagnetic interference shielding. Adv. Electron. Mater.1901094 (2019)

126. Luo, J.Q., Zhao, S., Zhang, H.B., Deng, Z., Li, L., Yu, Z.Z.: Flexible, stretchable and electrically conductive MXene/natural rubber nanocomposite films for efficient electromagnetic interference shielding. Compos. Sci. Technol. **182** (2019)
127. Sambyal, P., Iqbal, A., Hong, J., Kim, H., Kim, M.K., Hong, S.M., et al.: Ultralight and mechanically robust $Ti_3C_2T_x$ hybrid aerogel reinforced by carbon nanotubes for electromagnetic interference shielding. ACS Appl. Mater. Interfaces. **11**, 38046–38054 (2019)
128. Wu, X., Han, B., Zhang, H.B., Xie, X., Tu, T., Zhang, Y., et al.: Compressible, durable and conductive polydimethylsiloxane-coated MXene foams for high-performance electromagnetic interference shielding. Chem. Eng. J. **381** (2020)
129. Shahzad, F., Alhabeb, M., Hatter, C.B., Anasori, B., Soon, M.H., Koo, C.M., et al.: Electromagnetic interference shielding with 2D transition metal carbides (MXenes). Science **353**, 1137–1140 (2016)
130. Liu, J., Zhang, H.B., Sun, R.H., Liu, Y.F., Liu, Z.S., Zhou, A.G., et al.: Hydrophobic, flexible, and lightweight MXene foams for high-performance electromagnetic-interference shielding. Adv. Mater. **29**, 1702367 (2017)
131. Han, M.K., Yin, X.W., Wu, H., Hou, Z.X., Song, C.Q., Li, X.L., et al.: Ti_3C_2 MXenes with modified surface for high-performance electromagnetic absorption and shielding in the X-band. ACS Appl. Mater. Interfaces. **8**, 21011–21019 (2016)
132. Bian, R.J., He, G.L., Zhi, W.Q., Xiang, S.L., Wang, T.W., Cai, D.Y.: Ultralight MXene-based aerogels with high electromagnetic interference shielding performance. J. Mater. Chem. C. **7**, 474–478 (2019)
133. He, P., Wang, X.X., Cai, Y.Z., Shu, J.C., Zhao, Q.L., Yuan, J., et al.: Tailoring $Ti_3C_2T_x$ nanosheets to tune local conductive network as an environmentally friendly material for highly efficient electromagnetic interference shielding. Nanoscale. (2019) (in press)
134. Li, X.L., Yin, X.W., Liang, S., Li, M.H., Cheng, L.F., Zhang, L.T.: 2D carbide MXene Ti_2CT_X as a novel high-performance electromagnetic interference shielding material. Carbon **146**, 210–217 (2019)
135. Han, M.K., Yin, X.W., Hantanasirisakul, K., Li, X.L., Iqbal, A., Hatter, C.B., et al.: Anisotropic MXene aerogels with a mechanically tunable ratio of electromagnetic wave reflection to absorption. Adv. Opt. Mater. **7**, 1900267 (2019)
136. Sun, R.H., Zhang, H.B., Liu, J., Xie, X., Yang, R., Li, Y., et al.: Highly conductive transition metal carbide/carbonitride(MXene)@polystyrene nanocomposites fabricated by electrostatic assembly for highly efficient electromagnetic interference shielding. Adv. Func. Mater. **27**, 1702807 (2017)
137. Liu, R.T., Miao, M., Li, Y.H., Zhang, J.F., Cao, S.M., Feng, X.: Ultrathin biomimetic polymeric $Ti_3C_2T_X$ MXene composite films for electromagnetic interference shielding. ACS Appl. Mater. Interfaces. **10**, 44787–44795 (2018)
138. Qing, Y.C., Zhou, W.C., Luo, F., Zhu, D.M.: Titanium carbide (MXene) nanosheets as promising microwave absorbers. Ceram. Int. **42**, 16412–16416 (2016)
139. Wang, Q.W., Zhang, H.B., Liu, J., Zhao, S., Xie, X., Liu, L.X., et al.: Multifunctional and water-resistant MXene-decorated polyester textiles with outstanding electromagnetic interference shielding and Joule heating performances. Adv. Func. Mater. **29**, 1806819 (2019)
140. Xu, H.L., Yin, X.W., Li, X.L., Li, M.H., Liang, S., Zhang, L.T., et al.: Lightweight Ti_2CT_x MXene/poly(vinyl alcohol) composite foams for electromagnetic wave shielding with absorption-dominated feature. ACS Appl. Mater. Interfaces. **11**, 10198–10207 (2019)
141. Raagulan, K., Braveenth, R., Jang, H.J., Lee, Y.S., Yang, C.M., Kim, B.M., et al.: Electromagnetic shielding by MXene-graphene-PVDF composite with hydrophobic, lightweight and flexible graphene coated fabric. Materials. **11**, 1803 (2018)
142. Zhao, S., Zhang, H.B., Luo, J.Q., Wang, Q.W., Xu, B., Hong, S., et al.: Highly electrically conductive three-dimensional $Ti_3C_2T_X$ MXene/reduced graphene oxide hybrid aerogels with excellent electromagnetic interference shielding performances. ACS Nano **12**, 11193–11202 (2018)
143. Weng, G.M., Li, J.Y., Alhabeb, M., Karpovich, C., Wang, H., Lipton, J., et al.: Layer-by-layer assembly of cross-functional semi-transparent MXene-carbon nanotubes composite films for next-generation electromagnetic interference shielding. Adv. Func. Mater. **28**, 1803360 (2018)

144. Cao, W.T., Chen, F.F., Zhu, Y.J., Zhang, Y.G., Jiang, Y.Y., Ma, M.G., et al.: Binary strengthening and toughening of MXene/cellulose nanofiber composite paper with nacre-inspired structure and superior electromagnetic interference shielding properties. ACS Nano **12**, 4583–4593 (2018)
145. Cui, C., Xiang, C., Geng, L., Lai, X.X., Guo, R.H., Zhang, Y., et al.: Flexible and ultra-thin electrospun regenerate cellulose nanofibers and d-$Ti_3C_2T_x$ (MXene) composite film for electromagnetic interference shielding. J. Alloy. Compd. **788**, 1246–1255 (2019)
146. Wang, L., Qiu, H., Song, P., Zhang, Y.L., Lu, Y.J., Liang, C.B., et al.: 3D $Ti_3C_2T_x$ MXene/C hybrid foam/epoxy nanocomposites with superior electromagnetic interference shielding performances and robust mechanical properties. Compos. Part A-Appl. Sci. Manufs. **123**, 293–300 (2019)
147. Liu, P.J., Ng, V.M.H., Yao, Z.J., Zhou, J.T., Kong, L.B.: Ultrasmall Fe_3O_4 nanoparticles on MXenes with high microwave absorption performance. Mater. Lett. **229**, 286–289 (2018)
148. Liu, P.J., Yao, Z.J., Ng, V.M.H., Zhou, J.T., Kong, L.B., Yue, K.: Facile synthesis of ultrasmall Fe_3O_4 nanoparticles on MXenes for high microwave absorption performance. Compos. Part A-Appl. Sci. Manuf. **115**, 371–382 (2018)
149. Feng, W.L., Luo, H., Wang, Y., Zeng, S.F., Tan, Y.Q., Deng, L.W., et al.: Mxenes derived laminated and magnetic composites with excellent microwave absorbing performance. Sci. Rep. **9**, 3957 (2019)
150. Yang, H.B., Dai, J.J., Liu, X., Lin, Y., Wang, J.J., Wang, L., et al.: Layered PVB/$Ba_3Co_2Fe_{24}O_{41}$/Ti_3C_2 Mxene composite: Enhanced electromagnetic wave absorption properties with high impedance match in a wide frequency range. Mater. Chem. Phys. **200**, 179–186 (2017)
151. Zhao, G.L., Lv, H.P., Zhou, Y., Zheng, X.T., Wu, C., Xu, C.: Self-assembled sandwich-like MXene-derived nanocomposites for enhanced electromagnetic wave absorption. ACS Appl. Mater. Interfaces. **10**, 42925–42932 (2018)
152. Liu, L.X., Chen, W., Zhang, H.B., Wang, Q.W., Guan, F. and Yu, Z.Z.: Flexible and multifunctional silk textiles with biomimetic leaf-like MXene/silver nanowire nanostructures for electromagnetic interference shielding, humidity monitoring, and self-derived hydrophobicity. Advanced Functional Materials. **29**,1905197 (2019)
153. Weng, C.X., Wang, G.R., Dai, Z.H., Pei, Y.M., Liu, L.Q., Zhang, Z.: Buckled AgNW/MXene hybrid hierarchical sponges for high-performance electromagnetic interference shielding. Nanoscale. **11**, 22804–22812 (2019)
154. Bai, Y.L., Zhou, K., Srikanth, N., Pang, J.H.L., He, X.D., Wang, R.G.: Dependence of elastic and optical properties on surface terminated groups in two-dimensional MXene monolayers: a first-principles study. RSC Advances. **6**, 35731–35739 (2016)
155. Berdiyorov, G.R.: Optical properties of functionalized $Ti_3C_2T_2$ (T = F, O, OH) MXene: First-principles calculations. AIP Adv. **6**, 055105 (2016)
156. Halim, J., Lukatskaya, M.R., Cook, K.M., Lu, J., Smith, C.R., Näslund, L.A., et al.: Transparent conductive two-dimensional titanium carbide epitaxial thin films. Chem. Mater. **26**, 2374–2381 (2014)
157. Dillon, A.D., Ghidiu, M.J., Krick, A.L., Griggs, J., May, S.J., Gogotsi, Y., et al.: Highly conductive optical quality solution-processed films of 2D titanium carbide. Adv. Func. Mater. **26**, 4162–4168 (2016)
158. Hantanasirisakul, K., Zhao, M.Q., Urbankowski, P., Halim, J., Anasori, B., Kota, S., et al.: Fabrication of $Ti_3C_2T_x$ MXene transparent thin films with tunable opoelectronic properties. Adv. Electro. Mater. **2**, 1600050 (2016)
159. Mariano, M., Mashtalir, O., Antonio, F.Q., Ryu, W.H., Deng, B.C., Xia, F.N., et al.: Solution-processed titanium carbide MXene films examined as highly transparent conductors. Nanoscale. **8**, 16371–16378 (2016)
160. Ali, A., Belaidi, A., Ali, S., Helal, M.I., Mahmoud, K.A.: Transparent and conductive $Ti_3C_2T_x$ (MXene) thin film fabrication by electrohydrodynamic atomization technique. J. Mater. Sci.: Mater. Electron. **27**, 5440–5445 (2016)

161. Aïssa, B., Ali, A., Mahmoud, K.A., Haddad, T., Nedil, M.: Transport properties of a highly conductive 2D $Ti_3C_2T_x$ MXene/graphene composite. Appl. Phys. Lett. **109**, 043109 (2016)
162. Lipatov, A., Alhabeb, M., Lukatskaya, M.R., Boson, A., Gogotsi, Y., Sinitskii, A.: Effect of synthesis on quality, electronic properties and environmental stability of individual monolayer Ti_3C_2 MXene flakes. Advanced Electronic Materials. **2**, 1600255 (2016)
163. Ying, G.B., Dillon, A.D., Fafarman, A.T., Barsoum, M.W.: Transparent, conductive solution processed spincast 2D Ti_2CT_x (MXene) films. Materials Research Letters. **5**, 391–398 (2017)
164. Ying, G.B., Kota, S., Dillon, A.D., Fafarman, A.T., Barsoum, M.W.: Conductive transparent V_2CT_x (MXene) films. Flatchem. **8**, 25–30 (2018)
165. Kim, S.J., Choi, J., Maleski, K., Hantanasirisakul, K., Jung, H.T., Gogotsi, Y., et al.: Interfacial assembly of ultrathin, functional MXene films. ACS Appl. Mater. Interfaces. **11**, 32320–32327 (2019)
166. Li, W.Y., Song, Z.Q., Zhong, J.M., Qian, J., Tan, Z.Y., Wu, X.Y., et al.: Multilayer-structured transparent MXene/PVDF film with excellent dielectric and energy storage performance. J. Mater. Chem. C. **7**, 10371–10378 (2019)
167. Liu, J., Zhang, L., Li, C.Z.: Highly stable, transparent, and conductive electrode of solution-processed silver nanowire-Mxene for flexible alternating-current electroluminescent devices. Ind. Eng. Chem. Res. **58**, 21485–21492 (2019)

The manufacturer's authorised representative in the EU is Springer Nature Customer Service Centre GmbH, Europaplatz 3, 69115 Heidelberg, Germany. If you have any concerns regarding our products, please contact ProductSafety@springernature.com

Printed and bound by CPI Group (UK) Ltd, Croydon, CR0 4YY
30/09/2026
02225503-0015